Hydrology and Earth Systems

Hydrology and Earth Systems

Edited by Allison Sergeant

SYRAWOOD
PUBLISHING HOUSE

New York

Published by Syrawood Publishing House,
750 Third Avenue, 9th Floor,
New York, NY 10017, USA
www.syrawoodpublishinghouse.com

Hydrology and Earth Systems
Edited by Allison Sergeant

© 2018 Syrawood Publishing House

International Standard Book Number: 978-1-68286-521-7 (Hardback)

Cataloging-in-Publication Data

Hydrology and earth systems / edited by Allison Sergeant.
        p. cm.
    Includes bibliographical references and index.
    ISBN 978-1-68286-521-7
    1. Hydrology. 2. Earth sciences. I. Sergeant, Allison.
GB661.2 .H93 2018
551.48--dc23

TABLE OF CONTENTS

PREFACE

Every book is a source of knowledge and this one is no exception. The idea that led to the conceptualization of this book was the fact that the world is advancing rapidly; which makes it crucial to document the progress in every field. I am aware that a lot of data is already available, yet, there is a lot more to learn. Hence, I accepted the responsibility of editing this book and contributing my knowledge to the community.

Hydrology is the study of water and its occurrence on Earth as well as its movement and properties in each phase of the hydrologic cycle. Water is a scarce natural resource and hydrologists seek to intervene in the availability and distribution of water in various regions of Earth. Concerns related to hydrology include the supply of water to cities or for cultivation, solve soil erosion and runoff as well as calamities like flooding. Geologic and atmospheric properties of various ecologies that are studied under Earth sciences allow for precise models for hydrological prediction and forecasting. This book unfolds the innovative aspects of this subject which will be crucial for the progress of this field in the future. It consists of contributions made by international experts. For all readers who are interested in hydrology, the case studies included in this book will serve as an excellent guide to develop a comprehensive understanding.

While editing this book, I had multiple visions for it. Then I finally narrowed down to make every chapter a sole standing text explaining a particular topic, so that they can be used independently. However, the umbrella subject sinews them into a common theme. This makes the book a unique platform of knowledge.

I would like to give the major credit of this book to the experts from every corner of the world, who took the time to share their expertise with us. Also, I owe the completion of this book to the never-ending support of my family, who supported me throughout the project.

Editor

A question driven socio-hydrological modeling process

M. Garcia[1], K. Portney[2], and S. Islam[1,3]

[1]Civil & Environmental Engineering Department, Tufts University, 200 College Avenue, Medford, MA 02155, USA
[2]Bush School of Government & Public Service, Texas A&M University, 4220 TAMU, College Station, TX 77843, USA
[3]The Fletcher School of Law and Diplomacy, Tufts University, 160 Packard Avenue, Medford, MA 02155, USA

Correspondence to: M. Garcia (margaret.garcia@tufts.edu)

Abstract. Human and hydrological systems are coupled: human activity impacts the hydrological cycle and hydrological conditions can, but do not always, trigger changes in human systems. Traditional modeling approaches with no feedback between hydrological and human systems typically cannot offer insight into how different patterns of natural variability or human-induced changes may propagate through this coupled system. Modeling of coupled human–hydrological systems, also called socio-hydrological systems, recognizes the potential for humans to transform hydrological systems and for hydrological conditions to influence human behavior. However, this coupling introduces new challenges and existing literature does not offer clear guidance regarding model conceptualization. There are no universally accepted laws of human behavior as there are for the physical systems; furthermore, a shared understanding of important processes within the field is often used to develop hydrological models, but there is no such consensus on the relevant processes in socio-hydrological systems. Here we present a question driven process to address these challenges. Such an approach allows modeling structure, scope and detail to remain contingent on and adaptive to the question context. We demonstrate the utility of this process by revisiting a classic question in water resources engineering on reservoir operation rules: what is the impact of reservoir operation policy on the reliability of water supply for a growing city? Our example model couples hydrological and human systems by linking the rate of demand decreases to the past reliability to compare standard operating policy (SOP) with hedging policy (HP). The model shows that reservoir storage acts both as a buffer for variability and as a delay triggering oscillations around a sustainable level of demand. HP reduces the threshold for action thereby decreasing the delay and the oscillation effect. As

a result, per capita demand decreases during periods of water stress are more frequent but less drastic and the additive effect of small adjustments decreases the tendency of the system to overshoot available supplies. This distinction between the two policies was not apparent using a traditional noncoupled model.

1 Introduction

Humans both respond to and ignore changes in environmental conditions. While humans depend on the natural hydrological cycle to supply water for both personal and economic health (Falkenmark, 1977), they also depend on an array of other natural and human resources to maintain and grow communities. At times water availability can act as the limiting constraint, locally preventing or stalling the expansion of human activity. For example, water availability and variability constrained agricultural development in the Tarim River basin in western China before major water storage and transport infrastructure was constructed (Liu et al., 2014). At other times the water-related risks rise in the background, disconnected from decision making, while other priorities prevail. For instance, the level of the Aral Sea has continued to decline for decades imposing significant costs on adjacent communities but no coordinated effort to stop the decline emerged (Micklin, 2007). At still other times public policy decisions may work to exacerbate water problems, as when decisions are made to keep municipal water prices artificially low or when "senior water rights" encourage water usage in the face of shortages (Chong and Sunding, 2006; Hughes et al., 2013; Mini et al., 2014).

Human and hydrological systems are coupled. Many impacts of human activity on the hydrological system are now well documented (Tong and Chen, 2002; Wissmar et al., 2004; Vörösmarty et al., 2010; Vahmani and Hogue, 2014) and there is increasing evidence that how and when humans respond individually and collectively to hydrological change has important implications for water resources planning, management and policy (Srinivasan et al., 2010; Di Baldassarre et al., 2013; Elshafei et al., 2014). These observations have prompted a call to treat humans as an endogenous component of the water cycle (Wagener et al., 2010; Sivapalan et al., 2012). Representing water systems as coupled human–hydrological systems or socio-hydrological systems with two-way feedback allows new research questions and potentially transformative insights to emerge.

Traditional modeling approaches assume that there is no feedback between hydrological and human systems and, therefore, cannot provide insights into how different patterns of natural variability or human-induced change may propagate through the coupled system. Over short timescales, such as a year, many human and hydrological variables can be considered constant and their couplings may be ignored (Srinivasan, 2015). However, water resources infrastructure decisions have impacts on longer (decadal to century) timescales; therefore, there is a need for an approach that can handle not only long-term variability and nonstationarity in the driving variables (e.g., precipitation, temperature, population) but also addresses how these changes can propagate through the coupled system, affecting the structure and properties of the coupled system (Sivapalan et al., 2012; Thompson et al., 2013). Dynamic modeling of socio-hydrological systems recognizes the potential for humans to transform hydrological systems and for hydrological conditions to influence human behavior. While human behavior is usually incorporated into a model through scenarios, scenarios cannot include two-way feedback. Building effects of human behavior into a simulation model can enable testing of feedback cycles and can illuminate the impact of feedback and path dependencies that are not easily identifiable in scenario-based modeling.

Coupled modeling, on the other hand, introduces new challenges. First, it is not possible to exhaustively model complex systems such as the coupled human–hydrological system (Sterman, 2000; Schlüter et al., 2014). Bounds must be set to develop an effective model but researchers are challenged to objectively define the scope of coupled modeling studies. Second, by definition coupled models cross disciplines and modelers are unable to point to the theoretical framework of any single discipline to defend the relevant scope (Srinivasan, 2015). At the same time researchers must balance the scope and level of detail in order to create a parsimonious and communicable model. Finally, critical assessment of models is more challenging when the theories, empirical methods and vocabulary drawn upon to create and communicate a model span disciplinary boundaries (Schlüter et al., 2014). At the same time, critique is needed to move the field forward as the science is new and lacks established protocols. Transparency of the model aims, the development process, conceptual framework and assumptions are thus particularly important. A structured but flexible modeling process can address these challenges by encouraging modelers to clearly define model objectives, document reasoning behind choices of scale, scope and detail, and take a broad view of potentially influential system processes.

In this paper we present a question driven process for modeling socio-hydrological systems that builds on current modeling tools from both domains and allows the flexibility for exploration. We demonstrate this process by revisiting a classic question in water resources engineering on reservoir operation rules: the tradeoff between standard operating policy (SOP) and hedging policy (HP). Under SOP, demand is fulfilled unless available supply drops below demand; under HP, water releases are reduced in anticipation of a deficit to decrease the risk a large shortfall (Cancelliere et al., 1998). We add to this classic question a linkage between supply reliability and demand. As this question has been asked by numerous researchers before, it offers an excellent opportunity to test the utility of our proposed modeling framework using a hypothetical municipality called Sunshine City as a case study.

2   Modeling socio-hydrological systems

Modeling the interactions between human and hydrological systems exacerbates challenges found in modeling purely hydrological systems including setting the model boundary, determining the relevant processes and relationships and clearly communicating model framing and assumptions. Common approaches to hydrological modeling are reviewed to put socio-hydrological modeling in the context of hydrological modeling practice. Next, modeling approaches used in system dynamics and social-ecological systems science, both of which address coupled systems, are described. Then, socio-hydrological modeling approaches are reviewed and gaps identified. While no one approach is directly transferrable to socio-hydrological systems, practices from hydrological modeling, along with those from integrative disciplines, serve as a baseline for comparison and inform our socio-hydrological modeling process. We then present our recommendations for socio-hydrological model conceptualization.

2.1   Modeling hydrological systems

In hydrology the basic steps of model development are (a) data collection and analysis, (b) conceptual model development, (c) translation of the conceptual model to a mathematical model, (d) model calibration and (e) model validation (Blöschl and Sivapalan, 1995). While the basic steps of model development are generally accepted, in practice approaches diverge, particularly in conceptual model development. In hydrology, Wheater et al. (1993) identified four commonly used modeling approaches: physics-based, concept-based (also called conceptual), data driven and hybrid data–conceptual. Physics-based models represent a system by linking small-scale hydrological processes (Sivapalan et al., 2003). Concept-based models use prior knowledge to specify the influential processes and determine the structure. Data driven models are derived primarily from observations and do not specify the response mechanism. Hybrid data–conceptual models use data and prior knowledge to infer model structure (Wheater et al., 1993; Sivapalan et al., 2003).

Modeling purpose typically determines the modeling approach. Environmental models may be developed to formulate and test theories or to make predictions (Beven, 2002). Physics-based models can be used to test theories about small-scale processes or to predict catchment response by scaling up these processes. Concept-based models hypothesize the important elements and processes and their structure of interaction to answer a question or predict a certain property, although hypotheses are often not explicitly stated and tested (Wheater et al., 1993). A reliance on prior knowledge limits the applicability of concept-based modeling in fields lacking consensus on both the presence and relevance of feedback processes. Data driven models are effective in prediction. While they have potential for hypothesis testing, a focus on black box input–output models limits insight into system processes and the ability to extrapolate beyond observed data (Sivapalan et al., 2003). Hybrid data–conceptual models use data and other knowledge to generate and test hypotheses about the structure of the system (Wheater et al., 1993; Young, 2003). As socio-hydrology is a new area of research, prior knowledge alone is insufficient and the focus is on modeling to enhance understanding through hypothesis generation and testing; hybrid data–conceptual modeling tactics aimed at enhancing understanding therefore inform our proposed process.

2.2 Modeling coupled systems

While coupling of natural and human systems is in its infancy in hydrology, there is a strong tradition of studying coupled systems in the fields of system dynamics and social-ecological systems. These fields have developed approaches to understand and model complex systems and can inform a socio-hydrological modeling process. First, in both fields the research question or problem drives modeling decisions. Much of the work to date on socio-hydrological systems is exploratory and aims to explain evidence of system coupling seen in case data. Developing a model to answer a question or solve a problem allows a more structured and defensible framework to support the modeling decisions and provides a benchmark for model validation (Sterman, 2000; Hinkel et al., 2015). For example, Jones et al. (2002), in modeling

the sawmill industry in the northeastern United States focus on understanding if the system has the structural potential to overshoot sustainable yield. While the resulting model is a significant simplification of a complex system, the reason for inclusion of tree growth dynamics, mill capacity and lumber prices and the exclusion of other variables is clear. Second, system dynamics and social-ecological systems science use multiple data sources, both quantitative and qualitative, to specify and parameterize model relationships. Omitting influential relationships or decision points due to lack of quantitative data results in a greater error than their incorrect specification (Forrester, 1992). Third, system dynamics focuses on developing a dynamic hypothesis that explains the system behavior of interest in terms of feedback processes (Sterman, 2000). Finally, social-ecological systems science has found that the use of frameworks as part of a structured model development process can aid transparency and comparability across models (Schlüter et al., 2014).

2.3 Progress and gaps in socio-hydrological modeling

Several research teams have operationalized the concepts of socio-hydrology using approaches ranging from simple generic models to contextual data-driven models. Di Baldassarre et al. (2013) developed a simple generic model to explore the dynamics of human–flood interactions for the purpose of showing that human responses to floods can exacerbate flooding problems. Viglione et al. (2014) extended this work to test the impact of collective memory, risk-taking attitude and trust in risk reduction measures on human–flood dynamics. Kandasamy et al. (2014) analyzed the past 100 years of development in the Murrumbidgee River basin in eastern Australia and built a simple model of the transition from the dominance of agricultural development goals, through a slow realization of adverse environmental impacts, to emergence of serious ecological restoration efforts. Elshafei et al. (2014) proposed a conceptual socio-hydrological model for agricultural catchments and applied it to the Murrumbidgee and the Lake Toolibin basins; they then built upon this conceptual model to construct a detailed semi-distributed model of the Lake Toolibin basin (Elshafei et al., 2015). Srinivasan and collaborators analyzed water security in the city of Chennai, India. By modeling the feedback between household level coping mechanisms and regional-scale stressors, the team explained the counterintuitive effects of policy responses such as the observation that reduced groundwater recharge caused by fixing leaky pipelines decreased a household's ability to use wells to cope with water system interruptions (Srinivasan et al., 2010, 2013).

Researchers have also addressed the methodological questions of how to frame and model socio-hydrological systems. Blair and Buytaert (2015), provide a detailed review of the model types and modeling methods used in socio-hydrology and those that may have utility in the field. Sivapalan and Blösch (2015) offer guidance on framing and modeling socio-hydrological systems from stating framing assumptions to model validation techniques and highlight the specific challenges of scale interactions found in these coupled systems. Elshafei et al. (2014) and Liu et al. (2014) detailed the development of conceptual models, giving readers insight into the framing of their case study work.

These methodological advances have begun to address the many challenges of translating the concept of feedback between human and hydrological systems into actionable science. However, obstacles remain: principally, expanding the scope of modeling to include societal systems and human decision-making exacerbates the challenges of setting the model boundary and process detail, and of evaluating those choices. The source of this challenge is twofold. First, there are fundamental differences between natural and social systems. The laws governing physical, chemical and biological systems such as conservation of mass and energy are broadly applicable across contexts; the relevance of rules influencing social systems varies by context. Second, the modeling of coupled human–hydrological systems is new intellectual territory. At this intersection the norms and unstated assumptions instilled by disciplinary training must be actively questioned and examined within a transparent model development, testing and validation process.

There are no universally accepted laws of human behavior as there are for the physical and biological sciences (Loucks, 2015). While institutions (formal and informal rules) influence behavior, the impact of institutions on the state of the system depends on whether people follow the rules (Schlager and Heikkila, 2011). Additionally, these rules are not static. In response to outcomes of past decisions or changing conditions, actors change both the rules that shape the options available for practical decisions and the rules governing the collective choice process through which these operation rules are made (McGinnis, 2011). Furthermore, water policy decisions are not made in isolation of other policy decisions. Decisions are interlinked as the same actors may interact with and get affected differently depending on the contexts (McGinnis, 2011b). The outcome of a related policy decision may alter the choices available to actors or the resources available to address the current problem. The state of the hydrological system, particularly during extreme events, can spark institutional changes; yet, other factors such as political support and financial resources as well as the preparedness of policy entrepreneurs also play a role (Crow, 2010; Hughes et al., 2013). Given this complexity, Pahl-Wostl et al. (2007) argue that recognizing the unpredictability of policy making and social learning would greatly improve the conceptualization of water management. Nevertheless, some dynamics persist across time and space; water management regimes persist for decades or centuries and some transitions in different locations share characteristics (Elshafei et al., 2014; Kandasamy et al., 2014; Liu et al., 2014). Furthermore, modeling is a useful tool to gain insight into the impacts of these dynamics (Thompson et al., 2013; Sivapalan and

Blöschl, 2015). However, complex systems such as socio-hydrological systems cannot be modeled exhaustively (Sterman, 2000; Schlüter et al., 2014). Rather, model conceptualization must balance sufficient process representation and parsimony (Young et al., 1996; Ostrom, 2007).

Model conceptualization is based on general assumptions about how a system works. Often these assumptions are implicit and not challenged by others within the same research community (Kuhn, 1996). This works well when research stays within the bounds of the existing methods, theories and goals of one's research community; when working in new intellectual territory, research community norms cannot be relied upon to guide assumptions. Further disciplinary training is highly successful at teaching these community norms, and researchers working on interdisciplinary projects must actively question the framing assumptions they bring to the project (Lélé and Norgaard, 2005; McConnell et al., 2009). By its integrative nature, socio-hydrological modeling crosses disciplines and modelers are unable to point to the theoretical framework of any single discipline to make simplifying assumptions (Srinivasan, 2015). In absence of research community norms, we must return to modeling fundamentals. Models are simplifications of real systems that, in a strict sense, cannot be validated but the acceptability of model assumptions for the question at hand can be assessed (Sterman, 2000). Careful articulation of the research questions links the assessment of important variables and mechanisms to the question context. This allows the critique to focus on the acceptability of these choices relative to model goals and enables critical assessment of the range of applicability of identified processes through case and model comparison.

The recent Water Resources Research Debate Series offers an excellent illustration of this point. Di Baldassarre et al. (2015) catalyze the debate by presenting a generic model of human–flood interaction. This model incorporates both the "levee effect", in which periods of infrequent flooding (sometimes caused by flood protection infrastructure) increase the tendency for people to settle in the floodplain, and the "adaptation effect", in which the occurrence of flooding leads to an adaptive response. In the model they link flood frequency and adaptive action through a social memory variable which increases with the occurrence of floods and decays slowly overtime; flood occurrence directly triggers levee heightening in technological societies and indirectly, through the social memory, decreases floodplain population density (Di Baldassarre et al., 2015).

In the debate this modeling approach is both commended as an impressive innovation and critiqued for its simplification of social dynamics (Gober and Wheater, 2015; Loucks, 2015; Sivapalan, 2015; Troy et al., 2015). Gober and Wheater (2015) note that while social or collective memory is an important factor in flood resilience it does not determine flood response; flood awareness may or may not result in an adaptive response based on the way individuals, the media and institutions process the flood threat, the social capacity for adaptation and the preparedness of policy entrepreneurs, among other factors. Loucks (2015) observes that data on past behavior is not necessarily an indicator of future behavior and suggests that observing stakeholder responses to simulated water management situations may offer additional insight. Troy et al. (2015) and Di Baldassarre et al. (2015) note that the human–flood interaction model presented represents a hypothesis of system dynamics which allows for exploration, and that simple stylized models enable generalization across space and time. In sum, the debate presents different perspectives on the acceptability of the modeling assumptions.

A close look at how the debate authors critique and commend the human–flood interaction model illustrates that the acceptability of modeling assumptions hinges upon the model's intended use. For example, Gober and Wheater (2015) critique the simplicity of social memory as a proxy for social system dynamics but acknowledge the utility of the model in clarifying the tradeoffs of different approaches to meet water management goals. As we can never have comprehensive representation of a complex and coupled human–hydrological system, we need transparency of the abstracting assumptions and their motivation. This is not a new insight; however, a question driven modeling process allows the flexibility and transparency needed to examine the acceptability of model assumptions while acknowledging the role of context and the potential for surprise.

2.4 A question driven modeling process

Our proposed process begins with a research question. The research question is then used to identify the key outcome metric(s). A dynamic hypothesis is developed to explain the behavior of the outcome metric over time; a framework can be used to guide and communicate the development of the dynamic hypothesis. Remaining model processes are then specified according to established theory.

As emphasized by both system dynamics and social-ecological systems researchers, the research question drives the process of system abstraction. One way to think about this process of abstraction is through the lens of forward and backward reasoning. Schlüter et al. (2014) introduced the idea of forward and backward reasoning to develop conceptual models of social-ecological systems. In a backward-reasoning approach, the question is first used to identify indicators or outcome metrics; next, the analysis proceeds to identify the relevant processes and then the variables and their relationships, as seen in Fig. 1 (Schülter et al., 2014). These three pieces then form the basis for the conceptual model. In contrast, a forward-reasoning approach begins with the identification of variables and relationships and then proceeds toward outcomes. Forward reasoning is most successful when there is expert knowledge of the system, and backward-reasoning is useful primarily when prior knowl-

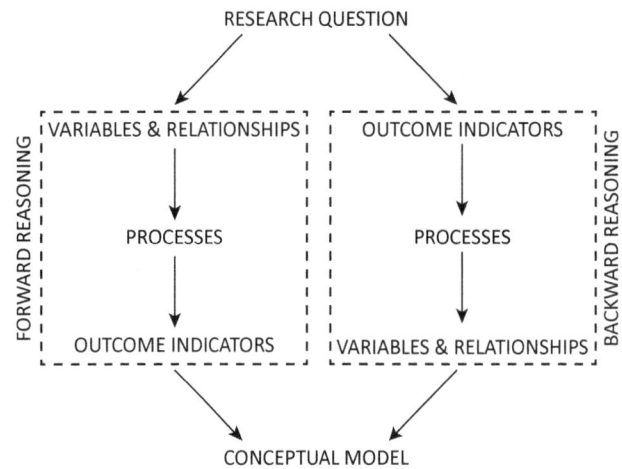

Figure 1. Backward-reasoning process (adapted from Schlüter et al., 2014).

edge is insufficient (Arocha et al., 1993). As few researchers have expert knowledge of all domains involved in socio-hydrological modeling and data is often sparse, a backward-reasoning approach is here used to conceptualize a socio-hydrological model. Additionally, this outcome-oriented approach will focus the scope of the model on the question's relevant variables and processes.

The research question helps to define the outcome metric(s) of interest; however, determining the relevant processes and variables requires further analysis. One tool to identify influential processes and variables is the dynamic hypothesis. A dynamic hypothesis is a working theory, informed by data, of how the system behavior in question arose (Sterman, 2000). It is dynamic in nature because it explains changes in behavior over time in terms of the structure of the system (Stave, 2003). The dynamic hypothesis could encompass the entire socio-hydrological model, but in practice many processes within a model will be based on established theory such as rainfall runoff or evaporation processes. The intent is to focus the dynamic hypothesis on a novel theory explaining observed behavior. Stating the dynamic hypothesis clarifies which portion of the model is being tested.

A framework can aid the development of the dynamic hypotheses and the communication of the reasoning behind it. The use of frameworks enhances the transparency of model development by clearly communicating the modeler's broad understanding of a system. Socio-hydrological modelers can develop their own framework (Elshafei et al., 2014) or draw on existing frameworks that address coupled human–hydrological systems such as the social-ecological systems (SES) framework, the management transition framework, or the integrated structure–actor–water framework (Ostrom, 2007; Pahl-Wostl et al., 2010; Hale et al., 2015).

To illustrate how a framework may be used in model conceptualization we will focus on the SES framework. The SES framework is a nested conceptual map that partitions the attributes of a social-ecological system into four broad classes: (1) resource system, (2) resource units, (3) actors and (4) the governance system (McGinnis and Ostrom, 2014). Each of the four top tier variables has a series of second tier (and potentially higher tier) variables; for example, storage characteristics and equilibrium properties are second tier attributes of the resource system (Ostrom, 2009). The SES framework prescribes a set of elements and general relationships to consider when studying coupled social and ecological systems (Ostrom, 2011). The variables defined in the SES framework were found to impact the interactions and outcomes of social-ecological systems in a wide range of empirical studies (Ostrom, 2007). In addition to specifying candidate variables, the SES framework specifies broad process relationships (Schlüter et al., 2014). At the broadest level, SES specifies that the state of the resource system, governance system, resource unit properties and actor characteristics influence interactions and are subsequently influenced by the outcomes of those interactions. To operationalize the SES framework for model conceptualization one must move down a level to assess the relevance of the tier two variables against case data and background knowledge. This review aims to check the dynamic hypothesis against a broader view of coupled system dynamics and to inform determination of remaining model processes.

The following case presents the development of a sociohydrological (coupled) and a traditional (noncoupled) model to illustrate this process. While this process is developed to study real-world cases a hypothetical case is used here for simplicity, brevity and proof of concept.

3   Sunshine City: a case study of reservoir operations

Sunshine City is located in a growing region in a semi-arid climate. The region is politically stable, technologically developed, with a market economy governed by a representative democracy. Sunshine City draws its water supply from the Blue River, a large river which it shares with downstream neighbors. The water users must maintain a minimum flow in the Blue River for ecological health. Sunshine City can draw up to 25% of the annual flow of the Blue River in any given year. A simple prediction of the year's flow is made by assuming that the flow will be equal to the previous year's flow; the resulting errors are corrected by adjusting the next year's withdrawal.

The city's Water Utility is responsible for diverting, treating and transporting water to city residents and businesses. It is also tasked with making infrastructure investment decisions, setting water prices. Water users receive plentiful supply at cost and there have been no shortages in recent years. While located in a semi-arid environment, the large size of Sunshine City's Blue River water availability and allocation created a comfortable buffer. The city's Water Utility is also

Table 1. Summary of Sunshine City properties.

Sunshine City properties		
Variable	Value	Units
Blue River mean flow	2	$km^3 \, yr^{-1}$
Blue River variance	0.5	$km^3 \, yr^{-1}$
Blue River lag 1 autocorrelation	0.6	–
Average evaporation rate	1	$m \, yr^{-1}$
Population	1 000 000	people
Average annual growth rate	3	%
Per capita water usage	400	$m^3 \, yr^{-1}$
Water price	0.25	$USD \, m^{-3}$
Reservoir capacity	0.2	km^3
Reservoir slope	0.1	–

responsible for setting water efficiency codes and other conservation rules. The current building code includes only basic efficiencies required by the national government. The Blue River, along with other regional sources, is fully allocated making future augmentation of supplies unlikely. See Table 1 above for a summary of key characteristics of Sunshine City.

Along with the rest of the region, Sunshine City's population, and its water demand, has grown rapidly over the past few years. Managers at the Water Utility are concerned they will no longer be able to meet its reliability targets as demands rise and have added a reservoir to increase future reliability. They now must decide how to operate the reservoir and are considering two options: standard operating policy (SOP) and hedging policy (HP). The selected operating policy must satisfy downstream user rights and maintain minimum ecological flows. In addition to meeting the legal requirements, the Water Utility managers are concerned with finding a policy that will enable the city to provide the most reliable water supply throughout the lifetime of the reservoir (50–100 years). From experience they have observed that both water price and reliability affect demand. A key puzzle that emerges for water managers from this experience is: *how do operational rules governing use of water storage influence long-term water supply reliability when consumers make water usage decisions based on both price and reliability?*

As the question implies, the Water Utility managers have a working hypothesis relating demand change with water shortages. Therefore, along with the research question the following dynamic hypothesis is considered: *the occurrence of water shortages increases the tendency of users to adopt water conservation technologies and to make long-term behavioral changes. HP triggers shortages sooner than SOP thus triggering earlier decreases in demand.*

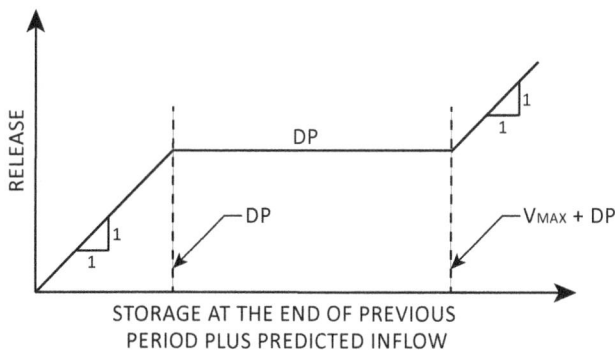

Figure 2. Standard operating policy, where D is per capita demand, P is population and V_{max} is reservoir capacity (adapted from Shih and ReVelle, 1994).

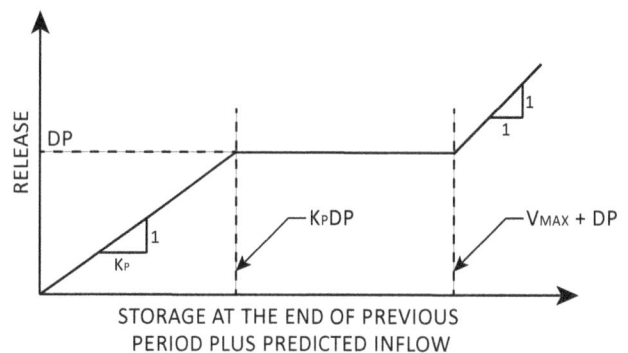

Figure 3. Hedging policy, where K_P is hedging release function slope (adapted from Shih and ReVelle, 1994).

3.1 Background

The decision of how much water to release for use each time period is deceptively complex due to the uncertainty of future streamflows and the nonlinear benefits of released water (Shih and ReVelle, 1994; Draper and Lund, 2004). In making release decisions, water utilities must fulfill their mandate to maintain a reliable water supply in a fiscally efficient manner. Reliability is the probability that the system is in a satisfactory state (Hashimoto et al., 1982). In this case, a satisfactory system state is one in which all demands on the system can be met. The definition of an unsatisfactory state is more nuanced. Water shortages have a number of characteristics that are important to water management including frequency, maximum shortage in a given time period, and length of shortage period (Cancelliere et al., 1998). Long-term reliability here refers to the projected reliability over several decades. The time frame used for long-term projections varies between locations and utilities (i.e., Boston uses a 25-year time frame, Denver uses a 40-year time frame, and Las Vegas uses a 50-year time frame) and a 50-year time frame is used here (MWRA, 2003; SNWA, 2009; Denver Water, 2015).

Two operational policies, SOP and HP, are commonly used to address this decision problem. Under SOP, demand is always fulfilled unless available supply drops below demand; under HP, water releases are limited in anticipation of an expected deficit (Cancelliere et al., 1998). Hedging is used as a way to decrease the risk of a large shortfall by imposing conservation while stored water remains available. Figures 2 and 3 illustrate SOP and HP, respectively. For this simple experiment only linear hedging, where K_P is the slope of the release function, is tested.

The traditional argument for hedging is that it is economical to allow a small deficit in the current time period in order to decrease the probability of a more severe shortage in a future time periods (Bower et al., 1962). This argument holds true if the loss function associated with a water shortage is

nonlinear and convex; in other words that a severe shortage has a larger impact than the sum of several smaller shortages (Shih and ReVelle, 1994). Gal (1979) showed that the water shortage loss function is convex, thereby proving the utility of hedging as a drought management strategy. Other researchers have shown that hedging effectively reduces the maximum magnitude of water shortages and increases total utility over time (Shih and ReVelle, 1994; Cancelliere et al., 1998). More recent work by Draper and Lund (2004) and You and Cai (2008) confirms previous findings and demonstrates the continued relevance reservoir operation policy selection.

Researchers and water system managers have for decades sought improved policies for reservoir operation during drought periods (Bower et al., 1962; Shih and ReVelle, 1994; You and Cai, 2008). We add to this classic question the observation that water shortages influence both household conservation technology adoption rates and water use behavior. In agreement with Giacomoni et al. (2013), we hypothesize that the occurrence of water shortages increases the tendency of users to adopt water conservation technologies and to make long-term behavioral changes. Household water conservation technologies include low flow faucets, shower heads and toilets, climatically appropriate landscaping, grey water recycling and rainwater harvesting systems (Schuetze and Santiago-Fandiño, 2013). The adoption rates of these technologies are influenced by a number of factors including price, incentive programs, education campaigns and peer adoption (Campbell et al., 2004; Kenney et al., 2008). A review of studies in the US, Australia and UK showed that the installation of conservation technologies results in indoor water savings of 9–12 % for fixture retrofits and 35–50 % for comprehensive appliance replacements (Inman and Jeffrey, 2006). In some cases offsetting behavior reduces these potential gains; however, even with offsetting, the adoption of conservation technologies still results in lower per capita demands (Geller et al., 1983; Fielding et al., 2012). Water use behavior encompasses the choices that individuals make related to water use ranging from length of showers

Table 2. Household conservation action by shortage experience (ISTPP, 2013).

Last experienced a water shortage	Percent of households, over the past year, that have		
	invested in efficient fixtures or landscapes	changed water use behavior	taken no action
Within a year	56 %	88 %	11 %
1–2 years ago	52 %	87 %	11 %
2–5 years ago	51 %	78 %	17 %
6–9 years ago	50 %	79 %	18 %
10 or more years ago	42 %	74 %	24 %
Never experienced	36 %	66 %	31 %

and frequency of running the dishwasher to timing of lawn watering and frequency of car washing. Water use behavior is shaped by knowledge of the water system, awareness of conservation options and their effectiveness, and consumer's attitudes toward conservation (Frick et al., 2004; Willis et al., 2011). Changes to water use behavior can be prompted by price increases, education campaigns, conservation regulations and weather (Campbell et al., 2004; Kenney et al., 2008; Olsmtead and Stavins, 2009).

As a city begins to experience a water shortage, the water utility may implement water restrictions, price increases, incentive programs or education campaigns to influence consumer behavior. While staff within the water utility or city may have planned these measures before, the occurrence of a water shortage event, particularly if it aligns with other driving forces, offers a window of opportunity to implement sustainable water management practices (Jones and Baumgartner, 2005; Hughes et al., 2013). In addition, water users are more likely to respond to these measures with changes in their water use behavior and/or adoption of conservation technologies during shortages. Baldassare and Katz (1992) examined the relationship between the perception of risk to personal well-being from an environmental threat and adoption of environmental practices with a personal cost (financial or otherwise). They found that the perceived level of environmental threat is a better predictor for individual environmental action, including water conservation, than demographic variables or political factors. Illustrating this effect, Mankad and Tapsuwan (2011) found that adoption of alternative water technologies, such as on-site treatment and reuse, is increased by the perception of risk from water scarcity.

Evidence of individual level behavior change can also be seen in the results of a 2013 national water policy survey conducted by the Institute for Science, Technology and Public Policy at Texas A&M University. The survey sampled over 3000 adults from across the United States about their attitudes and actions related to a variety of water resources and public policy issues. Included in the survey were questions that asked respondents how recently, if ever, they personally experienced a water shortage and which, if any, household efficiency upgrade or behavioral change actions their household had taken in the past year. Efficiency upgrade options offered included low-flow shower heads, low-flush toilets and changes to landscaping; behavioral options given included shorter showers, less frequent dishwasher or washing machine use, less frequent car washing and changes to yard watering (ISTPP, 2013). As seen in Table 2, respondents who had recently experienced a water shortage were more likely to have made efficiency investments and to have changed their water use behavior. This finding is corroborated by a recent survey of Colorado residents. Of the 72 % of respondents reporting increased attention to water issues, the most-cited reason for the increase (26 % of respondents) was a recent drought or dry year (BBC Research, 2013). Other reasons cited by an additional 25 % of respondents including news coverage, water quantity issues and population growth may also be related water shortage concerns or experiences.

The increased receptivity of the public to water conservation measures and the increased willingness of water users to go along with these measures during shortage events combine to drive changes in per capita demands. The combined effect of these two drivers was demonstrated in a study of the Arlington, Texas, water supply system (Giacomoni et al., 2013; Kanta and Zechman, 2014). Additional examples of city- and regional-scale drought response leading to long-term demand decreases include the droughts of 1987–1991 and the mid-2000s in California and of 1982–1983 and 1997–2009 in Australia (Zilberman et al., 1992; Turral, 1998; Sivapalan et al., 2012; Hughes et al., 2013). It is often difficult to separate the relative effects of the multiple price and nonprice approaches applied by water utilities during droughts (Olmstead and Stavins, 2009). The point is, however, that the response generally points to lower per capita water demands.

One example of lasting water use reductions after a shortage is the 1987–1992 drought in Los Angeles, California. An extensive public awareness and education campaign sparked both behavioral changes and the adoption of efficient fixtures such as low-flow shower heads and toilets and increasing block pricing introduced after the drought helped maintain conservation gains (LADWP, 2010). Evidence of the lasting effect can be seen in Fig. 4. Per capita water demands do not return to 1990 levels after the drought ends in 1992. Note that the data below also contains a counter example. The 1976–1977 drought caused a sharp drop in water consumption in Los Angeles; however, consumption quickly returned to predrought levels when the rainfall returned in 1978. While the 1976–1977 drought was more intense than any year in the 1987–1992 drought, the long duration of the later drought caused deeper draw downs in the city's water reserves ultimately prompting transformative action (LADWP, 2010). This may indicate that the impact of the 1976–1977 drought was below the threshold for significant action or that other priorities dominated public attention and resources at the

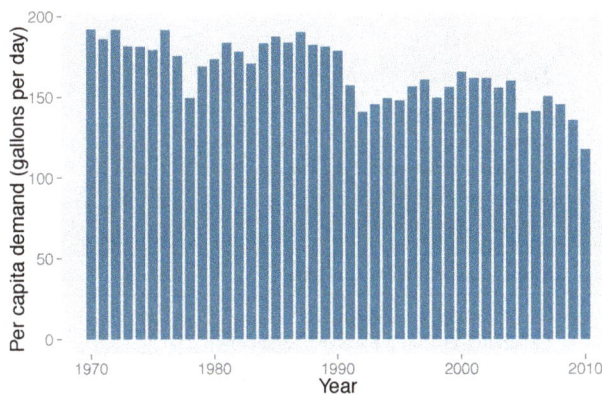

Figure 4. Historical city of Los Angeles water use (LADWP, 2010).

time. In sum, the Los Angeles case serves both to illustrate that hydrological change can prompt long-term changes in water demands and as a reminder that multiple factors influence water demands and hydrological events will not always dominate.

3.2 Model development

The Sunshine City water managers want to understand how the operational rules governing use of water storage influence long-term water supply reliability when consumers make water usage decisions based on price and reliability. A model can help the managers gain insight into system's behavior by computing the consequences of reservoir operation policy choice over time and under different conditions. As described in the background section, many supply side and demand side factors affect water system reliability. However, not all variables and processes are relevant for a given question. A question driven modeling process uses the question to determine model boundary and scope rather than beginning with a prior understanding of the important variables and processes. A question driven process is here used to determine the appropriate level of system abstraction for the Sunshine City reservoir operations model.

From the research question it is clear that reliability is the outcome metric of interest and that the model must test for the hypothesized link between demand changes and reliability. Reliability, as defined above, is the percent of time that all demands can be met. The SES framework is used to guide the selection of processes and variables, including the dynamic hypothesis. Given this wide range, the framework was then compared against the variables and processes found to be influential in urban water management and socio-hydrological studies (Brezonik and Stadelmann, 2002; Abrishamchi et al., 2005; Padowski and Jawitz, 2012; Srinivasan et al., 2013; Dawadi and Ahmad, 2013; Elshafei et al., 2014; Gober et al., 2014; Liu et al., 2014; Pande et al., 2013; van Emmerik et al., 2014). Based on this evaluation two second tier variables were added to the framework: land use to the resource system

characteristics and water demand to interactions; other variables were modified to reflect the language typically used in the water sciences (i.e., supply in place of harvesting). See Table 3 for urban water specific modification of the SES framework.

We then assess the relevance of the tier two variables against case data and background knowledge (summarized in Sects. 3 and 3.1, respectively) by beginning with the outcome metric, reliability. Within the framework reliability is an outcome variable, specifically a social performance metric, and it is the direct result of water supply and water demand interaction processes. Water supply encompasses the set of utility level decisions on reservoir withdrawals and discharges. As detailed in the case description, these decisions are shaped by the selected reservoir operating policy, streamflow, the existing environmental flow and downstream allocation requirements, reservoir capacity, water in storage and water demands. Streamflow is a stochastic process that is a function of many climatic, hydraulic and land surface parameters. However, given the driving question and the assumption that the city represents only a small portion of the overall watershed, a simple statistical representation is sufficient and streamflow is assumed independent of other model variables.

Total water demand is a function of both population and per capita demand. As described in the background section, per capita water demand changes over time in response to household level decisions to adopt more water efficient technologies and water use behavior change made by individuals in each time interval; these decisions may be influenced by conservation policies. As conditions change water users reassess the situation and, if they choose to act, decide between available options such as investment in efficient technology, changing water use behavior and, in extreme cases, relocation. Therefore, per capita demand is a function of price and historic water reliability as well as available technologies, and water user's perception of the water system. Since the focus of the question is on system wide reliability individual level decisions can be modeled in the aggregate as total demand, which is also influenced by population. Population increases in proportion to the current population, as regional economic growth is the predominant driver of migration trends. However, in extreme cases, perceptions of resource limitations can also influence growth rates. The SES variables used in the conceptual model are highlighted in Table 3 and the resulting processes are summarized in Fig. 5.

Only a subset of the variables and processes articulated in the SES framework are included in the conceptual model; other variables and processes were considered but not included. For example, economic development drives increasing per capita water demands in many developing regions but the relationship between economic growth and water demands in highly developed regions is weaker due to the increased cost of supply expansion and greater pressure for environmental protection (Gleick, 2000). The income elas-

Table 3. SES framework, modified for urban water systems.

First tier variables	Second tier variables	Third tier variables (examples)
Socio, economic and political settings	S1 – Economic development	Per capita income
	S2 – **Demographic trends**	Rapid growth
	S3 – Political stability	Frequency of government turnover
	S4 – **Other governance systems**	Related regulations
	S5 – Markets	Regional water markets
	S6 – Media organizations	Media diversity
	S7 – Technology	Infrastructure, communications
Resource systems[1]	RS1 – Type of water resource	Surface water, groundwater
	RS2 – Clarity of system boundaries	Groundwater–surface water interactions
	RS3 – Size of resource system	Watershed or aquifer size
	RS4 – **Human-constructed facilities**	Type, capacity, condition
	RS5 – Catchment land use	Urbanization, reforestation
	RS6 – **Equilibrium properties**	Mean streamflow, sustainable yield
	RS7 – **Predictability of system dynamics**	Data availability, historic variability
	RS8 – **Storage characteristics**	Natural/built, volume
	RS9 – Location	
Governance systems[2]	GS1 – Government organizations	Public utilities, regulatory agencies
	GS2 – Nongovernment organizations	Advocacy groups, private utilities
	GS3 – Network structure	Hierarchy of organizations
	GS4 – *Water*-rights systems	Prior appropriation, beneficial use
	GS5 – **Operational-choice rules**	Water use restrictions, operator protocol
	GS6 – **Collective-choice rules**	Deliberation rules, position rules
	GS7 – Constitutional-choice rules	Boundary rules, scope rules
	GS8 – Monitoring and sanctioning rules	Enforcement responsibility
Resource units[3]	RU1 – *Interbasin connectivity*	Infrastructure, surface–groundwater interactions
	RU2 – **Economic value**	Water pricing, presence of markets
	RU3 – *Quantity*	Volume in storage, current flow rate
	RU4 – Distinctive characteristics	Water quality, potential for public health impacts
	RU5 – **Spatial and temporal distribution**	Seasonal cycles, interannual cycles
Actors	A1 – **Number of relevant actors**	
	A2 – Socioeconomic attributes	Education level, income, ethnicity
	A3 – **History or past experiences**	Extreme events, government intervention
	A4 – Location	
	A5 – Leadership/entrepreneurship	Presence of strong leadership
	A6 – Norms (trust-reciprocity)/social capital	Trust in local government
	A7 – **Knowledge of SES/mental models**	Memory, mental models
	A8 – Importance of resource (dependence)	Availability of alternative sources
	A9 – **Technologies available**	Communication technologies, efficiency technologies
	A10 – *Values*	Preservation of cultural practices
Action situations: interactions → outcomes[4]	I1 – *Water supply*	Withdrawal, transport, treatment, distribution
	I2 – Information sharing	Public meetings, word of mouth
	I3 – Deliberation processes	Ballot initiatives, board votes, public meetings
	I4 – Conflicts	Resource allocation conflicts, payment conflicts
	I5 – Investment activities	Infrastructure construction, conservation technology
	I6 – Lobbying activities	Contacting representatives
	I7 – Self-organizing activities	Formation of NGOs
	I8 – Networking activities	Online forums
	I9 – Monitoring activities	Sampling, Inspections, self-policing
	I10 – *Water demand*	Indoor/Outdoor, residential/commercial/industrial
	O1 – **Social performance measures**	Efficiency, equity, accountability
	O2 – Ecological performance measures	Sustainability, minimum flows
	O3 – Externalities to other SESs	Ecosystem impacts
Related ecosystems	ECO1 – **Climate patterns**	El Niño impacts, climate change projections
	ECO2 – Pollution patterns	Urban runoff, upstream discharges
	ECO3 – **Flows into and out of focal SES**	Upstream impacts, downstream rights

Note: variables added are in italic, variables key to the conceptual model are in bold. Examples of third tier variables are given for clarification. [1] Resource system variables removed or replaced: productivity of system. [2] Governance system variables removed or replaced: property. [3] Resource unit variables removed or replaced: resource unit mobility, growth or replacement rate, interaction among resource units, number of units. [4] Interaction and outcome variables removed or replaced: harvesting.

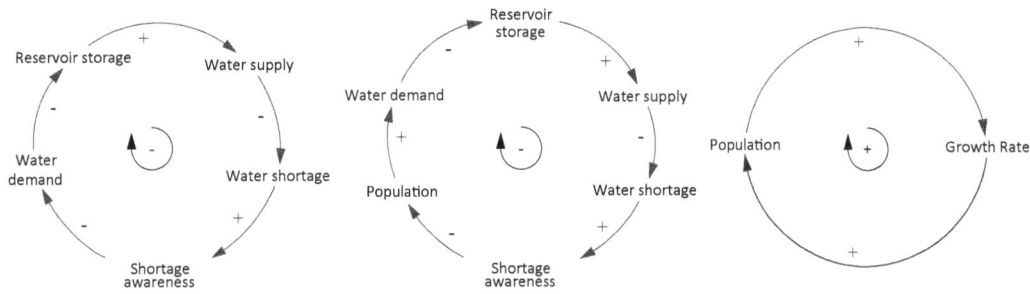

Figure 5. Causal loop diagrams: **(a)** water demand, shortage and conservation; **(b)** water demand, shortage and population; **(c)** population and growth rate.

Table 4. State and exogenous model variables.

Variable	Description	Units	Equation	Variable Type
Q	Streamflow	$km^3 \, yr^{-1}$	1	Exogenous
V	Reservoir storage volume	km^3	2	State
P	Population	Persons	3	State
W	Withdrawal	$km^3 \, yr^{-1}$	4	State
S	Shortage magnitude	$km^3 \, yr^{-1}$	5	State
M	Shortage awareness		6	State
D	Per capita demand	$m^3 \, yr^{-1}$	7	State

ticity of water can lead to increased water demands if rates do not change proportionally (Dalhuisen et al., 2003); here prices are assumed to keep pace with inflation. Given this assumption, and the focus on a city in a developed region, economic development likely plays a minor role. Similarly, group decision-making and planning processes such as public forums, voting and elections can shape the responses to reliability changes over time. This model aims to answer a question about the impact of a policy not the ease or likelihood of its implementation. Once the policy is established through whatever process that is used, the question here focuses on its efficacy. Therefore, group decision-making processes need not be included.

In addition to determining the appropriate level of detail of the conceptual model, we must determine which variables change in response to forces outside the model scope (exogenous variables), which variables must be modeled endogenously (state variables) and which can be considered constants (parameters). Again the nature of the question along with the temporal and spatial scale informs these distinctions. Variables such as stored water volume, per capita water demand and shortage awareness will clearly change over the 50-year study period. The population of the city is also expected to change over the study period. Under average hydrological conditions the population growth rate is expected to be driven predominately by regional economic forces exogenous to the system; however, under extreme conditions water supply reliability can influence the growth rate. Therefore, population is considered a state variable. Streamflow

characteristics may change over the 50-year timescale in response to watershed wide land use changes and global-scale climatic changes. Streamflow properties are first considered stationary parameters in order to understand the impact of the selected operating policy in isolation from land use and climate change. Climate scenarios or feedbacks between population and land use can be introduced in future applications of the model to test their impact on system performance. Reservoir operating policy, summarized as the hedging slope, K_P, is considered a parameter in the model. Alternate values of parameter K_P are tested but held constant during the study period to understand the long-term impacts of selecting a given policy. Reservoir properties such as capacity and slope are also held constant to hone in on the effect of operating policy. See Tables 4 and 5 for a summary of variable types. From these model relationships, general equations are developed by drawing from established theory, empirical findings and working hypotheses.

Streamflow, Q, is modeled using a first-order autoregressive model, parameterized by mean ($\mu_H \, km^3 \, yr^{-1}$), standard deviation ($\sigma_H \, km^3 \, yr^{-1}$) and lag one autocorrelation (ρ_H). The final term, a_t, is a normally distributed random variable with a mean zero and a standard deviation of 1.

$$Q_t = \rho_H (Q_{t-1} - \mu_H) + \sigma_H \left(1 - p_H^2\right)^{0.5} a_t + \mu_H \qquad (1)$$

At each time step the amount of water in storage, V, in the reservoir is specified by a water balance equation, where W is water withdrawal (km^3), η_H ($km \, yr^{-1}$) is evaporation, A is area (km^2), Q_D (km^3) is downstream demand and Q_E (km^3)

Table 5. Model parameters.

Parameters	Description	Value	Units	Equation
μ_H	Mean streamflow	2.0	$km^3\,yr^{-1}$	1
σ_H	Standard deviation of streamflow	0.5	$km^3\,yr^{-1}$	1
ρ_H	Streamflow lag one autocorrelation	0.6	–	1
η_H	Evaporation rate	0.001	$km\,yr^{-1}$	2
Q_D	Downstream allocation	$0.50Q$	km^3	2
Q_E	Required environmental flow	$0.25Q$	km^3	2
σ_T	Average slope of reservoir	0.1	–	Stage–storage curve
δ_I	Regional birth rate	0.04	yr^{-1}	3
δ_E	Regional death rate	0.03	yr^{-1}	3
δ_I	Regional immigration rate	0.05	yr^{-1}	3
δ_E	Regional emigration rate	0.03	yr^{-1}	3
τ_P	Threshold	0.4	–	3
V_{max}	Reservoir capacity	2.0	km^3	4
K_P	Hedging slope	Variable	–	5
μ_S	Awareness loss rate	0.05	yr^{-1}	6
α_D	Fractional efficiency adoption rate	0.15	–	7
β_D	Background efficiency rate	0.0001	–	7
D_{MIN}	Minimum water demand	200	$m^3\,yr^{-1}$	7

is the required environmental flow.

$$\frac{dV}{dt} = Q_t - W_t - \eta_H A_t - Q_D - Q_E \quad (2)$$

Population is the predominant driver of demand in the model. Population (P) changes according to average birth (δ_B, yr^{-1}), death (δ_D, yr^{-1}), emigration (δ_E, yr^{-1}) and immigration (δ_I, yr^{-1}) rates. However, immigration is dampened and emigration accelerated by high values of perceived shortage risk, as would be expected at extreme levels of resource uncertainty (Sterman, 2000). The logistic growth equation, which simulates the slowing of growth as the resource carrying capacity of the system is approached, serves as the basis for the population function. While the logistic function is commonly used to model resource-constrained population growth, the direct application of this function would be inappropriate for two reasons. First, an urban water system is an open system; resources are imported into the system at a cost and people enter and exit the system in response to reductions in reliability and other motivating factors. Second, individuals making migration decisions may not be aware of incremental changes in water shortage risk; rather, perceptions of water stress drive the damping effect on net migration. Finally, only at high levels does shortage perception influence population dynamics. To capture the effect of the open system, logistic damping is applied only to immigration driven population changes when shortage perception crosses a threshold, τ_P. To account for the perception impact, the shortage awareness variable, M, is used in place of the ratio of population to carrying capacity typically used; this modi-

fication links the damping effect to perceived shortage risk.

$$\frac{dP}{dt} = \begin{cases} P_t\,[\delta_B - \delta_D + \delta_I - \delta_E] \\ P_t\,[(\delta_B - \delta_D) + \delta_I(1 - M_t) - \delta_E(M_t)] & \text{for } M_t \geq \tau_P \end{cases} \quad (3)$$

Water withdrawals, W, are determined by the reservoir operating policy in use. As there is only one source, water withdrawn is equivalent to the quantity supplied. The predicted streamflow for the coming year is $0.25 \times Q_{t-1}$, accounting for both downstream demands and environmental flow requirements. Under SOP, K_P is equal to one which sets withdrawals equal to total demand, DP (per capita demand multiplied by population), unless the stored water is insufficient to meet demands. Under HP, withdrawals are slowly decreased once a pre-determined threshold, K_PDP, has been passed. For both policies excess water is spilled when stored water exceeds capacity, V_{max}.

$$W_t = \quad (4)$$
$$\begin{cases} V_t + 0.25Q_{t-1} - V_{max} & \text{for} \quad \begin{array}{l} V_t + 0.25Q_{t-1} \geq \\ D_t P_t + V_{max} \end{array} \\ D_t P_t & \text{for} \quad \begin{array}{l} D_t P_t + V_{max} > \\ V_t + 0.25Q_{t-1} \geq K_P D_t P_t \end{array} \\ \dfrac{V_t + 0.25Q_{t-1}}{K_P} & \text{for} \quad K_P D_t P_t > V_t + 0.25Q_{t-1} \end{cases}$$

When the water withdrawal is less than the quantity demanded by the users, a shortage, S, occurs.

$$S_t = \begin{cases} D_t P_t - W_t & \text{for } D_t P_t > W_t \\ 0 & \text{otherwise} \end{cases} \quad (5)$$

Di Baldassarre et al. (2013) observed that in flood plain dynamics awareness of flood risk peaks after a flood event. This model extends that observation to link water shortage events to the awareness of shortage risk. The first term in the equation is the shortage impact which is a convex function of the shortage volume. The economic utility of hedging hinges on the assumption that the least costly options to manage demand will be undertaken first. As both water utilities and water users have a variety of demand management and conservation options available and both tend to use options from most to least cost-effective, a convex shortage loss is also applicable to the water users (Draper and Lund, 2004). It is here assumed that the contribution of an event to shortage awareness is proportional to the shortage cost. At high levels of perceived shortage risk only a large shortage will lead to a significant increase in perceived risk. The adaptation cost is multiplied by one minus the current shortage awareness to account for this effect. The second term in the equation incorporates the decay of shortage, μ_S (yr^{-1}), awareness and its relevance to decision making that occurs over time (Di Baldassarre et al., 2013).

$$\frac{dM}{dt} = \left(\frac{S_t}{D_t P_t}\right)^2 (1 - M_t) - \mu_S M_t \qquad (6)$$

Historically, in developed regions per capita water demands have decreased over time as technology improved and as water use practices have changed. As described above, this decrease is not constant but rather is accelerated by shocks to the system. To capture this effect there are two portions to the demand change equation: shock-stimulated logistic decay with a maximum rate of α (yr^{-1}) and a background decay rate, β (yr^{-1}). Per capita water demand decrease accelerates in a time interval if water users are motivated by recent personal experience with water shortage (i.e., $M > 0$). As a certain amount of water is required for basic health and hygiene, there is ultimately a floor to water efficiencies, specified here as D_{min} (km^3 yr^{-1}). Reductions in per capita water usage become more challenging as this floor is approached; a logistic decay function is used to capture this effect. When no recent shortages have occurred (i.e., $M = 0$), there is still a slow decrease in per capita water demands. This background rate, β, of demand decrease is driven by both the replacement of obsolete fixtures with modern water efficient fixtures and the addition of new more efficient building stock. This background rate is similarly slowed as the limit is approached; this effect is incorporated by using a percentage-based background rate. Note that price is not explicitly included in this formulation of demand. As stated above, because price and nonprice measures are often implemented in concert it is difficult to separate the impacts of these two approaches and in this case unnecessary.

$$\frac{dD}{dt} = -D_t \left[M_t \alpha \left(1 - \frac{D_{min}}{D_t}\right) + \beta \right] \qquad (7)$$

As a comparison, a noncoupled model was developed. In this model, population and demand changes are no longer modeled endogenously. The shortage awareness variable is removed as it no longer drives population and demand changes. Instead the model assumes that population growth is constant at 3 % and that per capita demands decrease by 0.5 % annually. While these assumptions may be unrealistic they are not uncommon. Utility water management plans typically present one population and one demand projection. Reservoir storage, water withdrawals and shortages are computed according to the equations described above. A full list of model variables and parameters can be found in Tables 4 and 5, respectively.

3.3 Results

The model was run for SOP ($K_P = 1$) and three levels of HP where level one ($K_P = 1.5$) is the least conservative, level two ($K_P = 2$) is slightly more conservative and level three ($K_P = 3$) is the most conservative hedging rule tested. Three trials were conducted with a constant parameter set to understand the system variation driven by the stochastic streamflow sequence and to test if the relationship hypothesized was influential across hydrological conditions. For each trial streamflow, reservoir storage, shortage awareness, per capita demand, population and total demand were recorded and plotted. As a comparison, each trial was also run in the noncoupled model in which demand and population changes are exogenous.

In the first trial, shown in Fig. 6a, there were two sustained droughts in the study period: from years 5 to 11 and then from years 33 to 37. Higher than average flows in the years preceding the first drought allowed the utility to build up stored water as seen in Fig. 6b. The storage acts as a buffer and the impacts are not passed along to the water users until year 18 under SOP. Under HP the impacts, as well as water users' shortage awareness, increase in years 15, 13 and 12 based on the level of the hedging rule (slope of K_P) applied, as shown in Fig. 6c. The impact of this rising shortage awareness on per capita water demands is seen in the acceleration of the decline in demands in Fig. 6d. This demand decrease is driven by city level policy changes such as price increases and voluntary restrictions in combination with increased willingness to conserve.

The impacts of this decrease on individual water users will depend on their socio-economic characteristics as well as the particular policies implemented. While the aggregation hides this heterogeneity, it should be considered in the interpretation of these results. The increased shortage awareness also has a small dampening effect on population growth during and directly after the first drought (Fig. 6e). Changes to both per capita demands and population result in total demand changes (see Fig. 6f). After the first drought the system begins to recover under each of the three hedging policies as evidenced by the slow increase in reservoir storage. However, as streamflows fluctuate around average streamflow and total demands now surpass the average allocation, reservoir storage does not recover when no hedging restrictions are

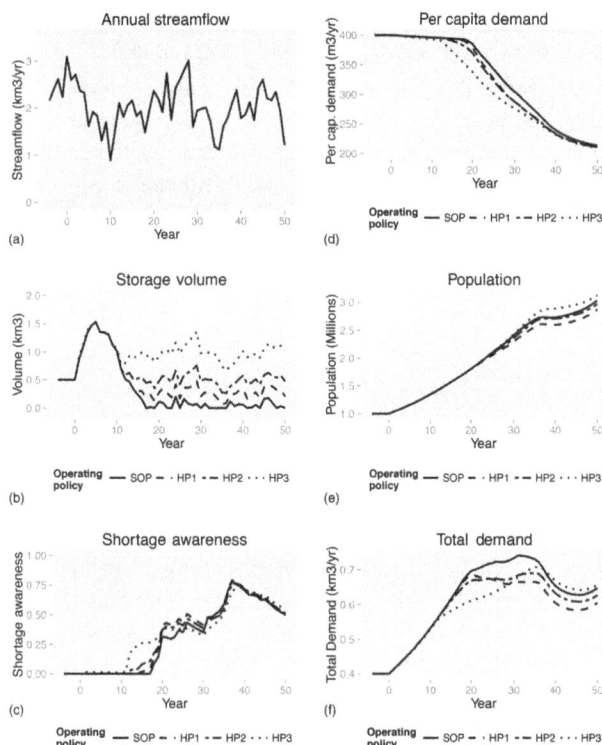

Figure 6. Model results, trial 1: (**a**) annual streamflow, (**b**) reservoir storage volume, (**c**) public shortage awareness, (**d**) per capita demand, (**e**) annual city population, (**f**) total demand.

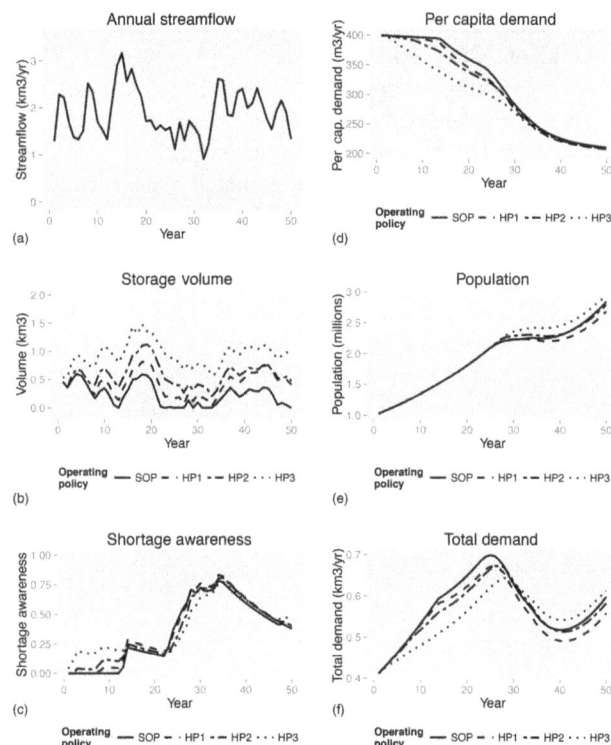

Figure 7. Model results, trial 2: (**a**) annual streamflow, (**b**) reservoir storage volume, (**c**) public shortage awareness, (**d**) per capita demand, (**e**) annual city population, (**f**) total demand.

imposed. Several years of above average flow ending in year 29 drive further recovery. The second prolonged drought has the most pronounced effect under the SOP scenario. Shortage impacts are drastic, driving further per capita demand decreases and a temporary decline in population. A slight population decrease is also seen under level one hedging but the results demonstrate that all hedging strategies dampen the effect.

In the second trial there are two brief droughts in the beginning of the study period, beginning in years 4 and 10, as seen in Fig. 7a. Under SOP and the first two hedging policies there is no change in operation for the first drought and the reservoir is drawn down to compensate as seen in Fig. 7a–b. Only under the level three HP are supplies restricted, triggering an increase in shortage awareness and a subsequent decrease in per capita demands, as found in Fig. 7c and d. When the prolonged drought begins in year 20, the four scenarios have very different starting points. Under SOP, there is less than $0.5\,km^3$ of water in storage and total annual demands are approximately $0.65\,km^3$. In contrast, under the level three HP there is $1.4\,km^3$ of water in storage and total annual demands are just under $0.6\,km^3$. Predictably, the impacts of the drought are both delayed and softened under HP. As the drought is quite severe, all scenarios result in a contraction of population. However, the rate of decrease and total population decrease is lowered by the use of HP.

In the third and final trial there is no significant low flow period until year 36 of the simulation when a moderate drought event occurs, as shown in Fig. 8a. Earlier in the simulation minor fluctuations in streamflow only trigger an acceleration of per capita demand declines under the level three HP, as seen in Fig. 8c and d. A moderate drought begins in year 36. However, the reservoir levels drop and shortage awareness rise starting before year 20, as seen in Fig. 8b and c. Then when the drought occurs the impacts are far greater than in the comparably moderate drought in trial 1 because a prolonged period of steady water supply enabled population growth and placed little pressure on the population to reduce demands. In the SOP scenario, the system was in shortage before the drought occurred and total demands peaked in year 30 at $0.82\,km^3$. The subsequent drought exacerbated an existing problem and accelerated changes already in motion.

Figure 9 presents results of the noncoupled model simulation. While the control model was also run for all three trials, the results of only trial three are included here for brevity. In the noncoupled model, HP decreases water withdrawals as reservoir levels drop and small shortages are seen early in the study period, as seen in Fig. 9b and c. In the second half of the study period significant shortages are observed, as in Fig. 9c. However, inspection of the streamflow sequence reveals no severe low flow periods indicating that the shortages are driven by increasing demands, as in Fig. 9a. As ex-

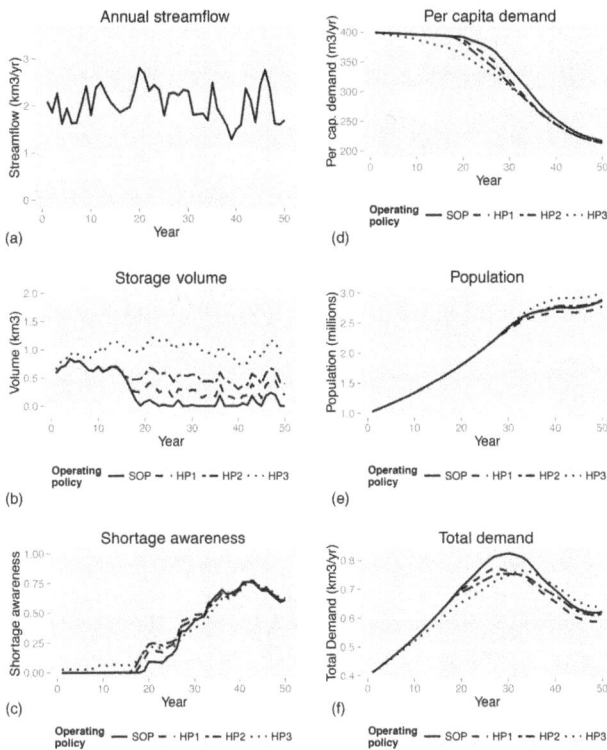

Figure 8. Model results, trial 3: (**a**) annual streamflow, (**b**) reservoir storage volume, (**c**) public shortage awareness, (**d**) per capita demand, (**e**) annual city population, (**f**) total demand.

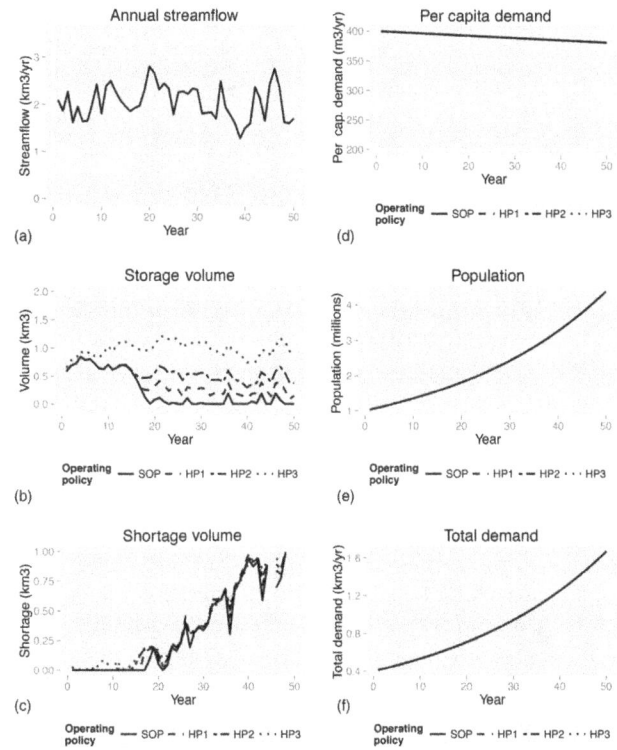

Figure 9. Noncoupled model results, trial 3: (**a**) annual streamflow, (**b**) reservoir storage volume, (**c**) shortage volume (demand–supply), (**d**) per capita demand, (**e**) annual city population, (**f**) total demand.

pected, changes to per capita demands, population, and total demands are gradual and consistent across the operating policy scenarios, found in Fig. 9e and f.

4 Discussion

The proposed question driven modeling process has three aims: to broaden the researcher's view of the system, to connect modeling assumptions to the model's purpose and to increase the transparency of these assumptions. A socio-hydrological model was developed to examine the difference in long-term reliability between two reservoir operating policies, SOP and HP. This question focused the conceptual model on processes influencing reliability at the city scale over the 50-year planning period. As part of the conceptual model development, the SES framework was used to check framing assumptions. The wide range of candidate variables included in the SES framework was reviewed against case data and background information. The model's intended use then informed decisions of which processes to include in the model, which processes were endogenous to the system and which variables could be held constant. The point here is not that the logic presented by the modeler using this process is unfailing but that it is clear and can inform debate. The questions raised about both the functional form of model re-

lationships and the variables excluded during the manuscript review process indicate that some transparency was achieved. However, the reader is in the best position to judge success on this third aim.

A socio-hydrological model of the Sunshine City water system was developed using the question driven modeling process and compared to a noncoupled model. The noncoupled model included assumes that both population growth and per capita demand change can be considered exogenous to the system. Both models show, as prior studies demonstrated, that by making small reductions early on HP reduces the chance of severe shortages. The socio-hydrological model also demonstrates that in the HP scenarios the moderate low flow events trigger an acceleration of per capita demand decrease that shifts the trajectory of water demands and in some instances slows the rate of population growth. In contrast, SOP delays impacts to the water consumers and therefore delays the shift to lower per capita demands. When extreme shortage events, such as a deep or prolonged drought occur, the impacts to the system are far more abrupt in the SOP scenario because per capita demands and population are higher than in hedging scenarios and there is less stored water available to act as a buffer. When we compare SOP and HP using a socio-hydrological model we see that HP decreases the magnitude of the oscillations in demand and population.

Hedging reduces the threshold for action, thereby decreasing the delay and the oscillation effect. This distinction between the two policies was not apparent when using a traditional noncoupled model. The significance of this observation is that a decrease in oscillation means a decrease in the magnitude of the contractions in population and per capita water demands required to maintain sustainability of the system. It is these abrupt changes in water usage and population that water utilities and cities truly want to avoid as they would hamper economic growth and decrease quality of life.

Examining the structure of the system can explain the differences in system response to SOP and HP. As seen in Fig. 5, there are one positive and two negative feedback loops in the system. Positive feedback loops, such as population in this model, exhibit exponential growth behavior but there are few truly exponential growth systems in nature and through interaction with other feedback loops most systems ultimately reach a limit (Sterman, 2000). Negative feedback loops generate goal seeking behavior. In its simplest form a negative feedback loop produces a slow approach to a limit or goal akin to an exponential decay function. In this case, the goal of the system is to match total demand with average supply. The fact that supply is driven by streamflow, a stochastic variable, adds noise to the system. Even if streamflow is correctly characterized with stationary statistics, as is assumed here, the variability challenges the management of the system. Reservoir storage helps utilities manage this variability by providing a buffer but it also acts as a delay. The delay between a change in the state of the system and action taken in response allows the system to overshoot its goal value before corrective action is taken, leading to oscillation around goal values. While water storage decreases the impact of a drought, changes to water consumption patterns are required to address demand driven shortages. Water storage simultaneously buffers variability and delays water user response by delaying impact. There are parallels between the feedback identified in this urban water supply system and the feedback identified by Elshafei et al. (2014) and Di Baldassarre (2013) in agricultural water management and human–flood interactions, respectively. Broadly, the three systems display the balance between the interaction between opposing forces, in this case articulated as positive and negative feedback loops.

The case of Sunshine City is simplified and perhaps simplistic. The limited number of available options for action constrains the system and shapes the observed behavior. In many cases water utilities have a portfolio of supply, storage and demand management policies to minimize shortages. Additionally, operating policies often shift in response to changing conditions. However, in this case no supply side projects are considered and the reservoir operating policy is assumed constant throughout the duration of the study period. As there are physical and legal limits to available supplies the first constraint reflects the reality of some systems. Constant operational policy is a less realistic constraint but can offer new insights by illustrating the limitations of main-

taining a given policy and the conditions in which policy change would be beneficial. Despite these drawbacks a simple hypothetical model is justified here to clearly illustrate the proposed modeling process.

There are several limitations to the hypothetical case of Sunshine City. First, the hypothetical nature of the case precludes hypothesis testing. Therefore, an important extension of this work will be to apply the modeling process presented here on a real case to fully test the resulting model against historical observations before generating projections. Second, only one set of parameters and functions was presented. Future extensions to this work on reservoir policy selection will test the impact of parameter and function selection through sensitivity analysis. Finally, we gain limited understanding of the potential of the model development process by addressing only one research question. We can further test the ability of the modeling process to generate new insights by developing different models in response to different questions. In this case, the narrow scope of the driving question leads to a model that just scratches the surface of socio-hydrological modeling as evidenced by the narrow range of societal variables and processes included. For example, this model does not address the ability of the water utility or city to adopt or implement HP. HP impacts water users in the short term. These impacts would likely generate a mix of reactions from water users and stakeholders making it impossible to ignore politics when considering the feasibility of HP. However, the question driving this model asks about the impact of a policy choice on the long-term reliability of the system not the feasibility of its implementation. A hypothesis addressing the feasibility of implementation would lead to a very different model structure.

While there is significant room for improvement, there are inherent limitations to any approach that models human behavior. The human capacity to exercise free will, to think creatively and to innovate means that human actions, particularly under conditions not previously experienced, are fundamentally unpredictable. Furthermore, as stated above we can never fully capture the complexity of the socio-hydrological system in a model. Instead we propose a modeling process that focuses socio-hydrological model conceptualization on answering questions and solving problems. By using model purpose to drive our modeling decisions we provide justification for simplifying assumptions and a basis for model evaluation.

5 Conclusions

Human and water systems are coupled. The feedback between these two subsystems can be, but are not always, strong and fast enough to warrant consideration in water planning and management. Traditional, noncoupled, modeling techniques assume that there is no significant feedback between human and hydrological systems. They therefore offer no insights into how changes in one part of the system may affect another. Dynamic socio-hydrological modeling recognizes and aims to understand the potential for feedbacks between human and hydrological systems. By building human dynamics into a systems model, socio-hydrological modeling enables testing of hypothesized feedback cycles and can illuminate the way changes propagate through the coupled system.

Recent work examining a range of socio-hydrological systems demonstrates the potential of this approach. However, there are significant challenges to modeling socio-hydrological systems. First, there are no widely accepted laws of human systems as there are for physical or chemical systems. Second, common disciplinary assumptions must be questioned due to the integrative nature of socio-hydrology. Transparency of the model development process and assumptions can facilitate the replication and critique needed to move this young field forward. We assess the progress and gaps in socio-hydrological modeling and draw lessons from adjacent fields of study, hydrology, social-ecological systems science and system dynamics, to inform a question driven model development process. We then illustrate this process by applying it to the hypothetical case of a growing city exploring two alternate reservoir operation rules.

By revisiting the classic question of reservoir operation policy, we demonstrate the utility of a socio-hydrological modeling process in generating new insights into the impacts of management practices over decades. This socio-hydrological model shows that HP offers an advantage not detected by traditional simulation models: it decreases the magnitude of the oscillation effect inherent in goal seeking systems with delays. Through this example we identify one class of question, the impact of reservoir management policy selection over several decades, for which socio-hydrological modeling offers advantages over traditional modeling. The model developed, and the resulting insights, are contingent upon the question context. The dynamics identified here may be more broadly applicable but this is for future cases and models to assess.

Acknowledgements. We would like to thank Brian Fath, Wei Liu, and Arnold Vedlitz for reviewing an early version of this paper. We would also like to thank the two reviewers for their careful reading and thoughtful comments. Financial support for this research comes from the NSF Water Diplomacy IGERT grant (0966093).

References

Abrishamchi, A. and Tajrishi, M.: Inter-basin water transfer in Iran, in: Water Conservation, Reuse, and Recycling, Proceeding of an Iranian American workshop, The National Academies Press, Washington, DC, 252–271, 2005.

Arocha, J. F., Patel, V. L., and Patel, Y. C.: Hypothesis generation and the coordination of theory and evidence in novice diagnostic reasoning, Med. Decis. Making, 13, 198–211, 1993.

Baldassare, M. and Katz, C.: The personal threat of environmental problems as predictor of environmental practices, Environ. Behav., 24, 602–616, 1992.

BBC Research: Public Opinions, Attitudes, and Awareness Regarding Water in Colorado, Final Report Prepared for the Colorado Water Conservation Board, available at: http://www.bbcresearch.com/images/Final_Report_072213_web.pdf (last access: 1 July 2015), 2013.

Beven, K.: Towards a coherent philosophy for modelling the environment, P. Roy. Soc. A, 458, 2465–2484, doi:10.1098/rspa.2002.0986, 2002.

Blair, P. and Buytaert, W.: Modelling socio-hydrological systems: a review of concepts, approaches and applications, Hydrol. Earth Syst. Sci. Discuss., 12, 8761–8851, doi:10.5194/hessd-12-8761-2015, 2015.

Blöschl, G. and Sivapalan, M.: Scale issues in hydrological modelling: a review, Hydrol. Process., 9, 251–290, doi:10.1002/hyp.3360090305, 1995.

Bower, B. T., Hufschmidt, M. M., and Reedy, W. W.: Operating procedures: their role in the design of water-resources systems by simulation analyses, in: Design of Water-Resource Systems, Harvard Univ. Press, Cambridge, MA, 443–458, 1962.

Brezonik, P. L. and Stadelmann, T. H.: Analysis and predictive models of stormwater runoff volumes, loads, and pollutant concentrations from watersheds in the Twin Cities metropolitan area, Minnesota, USA, Water Res., 36, 1743–1757, 2002.

Campbell, H. E., Johnson, R. M., and Larson, E. H.: Prices, devices, people, or rules: the relative effectiveness of policy instruments in water conservation, Rev. Policy Res., 21, 637–662, 2004.

Cancelliere, A., Ancarani, A., and Rossi, G.: Susceptibility of water supply reservoirs to drought conditions, J. Hydrol. Eng., 3, 140–148, 1998.

Chong, H. and Sunding, D.: Water markets and trading, Annu. Rev. Env. Resour., 31, 239–264, doi:10.1146/annurev.energy.31.020105.100323, 2006.

Crow, D. A.: Policy Entrepreneurs, Issue Experts, and Water Rights Policy Change in Colorado, Rev. Policy Research, 27, 299–315, doi:10.1111/j.1541-1338.2010.00443.x, 2010.

Dalhuisen, J. M., Florax, R. J. G. M., de Groot, H. L. F., and Nijkamp, P.: Price and Income Elasticities of Residential Water Demand: A Meta-Analysis, Land Econ., 79, 292–308, doi:10.2307/3146872, 2003.

Dawadi, S. and Ahmad, S.: Evaluating the impact of demand-side management on water resources under changing climatic conditions and increasing population, J. Environ. Manage., 114, 261–75, doi:10.1016/j.jenvman.2012.10.015, 2013.

Denver Water: Long-range Planning, available at: http://www.denverwater.org/SupplyPlanning/Planning/, last access: 1 July, 2015.

Di Baldassarre, G., Viglione, A., Carr, G., Kuil, L., Salinas, J. L., and Blöschl, G.: Socio-hydrology: conceptualising human-flood interactions, Hydrol. Earth Syst. Sci., 17, 3295–3303, doi:10.5194/hess-17-3295-2013, 2013.

Di Baldassarre, G., Viglione, A., Carr, G., Kuil, L., Yan, K., Brandimarte, L., and Blöschl, G.: Debates–Perspectives on socio-hydrology: Capturing feedbacks between physical and social processes, Water Resour. Res., 51, WR016416, doi:10.1002/2014WR016416, 2015.

Draper, A. J. and Lund, J. R.: Optimal hedging and carryover storage value, J. Water Res. Pl.-ASCE, 130, 83–87, 2004.

Elshafei, Y., Sivapalan, M., Tonts, M., and Hipsey, M. R.: A prototype framework for models of socio-hydrology: identification of key feedback loops and parameterisation approach, Hydrol. Earth Syst. Sci., 18, 2141–2166, doi:10.5194/hess-18-2141-2014, 2014.

Elshafei, Y., Coletti, J. Z., Sivapalan, M., and Hipsey, M. R.: A model of the socio-hydrologic dynamics in a semiarid catchment: Isolating feedbacks in the coupled human-hydrology system, Water Resour. Res., 6, WR017048, doi:10.1002/2015WR017048, 2015.

Falkenmark, M.: Water and mankind – complex system of mutual interaction, Ambio, 6, 3–9, 1977.

Fielding, K. S., Russell, S., Spinks, A., and Mankad, A.: Determinants of household water conservation: the role of demographic, infrastructure, behavior, and psychosocial variables, Water Resour. Res., 48, W10510, doi:10.1029/2012WR012398, 2012.

Forrester, J. W.: Policies, decisions, and information sources for modeling, Eur. J. Oper. Res., 59, 42–63, 1992.

Frick, J., Kaiser, F. G., and Wilson, M.: Environmental knowledge and conservation behavior: exploring prevalence and structure in a representative sample, Pers. Indiv. Differ., 37, 1597–1613, 2004.

Gal, S.: Optimal management of a multireservoir water supply system, Water. Resour. Res., 15, 737–749, 1979.

Geller, E. S., Erickson, J. B., and Buttram, B. A.: Attempts to promote residential water conservation with educational, behavioral and engineering strategies, Popul. Environ., 6, 96–112, 1983.

Giacomoni, M. H., Kanta, L., and Zechman, E. M.: Complex adaptive systems approach to simulate the sustainability of water resources and urbanization, J. Am. Water Resour. Assoc., 139, 554–564, 2013.

Gleick, P. H.: A Look at Twenty-first Century Water Resources Development, Water Int., 25, 127–138, doi:10.1080/02508060008686804, 2000.

Gober, P. and Wheater, H. S.: Debates–Perspectives on socio-hydrology: Modeling flood risk as a public policy problem, Water Resour. Res., 51, WR016945, doi:10.1002/2015WR016945, 2015.

Gober, P., White, D. D., Quay, R., Sampson, D. A., and Kirkwood, C. W.: Socio-hydrology modelling for an uncertain future, with examples from the USA and Canada, Geol. Soc. Spec. Publ., 408, SP408-2, 2014.

Hashimoto, T., Stedinger, J. R., and Loucks, D. P.: Reliability, resiliency, and vulnerability criteria for water resource system evaluation, Water Resour. Res., 18, 14–20, 1982.

Hale, R. L., Armstrong, A., Baker, M. A., Bedingfield, S., Betts, D., Buahin, C., Buchert, M., Crowl, T., Dupont, R. R., Ehleringer, J.R, Endter-Wada, J., Flint, C., Grant, J., Hinners, S., Jeffery, S., Jackson-Smith, D., Jones, A. S., Licon, C., and Null, S. E.: iSAW: integrating structure, actors, and water to

study socio-hydro-ecological systems, Earth's Future, 110–132, doi:10.1002/2014EF000295, 2015.

Hinkel, J., Schleuter, M., and Cox, M.: A diagnostic procedure for applying the social–ecological systems framework in diverse cases, Ecol. Soc., 20, 32, doi:10.5751/ES-07023-200132, 2015.

Hughes, S., Pincetl, S., and Boone, C.: Triple exposure: regulatory, climatic, and political drivers of water management changes in the city of Los Angeles, Cities, 32, 51–59, doi:10.1016/j.cities.2013.02.007, 2013.

Inman, D. and Jeffrey, P.: A review of residential water conservation tool performance and influences on implementation effectiveness, Urban Water J., 3, 127–143, 2006.

ISTPP: National Public Water Survey, Institute for Science, Technology and Public Policy, Bush School of Government and Public Service, Texas A & M University, College Station, TX, 2013.

Jones, B. D. and Baumgartner, F. R.: The Politics of Attention, University of Chicago Press, Chicago, IL, 2005.

Jones, A., Seville, D., and Meadows, D.: Resource sustainability in commodity systems: the sawmill industry in the Northern Forest, Syst. Dynam. Rev., 18, 171–204, doi:10.1002/sdr.238, 2002.

Kandasamy, J., Sounthararajah, D., Sivabalan, P., Chanan, A., Vigneswaran, S., and Sivapalan, M.: Socio-hydrologic drivers of the pendulum swing between agricultural development and environmental health: a case study from Murrumbidgee River basin, Australia, Hydrol. Earth Syst. Sci., 18, 1027–1041, doi:10.5194/hess-18-1027-2014, 2014.

Kanta, L. and Zechman, E.: Complex adaptive systems framework to assess supply-side and demand-side management for urban water resources, J. Water Res. Pl.-ASCE, 140, 75–85, 2014.

Kenney, D. S., Goemans, C., Klein, R., Lowrey, J., and Reidy, K.: Residential water demand management: Lessons from Aurora, Colorado, J. Am. Water Resour. As., 44, 192–207, 2008.

Kuhn, T. S.: The Structure of Scientific Revolutions, 3rd Edn., University of Chicago Press, Chicago, 1996.

LADWP – Los Angeles Department of Water and Power: Urban Water Management Plan 2010, Los Angeles, CA, 2010.

Lélé, S. and Norgaard, R. B.: Practicing Interdisciplinarity, BioScience, 55, 967-975, doi:10.1641/0006-3568(2005)055[0967:PI]2.0.CO;2, 2005.

Liu, Y., Tian, F., Hu, H., and Sivapalan, M.: Socio-hydrologic perspectives of the co-evolution of humans and water in the Tarim River basin, Western China: the Taiji-Tire model, Hydrol. Earth Syst. Sci., 18, 1289–1303, doi:10.5194/hess-18-1289-2014, 2014.

Loucks, D. P.: Debates–Perspectives on socio-hydrology: Simulating hydrologic-human interactions, Water Resour. Res., 51, WR017002, 10.1002/2015WR017002, 2015.

Mankad, A. and Tapsuwan, S.: Review of socio-economic drivers of community acceptance and adoption of decentralised water systems, J. Environ. Manage., 92, 380–91, doi:10.1016/j.jenvman.2010.10.037, 2011.

McConnell, W. J., Millington, J. D. A., Reo, N. J., Alberti, M., Asbjornsen, H., Baker, L. A., Brozović, N., Drinkwater, L. E., Scott, A., Fragoso, J., Holland, D. S., Jantz, C. A., Kohler, T. A., Herbert, D., Maschner, G., Monticino, M., Podestá, G., Pontius, R. G., Redman, C. L., Sailor, D., Urquhart, G., and Liu, J.: Research on Coupled Human and Natural Systems (CHANS): Approach, Challenges, and Strategies, B. Ecol. Soc. Am., Meeting Reports, 218–228, 2009.

McGinnis, M.: Networks of adjacent action situations in polycentric governance, Policy Stud. J., 39, 51–78, doi:10.1111/j.1541-0072.2010.00396.x/full, 2011.

McGinnis, M.: An Introduction to IAD and the Language of the Ostrom Workshop: A Simple Guide to a Complex Framework, Policy Stud. J., 39, 169–183, doi:10.1111/j.1541-0072.2010.00401.x, 2011b.

McGinnis, M. D. and Ostrom, E.: Social-ecological system framework: initial changes and continuing challenges, Ecol. Soc., 19, 30, doi:10.5751/ES-06387-190230, 2014.

Micklin, P.: The aral sea disaster, Annu. Rev. Earth Pl. Sc., 35, 47–72, doi:10.1146/annurev.earth.35.031306.140120, 2007.

Mini, C., Hogue, T. S., and Pincetl, S.: Patterns and controlling factors of residential water use in Los Angeles, California, Water Policy, 16, 1–16, doi:10.2166/wp.2014.029, 2014.

MWRA – Massachusetts Water Resources Authority: Summary of MWRA Demand Management Program, available at: http://www.mwra.state.ma.us/harbor/pdf/demandreport03.pdf (last access: 1 July 2015), 2003.

Olmstead, S. M. and Stavins, R. N.: Comparing price and nonprice approaches to urban water conservation, Water Resour. Res., 45, W04301, doi:10.1029/2008WR007227, 2009.

Ostrom, E.: A diagnostic approach for going beyond panaceas, P. Natl. Acad. Sci. USA, 104, 15181–15187, 2007.

Ostrom, E.: A general framework for analyzing sustainability of social-ecological systems, Science, 325, 419–422, doi:10.1126/science.1172133, 2009.

Ostrom, E.: Background on the Institutional Analysis and Development Framework, Policy Stud. J., 39, 7–27, 2011.

Pahl-Wostl, C., Craps, M., Dewulf, A., Mostert, E., Tabara, D., and Taillieu, T.: Social Learning and Water Resources Management, Ecol. Soc., 12, 5, 2007.

Pahl-Wostl, C., Holtz, G., Kastens, B., and Knieper, C.: Analyzing complex water governance regimes: the Management and Transition Framework, Environ. Sci. Policy, 13, 571–581, 2010.

Padowski, J. C. and Jawitz, J. W.: Water availability and vulnerability of 225 large cities in the United States, Water Resour Res, 48, W12529, doi:10.1029/2012WR012335, 2012.

Pande, S. and Ertsen, M.: Endogenous change: on cooperation and water availability in two ancient societies, Hydrol. Earth Syst. Sci., 18, 1745–1760, doi:10.5194/hess-18-1745-2014, 2014.

Schlager, E. and Heikkila, T.: Left High and Dry? Climate Change, Common-Pool Resource Theory, and the Adaptability of Western Water Compacts, Public Admin. Rev., 71, 461–470, doi:10.1111/j.1540-6210.2011.02367.x, 2011.

Schlüter, M., Hinkel, J., Bots, P., and Arlinghaus, R.: Application of the SES framework for model-based analysis of the dynamics of social–ecological systems, Ecol. Soc., 19, 36, doi:10.5751/ES-05782-190136, 2014.

Schuetze, T. and Santiago-Fandiño, V.: Quantitative assessment of water use efficiency in urban and domestic buildings, Water, 5, 1172–1193, 2013.

Shih, J. and Revelle, C.: Water-supply operations during drought: continuous hedging rule, J. Water Res. Pl.-ASCE, 120, 613–629, 1994.

Sivapalan, M.: Debates–Perspectives on socio-hydrology: Changing water systems and the "tyranny of small problems"– Socio-hydrology, Water Resour. Res., 51, WR017080, doi:10.1002/2015WR017080, 2015.

Sivapalan, M. and Blöschl, G.: Time scale interactions and the coevolution of humans and water, Water Resour. Res., 51, WR017896, doi:10.1002/2015WR017896, 2015.

Sivapalan, M., Blöschl, G., Zhang, L., and Vertessy, R.: Downward approach to hydrological prediction, Hydrol. Process., 17, 2101–2111, doi:10.1002/hyp.1425, 2003.

Sivapalan, M., Savenije, H. H. G., and Blöschl, G.: Socio-hydrology: A new science of people and water, Hydrol. Process., 26, 1270–1276, 2012.

SNWA – Southern Nevada Water Authority: Water Resources Management Plan, available at: http://www.snwa.com/assets/pdf/wr_plan.pdf (last access: 1 July 2015), 2009.

Srinivasan, V., Gorelick, S. M., and Goulder, L.: Sustainable urban water supply in south India: desalination, efficiency improvement, or rainwater harvesting?, Water Resour. Res., 46, W10504, doi:10.1029/2009WR008698, 2010.

Srinivasan, V., Seto, K. C., Emerson, R., and Gorelick, S. M.: The impact of urbanization on water vulnerability: A coupled human–environment system approach for Chennai, India, Global Environ. Chang., 23, 229–239, doi:10.1016/j.gloenvcha.2012.10.002, 2013.

Srinivasan, V.: Reimagining the past – use of counterfactual trajectories in socio-hydrological modelling: the case of Chennai, India, Hydrol. Earth Syst. Sci., 19, 785–801, doi:10.5194/hess-19-785-2015, 2015.

Stave, K. A.: A system dynamics model to facilitate public understanding of water management options in Las Vegas, Nevada, J. Environ. Manage., 67, 303–313, doi:10.1016/S0301-4797(02)00205-0, 2003.

Sterman, J.: Business Dynamics: Systems Thinking and Modeling for a Complex World, Irwin McGraw-Hill, Boston, 2000.

Thompson, S. E., Sivapalan, M., Harman, C. J., Srinivasan, V., Hipsey, M. R., Reed, P., Montanari, A., and Blöschl, G.: Developing predictive insight into changing water systems: use-inspired hydrologic science for the Anthropocene, Hydrol. Earth Syst. Sci., 17, 5013–5039, doi:10.5194/hess-17-5013-2013, 2013.

Tong, S. T. and Chen, W.: Modeling the relationship between land use and surface water quality, J. Environ. Manage., 66, 377–393, 2002.

Troy, T. J., Pavao-Zuckerman, M., and Evans, T. P.: Debates–Perspectives on socio-hydrology: Socio-hydrologic modeling: Tradeoffs, hypothesis testing, and validation, Water Resour. Res., 51, WR017046, doi:10.1002/2015WR017046, 2015.

Turral, H.: Hydro-Logic? Reform in Water Resources Management in Developed Countries with Major Agricultural Water Use – Lessons for Developing Nations, Overseas Development Institute, London, 1998.

Vahmani, P. and Hogue, T. S.: Incorporating an urban irrigation module into the Noah Land surface model coupled with an urban canopy model, J. Hydrometeorol., 15, 1440–1456, 2014.

van Emmerik, T. H. M., Li, Z., Sivapalan, M., Pande, S., Kandasamy, J., Savenije, H. H. G., Chanan, A., and Vigneswaran, S.: Socio-hydrologic modeling to understand and mediate the competition for water between agriculture development and environmental health: Murrumbidgee River basin, Australia, Hydrol. Earth Syst. Sci., 18, 4239–4259, doi:10.5194/hess-18-4239-2014, 2014.

Viglione, A., Di Baldassarre, G., Brandimarte, L., Kuil, L., Carr, G., Salinas, J. L., Scolobig, A., and Blöschl, G.: Insights from socio-hydrology modelling on dealing with flood risk – roles of collective memory, risk-taking attitude and trust, J. Hydrol., 518, 71–82, doi:10.1016/j.jhydrol.2014.01.018, 2014.

Vörösmarty, C. J., McIntyre, P. B., Gessner, M. O., Dudgeon, D., Prusevich, A., Green, P., Glidden, S., Bunn, S. E., Sullivan, C. A., Lierman, C. R., and Davies, P. M.: Global threats to human water security and river biodiversity, Nature, 467, 555–561, doi:10.1038/nature09549, 2010.

Wagener, T., Sivapalan, M., Troch, P. A., McGlynn, B. L., Harman, C. J., Gupta, H. V., and Wilson, J. S.: The future of hydrology: an evolving science for a changing world, Water Resour. Res, 46, W05301, doi:10.1029/2009WR008906, 2010.

Wheater, H. S., Jakeman, A. J., and Beven, K. J.: Progress and directions in rainfall–runoff modeling, in: Modeling Change in Environmental Systems, John Wiley and Sons, New York, 101–132, 1993.

Willis, R. M., Stewart, R. A., Panuwatwanich, K., Williams, P. R., and Hollingsworth, A. L.: Quantifying the influence of environmental and water conservation attitudes on household end use water consumption, J. Environ. Manage., 92, 1996–2009, 2011.

Wissmar, R. C., Timm, R. K., and Logsdon, M. G.: Effects of changing forest and impervious land covers on discharge characteristics of watersheds, Environ. Manage., 34, 91–98, 2004.

You, J. Y. and Cai, X.: Hedging rule for reservoir operations: 1. A theoretical analysis, Water Resour. Res., 44, 1–9, 2008.

Young, P., Parkinson, S., and Lees, M.: Simplicity out of complexity in environmental modelling: Occam's razor revisited, J. Appl. Stat., 23, 165–210, doi:10.1080/02664769624206, 1996.

Young, P.: Top-down and data-based mechanistic modelling of rainfall–flow dynamics at the catchment scale, Hydrol. Process., 17, 2195–2217, doi:10.1002/hyp.1328, 2003.

Zilberman, D., Dinar, A., MacDougall, N., Khanna, M., Brown, C., and Castillo, F.: Individual and institutional responses to the drought: The case of California agriculture. ERS Staff Paper, US Department of Agriculture, Washington, DC, 1992.

2

Integrated water system simulation by considering hydrological and biogeochemical processes: model development, with parameter sensitivity and autocalibration

Y. Y. Zhang[1], Q. X. Shao[2], A. Z. Ye[3], H. T. Xing[4], and J. Xia[1]

[1]Key Laboratory of Water Cycle and Related Land Surface Processes, Institute of Geographic Sciences and Natural Resources Research, Chinese Academy of Sciences, Beijing, 100101, China
[2]CSIRO Digital Productivity Flagship, Leeuwin Centre, 65 Brockway Road, Floreat Park, WA 6014, Australia
[3]College of Global Change and Earth System Science, Beijing Normal University, Beijing, 100875, China
[4]CSIRO Agriculture Flagship, GPO BOX 1666, Canberra, ACT 2601, Australia

Correspondence to: Y. Y. Zhang (zhangyy003@igsnrr.ac.cn) and Q. X. Shao (quanxi.shao@cisro.au)

Abstract. Integrated water system modeling is a feasible approach to understanding severe water crises in the world and promoting the implementation of integrated river basin management. In this study, a classic hydrological model (the time variant gain model: TVGM) was extended to an integrated water system model by coupling multiple water-related processes in hydrology, biogeochemistry, water quality, and ecology, and considering the interference of human activities. A parameter analysis tool, which included sensitivity analysis, autocalibration and model performance evaluation, was developed to improve modeling efficiency. To demonstrate the model performances, the Shaying River catchment, which is the largest highly regulated and heavily polluted tributary of the Huai River basin in China, was selected as the case study area. The model performances were evaluated on the key water-related components including runoff, water quality, diffuse pollution load (or nonpoint sources) and crop yield. Results showed that our proposed model simulated most components reasonably well. The simulated daily runoff at most regulated and less-regulated stations matched well with the observations. The average correlation coefficient and Nash–Sutcliffe efficiency were 0.85 and 0.70, respectively. Both the simulated low and high flows at most stations were improved when the dam regulation was considered. The daily ammonium–nitrogen (NH_4–N) concentration was also well captured with the average correlation coefficient of 0.67. Furthermore, the diffuse source load of NH_4–N

and the corn yield were reasonably simulated at the administrative region scale. This integrated water system model is expected to improve the simulation performances with extension to more model functionalities, and to provide a scientific basis for the implementation in integrated river basin managements.

1 Introduction

Severe water crises are global issues that have emerged as a consequence of the rapid development of social economy, and include flooding, water shortages, water pollution and ecological degradation. These crises have hindered the equitable development of regions by compromising the sustainability of vital water resources and ecosystems. It is impossible to address these crises within a single scientific discipline (e.g., hydrology, hydraulics, water quality or aquatic ecology) because of the complicated interactions among physical, chemical and ecological components of an aquatic ecosystem (Kindler, 2000; Paola et al., 2006). The paradigm of integrated river basin management may be a sensible solution at basin scale by focusing on the coordinated management of water resources in terms of social economy, water quality and ecosystems. Integrated water system models have been popular since the last decade due to the rapid development of water-related sciences, computer science, Earth observation technologies and the availability of open data.

The hydrological cycle has been known as a critical linkage among other water-related processes (e.g., physical, biogeochemical and ecological processes) and energy fluxes at the basin scale (Burt and Pinay, 2005). For example, physiological and ecological processes of vegetation affect evapotranspiration, soil moisture distribution and nutrient movement. In the meantime, soil moisture and nutrient constrain the vegetation growth. Overland flow is a carrier of pollutants to water bodies. Therefore, all the processes should be considered simultaneously to capture the interactions and feedbacks between individual cycles. Multidisciplinary research provides an effective way to enable breakthroughs in the integrated water system modeling by integrating the theories in water-related sciences (e.g., accumulated temperature law for phenological development, Darcy's law for groundwater flow, Saint-Venant equation for flow routing, balance equation for mass and momentum, Richards' equation for unsaturated zone, Horton theory for infiltration, Penman–Monteith equation for evapotranspiration). Abundant open data sources further support the implementation of an integrated water system model, e.g., high-resolution spatial information data, chemical and isotopic data from field experiments (Singh and Woolhiser, 2002; Kirchner, 2006).

Several models have been developed since the 1980s (Di Toro et al., 1983; Brown and Barnwell, 1987; Johnsson et al., 1987; Hamrick, 1992; Li et al., 1992; Abrahamsen and Hansen, 2000; Tattari et al., 2001; Singh and Woolhiser, 2002). Owing to the complexity of the integrated water system and the scale conflicts between different processes, most existing models focus on only one or two major water-related processes, and can be categorized into three major classes. (1) Hydrological models emphasize the rainfall–runoff relationship and link with some dominating water quality and biogeochemical processes. These models generally show satisfactory performances in simulating the hydrological processes. Some widely accepted models are TOPMODEL (Beven and Kirkby, 1979), SHE (Abbott et al., 1986), HSPF (Bicknell et al., 1993), VIC (Liang et al., 1994), ANSWERS (Bouraoui and Dillaha, 1996), HBV-N (Arheimer and Brandt, 1998, 2000), HYPE (Lindström et al., 2010) and its improved version S-HYPE (Strömqvist et al., 2012). (2) Water quality models focus on the migration and transformation processes of pollutants in water bodies. These models can simulate the water quality variables at high spatial and temporal resolutions in river networks by adopting multi-dimensional dynamic equations. However, they have difficulties in simulating the overland processes of water and pollutants. Typical models include WASP (Di Toro et al., 1983), QUAL2E (Brown and Barnwell, 1987) and EFDC (Hamrick, 1992). (3) Biogeochemical models have advantages in simulating the physiological and ecological processes of vegetation and the vertical movements of nutrients and water in soil layers at the field or experimental

catchment scales. However, these models lack accurate hydrological features (Deng et al., 2011) and are hard to simulate the movements of water, nutrients and their losses along flow pathways in the basin. Some biogeochemical models are SOILN (Johnsson et al., 1987), EPIC (Sharpley and Williams, 1990), DNDC (Li et al., 1992), Daisy (Abrahamsen and Hansen, 2000) and ICECREAM (Tattari et al., 2001). Overall, most models usually achieve good performances on their oriented processes and only approximate the results for other processes outside of the model's focus in the integrated river basin management. An important scientific question is "does including these extra processes in an integrated manner improve model results compared to models that are focused only on one component?"

SWAT is an integrated water system model that can simulate most water-related processes over a long period at large scales (Arnold et al., 1998). However, not all water-related processes can be well captured in practice because of the inaccurate descriptions of some processes, such as daily extreme flow events (Borah and Bera, 2004), soil nitrogen and carbon (Gassman et al., 2007) and regulation rules of dams or sluices in regulated basins (Zhang et al., 2013). Particularly, the simulation methods of surface runoff yield in SWAT have been questioned, e.g., the general applicability of the curve number (Rallison and Miller, 1981) and the scale limitations of the Green-Ampt infiltration model (King et al., 1999). Furthermore, SWAT has difficulties in accurately capturing the complicated dynamic processes of soil nitrogen and carbon by comparing with other biochemical models (Gassman et al., 2007). Several modified versions have been developed, such as SWIM (Krysanova et al., 1998) and SWAT-N (Pohlert et al., 2006).

In this study, we tended to develop an integrated water system model based on a hydrological model. The time variant gain model (TVGM) proposed by Xia (1991) is a lumped hydrological model based on the rainfall and runoff observations from many basins with different scales all over the world. In the TVGM, the rainfall–runoff relationship is considered to be nonlinear because the surface runoff coefficient varies over time and is significantly affected by antecedent soil moisture. The TVGM has a strong mathematical basis because this nonlinear relationship is transformed into a complex Volterra nonlinear formulation. Wang et al. (2002) extended the TVGM to the distributed time variant gain model (DTVGM) by taking advantage of better computing facilities and available data sources. Currently, the DTVGM performs well in many basins with different scales and climate zones to investigate the effect of human activities and climate change on runoff (Xia et al., 2005; Wang et al., 2009).

In the model development, we would like to produce reasonable simulations simultaneously in both hydrological and water quality processes and to include more water-related processes such as soil biogeochemistry and crop growth for better understandings of the complicated water-related processes and their interactions in the real basins. Our proposed

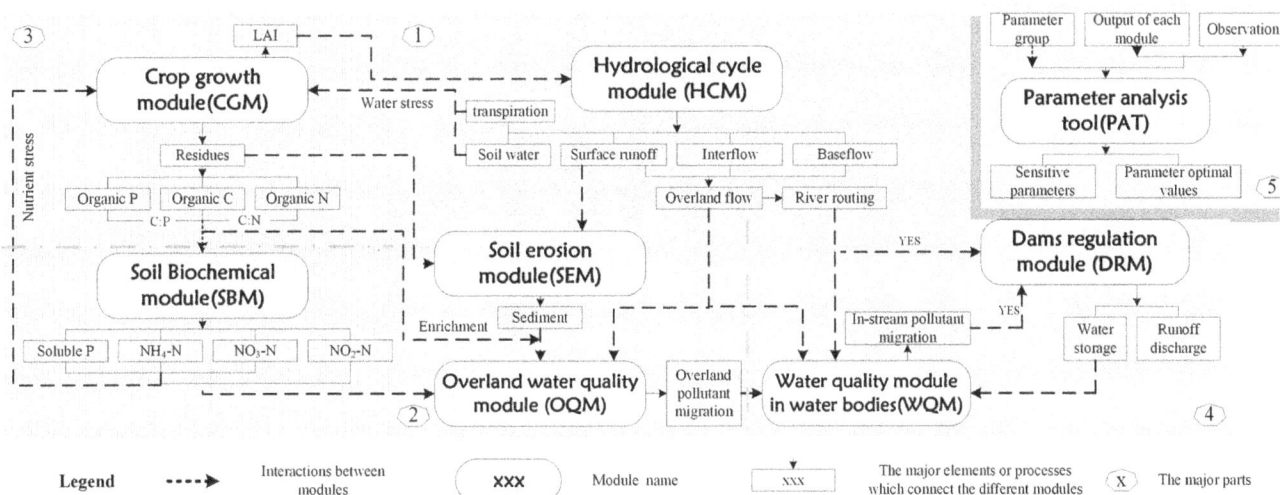

Figure 1. The model structure and the interactions among the major modules (1: hydrological part; 2: water quality part; 3: ecological part; 4: dam regulation part; 5: PAT).

model was built by extending the DTVGM through coupling of the detailed interactions and linkages among hydrological, water quality, soil biogeochemical and ecological processes, as well as considering the prevalent regulations of water projects (dams and sluices) at the basin scale. In order for readers to use the proposed model easily, a parameter analysis module, which included popular objective functions, autocalibration approaches and summary statistics, was also developed. To demonstrate the model performances, we simulated several key water-related components including flow regimes, diffuse source (or nonpoint source) pools of nutrients, water quality variables in water bodies and crop yield in a highly regulated and heavily polluted catchment (Shaying River catchment) in China.

2 Methods and material

2.1 Model framework

Our proposed model includes eight major modules, namely the hydrological cycle module (HCM), soil biochemical module (SBM), crop growth module (CGM), soil erosion module (SEM), overland water quality module (OQM), water quality module of water bodies (WQM) and dam regulation module (DRM). The parameter analysis tool (PAT) is also designed for model calibration. The model structure is shown in Fig. 1. More detailed descriptions of each module and its interactions with other modules are given in Sects. 2.1.1 to 2.1.5. The main equations of each module are deferred to the Appendix and Supplement for readers who are interested in the mathematical details.

Our model is based on the hypothesis that the cycles of water and nutrients (N, P and C) are inseparable and act as the critical linkages among all the modules. It takes full advan-

tage of the existing models, i.e., the powerful interconnections of the hydrological models with other processes at the spatial scale, the elaborative descriptions of the ecological models on nutrient vertical movement in soil layers, and the elaborative descriptions of the water quality models on nutrient movements along river networks. First, several key components, simulated by the hydrological cycle module (HCM) (e.g., evapotranspiration, soil moisture and flow), are treated as critical linkages in all the modules (Sect. 2.1.1). Second, the soil biochemical processes determine the nutrient loads absorbed in the crop growth process (CGM) and migrated into water bodies as the diffuse pollution source (OQM and WQM). The accurate descriptions of soil biochemical processes are helpful in improving the simulation of diffuse source processes in responding to agricultural management (Sect. 2.1.2). Third, the hydrological cycle module (HCM) provides a function for describing the connections between spatial calculation units to simulate the overland and in-stream movements of water and nutrients at the basin scale (Sects. 2.1.1 and 2.1.3).

2.1.1 Hydrological cycle module (HCM)

Surface runoff calculation is the core of hydrological simulation. The TVGM is adopted to calculate the surface runoff yields for different land-use/cover areas, such as forest, grassland, water body, urban area, unused land, paddy land and dryland agriculture. The potential evapotranspiration is calculated using the Hargreaves method (Hargreaves and Samani, 1982) because only the available daily maximum and minimum temperatures are used. The actual plant transpiration is expressed as a function of potential evapotranspiration and leaf area index, whereas soil evaporation is expressed as a function of potential evapotranspiration and

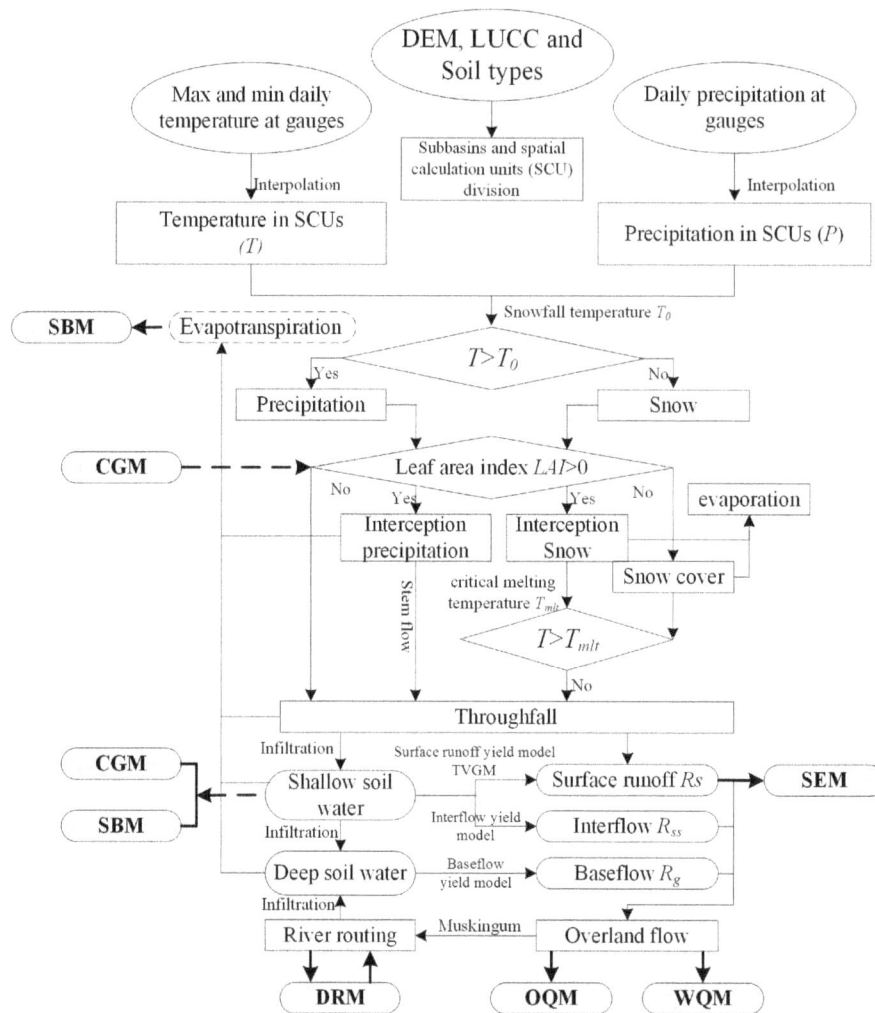

Figure 2. The flowchart of HCM and the interactions with other modules.

surface soil residues (Neitsch et al., 2011). The yields of interflow and baseflow have linear relationships with the soil moisture in the upper and lower layers, respectively (Wang et al., 2009). The infiltration from the upper to lower soil layers is calculated using the storage routing method (Neitsch et al., 2011). The Muskingum method or kinetic wave equation is used for river flow routing.

Figure 2 shows that the shallow soil moisture from the hydrological cycle module is a major factor that connects the crop growth module (to control crop growth) and the soil biochemical module (to control the vertical migration and reaction of nutrients in the soil layers). Plant transpiration is also linked to the soil biochemical module (to drive the vertical migration of nutrients in the soil layers). The surface runoff is linked to the soil erosion module, while the overland flow (surface runoff, interflow and baseflow) is connected to the overland water quality module (to drive the movements of nutrients and sediment along flow pathways) and the water quality module of water bodies (rivers and lakes) for runoff

routing. Moreover, the hydrological cycle module provides the inflows for individual dams or sluices in the dam regulation module.

2.1.2 Modules for ecological processes

The ecological processes are described in the soil biochemical module and the crop growth module. The crop growth and soil biochemical processes directly affect the soil moisture, evapotranspiration, nutrient transformation and loss from the soil layers. Therefore, our model incorporates the water cycle, nutrient cycle, crop growth and their key linkages.

Soil biochemical module (SBM)

The soil biochemical module simulates the key processes of carbon (C), nitrogen (N) and phosphorus (P) dynamics in the soil layers, including decomposition, mineralization, immobilization, nitrification, denitrification, leaching and plant uptake. Different forms of N and P outputted from the soil bio-

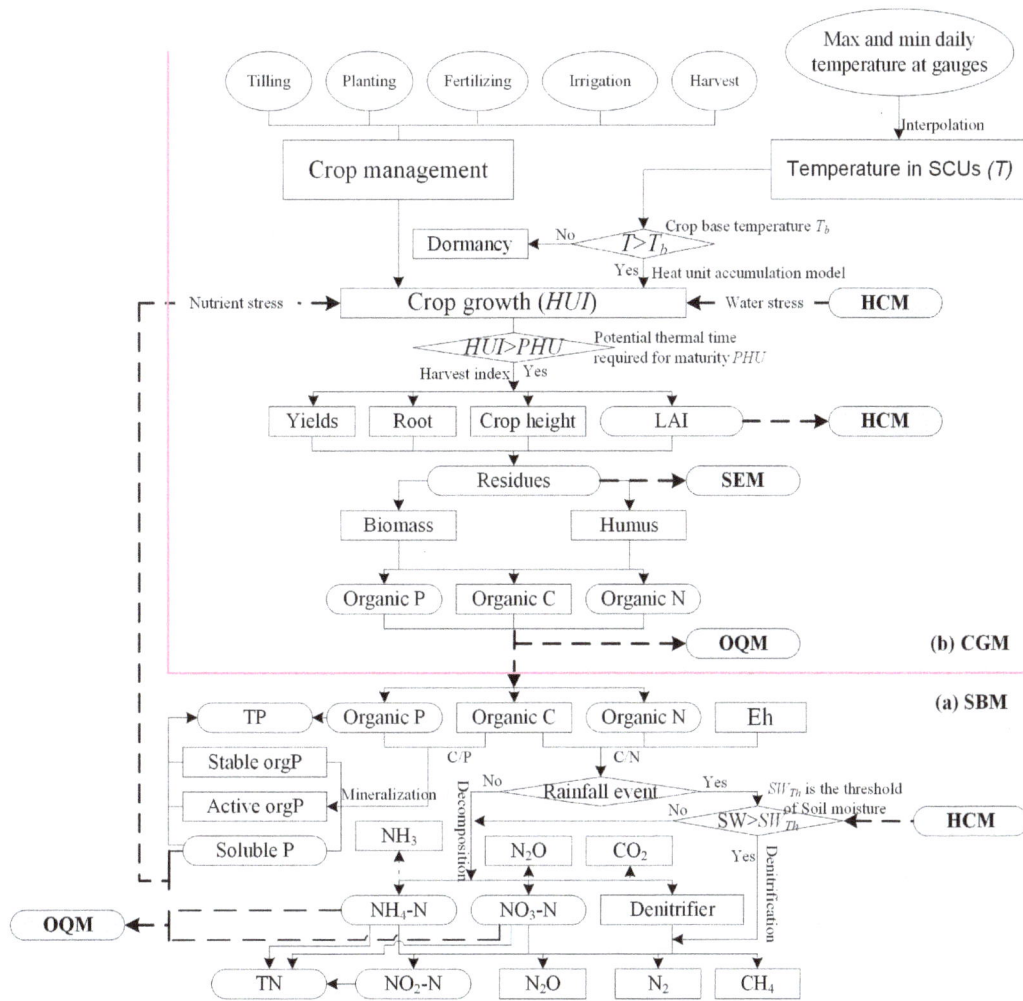

Figure 3. The flowchart of SBM (**a**) and CGM (**b**) in the ecological part and the interactions with other modules.

chemical module are connected to the crop growth module as the nutrient constraints of crop growth and to the overland water quality module as the main diffuse sources to water bodies (Fig. 3a).

Soil C and N cycle. The sub-models of daily step decomposition and denitrification in DNDC (Li et al., 1992) are adopted to simulate the soil biogeochemical processes of C and N at the field scale. The decomposition and other oxidation processes are the dominant microbial processes in the aerobic condition. The three conceptual organic C pools are the decomposable residue C pool, microbial biomass C pool and stable C pool. The decomposition of each C pool is treated as the first-order decay process with the individual decomposition rates constrained by the soil temperature and moisture, clay content and C : N ratio. The major simulated processes of decomposition under aerobic conditions are mineralization, immobilization, ammonia (NH_3) volatilization and nitrification. The mineralization and immobilization of mineral N (NH_4^+ and NO_3^-) are determined by the flow rates of soil organic carbon (SOC) pools. NH_3 volatilization

is controlled by the NH_4^+ concentration, clay content, pH, soil moisture and temperature. NH_4^+ is oxidized to NO_3^- during nitrification and nitrous oxide (N_2O) is emitted into the air during the nitrification. Denitrification occurs under the anaerobic condition, which is controlled by soil moisture, temperature, pH and dissolved SOC content. The detailed descriptions are given in Appendix B and Li et al. (1992).

Soil P cycle. The major processes of the soil P cycle are simulated according to the study of Horst et al. (2001). Six P pools are considered including three organic pools (stable and active pools for plant uptake, a fresh pool associated with plant residue) and three mineral pools (dissolved mineral, stable and active pools). The involved processes are the P release, mineralization and decomposition from fertilizer, manure, residue, microbial biomass, humic substances and the sorption by plant uptake (Horst et al., 2001; Neitsch et al., 2011).

The soil profile is divided into three layers, namely, surface (0–10 cm) and user-defined upper and lower layers, all

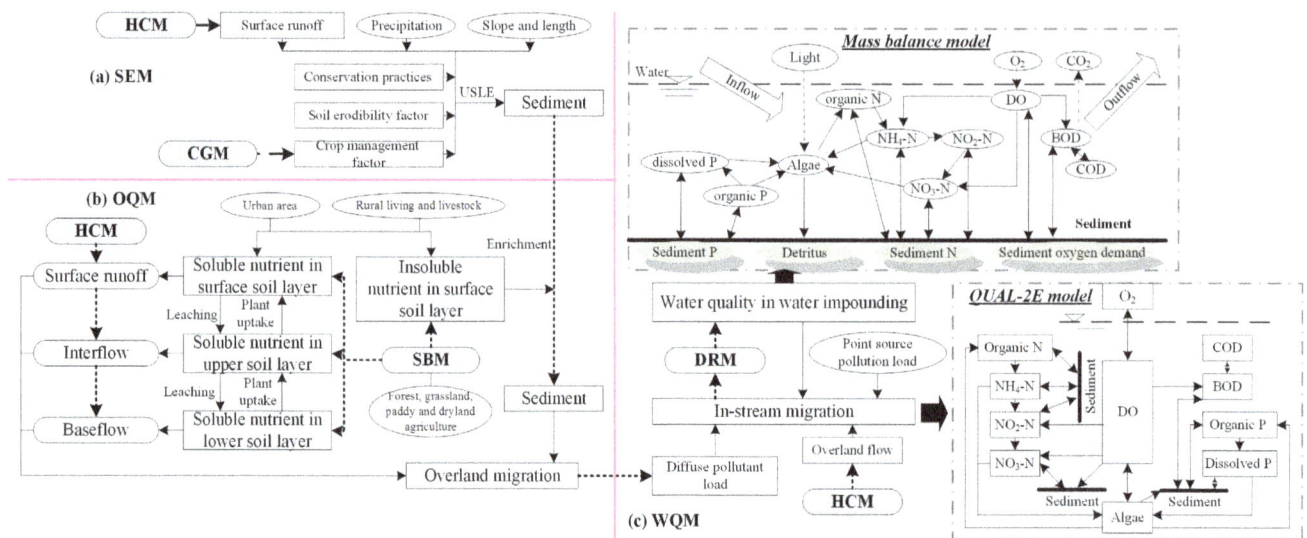

Figure 4. The flowchart of SEM (**a**), OQM (**b**) and WQM (**c**) in the water quality part and the interactions with other modules.

of which are consistent with the soil layers of the hydrological cycle module to smoothly exchange the values through the linkages (e.g., soil moisture) among different modules.

Crop growth module (CGM)

The crop growth module is developed based on the EPIC crop growth model (Hamrick, 1992). It simulates total dry matter, leaf area index, root depth and density distribution, harvest index, nutrient uptake, and so on (Williams et al., 1989; Sharpley and Williams, 1990). The crop respiration and photosynthesis drive the vertical movements of water and nutrients. The output of the leaf area index is a main factor connecting the hydrological cycle module (to control the transpiration), and the crop residue left in the fields is a main source of organic nutrients (C, N and P) connecting to the soil biochemical module for soil biochemical processes, to the overland water quality module and to the soil erosion module as one of the five constraint factors (Fig. 3b).

2.1.3   Modules for water quality processes

The water quality processes focus on the migration and transformation of water quality variables (e.g., sediment, different forms of nutrients, biochemical oxygen demand, BOD, and chemical oxygen demand, COD) along the flow pathways in the land surface and river network. The main modules are the soil erosion module for the sediment yield, the overland water quality module for the migration of overland diffuse source to water bodies and the water quality module for the migration and transformation of point and diffuse pollution sources in water bodies.

Soil erosion module (SEM)

The soil erosion by precipitation is estimated using the improved USLE equation (Onstad and Foster, 1975) based on runoff yields outputted from the hydrological cycle module and crop management factor outputted from the crop growth module. The soil erosion module simulates the sediment load for the overland water quality module to provide the carrier for the migration of insoluble organic matter along overland transport paths and water bodies (Fig. 4a).

Overland water quality module (OQM)

This module simulates the overland losses and migration loads of diffuse source pollutants (e.g., sediment, insoluble and dissolved nutrients, BOD and COD) (Fig. 4b). The main diffuse sources include the nutrient loss from the soil layers and urban areas, and the farm manure from livestock in rural areas. The nutrient loss from the soil layers, as the primary diffuse source in most catchments, is determined by the overland flow and sediment yield (Williams et al., 1989), and the other sources are estimated using the export coefficient method (Johnes, 1996). The overland migration processes contain the dissolved pollutant migration with overland flow and the insoluble pollutant migration with sediment. All the processes occur along the overland transport paths.

Water quality module of water bodies (WQM)

This module simulates the transformation and migration of water quality variables in different types of water bodies (in-stream and water impounding) (Fig. 4c). The simulated variables include water temperature, dissolved oxygen (DO), sediment, different forms of nutrients (N and P), BOD and COD. Point pollution sources are also considered. Point sources are directly added to the surface water in the model according to their geographic positions. Common point sources are urban water treatment plants and industrial plants.

Two modules are designed for the different types of water bodies, i.e., the in-stream water quality module and the water quality module for water impounding (reservoir or lake). The enhanced stream water quality model (QUAL-2E) (Brown and Barnwell, 1987) is adopted to simulate the longitudinal movement and transformation of water quality variables in the in-streams. The model is solved at the sub-basin scale rather than at the fine grid scale in order to maintain spatial consistency with the hydrological cycle module. The water quality outputs provide the water quality boundary of dams or sluices in the dam regulation module. The water quality module for water impounding assumes that water body is at the steady state and focuses on the vertical interaction of water quality processes. The main processes include water quality degradation and settlement, sediment resuspension and decay.

2.1.4 Dam regulation module (DRM)

Dams and sluices highly alter flow regimes and associated water quality processes in most river networks. Thus, the dam and sluice regulation should be considered in the water system models. The dam regulation module provides the regulated boundaries (e.g., water storage and outflow) to the hydrological cycle module for flow routing and to the water quality module of water bodies for pollutant migration.

Given that different types of dams and sluices are likely to show completely different regulation behaviors, we try to reproduce their common functionalities for either the flood control or water supply in this module. Three methods are proposed to calculate the water storage and outflow of dams or sluices, namely, the measured outflow, controlled outflow with target water storage and the relationship between outflow and water storage volume. The first method requires users to provide the measured outflow series during the simulation period. The second method simplifies the regulation rules of dams or sluices for long-term analysis based on the assumption that water is stored according to the usable water level during the non-flooding season and the flood control level during the flooding season, and the surplus water is discharged. This method requires the characteristic parameters of dam or sluice including water storage capacities of dead, usable, flood control and maximum flood levels and the corresponding water surface areas. The third method is based on the relationships among water level, water surface area, storage volume and outflow according to the designed dam data or long-term observed data (Zhang et al., 2013) (Appendix C).

2.1.5 Parameter analysis tool (PAT)

In our model, 66 lumped and 94 distributed parameters involve the hydrological, ecological and water quality processes. The distributed parameters are divided into 37 overland parameters, 17 stream parameters and 40 parameters of water projects (only for the sub-basin with reservoir or sluice) according to their spatial distribution. These parameter values are determined by the properties of overland landscape and soil, stream patterns and water projects, respectively. Different spatial calculation units share many common parameter values if their properties are the same.

Owing to a large number of parameters, it is hard to find optimal parameter values by manual tuning. The limited number of observed processes causes equifinality in the model calibration (Beven, 2006). Therefore, the parameter sensitivity analysis and calibration are important steps to alleviate equifinality in the applications of highly parameterized models, particularly for integrated water system models (Mantovan and Todini, 2006; Mantovan et al., 2007; McDonnell et al., 2007). The PAT is designed for parameter sensitivity analysis, autocalibration and model performance evaluation (Fig. 5).

To evaluate model performance, five traditionally used criteria are included in the PAT, i.e., bias (bias), relative error (re), root mean square error (RMSE), correlation coefficient (r) and Nash–Sutcliffe efficiency (NS defined by Nash and Sutcliffe, 1970). The detailed definitions of these criteria are given in Appendix D. Furthermore, flow duration curve and cumulative distribution function are also provided for capturing multiple signatures of calibrated processes. More criteria can also be proposed by the users. The objective function(s) to calibrate the model can be formed by single or multiple criteria or their function (such as weighted average).

The parameter analysis algorithms in the PAT include the parameter sensitivity method (Latin hypercube one factor at a time: LH-OAT) (van Griensven et al., 2006), the single objective auto-optimization methods such as particle swarm optimization (PSO) (Kennedy, 2010), the genetic algorithm (GA) (Goldberg, 1989) and shuffled complex evolution (SCE-UA) (Duan et al., 1994), as well as the multi-objective auto-optimization methods such as the weighted sum method and nondominated sorting genetic algorithm II (NSGA-II) (Deb et al., 2002). The method can be selected on the basis of the specific requirements of users.

In order to obtain the optimal parameter values, the following treatments are adopted in the PAT. First, the prior ranges of all the parameter values or their prior distributions (i.e., uniform or normal) are preset by referring to the literatures or similar basins. The constraints on parameters are also considered in both parameter sensitive analysis and autocalibration. In the hydrological cycle module, the constraints on soil moisture parameters are W_m (minimum moisture) $< W_w$ (moisture at permanent wilting point) $< W_{fc}$ (field capacity) $< W_{sat}$ (saturated moisture capacity). The basic surface

Prior information

Parameter constraints — Parameters and their prior ranges — Parameter values

(a) Integrated water system model

Integrated water system model

Observations — Simulations

Hydrological components — Water quality components — Ecological components

Model performance evaluation

Traditional indices — Statistical indices — User-defined indices

Bias — Relative error — Correlation coefficient — Root mean square error — Nash–Sutcliffe efficiency — Flow duration curve — Cumulative distribution function

Objective function

(b) PAT

Sensitivity analysis

LH-OAT

Maximum iterations — No

Yes

Sensitivity ranks of all the parameters

Selected parameters — Parameter ranges

Parameter autocalibration

User selection

Single objective calibration — Multi-objective calibration

GA — PSO — SCE-UA — Weighted sum method — NSGA-II

No

Best objectives or maximum iterations

Yes

Parameter optimal values — Optimal simulated components

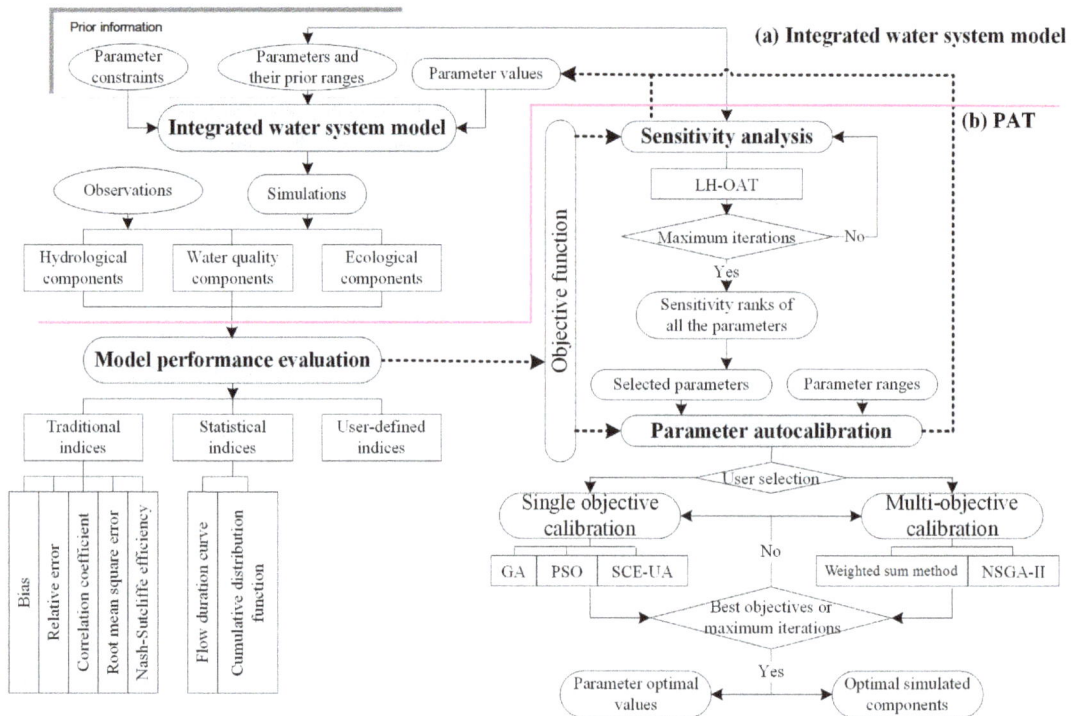

Figure 5. The flowchart of PAT and its interactions with other modules.

runoff coefficient (g_1) for different land uses/covers is set in ascending order (water body, paddy land, urban area, forest, dryland agriculture, unused land and grassland). The interflow yield coefficient (K_{ss}) is greater than the baseflow coefficient (K_{bs}). In the water quality module of water bodies, the settling rates of water quality variables (K_{set}) in the water impounding are greater than the resuspension rates (K_{scu}) and the settling rates (R_{set}) in channels. Second, the sensitive parameters are determined to reduce the parameter dimensions by sensitivity analysis. Third, the selected sensitive parameters are calibrated by the auto-optimization method, while the insensitive parameters remain as their default values that are given by referring to the literatures or other models (e.g., SWAT, EPIC and DNDC) in the same/similar basins.

The PAT connects with other modules through the parameter values that are used to simulate the processes of other modules and evaluate the objective functions in sensitivity analysis and autocalibration. Depending on the algorithm used, the parameter values are (randomly) sampled from the multi-dimensional parameter spaces to drive our model, and the objective function value of each parameter set is then obtained. For the parameter sensitivity analysis, the sensitivity index of each parameter set is evaluated by comparing the variation of the objective function value along with the change in parameter value. For the parameter autocalibration, the good parameter sets are kept or updated by the auto-optimization method until the convergence or the maximum number of iterations is achieved.

2.2 Model operation

2.2.1 Multi-scale solution

The spatial heterogeneities of basin attributes and the different timescales used in individual processes cause inconsistent spatial and temporal scales in model integration (Sivapalan and Kalma, 1995; Singh and Woolhiser, 2002). For the spatial scale, three levels of spatial calculation units are designed, namely, sub-basin, land-use/cover and crop from largest to smallest. These units are defined as the minimum polygons with similar hydrological properties, land uses/covers and agriculture crop cultivation patterns, respectively. The sub-basins are defined on the basis of a digital elevation model (DEM), the positions of gauges and water projects, and are used in the hydrological cycle module (e.g., flow routing in both land and in-stream), overland water quality module, water quality module of water bodies and dam regulation module. Seven specific land-use/cover units of each sub-basin are partitioned by the land-use/cover classification (i.e., forest, grassland, water, urban, unused land, paddy land and dryland agriculture) and are used in the hydrological cycle module (e.g., water yield, infiltration, interception and evapotranspiration) and the soil erosion module. Moreover, several specific land-use/cover units (paddy land, dryland agriculture, forest and grassland), where agricultural activities usually occur, are divided further into the crop units for the detailed analysis of the impact of agricultural management on water and nutrient cycles. In the current version of

Table 1. The data sets and their categories used in the model.

Category	Data	Objectives	Controlled processes
GIS	DEM	Elevation, area, longitude and latitude, slopes and lengths of each sub-basin and channel	Hydrology and water quality
	Land-use/cover map	Land-use/cover types and their corresponding areas in each sub-basin	Hydrology, water quality and ecology
	Soil map	Soil physical properties of each sub-basin such as bulk density and saturated conductivity	
Weather	Daily precipitation	Daily precipitation of each sub-basin	Hydrology
	Daily maximum and minimum temperature	Daily maximum and minimum temperature of each sub-basin	
Hydrology	Observed runoff or other hydrological components, etc.	Hydrological parameter calibration	Hydrology
Water quality	Urban wastewater discharge outlets and discharge load	Model input of point source pollutant load	Water quality
	Water quality observations (concentration or load), etc.	Water quality parameter calibration	
Ecology	Crop yield, leaf area index, etc.	Ecological parameter calibration	Ecology
Economy	Basic economic statistical indictors	Populations, breeding stock of large animals and livestock, water withdrawal in each sub-basin	Hydrology and water quality
Water projects	Design data attribute parameters	Regulation rules of dams or sluices	Hydrology
Agricultural management	Fertilization and irrigation types, timing and amount, time of seeding and harvest, and crop types	Agricultural management rules of each sub-basin	Water quality and ecology

our model, these four land-use/cover units are divided into 10 specific categories of crop units as fallow for all these land-use/cover units, grass for grassland unit, fruit tree and non-economic tree for forest unit, early rice and late rice for paddy unit, spring wheat, winter wheat, corn and mixed dry crop for dryland agriculture unit. The crop unit of a specific land-use/cover pattern varies depending on crop cultivation structure and timing. The related modules are the soil biochemical module and the crop growth module. All of the outputs of the crop unit are summarized at the land-use/cover scale or sub-basin scale based on the area percentages in different crop units.

For the temporal scale, it is practical to use a daily time step, as this is consistent with the underlying rainfall–runoff module and the data availability. The sub-daily scale may improve the performance in some modules (e.g., SEM and WQM). However, most observations (e.g., climate data sets, soil nutrient availability and water quality concentrations) are at the daily scale, leading to potential uncertainties or instabilities to disaggregate the observations into a sub-daily

scale. Linear or nonlinear aggregation functions are used to transform different timescales to daily scale (Vinogradov et al., 2011), such as exponential functions for flow infiltration and overland flow routing processes in the hydrological cycle module, for soil erosion processes in the soil erosion module (Eqs. A5, A6 and S32 in Appendix A and the Supplement), and accumulation functions for the crop growth process in the crop growth module (Eq. S7 in the Supplement).

2.2.2 Basic data sets and spatial delineation

The indispensable data sets for model setup are GIS data, daily meteorological data series, social and economic data series and dam attribute data. Several monitoring data series are needed for model calibration, such as runoff and water quality series in river sections, soil moisture and crop yield at the field scale. Table 1 shows all of the detailed data sets and their usages.

The hydrological toolset of the Arc GIS platform is used to delineate all the spatial calculation units based on a DEM and

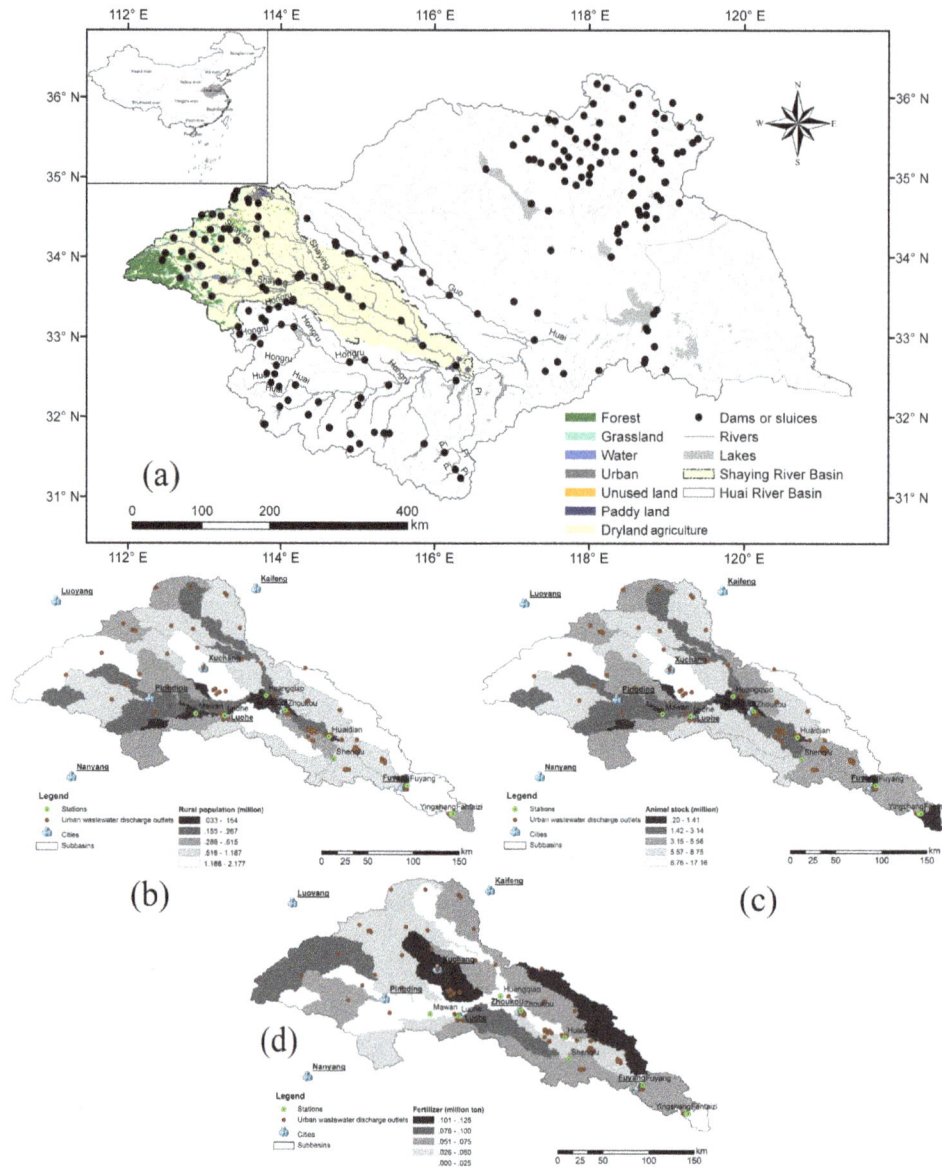

Figure 6. The location of the study area (**a**) and the digital delineation of the sub-basin, point source pollutant outlets, rural population (**b**), animal stock (**c**) and fertilization (**d**).

land-use/cover data. The sub-basin attributes (e.g., location, evaluation, area, land surface slope and slope length, land-use/cover areas) and flow routing relationship between sub-basins are obtained during this procedure.

2.3 Study area and model testing

In this study, our model was applied to a highly regulated and heavily polluted catchment (the Shaying River catchment) in China. The simulated water-related components contained daily runoff and water quality concentrations at river sections, spatial patterns of diffuse source pollution load and crop yield at sub-basin scale.

2.3.1 Study area

The Shaying River catchment (112°45′–113°15′ E, 34°20′–34°34′ N), which is the largest sub-basin of the Huai River basin in China, is selected as the study area (Fig. 6a). The drainage area is 36 651 km², with a mainstream of 620 km. The average annual population (2003–2008) (Fig. 6b) is 32.42 million, with a rural population of 23.70 million. The average annual stocks include 8.30 million big animals (cattle, pigs and sheep) and 178.42 million poultry (Fig. 6c). The average annual use of chemical fertilizer is 1.55 million ton (N: 38–51 %; P: 16–25 %; and others: 23–47 %) (Fig. 6d). The catchment is located in the typical warm temperate and semi-humid continental climate zone. The annual average temperature and rainfall are 14–16 °C and 769.5 mm, respectively. The Shaying River is the most seriously polluted tributary, with a pollutant load contribution of over 40 % in the whole Huai River, and is usually known as the water environment barometer of the Huai River mainstream. To reduce flood or drought disasters, 24 reservoirs and 13 sluices, whose regulation capacities are over 50 % of the total annual runoff, have been constructed, and fragmented the river into several impounding pools.

2.3.2 Model setup

All data sets for model setup and calibration were collected from the government bureaus, official books and scientific references. The detailed descriptions were presented in Tables S2 and S3 of the Supplement. The resolutions of GIS and weather input data were quite satisfactory for the model application. However, most data on water quality, ecology and agricultural management were at monthly or annual temporal scale. The data for economy, agricultural management and diffuse source load were collected from individual administrative regions. Both the temporal and spatial scales were larger than the required daily scale or spatial calculation units (sub-basin, land-use/cover and crop). In these cases, the data values were uniformly distributed to the required temporal and/or spatial scales, such as the input of point sources, and social and economic data.

The Shaying River catchment was divided into 46 sub-basins. According to the land-use/cover classification standard of China (CNS, 2007), the main land-use/cover types were dryland agriculture (84.04 %), forest (7.66 %), urban (3.27 %), grassland (2.68 %), water (1.43 %), paddy land (0.91 %) and unused land (0.01 %). The soil input parameters (the contents of sand, clay and organic matter) were calculated based on the percentage of soil types in each sub-basin. The main crops were early rice and late rice in the paddy land, and winter wheat and corn in the dryland agriculture. The main agricultural management schemes (fertilize, plant, harvest and kill) were summarized by field investigation in the studies of Wang et al. (2008) and Zhai et al. (2014) (Table S3). Crop rotations and management schemes were considered in the model by setting the start time, the duration of management and the fertilizer amounts. Two fertilizations (base and additional fertilization) were considered in the model during the complete growth cycle of a certain crop. The areas of sub-basin, land-use/cover and crop units ranged from 46.48 to 3771.15 km^2, from 0.04 to 2762.5 km^2, and from 3.73 to 2762.5 km^2, respectively.

The daily precipitation series from 2003 to 2008 at 65 stations were interpolated to each sub-basin using the inverse distance weighting method, while the daily temperature series at six stations were interpolated using the nearest-neighbor interpolation method. The social and economic data (e.g., population and livestock in the rural area, chemical fertilizer amounts) were calculated for each sub-basin based on the area percentage.

Moreover, 5 reservoirs, 12 sluices and over 200 wastewater discharge outlets were considered according to their geographical positions. The farm manure from rural living and livestock farming was considered as a diffuse source owing to its scattered characteristics and the deficient sewage treatment facilities in the rural areas.

2.3.3 Model evaluation

The observation series of daily runoff and NH$_4$–N concentration were used to calibrate the model parameters. There were five regulated stations (Luohe, Zhoukou, Huaidian, Fuyang and Yingshang) and one less-regulated station (Shenqiu), which is the downstream station situated far from water projects. Moreover, given that the observed yields of diffuse pollutant loads and crops were hard to collect for the whole catchment, only the statistical results from official reports or statistical yearbooks (Wang, 2011; Henan Statistical Yearbooks, 2003, 2004 and 2005) were collected to validate the model performances.

We selected LH-OAT for parameter sensitivity analysis and SCE-UA for parameter calibration in the PAT. To reduce the dimensions of the calibration problem, we restricted SCE-UA to calibrate only the sensitive parameters defined by LH-OAT, whereas the rest of the parameters remained constants. The selected evaluation indices of model performance were bias, r and NS. However, NS was sensitive to the extreme value, outlier and number of the data points, and was not commonly used in environmental sciences (Ritter and Muñoz-Carpena, 2013). Thus NS was not used to evaluate the NH$_4$–N concentration simulation.

The model calibration was conducted by the following steps. Hydrological parameters were calibrated first against the observed runoff series at each station from upstream to downstream, and then water quality parameters against the observed NH$_4$–N concentration series. The calibration and validation periods were from 2003 to 2005 and from 2006 to 2008, respectively. The weighted sum method was usually used to comprehensively handle multi-objectives (Efstratiadis and Koutsoyiannis, 2010). In this study, single-objective functions were formed by equally weighting the evaluation indices as (f_{runoff} and $f_{\text{NH4–N}}$) because the case study was only a demonstration of the model performance.

$$\begin{cases} f_{\text{runoff}} = \min[(|\text{bias}| + 2 - r - \text{NS})/3] \\ f_{\text{NH}_4\text{–N}} = \min[(|\text{bias}| + 1 - r)/2] \end{cases} \qquad (1)$$

Moreover, the effect of dam regulation was considered because of the high regulation in most rivers. The dam and sluice regulation usually altered the intra-annual distribution of flow events, such as flattening high flow and increasing low flow. The simulation performances of high and low flows were separately evaluated and the effectiveness of the DRM was tested by comparing the simulation with and with-

Table 2. Sensitive parameters, their value ranges and relative importance for runoff and NH$_4$–N simulations.

Variables	Range	Definition	Relative importance for runoff (%)	Relative importance for NH$_4$–N (%)
W_{fc}	0.20 to 0.45	Field capacity of soil	32.73	11.10
W_{sat}	0.45 to 0.75	Saturated moisture capacity of soil	11.68	11.83
g_1	0 to 3	Basic surface runoff coefficient	7.30	10.34
g_2	0 to 3	Influence coefficient of soil moisture	10.54	12.11
K_{ET}	0 to 3	Adjustment factor of evapotranspiration	23.21	10.71
K_{ss}	0 to 1	Interflow yield coefficient	9.55	3.20
T_g	1 to 100	Delay time for aquifer recharge	1.74	–
K_{bs}	0 to 1	Baseflow yield coefficient	2.91	–
K_{sat}	0 to 120	Steady-state infiltration rate	0.33	–
R_d(BOD)	0.02 to 3.4	BOD deoxygenation rate at 20 °C	–	6.62
R_{set} (BOD)	−0.36 to 0.36	BOD settling rate at 20 °C	–	3.60
R_d (NH$_4$)	0.1 to 1	Bio-oxidation rate of NH$_4$–N at 20 °C	–	1.97
K_{set} (NH$_4$)	0 to 100	Settling rate of NH$_4$–N in the reservoirs	–	14.17
K_d (BOD)	0.02 to 3.4	BOD deoxygenation rate in the reservoirs at 20 °C	–	2.12
K_d (NH$_4$)	0.1 to 1.0	Bio-oxidation rate of NH$_4$–N in the reservoirs at 20 °C	–	4.51
Total relative importance			100.00	92.27

out the consideration of dam regulation. The high and low flows were determined by the cumulative distribution function (CDF). A threshold of 50 % was used for easy presentation; i.e., the flow was treated as high flow (or low flow) if its percentile was greater than (or smaller than) the threshold.

3 Results

3.1 Parameter sensitivity analysis

Nine sensitive parameters were detected for runoff simulation by LH-OAT (Table 2), including soil-related parameters W_{fc} (field capacity), W_{sat} (saturated moisture capacity), K_r (interflow yield coefficient) and K_{sat} (steady-state infiltration rate); TVGM parameters g_1 (basic surface runoff coefficient) and g_2 (influence coefficient of soil moisture); baseflow parameters K_g (baseflow yield coefficient) and T_g (delay time for aquifer recharge); and evapotranspiration parameter K_{ET} (adjusted factor of actual evapotranspiration). All of these parameters controlled the main hydrological processes in which soil water and evapotranspiration processes were distinctly important and explained 54.3 and 23.2 % of the runoff variation, respectively.

For NH$_4$–N concentration simulation, over 90 % of observed NH$_4$–N concentration variations were explained by 14 sensitive parameters that were categorized into hydrological (59.28 % of variation), NH$_4$–N (20.65 % of variation) and COD (12.34 % of variation) related parameters. The main explanation was that hydrological processes provided the hydrological boundaries that affected the diffuse source load into rivers and the degradation and settlement processes of NH$_4$–N in water bodies NH$_4$–N concentration was further influenced by the settlement and biological oxidation. More-

over, it was a competitive relationship between COD and NH$_4$–N to consume DO of water bodies in a certain limited level (Brown and Barnwell, 1987).

3.2 Hydrological simulation

The runoff simulations fitted the observations well at all the stations (Fig. 7 and Table 3). The biases were very close to 0.0 at all the regulated stations except Zhoukou with an underestimation (bias: 0.24 for calibration and 0.41 for validation) and Luohe with an overestimation (bias: −0.52 for validation). The obvious biases were caused by the average objective function of all three evaluations rather than the bias only. The r values ranged from 0.75 (Luohe for validation) to 0.92 (Yingshang for calibration) with the average value of 0.85, whereas the NS values ranged from 0.51 (Luohe for validation) to 0.84 (Yingshang for calibration) with the average value of 0.70. The results of the regulated stations were a little worse than those of the less-regulated station (Shenqiu) owing to the regulation.

By comparing the simulations with the observations from 2003 to 2008, we saw that the high and low flows were always overestimated if the model did not consider the regulations (Fig. 8). Except for the high flows at Zhoukou, both high and low flows at all the stations were simulated well when the dam and sluice regulation was considered (Table 4). The best fitting was at Fuyang, particularly for the high flow simulation (bias = 0.10, $r = 0.89$ and NS = 0.78). From unregulation to regulation settings, the improvements measured by f_{runoff} ranged from −0.08 (Zhoukou) to −0.29 (Huaidian) for high flow simulations, from −0.05 (Zhoukou) to −0.31 (Huaidian) for average flow simulations, and from −1.97 (Fuyang) to −3.91 (Yingshang) for low flow simulations except Zhoukou (1.28). The improvements in the low

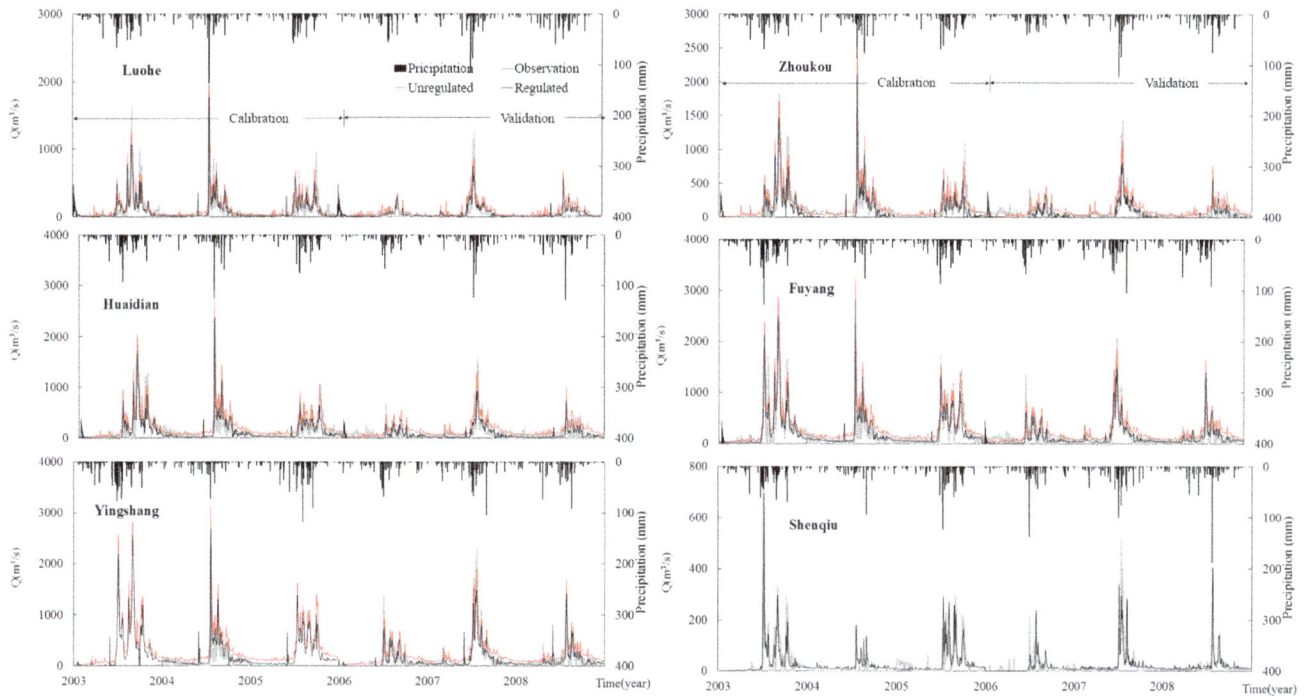

Figure 7. The daily runoff simulation at all stations.

Table 3. Runoff simulation results for regulated and less-regulated stations.

Stations	Periods	Daily flow				Monthly flow			
		Bias	r	NS	f	Bias	r	NS	f
Regulated stations									
Luohe	Calibration	0.00	0.84	0.70	0.15	0.00	0.87	0.71	0.14
	Validation	−0.52	0.75	0.51	0.42	−0.52	0.87	0.67	0.33
Zhoukou	Calibration	0.24	0.87	0.73	0.21	0.24	0.90	0.76	0.19
	Validation	0.41	0.79	0.55	0.36	0.41	0.91	0.70	0.26
Huaidian	Calibration	0.03	0.88	0.77	0.13	0.03	0.91	0.81	0.10
	Validation	0.12	0.76	0.54	0.27	0.12	0.87	0.70	0.18
Fuyang	Calibration	0.00	0.90	0.81	0.10	0.00	0.95	0.89	0.05
	Validation	0.14	0.88	0.76	0.17	0.14	0.94	0.86	0.11
Yingshang	Calibration	−0.13	0.92	0.84	0.12	−0.13	0.92	0.84	0.12
	Validation	0.16	0.87	0.74	0.18	0.16	0.93	0.82	0.13
Less-regulated stations									
Shenqiu	Calibration	0.00	0.91	0.82	0.09	0.00	0.94	0.88	0.06
	Validation	−0.13	0.83	0.67	0.21	−0.13	0.98	0.94	0.08

flow simulations were very obvious. However, their performances still needed to be improved further, particularly for the underestimation at Zhoukou and Huaidian. The possible reasons were as follows. On the one hand, the applied evaluation indices (r and NS) were known to emphasize the high flow simulation rather than the low flow simulation (Pushpalatha et al., 2012), and the objective of autocalibration was to obtain the optimal solution for the average of three evalu-

ation indices rather than the bias only. The slight sacrifice of bias improved the overall simulation performance evaluated by all three indices. One the other hand, the dam regulation module still could not fully capture the low flows.

Furthermore, the model performances on monthly flows were even better, particularly for r and NS. The r values ranged from 0.87 (Luohe for both calibration and validation) to 0.95 (Fuyang for calibration) with the average value of

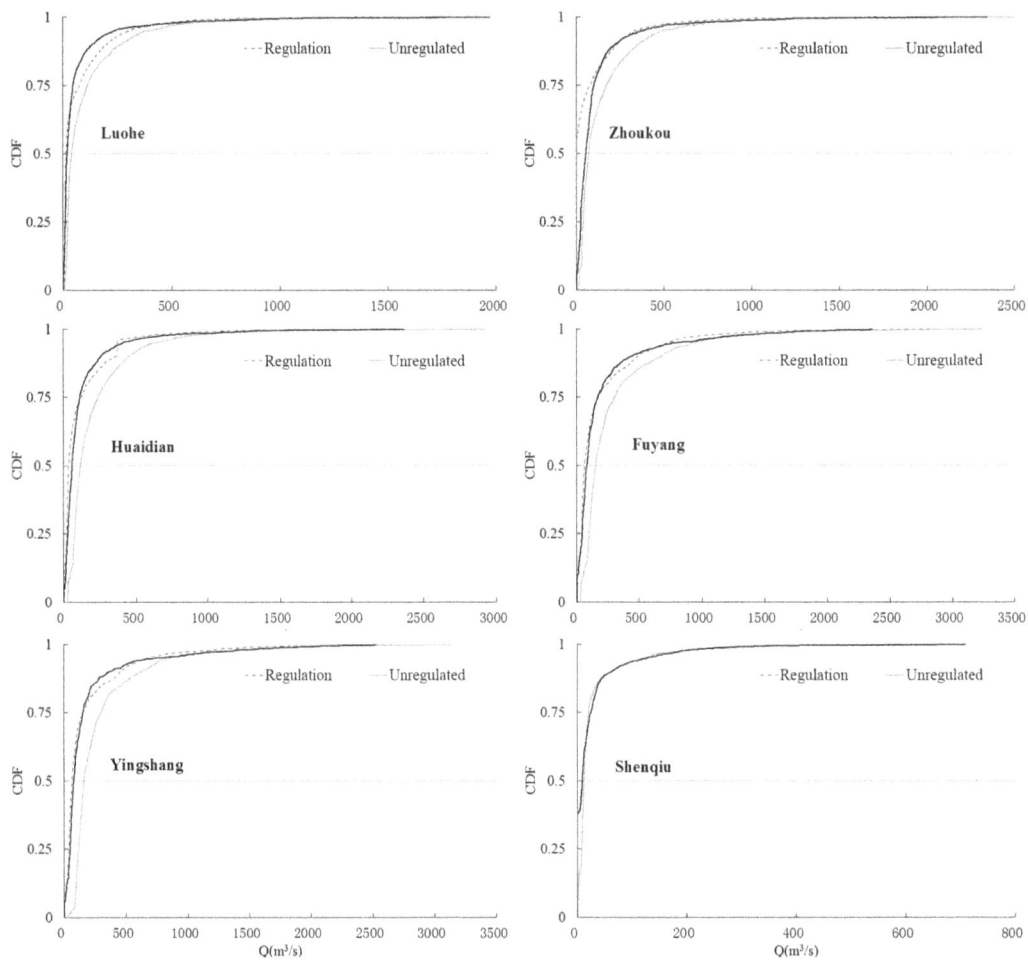

Figure 8. The cumulative distributions of simulated and observed daily runoff at all stations.

Table 4. The runoff simulation results at regulated stations with and without the dam regulation considered. Range means the difference of the objective function value between regulations considered and not considered. If the range value is less than 0.0, then the simulation with regulation is better than that without regulation. Otherwise, the simulation without regulation is better.

Stations	Regulated capacity (%)	Flow event	Regulation considered				Regulation not considered				Range
			Bias	r	NS	f	Bias	r	NS	f	
Luohe	0.26	High	−0.16	0.97	0.92	0.09	−0.62	0.97	0.80	0.29	−0.20
		Low	−0.02	0.98	0.69	0.12	−1.46	0.99	−5.53	2.67	−2.55
		Average	−0.15	0.97	0.93	0.08	−0.68	0.96	0.82	0.30	−0.22
Zhoukou	1.31	High	0.21	0.98	0.93	0.10	−0.38	0.98	0.87	0.18	−0.08
		Low	1.00	0.00	−2.57	1.86	−0.64	0.99	−0.08	0.58	1.28
		Average	0.30	0.99	0.93	0.13	−0.41	0.98	0.89	0.18	−0.05
Huaidian	1.37	High	0.02	0.98	0.95	0.03	−0.64	0.98	0.68	0.32	−0.29
		Low	0.36	0.97	0.43	0.32	−1.51	0.98	−5.88	2.80	−2.48
		Average	0.06	0.98	0.96	0.04	−0.74	0.98	0.72	0.35	−0.31
Fuyang	2.21	High	0.04	0.98	0.96	0.03	−0.39	0.99	0.86	0.18	−0.15
		Low	0.17	0.99	0.87	0.10	−1.43	0.99	−3.78	2.07	−1.97
		Average	0.05	0.99	0.97	0.03	−0.50	0.99	0.88	0.21	−0.18
Yingshang	1.76	High	0.03	0.98	0.95	0.03	−0.44	0.99	0.86	0.20	−0.17
		Low	0.18	0.99	0.82	0.12	−1.77	0.95	−9.26	4.03	−3.91
		Average	0.05	0.99	0.96	0.03	−0.60	0.98	0.86	0.25	−0.22

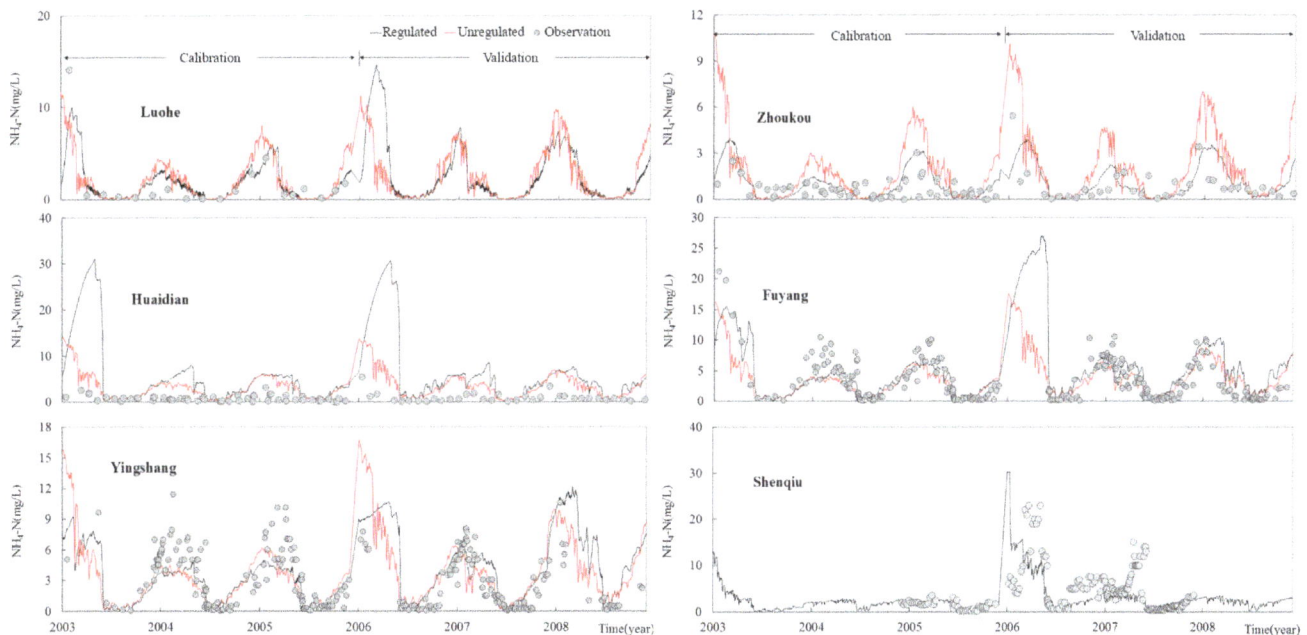

Figure 9. The simulated NH$_4$–N concentration variation at all stations.

0.92, whereas the NS values ranged from 0.67 (Luohe for validation) to 0.94 (Shenqiu for validation) with the average value of 0.80. Compared with the existing results at the same stations by SWAT (Zhang et al., 2013), the flow simulations at the downstream stations were improved, although they became a little worse at the upstream stations (Luohe and Zhoukou for calibration). In particular, the total water volume and agreements with the observations (i.e., bias and NS) were well captured.

3.3 Water quality simulation

The simulated concentrations of NH$_4$–N matched well with the observations according to the evaluation standard recommend by Moriasi et al. (2007) (Fig. 9 and Table 5). The r values were over 0.60 for all the stations except Zhoukou (0.56 for validation), Yingshang (0.49 for validation) and Shenqiu (0.41 for validation), and the average value was 0.67. The biases were considered to be "acceptable" with a range from −0.27 (Fuyang for validation) to 0.29 (Zhoukou for calibration). The best simulation was at Luohe station. The obvious discrepancies between the simulations and observations often appeared in the period from January to May because of the poor simulation performances on the low flows. Although the biases changed markedly from calibration to validation at Fuyang and Yingshang stations, the model performances were still acceptable. The possible explanation was that the biases for corresponding runoff simulations at these two stations also changed.

Compared with the results without the consideration of regulation, the simulation results were obviously improved

when the regulation was considered, except those at Fuyang station in the calibration period. The decreases in the $f_{\text{NH4−N}}$ value ranged from 0.10 (Huaidian for calibration) to 0.49 (Zhoukou for validation), although there was a slight increase at Fuyang for the calibration (0.02). Therefore, it was concluded that the consideration of dam and sluice regulation played an important role in the water quality simulation. In the upper stream of the Shaying River, the flow was small and the NH$_4$–N concentration decreased obviously because of the degradation and settlement of large water storage. In the downstream of the Shaying River, the NH$_4$–N concentration increased because of the pollutant accumulation and the decreasing flow from dams and sluices owing to the regulation (Zhang et al., 2010). Therefore, the simulated concentrations without regulation were usually overestimated or higher than the simulation with regulation at the upstream stations (Luohe and Zhoukou). However, the concentrations were underestimated at the downstream stations (Huaidian, Fuyang and Yingshang). The largest differences between the simulations with and without the consideration of regulation appeared at Zhoukou.

The spatial pattern of average annual load of diffuse source NH$_4$–N was shown in Fig. 10a. The estimated annual yield rates ranged from 0.048 to 11.00 t km^{-2} year^{-1} with the average value of 0.73 t km^{-2} year^{-1}. The yield in each administrative region was summarized from the results of each sub-basin according to the area percentage of sub-basins in each administrative region. Compared with the statistical load of each administrative region based on the soil erosion, land use/cover and fertilizer amount in the official report (Wang, 2011), the bias of simulated diffuse source load in the whole

Table 5. The comparison of NH$_4$–N simulation results with and without dam regulation considered.

Stations	Periods	Regulated			Unregulated			Range	Ratio of diffuse source load (%)
		Bias	r	f	Bias	r	f		
Regulated stations									
Luohe	Calibration	−0.02	0.93	0.05	−0.67	0.60	0.54	−0.49	46.10
	Validation	−	−	−	−	−	−		
Zhoukou	Calibration	0.29	0.61	0.34	−0.56	0.38	0.59	−0.25	44.54
	Validation	0.27	0.56	0.36	−1.35	0.66	0.85	−0.49	
Huaidian	Calibration	0.22	0.73	0.25	0.49	0.80	0.35	−0.10	31.72
	Validation	0.02	0.67	0.18	0.22	0.51	0.36	−0.18	
Fuyang	Calibration	0.28	0.78	0.25	0.26	0.80	0.23	0.02	33.12
	Validation	−0.27	0.76	0.26	−0.38	0.56	0.41	−0.15	
Yingshang	Calibration	0.24	0.79	0.23	0.25	0.58	0.34	−0.11	33.26
	Validation	−0.24	0.49	0.38	−0.76	0.62	0.57	−0.19	
Less-regulated stations									
Shenqiu	Calibration	0.13	0.62	0.26	−	−	−	−	47.13
	Validation	0.16	0.41	0.37	−	−	−	−	

Figure 10. The spatial pattern of diffuse source NH$_4$–N load **(a)** and its relationship with paddy area **(b)** and rice yield **(c)** at the sub-basin and regional scale in the Shaying River catchment.

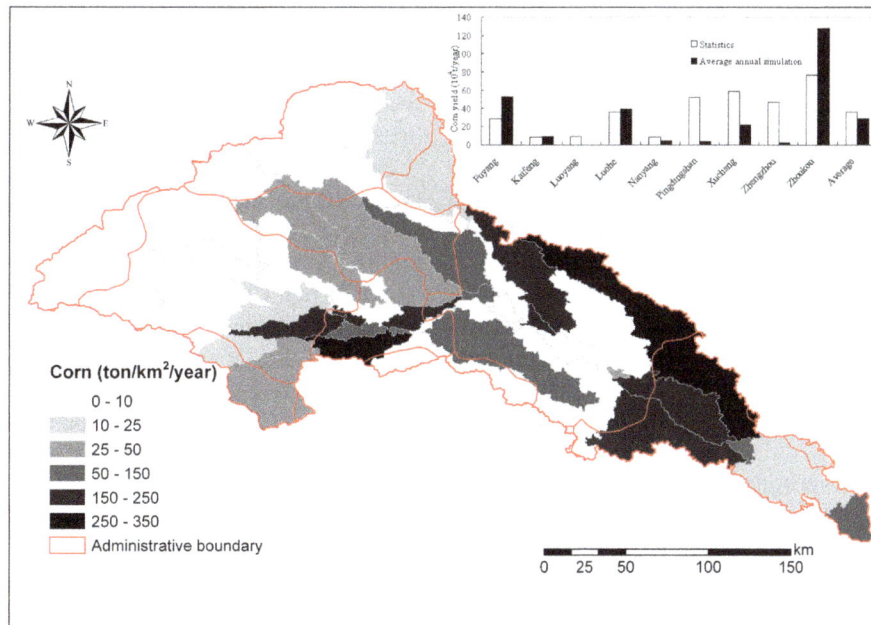

Figure 11. The spatial pattern of corn yield at the sub-basin and regional scale in the Shaying River catchment.

region was 21.31 % when the two regions with the biggest biases (Fuyang and Pingdingshan) were excluded as outliers. The high load regions were in the middle of the Pingdingshan, Xuchang, Zhengzhou, Fuyang and Zhoukou regions. The spatial pattern was significantly correlated with the distribution of paddy area ($r = 0.506$, $p < 0.001$) and rice yield ($r = 0.799$, $p < 0.001$) (Fig. 10b and c). The fertilizer losses in the paddy areas might be the primary contributor to the diffuse source NH_4–N load because the average nitrogen loss coefficient in China was just 30–70 % in the paddy areas, which was higher than that in the dryland agriculture (20–50 %) (Zhu, 2000; Xing and Zhu, 2000).

Summarized from the collected data for model input, the observed average load of point source NH_4–N into rivers was approximately 4.70×10^4 t year^{-1} in the Shaying River catchment. The diffuse source contributed 38.57 % of the overall NH_4–N load on average from 2003 to 2005, and this value was slightly higher than the statistical results (29.37 %) given in the official report (Wang, 2011). Moreover, the diffuse source contributions at the stations ranged from 31.72 % (Huaidian) to 47.13 % (Shenqiu). Compared with the diffuse source loads in the individual administrative regions in 2000, the simulated loads tended to increase from 2003 to 2005, except in the Kaifeng region. The yields in the Fuyang and Pingdingshan regions increased at the highest rates. The primary pollution source in the Shaying River catchment was still the point source, but the diffuse source was also an important concern. In terms of spatial variation, the contribution of diffuse source to the pollutant load was high in the upstream and low in the middle and downstream because the point source emission was usually concentrated in the mid-

dle and downstream. Therefore, compared with the results in Zhang et al. (2013), the overall simulation performance of NH_4–N concentration was also improved remarkably by considering the detailed nutrient processes in the soil layers.

3.4 Crop yield simulation

The simulated corn yield and its spatial pattern were shown in Fig. 11. The average annual yields were summarized at sub-basin scale and ranged from 0.08 to 326.95 t km^{-2} year^{-1} with the average value of 76.84 t km^{-2} year^{-1}. The yield of each administrative region was further summarized and compared with the data from statistical yearbooks from 2003 to 2005 (Henan Statistical Yearbook, 2003, 2004 and 2005). The high-yield regions were Luohe, Fuyang and Zhoukou in the middle and downstream where the primary land use/cover was the dryland agriculture (93.12, 95.87 and 93.18 %, respectively). The crop yields in the Luohe, Nanyang and Kaifeng regions were well simulated. The total yield was underestimated in the whole basin with a bias of 19.93 %. The discrepancies might be caused by the boundary mismatch between the administrative region and sub-basin, spatial heterogeneities of human agricultural activities and inaccurate cropping pattern used in such huge regions. A high-resolution remote sensing image and field investigation might be helpful to improve the model performance.

4 Discussion

4.1 Comparison with other models

It is a natural tendency that models grow in complexity in order to capture more interactions of complex water-related processes in the real basins because of more and more available observations and improved accuracies (Beven, 2006). Our proposed model was developed in this direction and tended to benefit integrated river basin management, although the model applicability needs to be further evaluated in different regions. In comparison with most existing models, our proposed model considered all the water-related processes as an integrated system rather than isolated systems for individual processes.

Our model provided competitive simulation results in the Huai River basin (Figs. 7–9; Tables 3–5). Several typical models were also applied in this basin, such as SWAT for the monthly runoff and water quality simulation at the regulated stations (Zhang et al., 2013), the SWAT and Xinanjiang models for the daily runoff simulation at the unregulated upstream stations (Shi et al., 2011) and the DTVGM for daily runoff simulation (Ma et al., 2014). Compared with the results of these models, our model generally performed better in the runoff or water quality simulations. In particular, our model performed even better than SWAT at the regulated stations, as more detailed dam regulation rules and soil biochemical processes were considered. For example, the average values of f_{runoff} at the monthly scale decreased from 0.32 (SWAT in Zhang et al., 2013) to 0.15 (our model) at the regulated stations. The average values of $f_{\text{NH4-N}}$ decreased from 0.47 (SWAT in Zhang et al., 2013) to 0.27 (our model). Moreover, both the Xinanjiang model and the DTVGM are limited to simulating the flow series at the unregulated or less-regulated stations because they do not consider the dam regulation in their current model frameworks (Shi et al., 2011; Ma et al., 2014).

4.2 Equifinality

Until now, our understandings of water-related processes have still been ambiguous, and it is hard to describe all these processes in the real-word systems from strong physical foundations (Beven, 2006; Hrachowitz et al., 2014). Empirical equations are usually adopted to approximate the physical processes with numerous unknown parameters, especially in the large-scale models. A single output variable of models is associated with multiple processes and many parameters. For examples, SWAT contains over 200 parameters (Arnold et al., 1998) and DNDC has nearly 100 parameters (Li et al., 1992). Pohlert et al. (2006) reported that six hydrological and 12 N-cycle sensitive parameters were detected in SWAT-N for the simulation of water flow and N leaching. In the case study, 9 and 14 sensitive parameters of our model were detected for runoff and NH_4–N simulation, respectively

(Table 2). Therefore, due to the large numbers of model parameters and limited observations, most existing models are subject to equifinality, which is more serious if more water-related processes are considered or more sub-basins are delineated for the distributed models.

Several strategies would be helpful to alleviate the equifinality, such as field experiments on the physical parameters (Kirchner, 2006), the utilization of more observed processes, multiple evaluation measures for a single predicted component (Her and Chaubey, 2015), parameter regularization and process constraints (Tonkin and Doherty, 2005; Pokhrel et al., 2008; Euser et al., 2013). Moreover, some attempts are made to move away from traditional curve fitting towards more process consistency and efficient model selection techniques (Hrachowitz et al., 2014; Fovet et al., 2015).

For our model, all the independent calibration and validation data sets were specified in Table 1, and most widely used measures of model performances were also provided in the PAT. In the case study, we also employed several observation sources (e.g., runoff and water quality observations at different stations, the diffuse pollution load and crop yield data) and used three measures to evaluate model performance for the individual components (e.g., bias, r and NS). To make full use of the existing data in practice, parameter sensitivity analysis would be an effective way to reduce dimensionality in model calibration and then focus only on the critical processes and parameters that are sensitive to model outputs (van Griensven et al., 2006). Model autocalibration would be efficient to obtain the optimal simulations from numerous samples in multi-dimensional parameter spaces.

4.3 Model limitations

It should be noted that our extended model still has several limitations.

1. The mathematical descriptions of groundwater, crop growth processes and agriculture management practices were still inaccurate. The current version focused on the detailed descriptions of hydrological and nutrient cycles in the soil layers and water bodies, and the consideration of dam regulation. Satisfactory performances on water quantity and quality simulation were achieved in our case study. However, the simulations for groundwater, diffuse pollution and crop yield in the agriculture regions could be improved further. The stratification of water impounding in the water quality module should be considered if the high-resolution bathymetric data of dams or lakes are available.

2. High parameterization is an inevitable issue because of its all-inclusive framework. Our model considered the main water-related processes in the hydrological, ecology and water quality subsystems, but numerous processes were still controlled by unmeasurable parameters because of their empirical and/or scale-dependent

nature (Her and Chaubey, 2015). Although the parameter sensitivity analysis and calibration are widely used to handle the high parameterization issue, the equifinality and parameter uncertainty are still inevitable because of the insufficient observations and the complex interactions among different subsystems.

5 Conclusions

In this study, the TVGM hydrological model was extended primarily to an integrated water system model to address the complex water issues emerging in the basins. The model performance was demonstrated in the Shaying River catchment, China. The model provided a reasonable tool for the effective water governance by simultaneously simulating several indicative components of water-related processes including the hydrological components (e.g., runoff, soil moisture, evaporation and plant transpiration, water storage in the dams and sluices), water quality components (e.g., diffuse pollution source load, water quality concentrations in water bodies) and ecological components (e.g., crop yield), which could be calibrated if observations were available. The case study showed that the simulated runoffs at most stations fitted the observations well in the highly regulated Shaying River catchment. All the evaluation criteria were acceptable for both the daily and monthly simulations at most stations. This model simulated the discontinuous daily NH_4–N concentration well and properly captured the spatial patterns of diffuse pollution load and corn yield.

Owing to the heterogeneity of spatial data in large basins and insufficient observations of individual subsystems, not all the results were acceptable and several processes were still not well calibrated (such as low flow events, diffuse pollution source load and crop yield). More available data and improved data quality will reduce the model uncertainty and equifinality problem, especially the higher-resolution data for surface conditions, water quality, agricultural management and socio-economic data. The model would be improved by further considering more accurate human activities in the agricultural management, calibrating multiple components by multi-objective optimization and model uncertainty analysis because of the interactions and tradeoffs among different processes. The over-parameterization and the reasonable prior parameter conditions should also be treated carefully in applications. Advanced analysis technologies would benefit the future model development, such as model selection techniques, parameter regularization. Moreover, an easily used operational software package can broaden the model's applications in different regions. More case studies are needed to further demonstrate its applicability.

Appendix A: Hydrological cycle module

The basic water balance equation is

$$P_i + SW_i = SW_{i+1} + Rs_i + E_{a_i} + Rss_i + Rbs_i + In_i, \quad (A1)$$

where P is the precipitation (mm); SW is the soil moisture (mm); E_a is the actual evapotranspiration (mm) including soil evaporation (E_s, mm) and plant transpiration (E_p, mm); Rs, Rss and Rbs are the surface runoff, interflow and baseflow (mm), respectively; In is the vegetation interception (mm) and i is the time step (day).

E_s and E_p are determined by the potential evapotranspiration (E_0, mm), leaf area index (LAI, $m^2\,m^{-2}$) and surface soil residues (rsd, $t\,ha^{-1}$) (Ritchie, 1972) as

$$\begin{cases} E_a = E_t + E_s \le E_0, \\ E_p = \begin{cases} LAI \cdot E_0/3 & 0 \le LAI \le 3.0, \\ E_0 & LAI > 3.0, \end{cases} \\ E_s = E_0 \cdot \exp(-5.0 \times 10^{-5} \times rsd), \end{cases} \quad (A2)$$

where E_0 is calculated by the Hargreaves method (Hargreaves and Samani, 1982).

The surface runoff (Rs, mm) yield equation (TVGM; Xia et al., 2005) is given as

$$Rs = g_1(SW_u/W_{sat})^{g_2} \cdot (P - In), \quad (A3)$$

where SW_u and W_{sat} are the surface soil moisture and saturation moisture (mm), respectively; g_1 and g_2 are the basic coefficient of surface runoff and the influence coefficient of soil moisture, respectively.

The interflow (Rss, mm) and baseflow (Rbs, mm) have linear relationships with the soil moistures in the upper and lower layers, respectively (Wang et al., 2009), as

$$\begin{cases} Rss = k_{ss} \cdot SW_u, \\ Rbs = k_{bs} \cdot SW_l, \end{cases} \quad (A4)$$

where k_{ss} and k_{bs} are the yield coefficients of interflow and baseflow, respectively; SW_l is the soil moisture in the lower layer (mm).

The infiltration from the upper to lower soil layers is calculated using the storage routing method (Neitsch et al., 2011) as

$$\begin{cases} W_{inf} = (SW_u - W_{fc}) \cdot [1 - \exp(-24/T_{inf})], \\ T_{inf} = (W_{sat} - W_{fc})/K_{sat}, \end{cases} \quad (A5)$$

where W_{inf} is the water infiltration amount on a given day (mm); W_{fc} is the soil field capacity (mm); and T_{inf} is the travel time for infiltration (h), respectively; K_{sat} is the saturated hydraulic conductivity ($mm\,h^{-1}$).

The calculation of overland flow routing is adopted from Neitsch et al. (2011) as

$$\begin{cases} Q_{overl} = (Q'_{overl} + Q_{stor,i-1}) \\ \quad \cdot [1 - \exp(-T_{retain}/T_{route})], \\ T_{route} = T_{overl} + T_{rch} = \dfrac{L_{overl}^{0.6} \cdot n_{overl}^{0.6}}{18 \cdot slp_{overl}^{0.3}} \\ \quad + \dfrac{0.62 \cdot L_{rch} \cdot n_{rch}^{0.75}}{A^{0.125} \cdot slp_{rch}^{0.375}}, \end{cases} \quad (A6)$$

where Q_{overl} is the overland flow discharged into the main channel (mm); Q'_{overl} is the lateral flow amount generated in the sub-basin (mm); $Q_{stor,i-1}$ is the lateral flow in the previous day (mm); T_{retain} is the residence time of flow (days); T_{route} is the flow routing time in sub-basin (days); T_{overl} and T_{rch} are the routing times of overland flow and river flow, respectively (days); L_{overl} and L_{rch} are the lengths of sub-basin slope and river, respectively (km); slp_{overl} and slp_{rch} are the slopes of sub-basin and river, respectively ($m\,m^{-1}$); n_{overl} and n_{rch} are the Manning roughness coefficients for sub-basin and river, respectively ($m\,m^{-1}$); and A is the sub-basin area (km^2).

Appendix B: Soil biochemical module

B1   Soil temperature (Williams et al., 1984)

$$T(Z,t) = \bar{T} + (AM/2 \cdot \cos[2\pi \cdot (t - 200)/365] \\ + TG - T(0,t)) \cdot \exp(-Z/DD), \quad (B1)$$

where Z is the soil depth (mm); t is the time step (days); $\bar{T}$ and TG are the average annual temperature and surface temperature (°C), respectively; AM is the annual variation amplitude of daily temperature; and DD is the damping depth (mm) of soil temperature given as

$$\begin{cases} DD = DP \cdot \exp\{(\ln(500/DP) \cdot [(1 - \xi)(1 + \xi)]^2\}, \\ DP = 1000 + 2500BD/[BD + 686\exp(-5.63BD)], \\ \xi = SW/[(0.356 - 0.144BD) \cdot Z_M], \\ TG_{IDA} = (1 - AB) \cdot (T_{mx} + T_{mn})/2 \cdot (1 - RA/800) \\ \quad + T_{mx} \cdot RA/800 + AB \cdot TG_{IDA-1}, \end{cases} \quad (B2)$$

where DP is the maximum damping depth of soil temperature (mm); BD is the soil bulk density ($t\,m^{-3}$); ζ is a scale parameter; IDA is the day of the year; AB is the surface albedo; and RA is the daily solar radiation (ly).

B2   C and N cycle (Li et al., 1992)

Decomposition. The decomposition of resistant and labile C is described by the first-order kinetic equation, viz.

$$dC/dt = \mu_{CLAY} \cdot \mu_{C:N} \cdot \mu_{t,n} \cdot [S \cdot k_1 + (1 - S) \cdot k_2], \quad (B3)$$

where μ_{CLAY}, $\mu_{C:N}$ and $\mu_{t,n}$ are the reduction factors of clay content, C : N ratio and temperature for nitrification, respectively; S is the labile fraction of organic C compounds; k_1

and k_2 are the specific decomposition rates of labile faction and resistant fraction, respectively (day^{-1}).

The NH$_4$ amount (FIX$_{NH4}$, kg ha^{-1}) absorbed by clay and organic matter is estimated by

$$FIX_{NH_4} = [0.41 - 0.47 \cdot \log(NH_4)] \cdot (CLAY/CLAY_{max}), \quad (B4)$$

where NH$_4$ is the NH$_4^+$ concentration in the soil liquid (g kg^{-1}). CLAY and CLAY$_{max}$ are the clay content and the maximum clay content, respectively.

$$\begin{cases} \log(K_{NH_4}/K_{H_2O}) = \log(NH_{4m}/NH_{3m}) + pH, \\ NH_{3m} \\ = 10^{\{\log(NH_4) - (\log(K_{NH_4}) - \log(K_{H_2O})) + pH\}} \cdot (CLAY/CLAY_{max}), \\ AM = 2 \cdot (NH_3) \cdot (D \cdot t/3.14)^{0.5}, \end{cases} \quad (B5)$$

where K_{NH_4} and K_{H_2O} are the dissociation constants for NH$_4^+$: NH$_3$ equilibrium and H$^+$: OH$^-$ equilibrium, respectively; NH$_{4m}$ and NH$_{3m}$ are the NH$_4^+$ and NH$_3$ concentrations (mol L^{-1}) in the liquid phase, respectively; AM and D are the accumulated NH$_3$ loss (mol cm^{-2}) and diffusion co-efficients (cm^2 d^{-2}), respectively.

The nitrification rate (dNNO, kg/ha/day) is a function of the available NH$_4^+$, soil temperature and moisture; N$_2$O emission is a function of soil temperature and soil NH$_4^+$ concentration, and is given as

$$\begin{cases} dNNO = NH_4 \cdot [1 - \exp(-K_{35} \cdot \mu_{t,n} \cdot dt)] \cdot \mu_{SW,n}, \\ N_2O = (0.0014 \cdot NH_4/30.0) \cdot (0.54 + 0.51 \cdot T)/15.8, \end{cases} \quad (B6)$$

where K_{35} is the nitrification rate at 35 °C (mg kg^{-1} ha^{-1}); $\mu_{SW,n}$ is the soil moisture adjusted factor for nitrification.

Denitrification. The growth rate of denitrifiers ((dB/dt)$_g$, kg ha^{-1} day^{-1}) is proportional to their respective biomass and is calculated by the double Monod kinetics equation as

$$\begin{cases} (dB/dt)_g = \mu_{DN} \cdot B(t), \\ \mu_{DN} = \mu_{t,dn} \cdot (u_{NO_3} \cdot \mu_{PH,NO_3} + u_{NO_2} \cdot \mu_{PH,NO_2} \\ \quad + u_{N_2O} \cdot \mu_{PH,N_2O}), \\ u_{N_xO_y} = u_{N_xO_y,max} \cdot (C/K_{C,1/2} + C) \\ \quad \cdot (N_xO_y/K_{N_xO_y,1/2} + N_xO_y), \end{cases} \quad (B7)$$

where B is the denitrifier biomass (kg); μ_{DN} is the relative growth rate of the denitrifiers; $u_{N_xO_y}$ and $u_{N_xO_y,max}$ are the relative and maximum growth rates of NO$_2^-$, NO$_3^-$ and N$_2$O denitrifiers, respectively. $K_{C,1/2}$ and $K_{N_xO_y,1/2}$ are the half velocity constants of C and N$_x$O$_y$, respectively; μ_{PH,N_xO_y} and $\mu_{t,dn}$ are the reduction factors of soil pH and temperature, respectively. The mathematical expressions are given as

$$\begin{cases} \mu_{PH,NO_3} = 7.14 \cdot (pH - 3.8)/22.8, \\ \mu_{PH,NO_2} = 1.0, \\ \mu_{PH,N_2O} = 7.22 \cdot (pH - 4.4)/18.8, \\ \mu_{t,dn} = \begin{cases} 2^{(T-22.5)/10} & \text{if } T < 60\,°C, \\ 0 & \text{if } T \geq 60\,°C. \end{cases} \end{cases} \quad (B8)$$

The death rate of denitrifier ((dB/dt)$_d$, kg ha^{-1} h^{-1}) is proportional to denitrifier biomass and is given as

$$(dB/dt)_d = M_C \cdot Y_C \cdot B(t), \quad (B9)$$

where M_C and Y_C are the maintenance coefficient of C (1 h^{-1}) and maximum growth yield of dissolved C (kg ha^{-1} hr^{-1}), respectively.

The consumption rates of dissolved C and CO$_2$ production are calculated as

$$\begin{cases} dC_{con}/dt = (\mu_{DN}/Y_C + M_C) \cdot B(t) \cdot \mu_{SW,d} \\ dCO_2/dC_{con,t} dt - (dB/dt)_d, \end{cases} \quad (B10)$$

where $\mu_{SW,d}$ is the soil moisture adjusted factor for denitrification.

The NO$_3^-$, NO$_2^-$, NO and N$_2$O consumption is calculated as

$$dN_xO_y/dt = (u_{N_xO_y}/Y_{N_xO_y} + M_{N_xO_y} \cdot N_xO_y/N) \\ \cdot B(t) \cdot \mu_{PHN_xO_y} \cdot \mu_{t,dn}, \quad (B11)$$

where $M_{N_xO_y}$ and $Y_{N_xO_y}$ are the maintenance coefficient (1 h^{-1}) and maximum growth yield on NO$_3^-$, NO$_2^-$, NO or N$_2$O (kg ha^{-1} h^{-1}), respectively.

N assimilation is calculated on the basis of the growth rates of denitrifiers and the C : N ratio (CNR$_{D:N}$) in the bacteria, viz.

$$(dN/dt)_{ass} = (dB/dt)_g \cdot (1/CNR_{D:N}). \quad (B12)$$

The emission rates are the functions of adsorption coefficients of the gases in soils and to the air-filled porosity of the soil, and are given as

$$\begin{cases} P(N_2) = 0.017 + ((0.025 - 0.0013 \cdot AD) \cdot PA \\ P(N_2O) = [30.0 \cdot (0.0006 + 0.0013 \cdot AD) \\ \quad + (0.013 - 0.005 \cdot AD)] \cdot PA \\ P(NO) = 0.5 \cdot [(0.0006 + 0.0013 \cdot AD) \\ \quad + (0.013 - 0.005 \cdot AD) \cdot PA] \end{cases} \quad (B13)$$

where $P(N_2)$, $P(NO)$ and $P(N_2O)$ are the emission rates of N$_2$, NO, and N$_2$O, respectively, during a day; PA and AD are the air-filled fractions of the total porosity and adsorption factor depending on the clay content in the soil, respectively.

Nitrate leaching. The NO$_3^-$ leaching rate is a function of clay content, organic C content and water infiltration in the soil layer, and is given as

$$Leach_{NO_3} = W_{inf} \cdot \mu_{CLAY} \cdot \mu_{soc}, \quad (B14)$$

where Leach$_{NO_3}$ is the NO$_3^-$ leaching rate; μ_{CLAY} and μ_{soc} are the influence coefficients of clay content and soil organic C, respectively.

B3 P cycle

The descriptions of P mineralization, decomposition and sorption are adopted from Neitsch et al. (2011) and are provided in the Supplement.

Appendix C: Dam regulation module (Zhang et al., 2013)

The water balance model of the dam or sluice is considered the inflow, outflow, precipitation, evapotranspiration, seepage and water withdrawal. The equation is

$$\Delta V = V_{\text{flowin}} - V_{\text{flowout}} + V_{\text{pcp}} - V_{\text{evap}} - V_{\text{seep}} - V_{\text{withd}}, \quad \text{(C1)}$$

where ΔV, V_{flowin} and V_{flowout} are the water storage variation, and water volumes of entering and flowing out, respectively (m^3), and are calculated by HCM; V_{pcp}, V_{evap} and V_{seep} are the volumes of precipitation, evaporation and seepage, respectively (m^3), and are the functions of surface water area and water storage. V_{withd} is the water withdraw volume (m^3) by humans and is given as a model input.

According to the design data of dams and sluices in China, there is a particular relationship among water level, storage and outflow. The outflow is determined by the water level or water storage volume. The relationships are described by equations.

$$\begin{cases} V_{\text{flowout}} = f'(V, H), \\ \text{SA} = f''(V, H), \end{cases} \quad \text{(C2)}$$

where V and H are the water storage volume (m^3) and water level (m) during a day, respectively; $f'()$ and $f''()$ are the functions that could be determined by statistical analysis methods (e.g., correlation analysis, linear or nonlinear regression analysis, polynomial regression analysis and least squares fitting).

Appendix D: Evaluation indices of model performance

Bias:

$$\text{bias} = \sum_{i=1}^{N}(O_i - S_i) \bigg/ \sum_{i=1}^{N} O_i \quad \text{(D1)}$$

Relative error:

$$\text{re} = \sum_{i=1}^{N} \frac{O_i - S_i}{O_i} \times 100\% \quad \text{(D2)}$$

Root mean square error:

$$\text{RMSE} = \sqrt{\sum_{i=1}^{N}(O_i - S_i)^2 / N} \quad \text{(D3)}$$

Correlation coefficient:

$$r = \sum_{i=1}^{N}(O_i - \bar{O}) \cdot (S_i - \bar{S}) \bigg/ \sqrt{\sum_{i=1}^{N}(O_i - \bar{O})^2 \cdot \sum_{i=1}^{N}(S_i - \bar{S})^2} \quad \text{(D4)}$$

Nash–Sutcliffe efficiency:

$$\text{NS} = 1 - \sum_{i=1}^{N}(O_i - S_i)^2 \bigg/ \sum_{i=1}^{N}(O_i - \bar{O})^2, \quad \text{(D5)}$$

where O_i and S_i are the ith observed and simulated values, respectively; $\bar{O}$ and $\bar{S}$ are the average observed and simulated values, respectively. N is the length of the series.

Acknowledgements. This study was supported by the Natural Science Foundation of China (no. 41271005), the China Youth Innovation Promotion Association CAS (no. 2014041), the Program for "Bingwei" Excellent Talents (no. 2015RC201) and the Key Project for the Strategic Science Plan (no. 2012ZD003) in IGSNRR, CAS, the Endeavour Research Fellowship, the China Visiting Scholar Project from the China Scholarship Council, and the CSIRO Computational and Simulation Sciences Research Platform. The authors would like to thank Yongqiang Zhang and James R. Frankenberger for their participation in our internal review procedure, Markus Hrachowitz and Christian Stamm for improving the quality and presentation of the manuscript, and the anonymous reviewers for their valuable comments and suggestions.

References

Abbott, M. B., Bathurst, J. C., Cunge, J. A., O'Connell, P. E., and Rasmussen, J.: An Introduction to the European System: Systeme Hydrologique Europeen (SHE), J. Hydrol., 87, 61–77, 1986.

Abrahamsen, P. and Hansen, S. D.: an open soil-crop-atmosphere system model, Environ. Model. Softw., 15, 313–330, 2000.

Arheimer, B. and Brandt, M.: Modelling nitrogen transport and retention in the catchments of southern Sweden, Ambio, 27, 471–480, 1998.

Arheimer, B. and Brandt, M.: Watershed modelling of non-point nitrogen pollution from arable land to the Swedish coast in 1985 and 1994, Ecol. Engin., 14, 389–404, 2000.

Arnold, J. G., Srinivasan, R., Muttiah, R. S., and Williams, J. R.: Large-area hydrologic modeling and assessment: Part I. Model development, J. Am. Water Resour. Assoc., 34, 73–89, 1998.

Beven, K. J.: A manifesto for the equifinality thesis, J. Hydrol., 320, 18–36, 2006.

Beven, K. J. and Kirkby, M. J.: A physically based variable contributing area model of basin hydrology, Hydrol. Sci. Bull., 24, 43–69, 1979.

Bicknell, B. R., Imhoff, J. C., Kittle, J. L., Donigian, A. S., and Johanson, R. C.: Hydrologic Simulation Program – FORTRAN (HSPF): User's Manual for Release 10, Report No. EPA/600/R–93/174, US EPA Environmental Research Lab, Athens, Ga, 1993.

Borah, D. K. and Bera, M.: Watershed-scale hydrologic and nonpoint-source pollution models: Review of application, Trans. ASAE, 47, 789–803, 2004.

Bouraoui, F. and Dillaha, T. A.: ANSWERS – 2000: Runoff and sediment transport model, J. Environ. Eng., 122, 493–502, 1996.

Brown, L. C. and Barnwell, T. O.: The enhanced stream water quality models QUAL2E and QUAL2E-UNCAS: documentation and user manual, Tufts University and Env. Res. Laboratory, US EPA, Athens, Georgia, 1987.

Burt, T. P. and Pinay, G.: Linking hydrology and biogeochemistry in complex landscapes, Prog. Phys. Geog., 29, 297–316, 2005.

China's national standard (CNS): Current land use condition classification (GB/T21010–2007), General administration of quality supervision, inspection and quarantine of China and Standardization administration of China, Beijing, China, 2007.

Deb, K., Pratap, A., Agarwal, S., and Meyarivan, T.: A fast and elitist multiobjective genetic algorithm: NSGA–II, IEEE T. Evolut. Comput., 6, 182–197, 2002.

Deng, J., Zhu, B., Zhou, Z. X., Zheng, X. H., Li, C. S., Wang, T., and Tang, J. L.: Modeling nitrogen loadings from agricultural soils in southwest China with modified DNDC, J. Geophys. Res., 116, G02020, doi:10.1029/2010JG001609, 2011.

Di Toro, D. M., Fitzpatrick, J. J., and Thomann, R. V.: Water quality analysis simulation program (WASP) and model verification program (MVP)-Documentation, Hydroscience, Inc., Westwood, NY, for US EPA, Duluth, MN, Contract No. 68–01–3872, 1983.

Duan, Q., Sorooshian, S., and Gupta, V. K.: Optimal use of the SCE-UA global optimization method for calibrating watershed models, J. Hydrol., 158, 265–284, 1994.

Efstratiadis, A. and Koutsoyiannis, D.: One decade of multi-objective calibration approaches in hydrological modelling: a review, Hydrol. Sci. J., 55, 58–78, 2010.

Euser, T., Winsemius, H. C., Hrachowitz, M., Fenicia, F., Uhlenbrook, S., and Savenije, H. H. G.: A framework to assess the realism of model structures using hydrological signatures, Hydrol. Earth Syst. Sci., 17, 1893–1912, doi:10.5194/hess-17-1893-2013, 2013.

Fovet, O., Ruiz, L., Hrachowitz, M., Faucheux, M., and Gascuel-Odoux, C.: Hydrological hysteresis and its value for assessing process consistency in catchment conceptual models, Hydrol. Earth Syst. Sci., 19, 105–123, doi:10.5194/hess-19-105-2015, 2015.

Gassman, P. W., Reyes, M. R., Green, C. H., and Arnold, A. G.: The soil and water assessment tool: historical development, applications, and future research directions, T. ASABE, 50, 1211–1250, 2007.

Goldberg, D. E.: Genetic algorithms in search, optimization, and machine learning, Reading Menlo Park: Addison-Wesley, Massachusetts, USA, 1989.

Hamrick, J. M.: A three-dimensional environmental fluid dynamics computer code: theoretical and computational aspects, Special Report, The College of William and Mary, Virginia Institute of Marine Science, Virginia, USA, 317, 1992.

Hargreaves, G. H. and Samani, Z. A.: Estimating potential evapotranspiration, J. Irrigat. Drain. Div., 108, 225–230, 1982.

Henan Statistical Yearbook in 2003, 2004 and 2005: China Statistics Press, Beijing, 2003, 2004, 2005.

Her, Y. and Chaubey, I.: Impact of the numbers of observations and calibration parameters on equifinality, model performance, and output and parameter uncertainty, Hydrol. Process., 29, 4220–4237, 2015.

Horst, W. J., Kamh, M., Jibrin, J. M., and Chude, V. O.: Agronomic measures for increasing P availability to crops, Plant. Soil., 237, 211–223, 2001.

Hrachowitz, M., Fovet, O., Ruiz, L., Euser, T., Gharari, S., Nijzink, R., Freer, J., Savenije, H. H. G., and GascuelOdoux, C.: Process consistency in models: The importance of system signatures, expert knowledge, and process complexity, Water Resour. Res., 50, 7445–7469, 2014.

Johnes, P. J.: Evaluation and management of the impact of land use change on the nitrogen and phosphorus load delivered to surface waters: the export coefficient modelling approach, J. Hydrol., 183, 323–349, 1996.

Johnsson, H., Bergstrom, L., Jansson, P. E., and Paustian, K.: Simulated nitrogen dynamics and losses in a layered agricultural soil, Agr. Ecosyst. Environ., 18, 333–356, 1987.

Kennedy, J.: Particle swarm optimization, Encyclopedia of Machine Learning, Springer USA, 760–766, 2010.

Kindler, J.: Integrated water resources management: the meanders, Water Int., 25, 312–319, 2000.

King, K. W., Arnold, J. G., and Bingner, R. L.: Comparison of Green-Ampt and curve number methods on Goodwin Creek watershed using SWAT, T. ASABE, 42, 919–925, 1999.

Kirchner, J. W.: Getting the right answers for the right reasons: Linking measurements, analyses, and models to advance the science of hydrology, Water Resour. Res., 42, W03S04, doi:10.1029/2005WR004362, 2006.

Krysanova, V., Mueller-Wohlfeil, D. I., and Becker, A.: Development and test of a spatially distributed hydrological/water quality model for mesoscale watersheds, Ecol. Model., 106, 261–289, 1998.

Li, C., Frolking, S., and Frolking, T. A.: A model of nitrous oxide evolution from soil driven by rainfall events: 1. Model structure and sensitivity, J. Geophys. Res., 97, 9759–9776, 1992.

Liang, X., Lettenmaier, D. P., Wood, E. F., and Burges, S. J.: A Simple hydrologically based model of land surface water and energy fluxes for GSMs, J. Geophys. Res., 99, 14415–14428, 1994.

Lindström, G., Pers, C. P., Rosberg, R., Strömqvist, J., and Arheimer, B.: Development and test of the HYPE (Hydrological Predictions for the Environment) model – A water quality model for different spatial scales, Hydrol. Res., 41, 295–319, 2010.

Ma, F., Ye, A., Gong, W., Mao, Y., Miao, C., and Di, Z.: An estimate of human and natural contributions to flood changes of the Huai River, Global Planet Change, 119, 39–50, 2014.

Mantovan, P. and Todini, E.: Hydrological forecasting uncertainty assessment: Incoherence of the GLUE methodology, J. Hydrol., 330, 368–381, 2006.

Mantovan, P., Todini, E., and Martina, M. L. V.: Reply to comment by Keith Beven, Paul Smith, and Jim Freer on "Hydrological forecasting uncertainty assessment: Incoherence of the GLUE methodology", J. Hydrol., 338, 319–324, 2007.

McDonnell, J. J., Sivapalan, M., Vache, K., Dunn, S., Grant, G., Haggerty, R., Hinz, C., Hooper, R., Kirchner, J., Roderick, M. L., Selker, J., and Weiler, M.: Moving beyond heterogeneity and process complexity: A new vision for watershed hydrology, Water Resour. Res., 43, W07301, doi:10.1029/2006WR005467, 2007.

Moriasi, D. N., Arnold, J. G., Van Liew, M. W., Binger, R. L., Harmel, R. D., and Veith, T.: Model evaluation guidelines for systematic quantification of accuracy in watershed simulations, T. ASABE, 50, 885–900, 2007.

Nash, J. E. and Sutcliffe, J. V.: River flow forecasting through conceptual models. Part I – A discussion of principles, J. Hydrol., 27, 282–290, 1970.

Neitsch, S., Arnold, J., Kiniry, J., and Williams, J. R.: SWAT2009 Theoretical Documentation, Texas Water Resources Institute, Temple, Texas, 2011.

Onstad, C. A. and Foster, G. R.: Erosion modeling on a watershed, T. ASAE, 18, 288–292, 1975.

Paola, C., Foufoula-Georgiou, E., Dietrich, W. E., Hondzo, M., Mohrig, D., Parker, G., Power, M. E., Rodriguez-Iturbe, I., Voller, V., and Wilcock, P.: Toward a unified science of the Earth's surface: opportunities for synthesis among hydrology, geomorphology, geochemistry, and ecology, Water Resour. Res., 42, W03S10, doi:10.1029/2005WR004336, 2006.

Pohlert, T., Breuer, L., Huisman, J. A., and Frede, H.-G.: Integration of a detailed biogeochemical model into SWAT for improved nitrogen predictions-model development, sensitivity and uncertainty analysis, Ecol. Model., 203, 215–228, 2006.

Pokhrel, P., Gupta, H. V., and Wagener, T.: A spatial regularization approach to parameter estimation for a distributed watershed model, Water Resour. Res., 44, W12419, doi:10.1029/2007WR006615, 2008.

Pushpalatha, R., Perrin, C., Le Moine, N., and Andréassian, V.: A review of efficiency criteria suitable for evaluating low-?ow simulations, J. Hydrol., 420–421, 171–182, 2012.

Rallison, R. E. and Miller, N.: Past, present and future SCS runoff procedure, in: Rainfall runoff relationship, edited by: Singh, V. P., Water Resources Publication, Littleton, CO, 353–364, 1981.

Ritchie, J. T.: A model for predicting evaporation from a row crop with incomplete cover, Water Resour. Res., 8, 1205–1213, 1972.

Ritter, A. and Muñoz-Carpena, R.: Performance evaluation of hydrological models: Statistical significance for reducing subjectivity in goodness-of-fit assessments, J. Hydrol., 480, 33–45, 2013.

Sharpley, A. N. and Williams, J. R.: EPIC-erosion/productivity impact calculator: 1. Model documentation. Technical Bulletin-United States Department of Agriculture, Agric. Res. Service, Washington D.C., USA, 1990.

Shi, P., Chen, C., Srinivasan, R., Zhang, X., Cai, T., Fang, X., Qu, S., Chen, X., and Li, Q.: Evaluating the SWAT model for hydrological modeling in the Xixian watershed and a comparison with the XAJ model, Water Resour. Manag., 25, 2595–2612, 2011.

Singh, V. P. and Woolhiser, D. A.: Mathematical modeling of watershed hydrology, J. Hydrol. Eng., 7, 270–292, 2002.

Sivapalan, M. and Kalma, J. D.: Scale problems in hydrology: contributions of the Robertson Workshop, Hydrol. Process., 9, 243–250, 1995.

Strömqvist, J., Arheimer, B., Dahné, J., Donnelly, C., and Lindström, G.: Water and nutrient predictions in ungauged basins: set-up and evaluation of a model at the national scale, Hydrol. Sci. J., 57, 229–247, 2012.

Tattari, S., Bärlund, I., Rekolainen, S., Posch, M., Siimes, K., Tuhkanen, H. R., and Yli-Halla, M.: Modeling sediment yield and phosphorus transport in Finnish clayey soils, T. ASABE, 44, 297–307, 2001.

Tonkin, M. J. and Doherty, J.: A hybrid regularized inversion methodology for highly parameterized environmental models, Water Resour. Res., 41, W10412, doi:10.1029/2005WR003995, 2005.

van Griensven, A., Meixner, T., Grunwald, S., Bishop, T., Diluzio, M., and Srinivasan, R.: A global sensitivity analysis tool for the parameters of multi-variable catchment models, J. Hydrol., 324, 10–23, 2006.

Vinogradov, Y. B., Semenova, O. M., and Vinogradova, T. A.: An approach to the scaling problem in hydrological modelling: the deterministic modelling hydrological system, Hydrol. Process., 25, 1055–1073, 2011.

Wang, G. S., Xia, J., Tan, G., and Lu, A. F.: A research on distributed time variant gain model: A case study on Chao River basin, Prog. Geogr., 21, 573–582, 2002 (in Chinese).

Wang, G., Xia, J., and Chen, J.: Quantification of effects of climate variations and human activities on runoff by a monthly water balance model: A case study of the Chaobai River basin in northern China, Water Resour. Res., 45, W00A11, doi:10.1029/2007WR006768, 2009.

Wang, J. Q., Ma, W. Q., Jiang, R. F., and Zhang, F. S.: Analysis about amount and ratio of basal fertilizer and topdressing fertilizer on rice, wheat, maize in China, Chin. J. Soil Sci., 39, 329–333, 2008 (in Chinese).

Wang, X.: Summary of Huaihe River Basin and Shandong Peninsula Integrated Water Resources Plan, China Water Resour., 23, 112–114, 2011.

Williams, J. R., Jones, C. A., and Dyke, P. T.: Modeling approach to determining the relationship between erosion and soil productivity, Trans. ASAE, 27, 129–144, 1984.

Williams, J. R., Jones, C. A., Kiniry, J. R., and Spanel, D. A.: The EPIC crop growth model, Trans. ASAE, 32, 497–511, 1989.

Xia, J.: Identification of a constrained nonlinear hydrological system described by Volterra Functional Series, Water Resour. Res., 27, 2415–2420, 1991.

Xia, J., Wang, G. S., Tan, G., Ye, A. Z., and Huang, G. H.: Development of distributed time-variant gain model for nonlinear hydrological systems, Sci. China: Earth Sci., 48, 713–723, 2005.

Xing, G. X. and Zhu, Z. L.: An assessment of N loss from agricultural fields to the environment in China, Nutr. Cycl. Agroecosys., 57, 67–73, 2000.

Zhai, X. Y., Zhang, Y. Y., Wang, X. L., Xia, J., and Liang, T.: Nonpoint source pollution modeling using Soil and Water Assessment Tool and its parameter sensitivity analysis in Xin'anjiang Catchment, China, Hydrol. Process., 28, 1627–1640, 2014.

Zhang, Y. Y., Xia, J., Liang, T., and Shao, Q. X.: Impact of water projects on River Flow Regimes and Water Quality in Huai River Basin, Water Resour. Manag., 24, 889–908, 2010.

Zhang, Y. Y., Xia, J., Shao, Q. X., and Zhai, X. Y.: Water quantity and quality simulation by improved SWAT in highly regulated Huai River Basin of China, Stoch. Env. Res. Risk A., 27, 11–27, 2013.

Zhu, Z. L.: Loss of fertilizer N from plants-soil system and the strategies and techniques for its reduction, Soil Environ. Sci., 9, 1–6, 2000 (in Chinese).

3

Improving flood forecasting capability of physically based distributed hydrological models by parameter optimization

Y. Chen[1], J. Li[1], and H. Xu[2]

[1]Department of Water Resources and Environment, Sun Yat-sen University, Room 108, Building 572, Guangzhou 510275, China
[2]Bureau of Hydrology and Water Resources of Fujian Province. Fuzhou, Fujian, China

Correspondence to: Y. Chen (eescyb@mail.sysu.edu.cn)

Abstract. Physically based distributed hydrological models (hereafter referred to as PBDHMs) divide the terrain of the whole catchment into a number of grid cells at fine resolution and assimilate different terrain data and precipitation to different cells. They are regarded to have the potential to improve the catchment hydrological process simulation and prediction capability. In the early stage, physically based distributed hydrological models are assumed to derive model parameters from the terrain properties directly, so there is no need to calibrate model parameters. However, unfortunately the uncertainties associated with this model derivation are very high, which impacted their application in flood forecasting, so parameter optimization may also be necessary. There are two main purposes for this study: the first is to propose a parameter optimization method for physically based distributed hydrological models in catchment flood forecasting by using particle swarm optimization (PSO) algorithm and to test its competence and to improve its performances; the second is to explore the possibility of improving physically based distributed hydrological model capability in catchment flood forecasting by parameter optimization. In this paper, based on the scalar concept, a general framework for parameter optimization of the PBDHMs for catchment flood forecasting is first proposed that could be used for all PBDHMs. Then, with the Liuxihe model as the study model, which is a physically based distributed hydrological model proposed for catchment flood forecasting, the improved PSO algorithm is developed for the parameter optimization of the Liuxihe model in catchment flood forecasting. The improvements include adoption of the linearly decreasing inertia weight strategy to change the inertia weight and the arccosine function

strategy to adjust the acceleration coefficients. This method has been tested in two catchments in southern China with different sizes, and the results show that the improved PSO algorithm could be used for the Liuxihe model parameter optimization effectively and could improve the model capability largely in catchment flood forecasting, thus proving that parameter optimization is necessary to improve the flood forecasting capability of physically based distributed hydrological models. It also has been found that the appropriate particle number and the maximum evolution number of PSO algorithm used for the Liuxihe model catchment flood forecasting are 20 and 30 respectively.

1 Introduction

Improving flood forecasting capability has long been the goal of the global hydrological community, and catchment hydrological models are the main tools for flood forecasting. The first model used for flood forecasting is commonly referred to as the Sherman's unit hydrograph method (Sherman, 1932). Early catchment hydrological models are usually referred to as lumped conceptual models (Refsgaard et al., 1996; Chen et al., 2011), and a large number of this kind of models have been proposed, such as the Stanford model (Crawford et al., 1966), the Xinanjiang model (Zhao, 1977), and many other lumped models included in the book *Computer Models of Watershed Hydrology* (Singh et al., 1995). Lumped conceptual models usually aggregate the hydrological forcings, state variables and model parameters over the whole catchment, so they could not represent the spatial distribution of the terrain characteristics and hydrological forcings finely, thus reducing their flood forecasting capabilities. With the development of remote sensing and GIS techniques, high-resolution terrain data such as those from the Shuttle Radar Topography Mission digital elevation model (DEM) database (Falorni et al., 2005; Sharma et al., 2014), the USGS land use type database (Loveland et al., 1991, 2000), the FAO soil type database (http://www.isric.org), and precipitation estimated by digital weather radar (Fulton et al., 1998; Chen et al., 2009) have been prepared and freely available globally. This largely facilitated the development of physically based distributed hydrological models (PBDHMs). PBDHMs divide the terrain of the whole catchment into a number of grid cells at fine resolution and assimilate different terrain data and precipitation to different cells, thus having the potential to improve the catchment hydrological process simulation and prediction capability (Ambroise et al., 2006). A dozen of PBDHMs have been proposed since the blueprint of PBDHMs was published by Freeze and Harlan (1969). The first full PBDHM is regarded as the SHE model published in 1987 (Abbott et al., 1986a, b); the others include WATERFLOOD model (Kouwen, 1988), THALES model (Grayson et al., 1992), VIC model (Liang et al., 1994), DHSVM model (Wigmosta et al., 1994), CASC2D model (Julien et al., 1995), WetSpa model (Wang et al., 1997), GBHM model (Yang et al., 1997), WEP-L model (Jia et al., 2001), Vflo model (Vieux and Vieux, 2002), WEHY model (Kavvas et al., 2004, 2006), Liuxihe model (Chen et al., 2011), and more. However, at the same time, the so-called semi-distributed hydrological models have also been proposed, such as the SWAT model (Arnold et al., 1994), TOPMODEL model (Beven et al., 1995), HRCDHM model (Carpenter et al., 2001), and others, with model complexity between the lumped model and distributed model.

Model parameters are very important to all kinds of models as they will determine the model performances in flood forecasting. Most of the model parameters could not be measured directly; therefore, they need to be estimated by some kind of model parameter estimation technique (Madsen, 2003; Laloy et al., 2010; Leta et al., 2015). As the lumped model has limited model parameters, the optimization technique has long been employed to calibrate the model parameters to improve the model's performance. For example, Dowdy et al. (1965) conducted a preliminary research on the parameter automatic optimization. Nash et al. (1970) and O'Connell et al. (1970) put forward a method to evaluate the accuracy of model simulation by utilizing efficiency coefficient. Ibbitt et al. (1971) designed a conceptual watershed hydrological model parameter fitting method. Duan et al. (1994) proposed the shuffled complex evolution (SCE) algorithm. Eberhart et al. (1995) proposed the particle swarm optimization method. Jasper et al. (2003) proposed the shuffled complex evolution metropolis algorithm-University of Arizona (SCEM-UA) method. Chu et al. (2011) proposed the shuffled complex evolution with principal components

analysis-University of California Irvine(SP-UCI) method. However, there are others. Now lots of parameter optimization methods for lumped hydrological models have been developed.

There are also many studies on parameter optimization for semi-distributed hydrologic models. Among them, the most studied model is SWAT due to its open-access codes and simple model structures. For example, the SCE-UA method was used to calibrate SWAT model for streamflow estimation (Ajami et al., 2004). The remote-sensing-derived evapotranspiration is used to calibrate the SWAT parameters by using Gauss–Marquardt–Levenberg algorithm (Immerzeel et al., 2008), and a multi-site calibration method with GA algorithm is also proposed for calibrating the SWAT parameters (Zhang et al., 2008). For estimating the parameters of Hydrology Laboratory Distributed Hydrologic Model, the regularization method was studied (Pokhrel et al., 2008).

PBDHMs usually have very complex model structures, and the hydrological processes are calculated by using physical meaning equations, so running a PBDHM is very time-consuming compared to the lumped model. In addition, PBDHM sets different model parameters to different cells, so the total model parameters of a PBDHM are huge even for a small catchment. This makes it difficult to calibrate the PBDHM parameters like calibration widely exercised in lumped models. In the early stage of PBDHMs, the PBDHMs are assumed to derive model parameters from the terrain properties directly, so there is no need to calibrate model parameters. This is true, and all the proposed PBDHMs could determine the model parameters with their own methods (Refsgaard, 1997; De Smedt et al., 2000; Vieux et al., 2002; Chen, 2009). It is fair when they are used to study the future impacts of the hydrological processes caused by climate changes, or by terrain changes due to human activities, in which there are no observation data to evaluate the model performance or to calibrate the model parameters. Here, the hydrological process simulation/prediction accuracy is not so important; detection of the changing trends is the key issue. However, like the lumped model, parameter uncertainty still exists in PBDHMs, and parameter optimization is still needed to reduce this uncertainty (Gupta et al., 1998; Madsen, 2003; Vieux and Moreda, 2003; Reed et al., 2004; Smith et al., 2004; Pokhrel et al., 2012), particularly for those applications with high prediction accuracy requirement, such as the catchment flood forecasting. The scalar method (Vieux et al., 2004; Vieux, 2004) proposed to adjust Vflo model parameters in its application to flood forecasting could be regarded as the first exploration of PBDHM parameter optimization. In this method, all parameters are adjusted manually with a factor or a multiplicator (scalar) based on the initially derived parameters from the terrain properties. The scalars for the same parameter in different cells take the same values, so the parameters to be adjusted are only a few. This is feasible computationally and proven to be effective. For MIKE SHE model, an automatic parameter optimization method with SCE (Duan et al.,

1994) was employed in simulating catchment runoff (Madsen, 2003), which considers two objectives: fitting the surface runoff at the catchment outlet and minimizing the error on simulated underground water level at different wells. In the Liuxihe model, a half-automated method was proposed to adjust the model parameter (Chen, 2009; Chen et al., 2011). In simulating a medium-sized catchment runoff processes with WetSpa Model, a multi-objective genetic algorithm was used to optimize the WetSpa parameter (Shafii and De Smedt, 2009). Compared with lumped model and semi-distributed model, studies on parameter optimization of PBDHMs are very few, particularly for their uses in flood forecasting. Further work needs to be done in this regard.

Current optimization methods are mainly used in lumped hydrological model parameter calibration, which could be divided into two categories: global optimization and local optimization (Sorooshian et al., 1995). Local optimization method searches the parameter starting from a given initial parameter value with a fixed step length step by step, such as the simplex method (Nelder et al., 1965), Rosenbrock method (Rosenbrock, 1960), pattern search method (Hooke and Jeeves, 1961), among others. Local optimization methods are widely applied in the early stage (Sorooshian et al., 1983; Hendrickson et al., 1988; Franchini et al., 1996), but using local optimization method it is difficult to find the global optimum parameters. Lots of global optimization methods have been proposed since then for lumped models in the past decades after realizing the disadvantages of the local optimization method, such as the genetic algorithm (Holland et al., 1975; Goldberg et al., 1989), adaptive random search (Masri et al., 1980), simulated annealing (Kirkpatrick et al., 1983), ant colony system (Dorigo et al., 1996), shuffled complex evolution algorithm (SCE) (Duan et al., 1994), differential evolution (DE) (Storn and Price, 1997), particle swarm optimization (PSO) algorithm (Eberhart et al., 2001), SCEM-UA (Jasper et al., 2003), SP-UCI (Chu et al., 2011), AMALGAM (Vrugt and Robinson, 2007), among others. Global optimization methods have been widely studied and applied in lumped model parameter calibration, with SCE and PSO the most widely used algorithms. SCE has been used for parameter optimization of Mike SHE (Madsen, 2003; Shafii and De Smedt, 2009), but PSO has never been used for PBDHM parameter optimization. PSO algorithm has the advantages of flexibility, easy implementation and efficiency (Poli et al., 2007; Poli, 2008); it has the potential to be employed to optimize the PBDHMs parameters.

There are two main purposes for this study: the first is to propose a parameter optimization method for PBDHMs in catchment flood forecasting by using PSO algorithm and to test its competence and improve its performances; the second is to explore the possibility of improving PBDHM capability in catchment flood forecasting by parameter optimization (i.e., whether PBDHM parameter optimization could improve model performance significantly and become achievable). In this paper, based on the scalar concept, a general framework for parameter optimization of the PBDHMs for catchment flood forecasting is first proposed that could be used for all PBDHMs. Then, with the Liuxihe model as the study model, which is a physically based distributed hydrological model proposed for catchment flood forecasting, the improved particle swarm optimization (PSO) algorithm is developed for the parameter optimization of the Liuxihe model in catchment flood forecasting. The method has been tested in two catchments in southern China with different sizes, and the results show that the improved PSO algorithm could be used for the Liuxihe model parameter optimization effectively and could improve the model capability largely in catchment flood forecasting.

2   Methodology

Based on the scalar concept, a general methodology for parameter optimization of the physically based distributed hydrological model for catchment flood forecasting is proposed, which is applicable to all physically based distributed hydrological models. This methodology has three steps: parameter classification, parameter initialization and normalization, and automated parameter optimization.

2.1   Parameter classification

In physically based distributed hydrological models, the whole terrain is divided into large numbers of grid cells. The model parameters in each cell are different, so the total parameter number is huge. The methodology proposed in this paper classifies the parameters into a few types, so as to reduce the parameter numbers needed to be optimized.

It is assume that all model parameters of a PBDHM are related and only related to one physical property of the terrain they belong – including the topography, soil type and vegetation type. Then the parameters of a PBDHM could be classified as four types: the climate-related parameters, the topography-related parameters, the vegetation-related (land-use-related) parameters and soil-related parameters. This classification could be used for all PBDHMs. With this classification, the parameters in different cells will have the same values if they have the same terrain properties. The independent parameters are defined based on this classification (i.e., the independent parameters are the parameters with the same terrain properties in each cell), and only the independent parameters need to be estimated and optimized. With this treatment, the number of model parameters with their values needed to be estimated will be largely reduced (i.e., from millions to tens), so the independent parameters could be optimized by employing optimization methods.

2.2   Parameter initialization and normalization

After classifying the model parameters into independent parameters, the feasible values of all the independent parameters will be derived from the terrain properties directly. These values, in this paper, are called the initial values of the model parameters. As mentioned above, all proposed PB-DHMs have their own methods to determine the initial model parameters.

Then the parameters are normalized with the initial values as follows:

$$x_i = x_i'/x_{i0}, \qquad (1)$$

where x_i' is the original value of parameter i, x_{i0} is the initial value of parameter i, and x_i is the normalized value of parameter i. With this normalization, all parameters become no-unit variables.

2.3 Automated parameter optimization

The normalized independent parameters will be automatically optimized with optimization methods. To do this, two important things need to be determined. The first one is to choose an optimization technique. In this study as mentioned above, the PSO algorithm will be employed. The second thing is to choose the optimization criterion (objective function). Different objective functions will result in different model parameters, thus different model performances. There are two main practices: the single-objective function and multiple-objective functions (Tang et al., 2006). Single-objective optimization uses one objective function in the parameter optimization. This is the prevailing practice for both lumped model and distributed model parameter optimization. Multiple-objective optimization considers simultaneously two or more objective functions. The different objectives could have same measures quantitatively, such as to minimize the model efficiency and model efficiency for logarithmic transformed discharges simultaneously (Shafii and De Smedt, 2009), or even have different measures quantitatively, such as to minimize the streamflow simulation error and the well water lever simulation error simultaneously (Madsen, 2003). Not producing one set of optimal parameters like in single-objective optimization, multiple-objective optimization produces Pareto-optimal parameter sets. Each Pareto-optimal parameter is a feasible parameter, which provides the user the opportunity to trade off among different simulation purposes. For example, if the user wants to have a better simulation to the high flow of the streamflow, then the high weight will be given to the model efficiency. However, if a better simulation to the low flow is expected, then the priority should be put on the model efficiency for logarithmic transformed discharges (Shafii and De Smedt, 2009). Multiple-objective optimization is more flexible than single-objective optimization, but it requires much more computation; if the model simulation purpose is determined (i.e., the objective is known), then the single-objective optimization is enough. In this study, the purpose is to optimize the model parameter for flood forecasting, so the purpose is obvious. The one objective function to minimize the peak flow rela-

tive error of the catchment discharge at outlet is chosen, and the single-objective optimization is carried out.

2.4 Liuxihe model and parameter classification

The Liuxihe model (Chen, 2009; Chen et al., 2011) is a physically based distributed hydrological model mainly for catchment flood forecasting. In the Liuxihe model, the studied area is divided into a number of cells horizontally by using a DEM. The cells are called a unit basin, and they are treated as a uniform basin in which elevation, vegetation type, soil characteristics, rainfall, and thus model parameters are considered to take the same value. The unit basin is then divided into three layers vertically: the canopy layer, the soil layer and the underground layer. The boundary of the canopy layer is from the terrain surface to the top of the vegetation. The evapotranspiration takes place in this layer, and the evapotranspiration model is used to determine the evapotranspiration at the unit-basin scale. In the soil layer, soil water is filled by the precipitation and depleted via evapotranspiration. The underground layer is beneath the soil layer with a steady underground flow that is recharged by percolation. All cells are categorized into three types, namely hillslope cell, river cell and reservoir cell.

There are five different runoff routings in the Liuxihe model: hillslope routing, river channel routing, interflow routing, reservoir routing and underground flow routing. Hillslope routing routes the surface runoff produced in one hillslope cell to its neighboring cell, and the kinematic wave approximation is employed to make this routing. For the river channel routing, the shape of the channel cross section is assumed to be trapezoid, which makes it estimated by satellite images. The one-dimensional diffusive wave approximation is employed to make this routing.

The parameters in the Liuxihe model are divided into unadjustable parameters and adjustable parameters. The flow direction and slope are unadjustable parameters which are derived from the DEM directly and remain unchanged. The other parameters are adjustable parameters and could be adjusted to improve the model performance. The adjustable parameters are classified as four types: climate-based parameters, topography-based parameters, vegetation-based parameters and soil-based parameters. Currently in the Liuxihe model, there is method for determining initial values of adjustable parameters, and then the adjustable parameters are optimized by a half-automated parameter adjusting method (i.e., based on the initial parameter values, the parameter values are adjusted by hand to improve the model performance, and the parameter adjustment is done one parameter by one parameter). In this way, it is very tedious and time-consuming. It takes months to adjust the parameters even in a very small catchment, so it is not highly proficient though it could improve the model performance. It is also not a global optimization method. An automatic, global optimization method of the Liuxihe model is needed. In this study, the

Liuxihe model will be employed as the representative PB-DHM.

2.5 Improved PSO algorithm for the Liuxihe model

2.5.1 Principles of particle swarm optimization (PSO)

Particle swarm optimization (PSO) algorithm was first proposed by American psychologist James Kennedy and electrical engineer Russell Eberhart (1995) during their study on the social and intelligent behaviors of a school of birds in their search for food and better living conditions. Now it is widely used in parameter calibration of lumped hydrological model. Resffa et al. (2013) used the PSO algorithm to optimize strategies for designing the membership functions of fuzzy control systems for the water tank and inverted pendulum. Mauricio et al. (2013) used the PSO optimization software for SWAT model calibration. Zambrano-Bigiarin et al. (2013) developed a hydroPSO software for model parameter optimization. Bahareh et al. (2013) used single-objective and multi-objective PSO algorithms to optimize parameters of Hydrologic Engineering Center-Hydrologic Modeling System(HEC-HMS) model. Leila et al. (2013) employed a multi-swarm version of particle swarm optimization (MSPSO) in connection with the well-known Hydrologic Engineering Center-Hydrologic Modeling System(HEC-HMS) simulation model in a parameterization–simulation–optimization (parameterization SO) approach. Richard et al. (2014) compared the PSO algorithm with other algorithms in hydrological model calibration. Jeraldin et al. (2014) used PSO in the tank system. These PSO applications are for lumped models only.

PSO is a global searching algorithm in which each particle represents a feasible solution to the model parameters, and usually an appropriate number of particles is chosen to act like a school of birds. The appropriate number of particles is a very important PSO parameter that will impact the PSO's performance. In the optimization process, these particles move forward over the searching space at the same time following certain rules – which include each particle's moving direction and moving speed – that can be determined with the following equations.

$$V_{i,k} = \omega \times V_{i,k-1} + C_1 \times \text{rand} \times \left(X_{i,\text{pBest}} - X_{i,k-1}\right)$$
$$+ C_2 \times \text{rand} \times \left(X_{\text{gBest}} - X_{i,k-1}\right) \quad (2)$$
$$X_{i,k} = X_{i,k-1} + V_{i,k}, \quad (3)$$

where $V_{i,k}$ is the moving speed of ith particle at kth step, $X_{i,k}$ is the position of ith particle at kth step, $X_{i,\text{pBest}}$ is the best position of ith particle at kth step (current), X_{gBest} is the best position of all particles at kth step, ω is inertia acceleration speed, C1 and C2 are learning factors, and rand is a random number between 0 and 1. Here ω, C1 and C2 are also important PSO parameters that will impact the PSO's performance.

For one-step optimization, also called one evolution, all particles move forward one step. All particles will then have their best positions up to now, and the best position of all particles represents the global optimal positions of all particles. With step-by-step evolution, the global positions of all the particles will be approached, and the corresponding parameter values are the optimal parameter values. In the evolution process, a maximum number of evolution is usually set to keep the optimization process to a reasonable time limit.

2.5.2 Improved PSO algorithm

In the early PSO algorithm, particle number, ω, C1 and C2 are fixed. Studies show that changing the values of ω, C1 and C2 in the PSO search process will improve the PSO's performance (El-Gohary et al., 2007; Song et al., 2008; Acharjee et al., 2010; Chuang et al., 2011). In this study, current research progress in improving PSO's performance will be introduced to improve PSO algorithm. The strategies employed in changing ω, C1 and C2 are stated below and will be tested in the studied catchments. In this paper, the appropriate PSO particle number, ω, C1 and C2 are called PSO parameters.

Inertia weight ω

The inertia weight ω is a PSO parameter impacting the global search capability (Shi and Eberhart, 1998). In the earlier studies, ω takes a fixed value of less than 1. Current studies show that changing ω could improve the PSO performance, and a few methods for dynamically adjusting ω have been proposed, such as linearly decreasing inertia weight strategy (LDIW) (Shi and Eberhart, 2001), adaptive adjustment strategy (Ratnaweera et al., 2004), random inertia weight (RIW) (Shu et al., 2009) and fuzzy inertia weight (Eberhart and Shi, 2001). In this study, the LDIW strategy is employed to dynamically determining the value of ω with the following equation:

$$\omega = \omega_{\max} - \frac{i\left(\omega_{\max} - \omega_{\min}\right)}{\text{MaxN}}, \quad (4)$$

where i is the current evolution number, MaxN is the maximum evolution number, $\omega_{\max}$ takes the value of 0.9 and $\omega_{\min}$ takes the value of 0.1.

Acceleration coefficients C1 and C2

Acceleration coefficients C1 and C2 also impact PSO's performance. In early studies, acceleration coefficients C1 and C2 usually take the same value of 2, and they are fixed in the evolution process. Studies show that dynamically adjusting C1 and C2 and taking different values for C1 and C2 could improve PSO's performances, and a few methods have been proposed, such as the linear strategy (Ratnaweera et al., 2004), concave function strategy (Chen et al., 2006) and arccosine function strategy (Chen et al., 2007). In this study, the arccosine function strategy is employed to determine the values of C1 and C2. The equations are listed below.

$$c_1 = c_{1\min} + (c_{1\max} - c_{1\min}) \left(1 - \frac{\arccos\left(\frac{-2 \times i}{\mathrm{MaxN}} + 1\right)}{\pi} \right) \qquad (5)$$

$$c_2 = c_{2\max} - (c_{2\max} - c_{2\min}) \left(1 - \frac{\arccos\left(\frac{-2 \times i}{\mathrm{MaxN}} + 1\right)}{\pi} \right)., \qquad (6)$$

where $C_{1\max}$ and $C_{1\min}$ are the maximum and minimum value of C_1. The values of 2.75 and 1.25 are recommended. $C_{2\max}$ and $C_{2\min}$ are the maximum and minimum values of C_2, and the values of 2.5 and 0.5 are recommended. i is the current evolution number. MaxN is the maximum evolution number.

.3 PSO procedure

The parameter optimization method based on PSO is summarized below.

1. Choose the independent parameters to be optimized. In the case that the computation load is a great challenge, only highly sensitive parameters will be optimized; otherwise, all parameters could be optimized.

2. Initialize independent parameters to be optimized and normalize them.

3. Choose optimization criterion, particle number, maximum evolution number, ω, C1 and C2.

4. Initialize all particles (i.e., determine their initial positions, and calculate the value of the current objective function).

5. For every evolution, first determine the best position of every particle and the global positions of all particles; then calculate the moving directions and speeds of every particles at current evolution by using Eqs. (2) and (3). Finally, check the optimization criterion. If it is satisfied, then the optimization ends. Otherwise, continue to the next evolution.

3 Studied catchment and the Liuxihe model setup

3.1 Studied catchment and hydrological data

Two catchments in southern China have been selected as the case study catchments. The first catchment is Tiantoushui catchment in Lechang County of Guangdong Province. It is a small watershed with a drainage area of 511 km^2 and channel length of 70 km, which is a typical mountainous catchment with frequent flash flooding in southern China. Tiantoushui catchment will mainly be used to test the PSO parameter impacts on the algorithm performance, so as to propose the optimal PSO parameters for the Liuxihe model parameter optimization. As this work needs lots of model runs, a small

Table 1. Initial values of land-use-based parameters in Tiantoushui catchment.

ID	Name	Evaporation coefficient	Roughness coefficient
2	evergreen coniferous forest	0.7	0.4
3	evergreen broadleaved forest	0.7	0.6
5	shrub	0.7	0.4
15	cultivated land	0.7	0.35

catchment helps to keep the running time to a feasible limit. There are 50 rain gauges within the catchment and one river flow gauge in the catchment outlet. The high-density rain gauge network is built not only for flash flood forecasting but also for some kinds of scientific experiments. This will also help to reduce the uncertainties caused by the uneven precipitation spatial distribution. Figure 1a is the sketch map of Tiantoushui catchment with locations of rain gauges and the tributaries.

Hydrological data of nine flood events have been collected for this study, including the river flow at the catchment outlet and precipitation at each rain gauges at an hourly interval. The precipitation measured by the rain gauges will be interpolated to the grid cells by employing Thiessen polygon method (Derakhshan et al., 2011).

The second studied catchment is the upper portion of Wujiang catchment in southern China. It is called in this paper the upper and middle Wujiang catchment (UMWC). UMWC is in the upper and middle stream of Wujiang catchment with a drainage area of 3622 km^2. Flooding in the catchment is also very frequent and heavy. The purpose of studying this big catchment is to show that PSO could still work in a large catchment. There is one river flow gauge in the outlet of UMWC and 17 rain gauges within the catchment. Figure 1b shows the sketch map of the catchment with locations of rain gauges and the tributaries. Hydrological data of 14 flood events from UMWC have been collected, including the river flow at the catchment outlet and precipitation at each rain gauges at 1 h interval. The precipitation measured by the rain gauges will also be interpolated to the grid cells employing Thiessen polygon method.

3.2 Property data for the Liuxihe model setup

Catchment property data used for model setup in this study are DEM, land use types and soil types. These data of the studied catchments were downloaded from an open-access databases. The DEM was downloaded from the Shuttle Radar Topography Mission database at http://srtm.csi.cgiar.org. The land use type was downloaded from http://landcover.usgs.gov, and the soil type was downloaded from http://www.isric.org. The downloaded DEM has a spatial resolution of 90 m × 90 m, but the other two data sets have a spatial resolution of 1000 m × 1000 m, so they are rescaled to

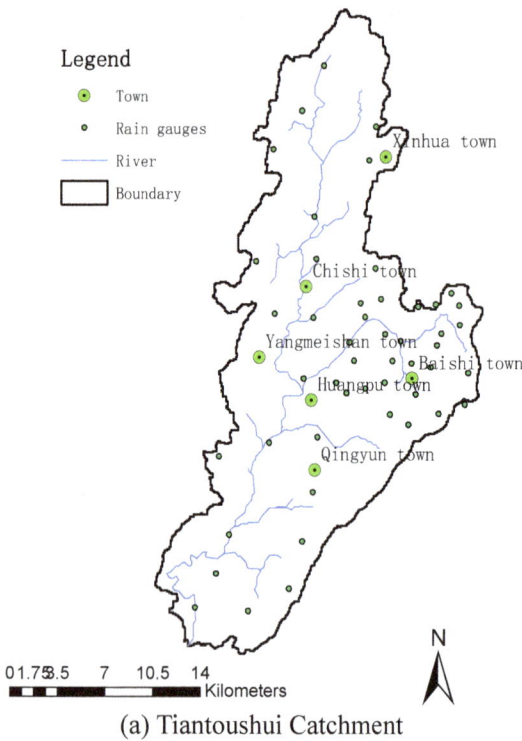

(a) Tiantoushui Catchment

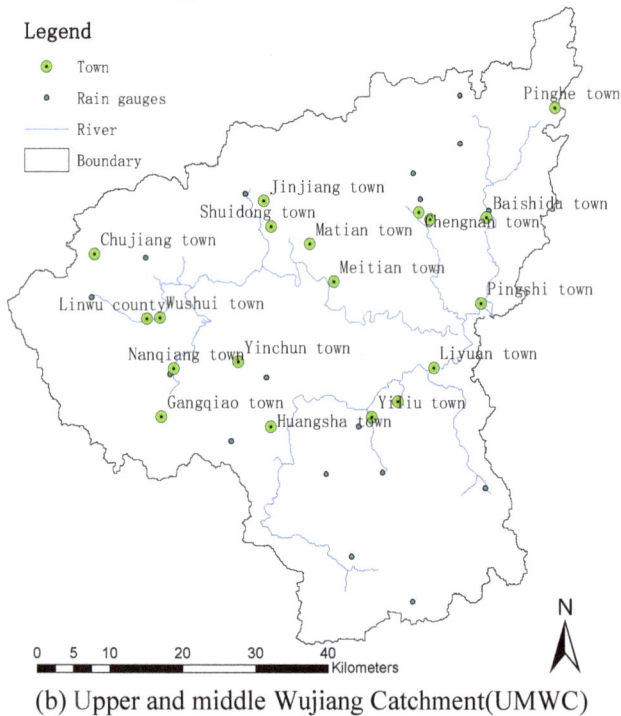

(b) Upper and middle Wujiang Catchment(UMWC)

Figure 1. Sketch map of the studied catchments: **(a)** Tiantoushui catchment and **(b)** upper and middle Wujiang catchment (UMWC).

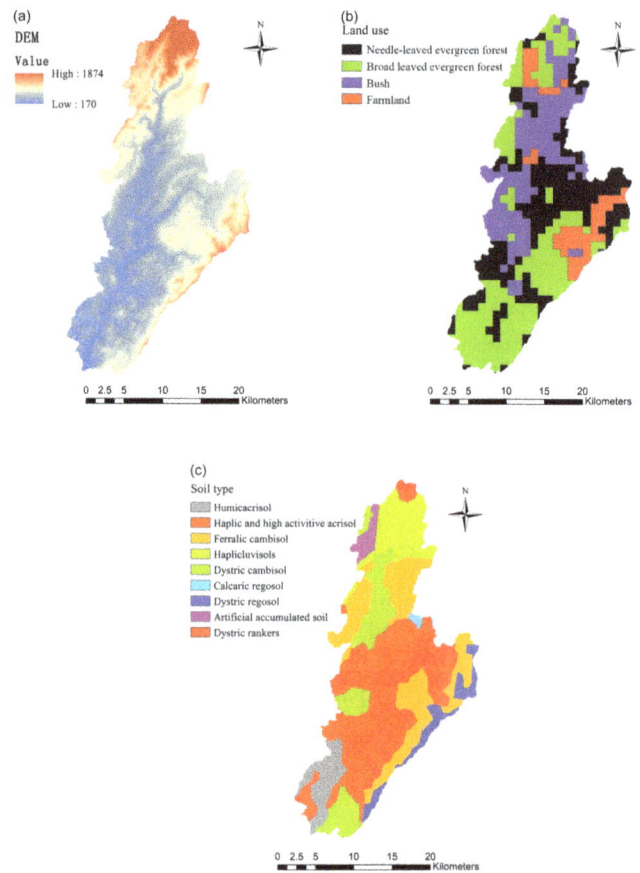

Figure 2. Terrain properties of Tiantoushui catchment: **(a)** DEM, **(b)** land use type and **(c)** soil type.

In the Tiantoushui catchment, the highest, lowest and average elevation are 1874, 174 and 782 m respectively. There are four land use types – evergreen coniferous forest, evergreen broadleaved forest, bush and farmland – accounting for 27.6, 36.5, 25.5, and 10.4 % of the total catchment area respectively. There are 10 soil types – water body, Humic Acrisol, Haplic and highly active Acrisol, Ferralic Cambisol, Haplic Luvisols, Dystric Cambisol, Calcaric Regosol, Dystric Regosol, Artificial accumulated soil and Dystric rankers – accounting for 4.8, 56.5, 1.7, 3.4, 6.5, 4.5, 0.7, 5.6, 9.8 and 6.5 % of the total catchment area respectively.

In the UMWC catchment, the highest, lowest and average elevation are 1793, 170 and 982 m respectively. There are eight land use types – evergreen coniferous forest, evergreen broadleaved forest, shrub, sparse wood, mountains and alpine meadow, slope grassland, lakes, and cultivated land – accounting for 26.4, 24.3, 35, 2.1, 0.1, 2.6, 0.5 and 9.1 % of the total catchment area respectively. There are 12 soil types – water body, Humic Acrisol, Haplic and highly active Acrisol, Ferralic Cambisol, Haplic Luvisols, Dystric Cambisol, Calcaric Regosol, Dystric Regosol, Haplic and weakly active Acrisol, artificial accumulated soil, Eutric Regosols and black limestone soil and dystric rankers – accounting for

the spatial resolution of 90 m × 90 m. Figures 2 and 3 show the property data of DEM, land use types and soil types of the two catchments respectively.

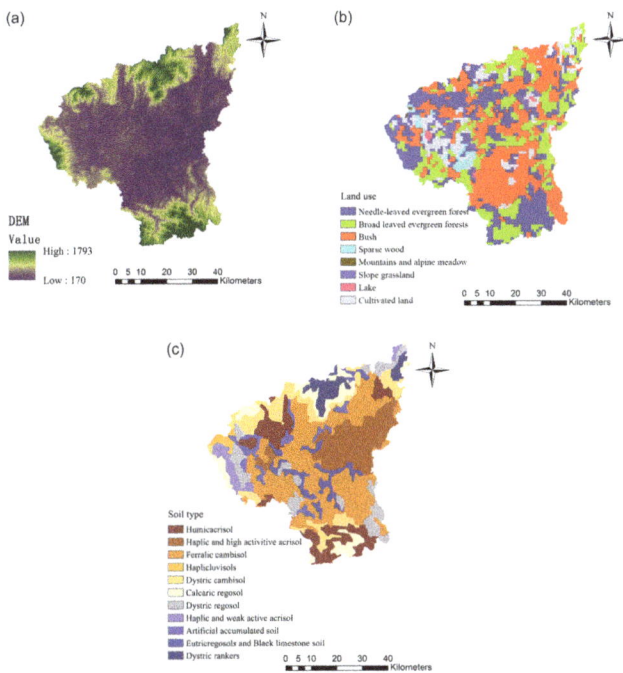

Figure 3. Terrain property data of UMWC: (**a**) DEM, (**b**) land use type and (**c**) soil type.

4.8, 56.5, 0.5, 3.4, 6.5, 4.5, 0.7, 5.6, 9.8, 6.6, 1.0 and 0.2 % of the total catchment area respectively.

3.3 Liuxihe model setup

Setting up the Liuxihe model in the studied catchments consists of dividing the whole catchment into grids with DEM. In this study, the Tiantoushui catchment is divided into 65 011 grid cells using the DEM with grid cell size of 90 m × 90 m; then they are categorized into reservoir cell, river channel cell and hillslope cell. In the studied catchments, there are no significant reservoirs, so there are no reservoir cells set. Based on the method for cell type classification proposed in the Liuxihe model, the river channel system is treated as a third-order channel system, and 1364 river channel cells and 63 647 hillslope cells have been produced in Tiantoushui catchment respectively. Further, 10 nodes have been set on the Tiantoushui catchment, and the river channel system is divided into 14 virtual sections. Their cross section sizes have been estimated by referencing to satellite remote-sensing images. The Liuxihe model structure of Tiantoushui catchment is shown in Fig. 4a.

The Liuxihe model is also set up in UMWC. The catchment is first divided into 460 695 grid cells using the DEM with grid cell size of 90 m × 90 m. The river channel system is treated as a third-order channel system, and 3295 river channel cells and 457 400 hillslope cells have been produced respectively; 32 nodes have been set on UMWC, and their cross-section sizes have been estimated by referencing to

satellite remote sensing images. The Liuxihe model structure of UMWC is shown in Fig. 4b.

3.4 Determination of initial parameter values

In the Liuxihe model, the flow direction and slope are two unadjustable parameters which will be derived from the DEM and will remain unchanged. Based on the DEM shown in Fig. 1a, the flow direction and slope of the studied catchments are derived. The other parameters are adjustable parameters, which need initial values for further optimization. Evaporation capacity is a climate-based parameter, and its initial value is set to $5\,\mathrm{mm\,d^{-1}}$ at both catchment based on the observation near the catchment outlet. Evaporation coefficient and roughness are land-use-based parameters and are less sensitive parameters in the Liuxihe model. The initial values of evaporation coefficient are set to 0.7 at both catchments as recommended by the Liuxihe model (Chen, 2009), while the initial values of roughness are derived based on Wang et al. (1997) and are listed in Tables 1 and 2 respectively for the two catchments.

The other parameters are soil-based parameters. In the Liuxihe model, b is recommended to take the value of 2.5. Soil water content under wilting conditions takes 30 % of the soil water content under saturated conditions. The initial values of other soil-based parameters are calculated by using the Soil Water Characteristics Hydraulic Properties Calculator (Arya et al., 1981), which calculates soil water content at saturation and field condition and the hydraulic conductivity at saturation based on the soil texture, organic matter, gravel content, salinity and compaction. The initial values of soil-based parameters are determined by using the program developed by Keith E. Saxton that can be downloaded for free at http://hydrolab.arsusda.gov/soilwater/Index.htm. The initial values of the soil-based parameters at the two studied catchments are listed in Tables 3 and 4 respectively.

4 Discussion and results

4.1 Impacting of particle number to PSO performance and the determination of appropriate particle number

Particle number is an important parameter of PSO, to understand the impact of the particle number on the PSO performance and to determine the appropriate particle number. Six values of particle number – 10, 15, 20, 25, 50 and 100 – have been used to optimize the model parameters of the Liuxihe model setup in Tiantoushui catchment. While maximum evolution number is set to 50, ω, C1 and C2 are dynamically adjusted with Eqs. (4)–(6). Flood event flood2006071409 is used to do this calculation. Five evaluation indices – Nash–Sutcliffe coefficient C, correlation coefficient R, process relative error P, peak flow relative error E and the coefficient of water balance W – have been computed and are listed in Ta-

Table 2. Initial values of land-use-based parameters in UMWC.

ID	Name	Evaporation coefficient	Roughness coefficient
2	Evergreen coniferous forest	0.7	0.4
3	Evergreen broadleaved forest	0.7	0.6
5	Shrub	0.7	0.4
6	Sparse wood	0.7	0.5
7	Mountains and alpine meadow	0.7	0.2
8	Slope grassland	0.7	0.3
10	Lakes	0.7	0.05
15	Cultivated land	0.7	0.35

Table 3. Initial values of soil-based parameters in Tiantoushui catchment.

Soil type	Thickness (mm)	Saturated water content	Field capacity	Saturated hydraulic conductivity (mm h^{-1})	b (percentage)	wilting
Humic Acrisol	700	0.515	0.362	3	2.5	0.2
Haplic and highly active Acrisol	1000	0.517	0.369	3	2.5	0.206
Ferralic Cambisol	700	0.419	0.193	15	2.5	0.1
Haplic Luvisols	1000	0.55	0.501	2	2.5	0.357
Dystric Cambisol	820	0.385	0.164	34	2.5	0.076
Calcaric Regosol	1000	0.5	0.324	3	2.5	0.172
Dystric Regosol	950	0.388	0.169	33	2.5	0.077
Artificial accumulated soil	1000	0.459	0.25	8	2.5	0.121
Dystric rankers	150	0.43	0.203	10	2.5	0.113

ble 5. The computation times for each optimization are also listed in Table 5.

We first analyze the impact of particle number on the computation time. From the results of Table 5 we found that with the increase of the particle number from 10 to 100, the computation time used decreases first. However, when the particle number is bigger than 20, the computation time increases then, and when the particle number is 20, the computation time is 12.1 h, which is the shortest among others. This means that particle number impacts the computation time used in optimization. The small and big particle number is not the best particle number. There exists an appropriate particle number to make the optimization in the least amount of time. In the Tiantoushui catchment, 20 is an appropriate particle number from the view of computational efficiency.

We further analyze the impact of particle number on the model performances by comparing the five evaluation indices. From the results, an obvious trend could be found: with the increase of the particle number, the Nash–Sutcliffe coefficient C, the correlation coefficient R and water balance coefficient increase first, but when the particle number reaches 20, the three indices decrease. However, for the process relative error W and peak flow relative error E, the trend is inversed (i.e., with the increase of the particle number, the process relative error W and peak flow relative error E decrease first, but when the particle number reaches 20, the two

indices increase). This also means that, with the increase of the particle number, the model performance increases first and then decreases. So from the view of model performance, we could assume 20 is the appropriate particle number in the Tiantoushui catchment. So in this paper, from the results above, we could suggest that 20 is the appropriate particle number of PSO algorithm for the Liuxihe model in catchment flood forecasting in Tiantoushui catchment.

The particle number of 20 is also used in the parameter optimization of UMWC catchment, and the model performance is also very satisfactory. The computation time is acceptable, so in this study we assume that 20 is the appropriate particle number for the Liuxihe model parameter optimization when employing the PSO algorithm for catchment flood forecasting no matter the size of the catchment. This conclusion can also be derived from the results of PSO's convergence in the next section.

4.2 PSO's convergence

PSO algorithm is an evolution algorithm; its searching process is an iteration process, so the convergence is a key issue (i.e., the algorithm should converge to its optimal state in a limited iteration number). Otherwise, it could not be used practically. In PSO, the iteration is called evolution; one iteration is called one evolution. To explore PSO's convergence, we first draw the optimization evolution process of PSO in

Table 4. Initial values of soil-based parameters in UMWC.

Soil type	Thickness (mm)	Saturated water content	Field capacity	Saturated hydraulic conductivity (mm h^{-1})	b (percentage)	wilting
Humic Acrisol	700	0.515	0.362	3	2.5	0.2
Haplic and highly active Acrisol	1000	0.517	0.369	3	2.5	0.206
Ferralic Cambisol	700	0.419	0.193	15	2.5	0.1
Haplic Luvisols	1000	0.55	0.501	2	2.5	0.357
Dystric Cambisol	820	0.385	0.164	34	2.5	0.076
Calcaric Regosol	1000	0.5	0.324	3	2.5	0.172
Dystric Regosol	950	0.388	0.169	33	2.5	0.077
Haplic and weakly active Acrisol	1000	0.55	0.501	2	2.5	0.357
Artificial accumulated soil	1000	0.459	0.25	8	2.5	0.121
Eutric Regosols and black limestone soil	430	0.495	0.312	4	2.5	0.156
Dystric rankers	150	0.43	0.203	10	2.5	0.113

Table 5. Performances of PSO algorithm in Tiantoushui catchment.

Particle number	Computation time (h)	Nash–Sutcliffe coefficient C	Correlation coefficient R	Process relative error P	Peak flow relative error E	Water balance coefficient W
10	21	0.793	0.896	0.319	0.086	0.894
15	13	0.849	0.925	0.235	0.077	0.903
20	12.1	0.962	0.951	0.13	0.07	0.917
25	18.6	0.852	0.927	0.237	0.056	0.884
50	45	0.862	0.932	0.242	0.043	0.885
100	86.8	0.838	0.92	0.256	0.054	0.867

Tiantoushui catchment in Fig. 5. Both the objective and parameter evolution processes are included.

From Fig. 5 we found that, during the evolution process, the objective function steadily decreases, which means the model performance is constantly improved. But for all the parameters, they do not change in the same direction: the parameters may increase in one evolution and decrease in the next evolution. However, after more than 25 evolutions, most of the parameters converge to their optimal values. With about 30 evolutions, all of the parameters converge to their optimal values; after that, there are almost no parameter changes. This means 30 is the maximum evolution number for PSO in Tiantoushui catchment.

From Fig. 5, we also found that the optimal parameter values of several parameters are quite different with the initial parameters, but some remain little changes. This also implies that the PSO algorithm has very good performance in convergence. Even the initial values of the parameters are far from their optimal values.

We further analyze PSO's performance in UMWC, but this time we only draw the parameter evolution process of PSO in UMWC in Fig. 6. The objective evolution process of PSO in UMWC is similar to that in the Tiantoushui catchment.

From Fig. 6 we also found that, during the evolution process, the objective function steadily decreases, but the parameters do not increase or decrease in a constant way. The changing patten is similar to that shown in Fig. 5. After 25 evolutions, most of the parameters converge to their optimal values. With about 30 evolutions, all of the parameters converge to their optimal values. The patten in UMWC is the same as that in Tiantoushui catchment.

From Fig. 6, we also found that the optimal parameter values of several parameters are quite different from the initial values, but some remain little changes. This patten in UMWC is the same as that in Tiantoushui catchment also.

From the above results both in UMWC and Tiantoushui catchment, we could assume that PSO algorithm has a very good performance in convergence in catchments with different sizes, and we could assume that the maximum evolution number could be set to 30 no matter the size of the studied catchments. This conclusion also supports the conclusion that 20 is the appropriate particle number for the Liuxihe model parameter optimization when employing PSO algorithm for catchment flood forecasting no matter the size of the catchment.

4.3 Computational efficiency

The computation time needed for physically based distributed hydrological model run is huge. For the parameter optimization, many model runs are needed, so the computation time needed for the parameter optimization is also a key factor impacting the performance of the PSO. From Table 5,

(a) Tiantoushui Catchment

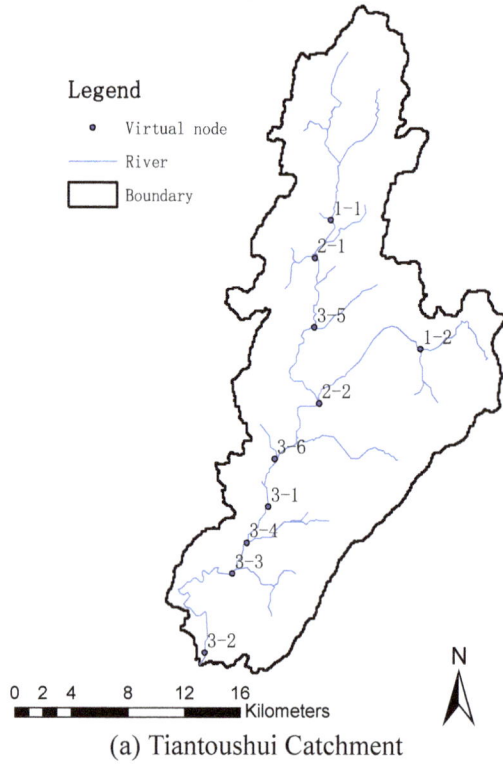

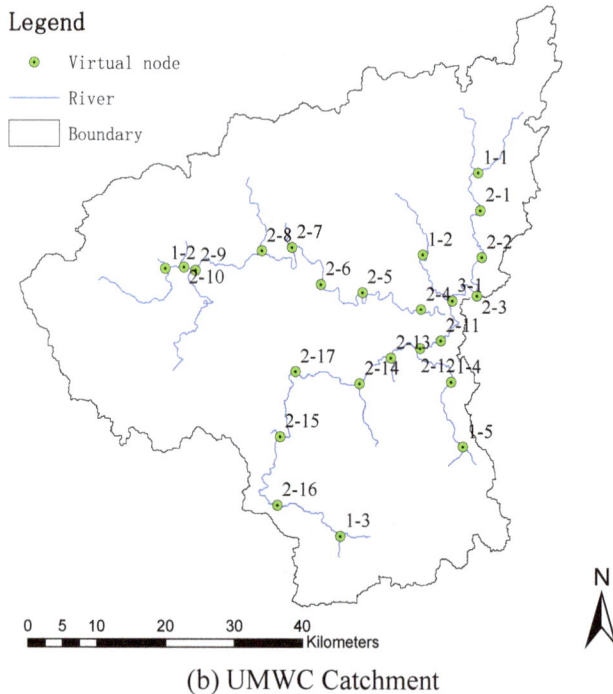

Figure 4. Model setup results: (a) Tiantoushui catchment and (b) UMWC catchment.

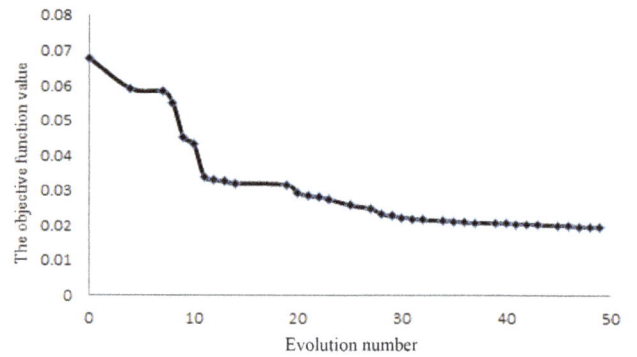

(a) Evolution of objective function

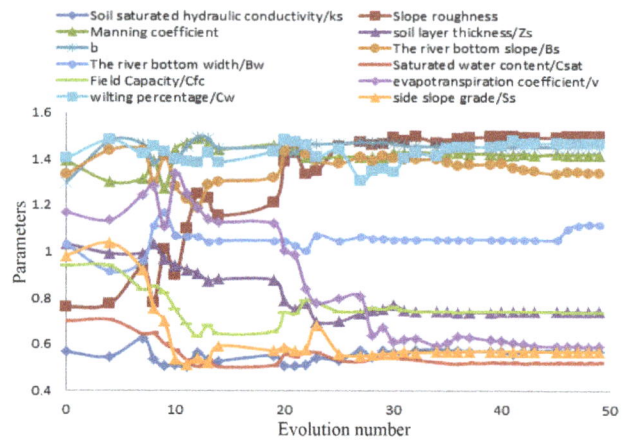

(b)Evolution of parameters

Figure 5. The evolution process of parameter optimization with PSO in Tiantoushui catchment: (a) evolution of objective function and (b) evolution of parameters.

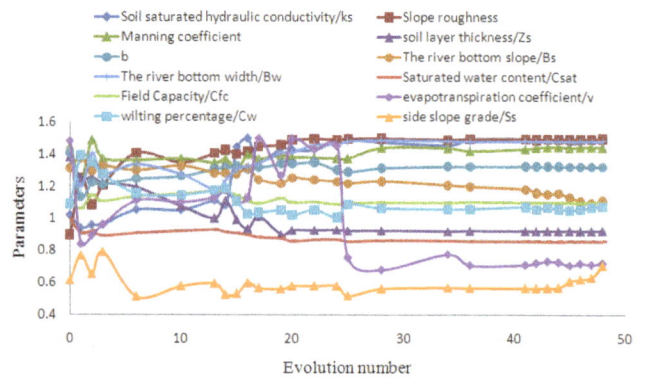

Figure 6. The evolution processes of parameter optimization with PSO in UMWC.

we know, in Tiantoushui catchment, the computation time for parameter optimization is about 12 h; this is acceptable. The time needed for parameter optimization in UMWC is about 82.6 h; it is also acceptable. The computer used for this study

(a) Flood1996071012

(c) flood2008061114

(b) flood2001061206

(d) flood2012060901

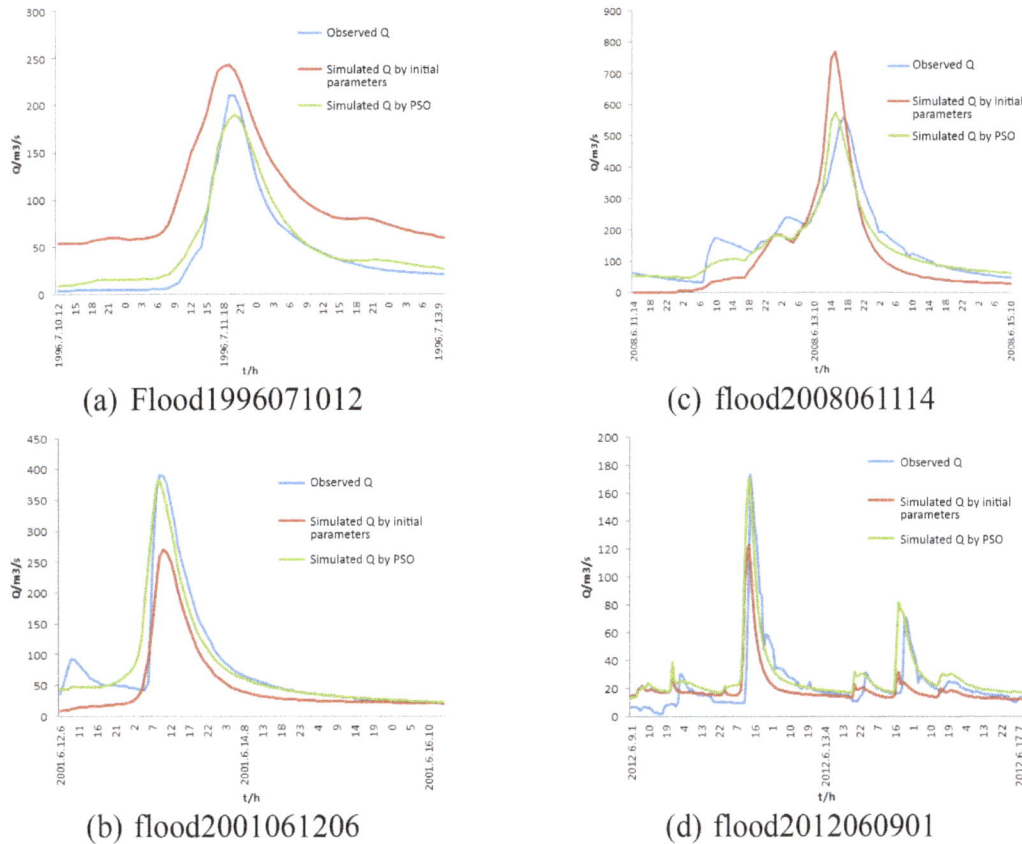

Figure 7. Simulated flood events of Tiantoushui Catchmen. **(a)** flood1996071012, **(b)** flood2001061206, **(c)** flood2008061114, **(d)** flood2012060901.

is a general server. If we used an advanced computer, the time needed could be reduced largely.

4.4 Model validation in Tiantoushui catchment

The parameters of the Liuxihe model in Tiantoushui catchment have been optimized by employing PSO algorithm proposed in this paper. The particle number used is 20. Maximum evolution number is set to 50; ω, C1 and C2 are dynamically adjusted with Eqs. (4)–(6). Flood event flood2006071409 is used to optimize the parameters.

The other eight observed flood events of Tiantoushui catchment are simulated by the model with parameters optimized above to validate the model performance for catchment flood forecasting. To analyze the effect of parameter optimization to model performance improvement, Fig. 7 shows four of the simulated hydrographs. The hydrographs simulated by the model with initial parameter values are also drawn in Fig. 7.

From the results, it has been found that the eight simulated hydrographs fit the observed hydrographs well. Particularly the simulated peak flow is quite good. From the results we also found that the model with initial parameter values does

not simulate the observed flood events satisfactorily (i.e., the uncertainties are high).

To further analyze the model performance with parameter optimization, the five evaluation indices of the eight simulated flood events have been calculated and are listed in Table 6.

From Table 6 we found that the five evaluation indices have been improved by parameter optimization at different extents. For the results simulated by the model with initial parameters, the five evaluation indices – the Nash–Sutcliffe coefficient, correlation coefficient, process relative error, peak flow relative error and water balance coefficient – have average values of 0.66, 0.85, 72 %, 21 % and 1.03 respectively. For the results simulated by the model with optimized parameters, the five evaluation indices have average values of 0.88, 0.94, 25 %, 6 % and 0.97 respectively. The average Nash–Sutcliffe coefficient has a 33 % increase, the correlation coefficient a 9.6 % increase, process relative error a 65.28 % decrease, peak flow relative error a 71.43 % decrease, and the water balance coefficient a 5.83 % decrease. Among the five evaluation indices, the peak flow relative error and the process relative error have the biggest improvement.

The above results imply that with parameter optimization using the PSO algorithm proposed in this paper, the model

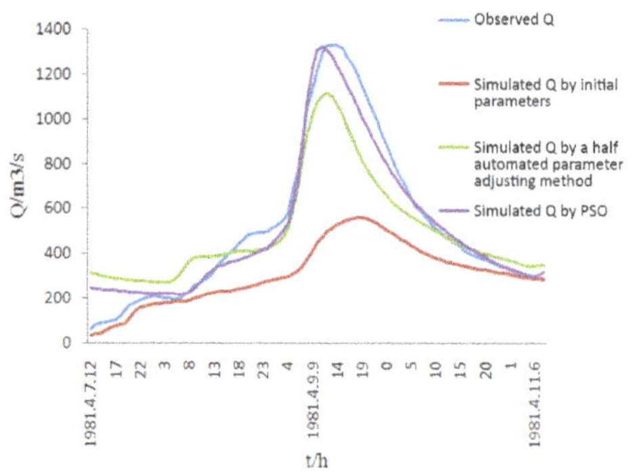

(a) flood1981040712

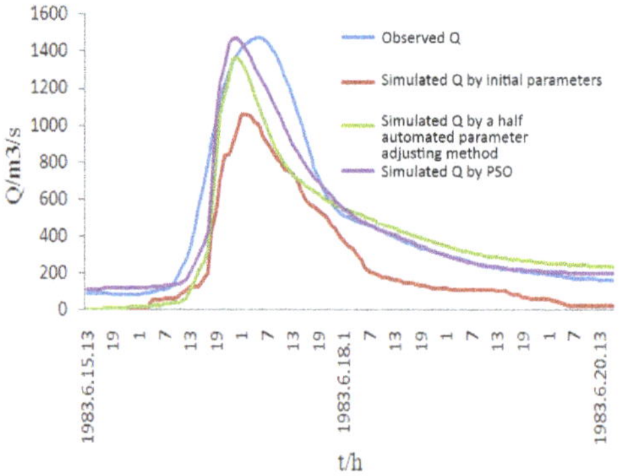

(c) flood1983022720

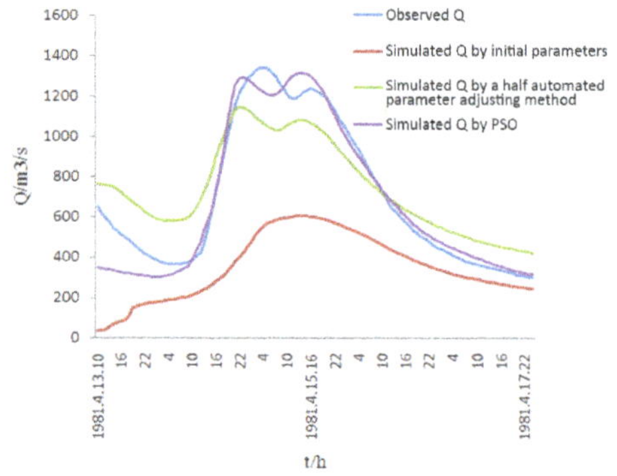

(b) flood1981041310

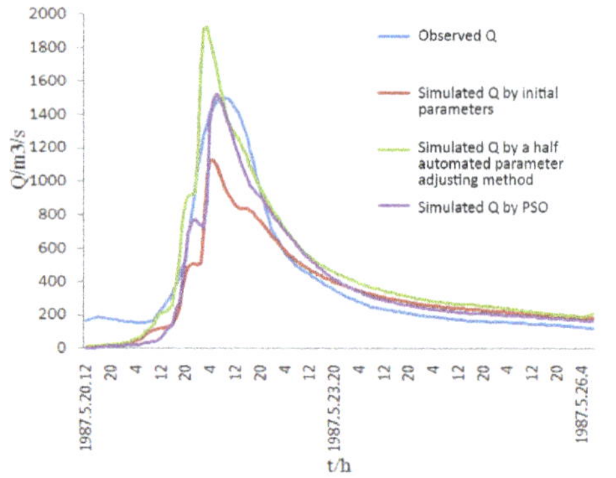

(d) flood1987052012

Figure 8. Simulated flood events of UMWC: (**a**) flood1981040712, (**b**) flood1981041310, (**c**) flood1983022720 and (**d**) flood1987052012.

Table 6. The evaluation index of the simulated flood events in Tiantoushui catchment.

Flood events	Nash–Sutcliffe coefficient C		Correlation coefficient R		Process relative error $P(\%)$		Peak flow relative error $E(\%)$		Water balance coefficient W	
	(1)*1	(2)*2	(1)*1	(2)*2	(1)*1	(2)*2	(1)*1	(2)*2	(1)*1	(2)*2
flood1996071012	0.964	0.85	0.990	0.79	16.3	30	11.2	15.6	1.102	2.19
flood1998061811	0.862	0.613	0.930	0.876	21.4	194.6	20.8	39.7	0.963	1.194
flood2001061206	0.836	0.758	0.926	0.969	31.8	35	0.9	31.1	0.841	0.64
flood2007082100	0.866	0.343	0.942	0.775	13.9	40.9	0.7	32.9	0.966	0.581
flood2008061114	0.882	0.74	0.943	0.883	20.8	71	2.5	31	0.930	0.36
flood2012040607	0.792	0.766	0.893	0.891	27.0	76.4	5.0	11.5	0.913	1.058
flood2012060901	0.912	0.454	0.958	0.752	37.0	74.5	3.2	1.5	1.072	1.238
flood2012062113	0.91	0.778	0.955	0.896	0.301	49.8	0.005	8.4	0.972	0.987
Average	0.88	0.66	0.94	0.85	25	72	6	21	0.97	1.03

*1: results simulated by model with optimized parameters, *2: results simulated by model with initial parameters.

performance of the Liuxihe model for catchment flood forecasting has been improved in Tiantoushui catchment. Optimizing the parameters of the Liuxihe model is necessary.

4.5 Model validation in UMWC

The parameters of the Liuxihe model in UMWC have been optimized by employing PSO algorithm proposed in this paper. The particle number and maximum evolution number are also set to 20 and 50 respectively; ω, C1 and C2

Table 7. The evaluation index of the simulated flood events in UMWC.

Flood events	Nash–Sutcliffe coefficient C			Correlation coefficient R			Process relative error P		
	(1)*1	(2)*2	(3)*3	(1)*1	(2)*2	(3)*3	(1)*1	(2)*2	(3)*3
flood1980050620	0.906	0.610	0.810	0.958	0.831	0.931	0.168	0.480	0.288
flood1980042313	0.892	0.724	0.824	0.972	0.768	0.968	0.282	0.270	0.307
flood1981041014	0.917	0.700	0.451	0.967	0.830	0.883	0.141	0.417	0.317
flood1981040712	0.805	0.686	0.686	0.964	0.738	0.938	0.154	0.550	0.255
flood1981041310	0.739	0.796	0.796	0.938	0.758	0.958	0.221	0.260	0.265
flood1982051014	0.831	0.793	0.793	0.924	0.852	0.952	0.271	0.440	0.174
flood1983061513	0.904	0.810	0.839	0.954	0.850	0.925	0.327	0.530	0.363
flood1983022720	0.896	0.750	0.850	0.974	0.740	0.934	0.152	0.220	0.102
flood1984050310	0.971	0.800	0.816	0.989	0.684	0.980	0.085	0.380	0.388
flood1985092216	0.967	0.840	0.940	0.986	0.785	0.978	0.375	0.480	0.380
flood1987051422	0.961	0.853	0.913	0.986	0.731	0.973	0.266	0.241	0.281
flood1987052012	0.902	0.727	0.927	0.951	0.628	0.968	0.332	0.362	0.262
flood2008060902	0.850	0.756	0.800	0.923	0.825	0.820	0.140	0.414	0.214
Average	0.888	0.757	0.8	0.960	0.771	0.94	0.248	0.388	0.28

Flood events	Peak flow relative error E			Water balance coefficient W		
	(1)*1	(2)*2	(3)*3	(1)*1	(2)*2	(3)*3
flood1980050620	0.004	0.230	0.013	0.913	0.760	0.796
flood1980042313	0.003	0.270	0.008	0.867	0.620	0.792
flood1981041014	0.043	0.180	0.185	0.973	0.729	0.729
flood1981040712	0.159	0.228	0.228	0.990	0.850	1.328
flood1981041310	0.006	0.146	0.146	0.830	1.160	1.061
flood1982051014	0.013	0.230	0.230	0.922	1.230	1.010
flood1983061513	0.007	0.350	0.072	0.944	0.680	0.967
flood1983022720	0.018	0.420	0.078	1.017	0.650	1.045
flood1984050310	0.010	0.210	0.010	0.951	0.720	0.820
flood1985092216	0.022	0.320	0.055	1.071	1.350	1.034
flood1987051422	0.012	0.280	0.013	0.925	1.510	0.892
flood1987052012	0.015	0.160	0.034	0.955	0.840	0.979
flood2008060902	0.004	0.240	0.104	0.985	0.910	0.850
Average	0.024	0.251	0.09	0.949	0.924	0.95

*1: results simulated by model with optimized parameters, *2: results simulated by model with initial parameters, *3: results simulated by model with half-automated optimized parameters.

are dynamically adjusted with Eqs. (4)–(6). Flood event flood1985052618 is used to optimize the parameters.

The other 13 observed flood events of UMWC are simulated by the model with parameters optimized above. Figure 8 shows four of the simulated hydrographs. To compare, the flood events also have been simulated with the parameters optimized with a half-automated parameter adjusting method (Chen, 2009), and the results are also shown in Fig. 8. From the simulated results, it has been found that the 13 simulated hydrographs fit the observed hydrographs well. Particularly the simulated peak flow is quite good. This conclusion is the same as the results in the Tiantoushui catchment. From the results we also found that the model with initial parameter values does not simulate the observed flood event satisfactorily. The simulated results with parameters optimized with a half-automated parameter adjusting method are a big im-

provement to those simulated with the initial model parameters, but the simulated results with the PSO optimized model parameters are the best among the three results.

To further analyze the model performance with parameter optimization, the five evaluation indices of the 13 simulated flood events have been calculated and are listed in Table 7.

From Table 7 we found that the five evaluation indices have been improved by parameter optimization at different extents. For the results simulated by the model with initial parameters, the five evaluation indices – the Nash–Sutcliffe coefficient, correlation coefficient, process relative error, peak flow relative error and water balance coefficient – have average values of 0.757, 0.771, 38.8 %, 25.1 % and 0.924 respectively. While for the results simulated by the model with optimized parameters, the five evaluation indices have average values of 0.888, 0.960, 24.8 %, 2.4 % and 0.949 respectively. The peak flow relative error has been reduced from 25.1 to 2.4 % after parameter optimization, which is 90.44 % down and also the biggest improvement among the five evaluation indices. The average Nash–Sutcliffe coefficient has a 17.31 % increase, the correlation coefficient a 24.51 % increase, process relative error a 36.08 % decrease and water balance coefficient a 2.71 % increase. The results have a similar trend to that in the Tiantoushui catchment. This also implies that with parameter optimization by using the PSO algorithm proposed in this paper, the model performance of the Liuxihe model for catchment flood forecasting has been improved in UMWC catchment: even for a larger catchment, PSO works well for the Liuxihe model. The Liuxihe model's capability for catchment flood forecasting could be improved by parameter optimization with PSO algorithm, and the Liuxihe model parameter optimization is necessary.

5 Conclusion

In this study, based on the scalar concept, a general framework for automatic parameter optimization of the physically based distributed hydrological model is proposed, and the improved particle swarm optimization algorithm is employed for the Liuxihe model parameter optimization for catchment flood forecasting. The proposed methods have been tested in two catchments in southern China with different sizes: one small and one large. Based on the study results, the following conclusions can be drawn:

1. When employing physically based distributed hydrological model for catchment flood forecasting, uncertainty in deriving model parameters physically from the terrain properties is high. Parameter optimization is still necessary to improve the model's capability for catchment flood forecasting.

2. Capability of physically based distributed hydrological model for catchment flood forecasting, specifically the Liuxihe model studied in this paper, could be improved largely by parameter optimization with PSO algorithm, and the model performance is quite good with the optimized parameters to satisfy the requirement of real-time catchment flood forecasting.

3. Improved particle swarm optimization (PSO) algorithm proposed in this paper for physically based distributed hydrological model for catchment flood forecasting, specifically the Liuxihe model studied in this paper, has very good optimization performance. The optimized model parameters are global optimal parameters and could be used for the Liuxihe model parameter optimization for catchment flood forecasting at different size catchments.

4. The appropriate particle number of PSO algorithm used for the Liuxihe model parameter optimization for catchment flood forecasting is 20.

5. The maximum evolution number of PSO algorithm used for the Liuxihe model parameter optimization for catchment flood forecasting is 30.

6. The PSO algorithm has high computational efficiency and could be used in large-scale catchment flood forecasting.

Acknowledgements. This study is supported by the Special Research Grant for the Water Resources Industry (funding no. 201301070), the National Science & Technology Pillar Program during the Twentieth Five-year Plan Period (funding no. 2012BAK10B06), the Science and Technology Program of Guangdong Province (funding no. 2013B020200007) and Water Resources Science Program of Guangdong Province (funding no. 2009-16).

References

Abbott, M. B., Bathurst, J. C., Cunge, J. A., O'Connell, P. E., and Rasmussen, J.: An Introduction to the European Hydrologic System-System Hydrologue Europeen, 'SHE', a: History and Philosophy of a Physically-based, Distributed Modelling System, J. Hydrol., 87, 45–59, 1986a.

Abbott, M. B.,Bathurst, J. C.,Cunge, J. A.,O'Connell, P. E., and Rasmussen, J.: An Introduction to the European Hydrologic System-System Hydrologue Europeen, 'SHE', b: Structure of a Physically based, distributed modeling System, J. Hydrol., 87, 61–77, 1986b.

Acharjee, P. and Goswami, S. K.: Chaotic particle swarm optimization based robust load flow, Int. J. Electr. Power Energ. Syst., 32, 141–146, 2010.

Ajami, N. K., Gupta, H., Wagener, T., and Sorooshian, S.: Calibration of a semi-distributed hydrologic model for streamflow estimation along a river system, J. Hydrol., 298, 112–135, 2004.

Ambroise, B., Beven, K., and Freer, J.: Toward a generalization of the TOPMODEL concepts: Topographic indices of hydrologic similarity, Water Resour. Res., 32, 2135–2145, 1996.

Arnold, J. G., Williams, J. R., and Srinivasan, R.: SWAT: Soil water assessment tool, US Department of Agriculture, Agricultural Research Service, Grassland, Soil and Water Research Laboratory, Temple, Texas, USA, 1994.

Arya, L. M. and Paris, J. F.: An empirical model to predict the soil moisture characteristic from particle-size distribution and bulk density data, Soil Sci. Soc. Am. J., 45, 1023–1030, 1981.

Bahareh, K. S., Mousavi, J., and Abbaspour, K. C.: Automatic calibration of HEC-HMS using single-objective and multi-objective PSO algorithms, Hydrol. Process., 27, 4028–4042, 2013.

Beven, K., Lamb, R., Quinn, P., Romanowicz, R., and Freer, J.: TOPMODEL. In. Computer Models of Watershed Hydrology, edited by: Singh, V., 627–668, Baton Rouge, Florida, USA, 1995.

Carpenter, T. M., Georgakakos, K. P., and Sperfslagea, J. A.: On the parametric and NEXRAD-radar sensitivities of a distributed hydrologic model suitable for operational use, J. Hydrol., 253, 169–193, 2001.

Chen, G., Jia, J., and Han, Q.: Study on the Strategy of Decreasing Inertia Weight in Particle Swarm Optimization Algorithm, Journal of Xi'an Jiantong University, 40, 53–56, 2006.

Chen, S., Cai, G. R., Guo, W. Z., and Chen, G. L.: Study on the Nonlinear Strategy of Acceleration Coefficient in Particle Swarm Optimization (PSO) Algorithm, Journal of Yangtze University (Nat. Sci. Edit), 1–4, 2007.

Chen, Y.: Liuxihe Model, Beijing, Science Press, 198 pp., 2009.

Chen, Y., Zhu, X., Han, J., and Cluckie, I.: CINRAD data quality control and precipitation estimation, Water Manage., 162, 95–105, 2009.

Chen, Y., Ren, Q. W., Huang, F. H., Xu, H. J., and Cluckie, I.: Liuxihe Model and its modeling to river basin flood, J. Hydrol. Eng., 16, 33–50, 2011.

Chu, W., Gao, X., and Sorooshian, S.: A new evolutionary search strategy for global optimization of high-dimensional problems, Inf. Sci., 181, 4909–4927, 2011.

Chuang, L. Y., Hsiao, C. J., and Yang, C. H.: Chaotic particle swarm optimization for data clustering, Expert Syst. Appl., 38, 14555–14563, 2011.

Crawford, N. H. and Linsley, R. K.: Digital simulation in hydrology, Stanford Watershed Model IV, Stanford Univ. Dep. Civ. Eng, Tech. Rep., 39, 1966.

De Smedt, F., Liu, Y. B., and Gebremeskel, S.: Hydrological modeling on a watershed scale using GIS and remote sensed land use information, edited by: Brebbia, C. A., in: Risk Analyses, WIT press, Southampton, Boston, p. 10, 2000.

Derakhshan, H. and Talebbeydokhti, N.: Rainfall disaggregation in non-recording gauge stations using space-time information system, Sci. Iran., 18, 995–1001, 2011.

Dorigo, M., Maniezzo, V., and Colorni, A.: Ant system: optimization by a colony of cooperating agents. Systems, Man, and Cybernetics, Part B: Cybernetics, IEEE Trans., 26, 29–41, 1996.

Duan, Q., Sorooshian, S., and Gupta, V. K.: Optimal use of the SCE-UA global optimization method for calibrating watershed models, J. Hydrol., 158, 265–284, 1994.

Eberhart, R. C. and Shi, Y.: Tracking and optimizing dynamic systems with particle swarms, IEEE, 1, 94–100, doi:10.1109/CEC.2001.934376, 2001.

Eberhart, R. C. and Shi, Y.: Particle swarm optimization: developments, applications and resources, IEEE, 1, 81–86, doi:10.1109/CEC.2001.934374, 2001.

El-Gohary, A., Al-Ruzaiza, A. S.:Chaos and adaptive control in two prey, one predator system with nonlinear feedback, Chaos, Solitons & Fractals, 34, 443–453, 2007.

Falorni, G., Teles, V., Vivoni, E. R., Bras, R. L., and Amaratunga, K. S.: Analysis and characterization of the vertical accuracy of digital elevation models from the Shuttle Radar Topography Mission, J. Geophys. Res. F-Earth Surf., 110, F02005, doi:10.1029/2003JF000113, 2005.

Franchini, M.: Use of a genetic algorithm combined with a local search for the automatic calibration of conceptual rainfall-runoff models, Hydrological Sciences Journal, 41, 21–39, 1996.

Freeze, R. A. and Harlan, R. L.: Blueprint for a physically-based, digitally simulated, hydrologic response model, J. Hydrol., 9, 237–258, 1969.

Fulton R. A., Breidenbach J. P. and Seo D-J., Miller, D. A.: The WSR-88D rainfall algorithm, Weather Forecast., 13, 377–395, 1998.

Goldberg, D. E.: Genetic algorithms in search, optimization and machine learning, Reading, MA, Addison-Wesley, 95–99, 1989.

Grayson, R. B., Moore, I. D., and McMahon, T. A.: Physically based hydrologic modeling: 1.A Terrain-based model for investigative purposes, Water Resour. Res., 28, 2639-2658, 1992.

Gupta, H. V., Sorooshian, S., and Yapo, P. O.: Toward improved calibration of hydrological models: multiple and non-commensurable measures of information, Water Resour. Res., 34, 751–763, 1998.

Hendrickson, J. D., Sorooshian, S., and Brazil, L. E.: Comparison of Newton-type and direct search algorithms for calibration of conceptual rainfall-runoff models, Water Resour. Res., 24, 691–700, 1988.

Holland, J. H.: Adaptation in natural and artificial systems:An introductory analysis with applications to biology, control, and artificial intelligence, Cambridge, MA, University of Michigan Press, ISBN:0262082136, 1992.

Hooke, R. and Jeeves, T. A.: "Direct Search" Solution of Numerical and Statistical Problems, JACM, 8, 212–229, 1961.

Ibbitt, R. P. and O'Donnell, T.: Designing conceptual catchment models for automatic fitting methods, IAHS Publication, 101, 462–475, 1971.

Immerzeel, W. W. and Droogers, P.: Calibration of a distributed hydrological model based on satellite evapotranspiration, J. Hydrol., 349, 411–424, 2008.

Jasper , A.. Vrugt. H. V., and Gupta, W. B.:A Shuffled Complex Evolution Metropolis algorithm for optimization and uncertainty assessment of hydrologic model parameters, Water Resour. Res., 39, 1201, doi:10.1029/2002WR001642, 2003.

Jeraldin, A. D. and Anitta, T.: PSO tuned PID-based Model Reference Adaptive Controller for coupled tank system, Applied Mechanics and Materials Trans Tech Publications, Switzerland, doi:10.4028/www.scientific.net/AMM.626.167, 626 pp., 167–171, 2014.

Jia, Y., Ni, G., and Kawahara, Y.: Development of WEP model and its application to an urban watershed, Hydrol. Process., 15, 2175–2194, 2001.

Julien, P. Y., Saghafian, B., and Ogden, F. L.: Raster-Based Hydrologic Modeling of spatially-Varied Surface Runoff, Water Resour. Bulletin, 31, 523–536, 1995.

Kavvas, M., Chen, Z., Dogrul, C., Yoon, J., Ohara, N., Liang, L., Aksoy, H., Anderson, M., Yoshitani, J., Fukami, K., and Matsuura, T.: Watershed Environmental Hydrology (WEHY) Model Based on Upscaled Conservation Equations: Hydrologic Module, J. Hydrol. Eng., 9, 450–464, doi:10.1061/(ASCE)1084-0699, 2004.

Kavvas, M., Yoon, J., Chen, Z., Liang, L., Dogrul, E., Ohara, N., Aksoy, H., Anderson, M., Reuter, J., and Hackley, S.: Watershed Environmental Hydrology Model: Environmental Module and Its Application to a California Watershed, J. Hydrol. Eng., 11, 261–272, doi:10.1061/(ASCE)1084-0699, 2006.

Kennedy, J. and Eberhart, R.: Particle swarm optimization: Proceedings, IEEE International Conference on Neural Networks, Picataway NJ, IEEE Service Center, 1942–1948, 1995.

Kirkpatrick, S., Gelatt, C. D., and Vecchi, M.: Optimization by simulated annealing, Science, 220, 671–680, 1983.

Kouwen, N.: WATFLOOD: A Micro-Computer based Flood Forecasting System based on Real-Time Weather Radar, Canad. Water Resour. J., 13, 62–77, 1988.

Laloy, E., Fasbender, D., and Bielders, C. L.: Parameter optimization and uncertainty analysis for plot-scale continuous modeling of runoff using a formal Bayesian approach, J. Hydrol., 380, 82–93, 2010.

Leila, O. , Miguel, A., and Mariño, A. A.: Multi-reservoir Operation Rules: Multi-swarm PSO-based Optimization Approach, Water Resour. Manage., 26, 407–427, 2012.

Leta O. T, Nossent J., Velez C., Shrestha N. K., Griensven, A., and Bauwens W.: Assessment of the different sources of uncertainty in a SWAT model of the River Senne (Belgium), Environ. Model. Softw., 68, 129–146, 2015.

Li, X., Chun, C., Xin, W., and Jian, L.: Study on Fuzzy Multi-objective SCE-UA Optimization Method for Rainfall-Runoff Models, Eng. Sci., 3, 52–57, 2007.

Liang, X., Lettenmaier, D. P., Wood, E. F., and Burges, S. J.: A simple hydrologically based model of land surface water and energy fluxes for general circulation models, J. Geophys. Res, 99, 14415–14428, 1994.

Loveland, T. R., Merchant, J. W., Ohlen, D. O., and Brown, J. F.: Development of a Land Cover Characteristics Data Base for the Conterminous U.S., Photogram. Eng. Remote Sens., 57, 1453–1463, 1991.

Loveland, T. R., Reed, B. C., Brown, J. F., Ohlen, D. O., Zhu, J., Yang, L., and Merchant, J. W.: Development of a Global Land Cover Characteristics Database and IGBP DISCover from 1-km AVHRR Data, Int. J. Remote Sens., 21, 1303–1330, 2000.

Madsen, H.: Parameter estimation in distributed hydrological catchment modelling using automatic calibration with multiple objectives, Adv. Water Resour., 26, 205–216, 2003.

Masri, S. F., Bekey, G. A., and Safford, F. B.: A global optimization algorithm using adaptive random search, Appl. Math. Comput., 7, 353–375, 1980.

Nash, J. E. and Sutcliffe, J. V.: River flow forecasting through conceptual models part – A discussion of principles, J. Hydrol., 10, 282–290, 1970.

Nelder, J. A. and Mead, R.: A simple method for function minimization, Comp. Journey, 7, 308–313, 1965.

O'Connell, P. E, Nash, J. E., and Farrell, J. P.: River flow forecasting through conceptual models part – The Brosna catchment at Ferbane, J. Hydrol., 10, 317–329, 1970.

Pokhrel, P., Gupta, H. V., and Wagener, T.: A spatial regularization approach to parameter estimation for a distributed watershed model, Water Resour. Res., 44, W12419, doi:10.1029/2007WR006615, 2008.

Pokhrel, P., Yilmaz, K. K., Gupta, H. V.: Multiple-criteria calibration of a distributed watershed model using spatial regularization and response signatures, J. Hydrol., 418–419, 49–60, 2012.

Poli, R.: Analysis of the publications on the applications of particle swarm optimisation. Journal of Artificial Evolution and Applications, 1-10, 2008.

Poli, R., Kennedy, J., and Blackwell, T.: Particle swarm optimization. Swarm Intelligence, 1, 33–57, 2007.

Ratnaweera, A., Halgamuge, S. K., and Watson, H. C.: Self-organizing hierarchical particle swarm optimizer with time-varying acceleration coefficients, Evolutionary Computation, IEEE Trans., 8, 240–255, 2004.

Reed, S., Koren, V., Smith, M., Zhang, Z., Moreda, F., and Seo, D.-J.: DMIP participants: Overall distributed model intercomparison project results, J. Hydrol., 298, 27–60, 2004.

Refsgaard, J. C. and Storm, B.: Construction, calibration and validation of hydrological models, in: Distributed Hydrological Modelling, edited by: Abbott, M. B. and Refsgaard, J. C., Kluwer Academic, Springer Netherlands, 41–54, 1996.

Refsgaard, J. C.: Parameterisation, calibration and validation of distributed hydrological models, J. Hydrol., 198, 69–97, 1997.

Resffa, F., O' Castillo., Fevrier, V., and Leticia, C.: Design of Optimal Membership Functions for Fuzzy Controllers of the Water Tank and Inverted Pendulum with PSO Variants, IFSA World Congress and NAFIPS Annual Meeting (IFSA/NAFIPS), 1068–1073, 2013.

Rosenbrock, H. H.: An automatic method for finding the greatest or least value of a function, Comp. Journey, 3, 175–184, 1960.

Richard, A., Annie, P., Pascal, C., and François, B.: Comparison of Stochastic Optimization Algorithms in Hydrological Model Calibration, American Society of Civil Engineers, doi:10.1061/(ASCE)HE.1943-5584.0000938, 1374–1384, 2014.

Shafii, M. and De Smedt, F.: Multi-objective calibration of a distributed hydrological model (WetSpa) using a genetic algorithm, Hydrol. Earth Syst. Sci., 13, 2137–2149, doi:10.5194/hess-13-2137-2009, 2009.

Sharma, A. Tiwari, K. N.: A comparative appraisal of hydrological behavior of SRTM DEM at catchment level, J. Hydrol., 519, 1394–1404, 2014.

Sherman, L. K.: Streamflow from Rainfall by the Unit-Graph Method, Eng. News-Rec., 108, 501–505 1932.

Shi, Y. and Eberhart, R. C.: A modified particle swarm optimizer, IEEE, 69–73, doi:10.1109/ICEC.1998.699146, 1998.

Shi, Y. and Eberhart, R. C.: Fuzzy adaptive particle swarm optimization, IEEE, 1, 101–106, doi:10.1109/CEC.2001.934377, 2001.

Shu, X. J., Chen, Y. B., Huang, F. H., and Zhou, H. L.: Application of PEST in the Parameter Calibration of Wetspa Distributed Hydrological Model, J. China Hydrol., 29, 45–49, 2009.

Singh, V. P.: Computer Models of Watershed Hydrology, Water Resources Publications, Colorado, 1130, ISBN:0-918334-91-8, 1995.

Smith, M. B., Seo, D.-J., Koren, V. I., Reed, S., Zhang, Z., Duan, Q.-Y., Cong, S., Moreda, F., and Anderson, R.: The distributed model intercomparison project (DMIP): motivation and experiment design, J. Hydrol., 298, 4–26, 2004.

Song, S. L., Kong, L., Gan, Y., and Rijian, S. B.: Hybrid particle swarm cooperative optimization algorithm and its application to MBC in alumina production, Prog. Natural Sci., 18, 1423–1428, 2008.

Sorooshian, S., Gupta, V. K., and Fulton, J. L.: Evaluaion of maximum likelihood parameter estimation techniques for conceptual rainfall-runoff models:Influence of calibration data variability and length on model credibility, Water Resour. Res., 19, 251–259, 1983.

Sorooshian, S., Gupta, V. K.:Model calibration. In: Singh VP, editor. Computer models of watershed hydrology, Colorado, Water Resources Publications, 23–68, 1995.

Storn, R. and Price, K.: Differential evolution e a simple and efficient heuristic for global optimization over continuous spaces, J. Global Opt., 11, 341–359, 1997.

Tang, Y., Reed, P., and Wagener, T.: How effective and efficient are multiobjective evolutionary algorithms at hydrologic model calibration?, Hydrol. Earth Syst. Sci., 10, 289–307, doi:10.5194/hess-10-289-2006, 2006.

Vieux, B. E.: Distributed Hydrologic Modeling Using GIS, second ed. Water Science Technology Series, vol. 48. ISBN:1-4020-2459-2, Kluwer Academic Publishers, Norwell, Massachusetts, p. 289, 2004.

Vieux, B. E. and Moreda, F. G.: Ordered physics-based parameter adjustment of a distributed model, in: Advances in Calibration of Watershed Models, edited by: Duan, Q., Sorooshian, S., Gupta, H. V., Rousseau, A. N., Turcotte, R., Water Science and Application Series, vol. 6. American Geophysical Union, 267–281, ISBN:0-87590-335-X (Chapter 20), 2003.

Vieux, B. E. and Vieux, J. E.: Vflo^{TM}: A Real-time Distributed Hydrologic Model[A], in: Proceedings of the 2nd Federal Interagency Hydrologic Modeling Conference, 28 July–1 August, Las Vegas, Nevada, Abstract and paper on CD-ROM, 2002.

Vieux, B. E., Cui, Z., and Gaur, A.: Evaluation of a physics-based distributed hydrologic model for flood forecasting, J. Hydrol., 298, 155–177, 2004.

Wigmosta, M. S., Vai, L. W., and Lettenmaier, D. P.: A Distributed Hydrology-Vegetation Model for Complex Terrain, Water Resour. Res., 30, 1665–1669, 1994.

Vrugt, J. and Robinson, B.: Improved evolutionary optimization from genetically adaptive multimethod search, P. Natl. Acad. Sci. USA, 104, 708–711, 2007.

Wang, Z., Batelaan, O., and De Smedt, F.: A distributed model for water and energy transfer between soil, plants and atmosphere (WetSpa), J. Phys. Chem. Earth, 21, 189–193, 1997.

Yang, D., Herath, S., and Musiake, K.: Development of a geomorphologic properties extracted from DEMs for hydrologic modeling, Ann. J. Hydr. Eng., JSCE, 47, 49–65, 1997.

Zambrano-Bigiarini, M. and Rojas, R.: A model-independent Particle Swarm Optimisation software for model calibration, Environ. Model. Softw., 43, 5–25, 2013.

Zhang, X., Srinivasan, R., and Liew, M. V.: Multi-site calibration of the SWAT model for hydrologic modeling, Transactions of the ASABE, 51, 2039–2049, 2008.

Zhao, R. J.: Flood Forecasting Method for Humid Regions of China. East China College of Hydraulic Engineering, Nanjing, China, 1977.

4

An index of floodplain surface complexity

M. W. Scown[1]**, M. C. Thoms**[1]**, and N. R. De Jager**[2]

[1]Riverine Landscapes Research Laboratory, University of New England, Armidale, Australia
[2]Upper Midwest Environmental Sciences Center, United States Geological Survey, La Crosse, Wisconsin, USA

Correspondence to: M. W. Scown (mscown2@myune.edu.au)

Abstract. Floodplain surface topography is an important component of floodplain ecosystems. It is the primary physical template upon which ecosystem processes are acted out, and complexity in this template can contribute to the high biodiversity and productivity of floodplain ecosystems. There has been a limited appreciation of floodplain surface complexity because of the traditional focus on temporal variability in floodplains as well as limitations to quantifying spatial complexity. An index of floodplain surface complexity (FSC) is developed in this paper and applied to eight floodplains from different geographic settings. The index is based on two key indicators of complexity, variability in surface geometry (VSG) and the spatial organisation of surface conditions (SPO), and was determined at three sampling scales. FSC, VSG, and SPO varied between the eight floodplains and these differences depended upon sampling scale. Relationships between these measures of spatial complexity and seven geomorphological and hydrological drivers were investigated. There was a significant decline in all complexity measures with increasing floodplain width, which was explained by either a power, logarithmic, or exponential function. There was an initial rapid decline in surface complexity as floodplain width increased from 1.5 to 5 km, followed by little change in floodplains wider than 10 km. VSG also increased significantly with increasing sediment yield. No significant relationships were determined between any of the four hydrological variables and floodplain surface complexity.

1 Introduction

The floodplain surface is an important component of floodplain ecosystems. It provides the primary physical template (sensu Southwood, 1977) upon which floodplain ecosystem processes are acted out (Salo, 1990). For example, the floodplain surface provides a succession of geomorphic features upon which vegetation can establish and different communities can develop (Hughes, 1997; Pollock et al., 1998), influencing inundation patterns, soil moisture, and nutrient dynamics (Pinay et al., 2000; De Jager et al., 2012). Topographic complexity of floodplain surfaces contributes to the abundance of different physical habitats (Hamilton et al., 2007), high biodiversity (Ward et al., 1999), and elevated levels of ecosystem productivity (Thoms, 2003), as well as complex nonlinear ecosystem responses to inundation (Murray et al., 2006; Thapa et al., 2015). The majority of floodplain research has focused on temporal variability, in particular how hydrological variability drives floodplain structure and function (Junk et al., 1989; Hughes, 1990; Bayley, 1995; Whited et al., 2007). Such a focus has contributed to a limited appreciation of the spatial complexity of floodplain surfaces.

There are two main components to the spatial complexity of floodplain surfaces (Scown et al., 2015a). The first component relates to the presence/absence, abundance, and diversity of geomorphic features present. This influences the number and range of distinct habitats and potential interactions between those habitats, both of which contribute to complexity (Levin, 1998; Phillips, 2003). The second component is concerned with the spatial organisation of those geomorphic features present within a floodplain surface. Spatial organisation affects local interactions and feedbacks between physical features of any landscape as well as the flux of matter and energy throughout the ecosystems present (Wiens, 2002).

Any measurement of spatial complexity must incorporate both components, something that does not generally occur (Cadenasso et al., 2006). In addition, riverine landscapes and their floodplains are hierarchically organised ecosystems (Dollar et al., 2007; Thorp et al., 2008), being composed of discrete levels of organisation distinguished by different process rates (O'Neill et al., 1989). Each level of organisation, or holon, has a spatial and temporal scale over which processes occur and patterns emerge (Holling, 1992). Thus, any measurement of spatial complexity must also acknowledge the effects of measurement scale (Scown et al., 2015a).

Studies of floodplain surface complexity have been limited because they tend to only measure one of the components of spatial complexity and often only at a single scale (Scown et al., 2016). Moreover, many of the measures of spatial complexity that have been proposed are based on categorical "patch" data (e.g. Papadimitriou, 2002). Such data have limitations because of the qualitative delineation of patch boundaries, loss of information within patches, and subsequent analyses of these data being restricted to the minimum scale at which patches were initially defined (McGarigal et al., 2009). Continuous numerical data have been used in some studies, and single metrics of surface complexity have been developed, such as rugosity or fractal dimension (see review by Kovalenko et al., 2012). These single-metric-based indices do not fully encompass the multivariate nature of spatial complexity; thus, multiple indicators are required to get the full measure of surface complexity (Dorner et al., 2002; Frost et al., 2005; Tokeshi and Arakaki, 2012). While frameworks encompassing the multiple dimensions of complexity have also been proposed (e.g. Cadenasso et al., 2006), they have not provided a quantitative measure of spatial complexity (Scown et al., 2016).

Environmental conditions that contribute to floodplain surface complexity have remained largely overlooked in floodplain research because of the limited application of quantitative measures of spatial complexity. However, several geomorphological and hydrological drivers are known to influence other floodplain patterns and processes. The valley trough or floodplain width has been identified as a primary controller of floodplain flow and sediment patterns in several previous studies. Spatial patterns of flow depth, velocity, and shear stress in overbank flows were all found by Miller (1995) to be influenced by valley width, and this influence was particularly noticeable at locations of valley widening or narrowing. Similarly, Thoms et al. (2000) found that valley width had a significant effect on sediment texture and associated heavy metal concentrations within different morphological units of the Hawkesbury River valley, New South Wales. The textural character of sediments delivered to the floodplain and local energy conditions during inundation have also been postulated as important controls of floodplain morphology (Nanson and Croke, 1992). In addition to these geomorphological drivers of pattern, hydrological variability is considered a major determinant of floodplain ecosystem processes (Junk et al., 1989; Hughes, 1990; Bayley, 1995; Whited et al., 2007). The influences of environmental drivers on floodplain pattern and process likely extend to floodplain surface complexity; however, determining such relationships requires a quantitative measure of surface complexity.

New technologies are available for intensive data capture, such as light detection and ranging (lidar), and the analysis of these data using geographic information systems (GIS) overcomes many of the limitations that have inhibited the quantification of spatial complexity. Lidar provides high-resolution, quantitative topographic data over large areas for many landscapes including floodplains. These data are useful for measuring floodplain surface complexity. Lidar-derived digital elevation models (DEMs) of floodplain surfaces can be used to measure the character and variability of surface features using a suite of surface metrics (McGarigal et al., 2009) and moving window analyses (Bar Massada and Radeloff, 2010; De Jager and Rohweder, 2012). The spatial organisation of these features can then be measured using spatial correlograms and geostatistical models (Rossi et al., 1992). These quantitative measurements of the two components of spatial complexity can be incorporated into a single multivariate index. The advantages of using single indices that can be decomposed into subindices (e.g. for use in assessing ecosystem health; Norris et al., 2007) have been widely favoured in ecosystem research.

A quantitative index of floodplain surface complexity is developed in this study and applied to eight floodplains from different geographic settings. The primary data source is a lidar-derived DEM for each floodplain. The character and variability of surface features and conditions and their spatial organisation are incorporated into a single quantitative index to enable a comparison of surface complexity between floodplains. The different environmental settings of each floodplain provide an opportunity to determine the influence of environmental controls on floodplain surface complexity. In addition, the index is measured over three sampling scales (moving window sizes) to investigate the effects of scale on floodplain surface complexity. In this study we ask two questions: (1) does the surface complexity of the eight floodplains differ and is this consistent among sampling scales? (2) What environmental factors influence floodplain surface complexity?

2 Study area

Eight floodplain surfaces from different geographic settings were examined in this study (Fig. 1, Table 1). The Bidgee, Gwydir, Macquarie, Narran, and Yanga floodplains are all located within the Murray–Darling Basin in S.E. Australia, whereas the floodplain of the Woodforde is located in central Australia approximately 150 km north of the town of Alice Springs. The floodplain of the Shingwedzi is located in N.E. South Africa, in the northern regions of Kruger Na-

Table 1. Summary of the geographical and climatic settings of the eight study floodplains.

Floodplain name	Valley setting	Climate	Stream network setting
Bidgee	Confined	Semi-arid/temperate	Lowland continuous
Gwydir	Unconfined	Semi-arid/temperate	Lowland terminal
Macquarie	Unconfined	Semi-arid/temperate	Lowland continuous
Mississippi	Confined	Continental	Upland continuous
Narran	Unconfined	Semi-arid	Lowland terminal
Shingwedzi	Confined	Subtropical	Upland continuous
Woodforde	Confined	Arid	Headwaters continuous
Yanga	Unconfined	Semi-arid/temperate	Lowland continuous

Table 2. Summary of the indicators used to calculate the index of floodplain surface complexity (FSC). Averages and standard deviations of the surface metrics (left columns) are calculated from 50 random sample locations throughout each floodplain. The nugget and range from Moran's I spatial correlograms (right columns) are extracted from the exponential isotropic models fit to these. See Scown et al. (2015a) for detailed calculation procedures.

Indicators of variability in surface geometry		Indicators of spatial organisation of surface conditions	
Average standard deviation of surface heights	Indicates variability in surface elevation within an area	Spatial correlogram exponential isotropic model nugget ($\times$ 4 metrics)	Indicates strength of spatial organisation
Average coefficient of variation of surface heights	Indicates variability in surface elevation relative to the mean elevation within an area	Inverse of the spatial correlogram exponential isotropic model range ($\times$ 4 metrics)	Indicates patchiness or fragmentation in spatial organisation
Standard deviation of skewness of surface heights	Indicates variability in erosional and depositional features within an area		
Average standard deviation of surface curvature	Indicates how convoluted the surface is		

tional Park, and the floodplain of the upper Mississippi is located within navigation pool 9 and forms the boundary of the states of Minnesota, Wisconsin, and Iowa in the USA. Details of the eight floodplains are provided in Table 1 and, in summary, they differ in terms of their degree of valley confinement, climate, and position within the stream network. Four floodplains (the Bidgee, Mississippi, Shingwedzi, and Woodforde) are contained within relatively confined river valley troughs with floodplain widths ranging between 1 and 5 km. The other four floodplains (the Gwydir, Macquarie, Narran, and Yanga) are all contained within relatively unconfined river valleys with floodplain widths of up to 60 km. The eight floodplains also differ in their hydrology and geomorphology, exhibiting a variety of morphological features such as flood channels, oxbows, natural levees, crevasse splays, and back swamps. Detailed descriptions of each of the eight floodplains are provided by Scown et al. (2015a).

3 Methods

The index of floodplain surface complexity (FSC) developed for this study was calculated from data extracted from lidar-derived DEMs for each floodplain. Floodplain extents were delineated using multiple lines of evidence. This delineation was based on examination of breaks of slope in the DEM, contours, changes in vegetation from aerial photography, soil conditions from local soil conservation surveys, and floodwater extents derived from Landsat TM imagery. A buffer within this manually delineated extent was also removed to ensure nothing other than what was deemed to be part of the floodplain was included. Permanently inundated areas were also removed because attaining accurate subsurface land elevations using lidar is difficult. Each DEM was then detrended to remove the overall downstream slope to ensure it had no effect on topographic measurements. Details of the detrending procedures for each of the floodplains are provided by Scown et al. (2015a, b). Each detrended DEM was subsequently resampled to a 5 m $\times$ 5 m grid size using the cubic

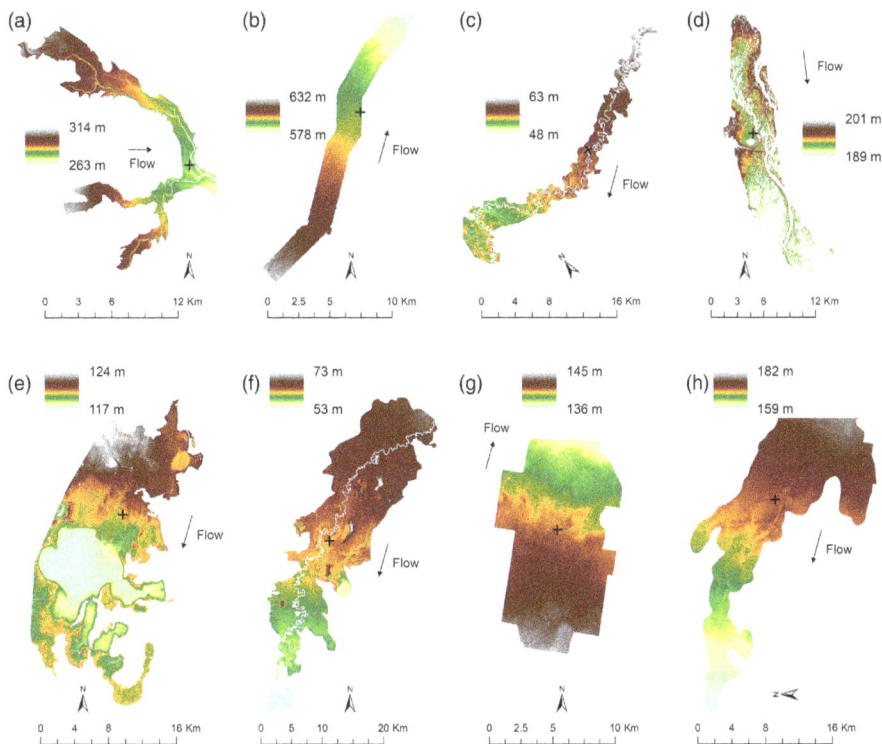

Figure 1. Digital elevation models displaying the floodplain surface in metres above sea level for each study site (crosses indicate coordinates listed): (**a**) Shingwedzi (31°24′ E, 23°05′ S), (**b**) Woodforde (133°20′ E, 22°21′ S), (**c**) Bidgee (143°24′ E, 34°42′ S), (**d**) Mississippi (91°15′ W, 43°29′ N), (**e**) Narran (147°23′ E, 29°48′ S), (**f**) Yanga (143°42′ E, 34°30′ S), (**g**) Macquarie (147°33′ E, 30°41′ S), and (**h**) Gwydir (149°20′ E, 29°16′ S).

method in ArcGIS 10.2 because this was the finest resolution available for one of the floodplains.

The FSC index is comprised of two subindices, which record the two components of spatial complexity: the variability in surface geometry (VSG) and the spatial organisation of surface conditions (SPO). VSG is a composite of four surface metrics (Table 2), measured at 50 random sample locations throughout each of the floodplains, while SPO is calculated from spatial correlogram models of Moran's I over increasing lag distances for each of the four surface metrics from 1000 random sample locations (Table 2). Details of the procedures for calculating each indicator are provided in Scown et al. (2015a). In summary, the surface metrics are used to indicate increasing surface variability, while the spatial correlogram model parameters (range and nugget) are used to indicate increasing "patchiness" or organisation in the surface (Table 2). It is argued here, and elsewhere (Scown et al., 2015a), that increasing variability and spatial organisation results in increasing spatial complexity. All surface metrics were measured within sampling windows of 50, 200, and 1000 m radius. These window sizes were chosen based on the identification of scale thresholds between them by Scown et al. (2015b). This enabled us to determine whether any effect of sampling-scale occurred.

The individual indicators were combined and weighted, using the standardised Euclidean distance procedure, to calculate the overall FSC index. This index was used for an overall assessment of floodplain surface complexity and the subindices of VSG and SPO were derived to provide specific interpretations of the two components of spatial complexity for each floodplain surface. An example of FSC calculation is given in Eq. (1), where I is the overall index and $A, B, C, \ldots, N$ are the n individual indicators of surface complexity, the details of which are provided in Table 2.

$$I = 1 - \frac{\sqrt{(1-A)^2 + (1-B)^2 + (1-C)^2 + \ldots + (1-N)^2}}{\sqrt{n}}. \quad (1)$$

Calculating the FSC index required the SPO indicators to have an additional weighting of 0.5, as there were twice as many indicators of SPO compared to VSG. All indicators were range-standardised and scaled between 0 and 1, hence this index provides a relative measure among those floodplains studied. An index value approaching one indicates the floodplain surface is among the most spatially complex of all floodplains observed, while an index value approaching zero indicates the floodplain surface is among the least spatially complex. The approach used has been applied successfully in developing a large-scale index of river condition (Norris et al., 2007).

Relationships between VSG, SPO, and FSC and seven environmental variables were also investigated. The environmental variables were mean daily discharge (in $ML\,day^{-1}$, Q), the coefficient of variation (CV) of daily discharge (Q_{CV}), CV mean annual discharge (Q_{CVAnn}), CV maximum annual discharge (Q_{CVMax}), sediment yield (in $t\,km^{-2}\,yr^{-1}$, SY), average valley slope (in $m\,m^{-1}$, Vs), and average floodplain width (in km, Fpw). Detailed calculations of environmental variables are provided by Scown et al. (2015a). Each of these environmental variables reflect an aspect of the flow, sediment, energy, and valley conditions, which have previously been shown to influence floodplain surface morphology (Nanson and Croke, 1992; Warner, 1992). Curve estimation between VSG, SPO, and FSC and each environmental variable at each sampling scale was conducted in SPSS. Q, SY, and Vs were normalised using a logarithmic transformation before analysis.

4   Results

4.1   Floodplain surface complexity (FSC)

Floodplain surface complexity, as measured by the FSC index, was highly variable among the eight floodplains and across sampling scales. The Gwydir floodplain had the least complex of surfaces across all sampling scales (mean FSC of 0.17), while the Shingwedzi floodplain had the most complex surface (mean FSC of 0.69) across all scales (Fig. 2). This presumably reflects differences in the geomorphology of these two floodplains. The Shingwedzi floodplain is dissected by numerous channels and gullies, which create highly organised patches of increased topographic relief, whereas the Gwydir floodplain has a relatively flat, featureless surface over larger continuous areas and limited organisation around any of the significant surface features. The effect of sampling scale on FSC was not consistent across the eight floodplains (Fig. 2), indicating that differences among floodplains are scale-dependent. For example, the Gwydir and Narran floodplain surfaces became more complex with increasing window size, whereas the Shingwedzi, Macquarie, and Mississippi floodplains became less complex.

4.2   Variability in surface geometry (VSG)

The VSG index was also highly variable among the eight floodplains and across sampling scales (Fig. 3). Again, the Gwydir floodplain consistently had the lowest values for this index over all window sizes (mean VSG of 0.06), while the Shingwedzi floodplain consistently had the highest (mean VSG of 0.65). This reflects the large differences in topographic relief and variability between these two floodplains. The VSG score of 0.00 for the Gwydir floodplain at the 50 m window size indicates that this floodplain had the lowest scores for all four indicators of variability in surface ge-

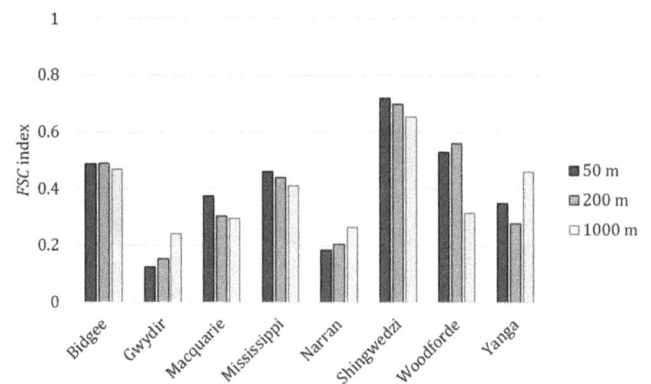

Figure 2. Index of floodplain surface complexity (FSC) for the eight floodplains at each of the three window sizes.

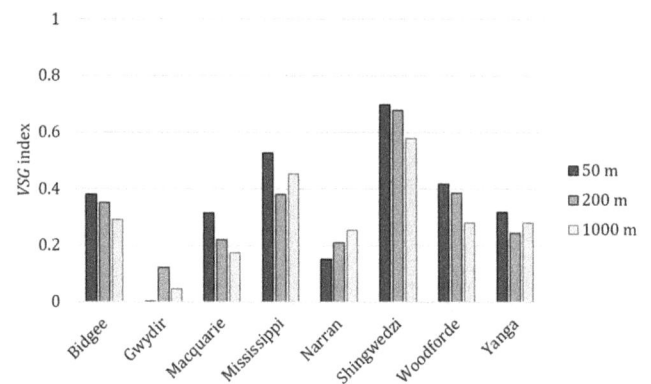

Figure 3. Index of variability in surface geometry (VSG) for the eight floodplains at each of the three window sizes.

ometry of the eight floodplains studied at this scale. Similar to FSC, the effect of sampling scale on VSG was not consistent across floodplains (Fig. 3). VSG increased with sampling scale for the Narran floodplain but decreased for the Shingwedzi, Bidgee, Macquarie, and Woodforde floodplains. VSG was highest at the 50 m window size and lowest at 200 m for the Mississippi and Yanga floodplains, while it was highest at 200 m and lowest at 50 m for the Gwydir. This indicates that the scale at which surface geometry is most variable depends on the floodplain.

4.3   Spatial organisation of surface conditions (SPO)

The SPO index was also highly variable among the eight floodplains and across sampling scales (Fig. 4). Unlike FSC and VSG, there was no consistency as to which floodplain had the highest and lowest SPO across sampling scales. This indicates that no floodplain has consistently the highest or lowest degree of spatial organisation of surface conditions among the eight floodplains studied. The effect of sampling scale on SPO was inconsistent across floodplains (Fig. 4). For five of the eight floodplains, SPO was lowest at the 200 m window size and highest at 1000 m. For the Mississippi and

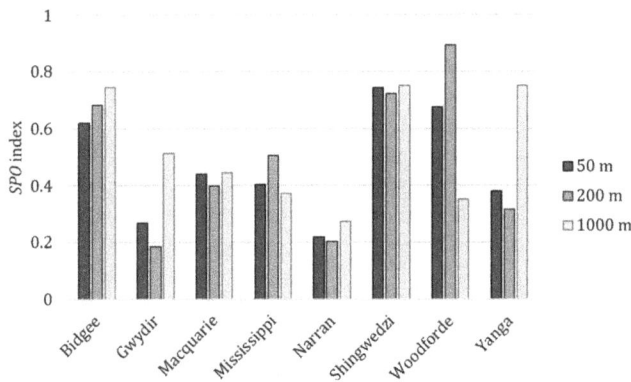

Figure 4. Index of spatial organisation of surface conditions (SPO) for the eight floodplains at each of the three window sizes.

Woodforde floodplains the opposite was observed, with SPO being highest at 200 m and lowest at 1000 m. The Bidgee floodplain was the only floodplain for which SPO increased consistently across all sampling scales. This indicates that the degree of spatial organisation of surface conditions is highest at large sampling scales for most floodplains, but at intermediate scales for some. SPO was highly variable across window sizes for the Yanga, Woodforde, and Gwydir floodplains. SPO was 178 % higher at the 1000 m window size than at 200 m for the Gwydir floodplain and 138 % higher for the Yanga floodplain, while for the Woodforde floodplain it was 61 % lower. This indicates a significant change in the spatial organisation of these floodplain surfaces between these two sampling scales. The results also showed that floodplain and window size have a greater combined effect on SPO among the eight floodplains than on relative FSC and VSG (Figs. 2, 3, 4).

4.4 Relationships between floodplain surface complexity and environmental variables

Floodplain width (Fpw) was the only environmental variable statistically related to any of the three indices of spatial complexity ($p < 0.05$). This variable was significantly related to FSC and VSG over all window sizes and to SPO over all but the 1000 m window size (Table 3). The decrease in all three complexity indices with increasing Fpw was best explained by either a power, logarithmic, or exponential function (Table 3). In terms of the decrease in FSC with increasing Fpw, this was best explained by a power function at all window sizes (Fig. 5a), indicating FSC undergoes rapid decline with increases in Fpw, approaching an asymptote at approximately 10 km in Fpw. The modelled change in FSC with increasing Fpw was almost identical between the 50 and 200 m window sizes. At the 1000 m window size, FSC was generally lower compared to that at 50 and 200 m windows sizes in narrow floodplains, before approaching a higher asymptote at larger Fpw. This indicates that broad floodplains generally have higher FSC when measured at larger sampling scales, whereas narrow floodplains generally have higher FSC when measured at smaller sampling scales.

Decreases in VSG with increasing Fpw was best explained by a logarithmic function at the 50 m window size, a power function at the 200 m window size, and an exponential function at 1000 m (Fig. 5b). These models indicate a more rapid initial decline in VSG with increasing Fpw at the 200 m window size than at the 50 and 1000 m window sizes. This is followed by approach to a higher asymptote at the 200 m window size above an Fpw of approximately 10 km, whereas modelled VSG continues to decline between Fpw values of 10 and 25 km at the 50 and 1000 m window sizes. This indicates that Fpw has a greater effect on VSG in wider floodplains when measured at small and large sampling scales than it does at intermediate scales. The relationship was strongest at the 200 m window size, with more than 80 % of the variance in VSG being explained by Fpw.

The decrease in SPO with increasing Fpw was best explained by a logarithmic function at the 50 and 200 m window sizes (Fig. 5c). The modelled decline in SPO was initially more rapid at the 50 m window size than at 200 m, before approaching a higher asymptote at narrower Fpw. This indicates that Fpw has more of an effect on SPO in wider floodplains when measured at the 200 m window size than at 50 m. The relationship was strongest at the 200 m window size, with more than 77 % of the variance in SPO being explained by Fpw. This was reduced to 71 % at the 50 m window size. There was no significant relationship between Fpw and SPO at the 1000 m window size (Fig. 5c). This suggests that Fpw exerts little or no control over the spatial organisation of surface conditions when measured at large sampling scales.

A weak statistical relationship was recorded between SY and VSG. An increase in VSG with increasing SY was observed at the 200 m window size ($r^2 = 0.44$; $p = 0.07$). The relatively lower level of significance of this result was attributable to the Gwydir having a high SY but a very low VSG. When the Gwydir floodplain was removed from the analysis, there was a significant and strong linear relationship between log-transformed SY and VSG across all window sizes for the remaining seven floodplains (Table 4, Fig. 6). This relationship was almost identical across all window sizes.

5 Discussion

5.1 The FSC index

The Euclidean index of FSC used in this study is comprised of two key components of spatial complexity: the character and variability of features or conditions, and their spatial organisation. This index appears to discriminate between floodplains with distinctly different geomorphological features.

Table 3. Results from regression analyses of FSC, VSG, and SPO against Fpw at each of the three window sizes. * and italics indicate no significant relationship.

		Best model	F	d.f.	p	r^2
FSC	50 m	$y = 0.765x^{-0.414}$	10.344	1, 7	0.02	0.63
	200 m	$y = 0.762x^{-0.420}$	25.523	1, 7	0.00	0.81
	1000 m	$y = 0.549x^{-0.213}$	5.871	1, 7	0.05	0.50
VSG	50 m	$y = -0.151 \ln x + 0.630$	9.642	1, 7	0.02	0.62
	200 m	$y = 0.627x^{-0.418}$	26.319	1, 7	0.00	0.81
	1000 m	$y = 0.472e^{-0.064}$	13.574	1, 7	0.01	0.69
SPO	50 m	$y = -0.145 \ln x + 0.737$	14.515	1, 7	0.01	0.71
	200 m	$y = -0.204 \ln x + 0.866$	20.586	1, 7	0.00	0.77
	1000 m		*0.570*	*1, 7*	*0.48**	*0.09*

Table 4. Results from regression analyses of VSG against each of the three window sizes with Gwydir removed.

	Best model	F	d.f.	p	r^2
50 m	$y = 0.183x + 0.088$	50.497	1, 6	0.00	0.91
200 m	$y = 0.158x + 0.084$	18.179	1, 6	0.00	0.78
1000 m	$y = 0.142x + 0.088$	36.076	1, 6	0.00	0.88

The multivariate nature of the index, comprised of 12 indicators of surface complexity (Table 2), has advantages over univariate indices that have been applied to measure floodplain surface complexity. Univariate indices fail to incorporate multiple aspects of surface structure that contribute to surface complexity (Dorner et al., 2002; Frost et al., 2005; Tokeshi and Arakaki, 2012). Having a single, multivariate-based index is also favourable, rather than multiple individual indicators of floodplain surface complexity, as it allows for a quantitative measure that can be compared for multiple riverine landscapes. Norris et al. (2007) provide a comparable example of such an application in their assessment of river condition, as do Flotemersch et al. (2015) in their watershed integrity index. It is important to note that, the standardisation of indicator scores from 0 to 1 is necessary for the Euclidean index equation (Norris et al., 2007); as such, the FSC index is a relative index of floodplain surface complexity across a group of floodplains all of which were included in the standardisation of the indicators. This is appropriate for examining relationships between floodplain surface complexity and environmental controls, given adequate replication over a range of floodplain settings is achieved. However, it should not be used to compare against indices of other studies, unless all floodplains being compared are included in the calculation of the index.

Recent approaches to examining and understanding ecosystem complexity and the emergent properties that arise from interactions within systems emphasise the importance of heterogeneity, connectivity, and contingency within the landscape (Loreau et al., 2003; Cadenasso et al., 2006). We have presented an index of floodplain surface complexity within such a framework that incorporates measures of variability and spatial organisation. These two components of spatial complexity are directly associated with heterogeneity and connectivity (Wiens, 2002), although no direct measure of historical contingency is given in this spatial approach. Metrics and indicators used to measure properties of landscape and ecosystem complexity in the past have largely been based on discrete units and the familiar concept of "patches" (Forman and Godron, 1981). The surface metrics employed in this study are conceptually equivalent to certain patch metrics and a comprehensive comparison of surface and patch metrics is provided by McGarigal et al. (2009). Thus, the approach presented in this study should be considered complimentary to other ecosystem complexity frameworks, such as the meta-ecosystem approach (Loreau et al., 2003), which are based on patches.

5.2 Environmental drivers of floodplain surface complexity

The results of this research demonstrate that floodplain surface complexity is highly variable among the eight floodplains studied, and that floodplain width exerts a significant top-down control (sensu Thorp et al., 2008) on differences in floodplain surface complexity. These results clearly support geomorphological and ecological thinking that "...the valley rules the stream...", as argued first by Hynes (1975) and strongly supported since (e.g. Schumm, 1977; Miller, 1995; Panin et al., 1999; Thoms et al., 2000). In this case, the valley rules the floodplain surface complexity, at least in terms of the top-down influences investigated here. The influence of floodplain width on floodplain surface complexity decreases once widths are greater than 10 km. This is likely due to the dissipation of flood energy in wide floodplains, limiting the construction of large topographic features that contribute to surface complexity. However, subtle topographic features in wide floodplains are also importance surface features (Fa-

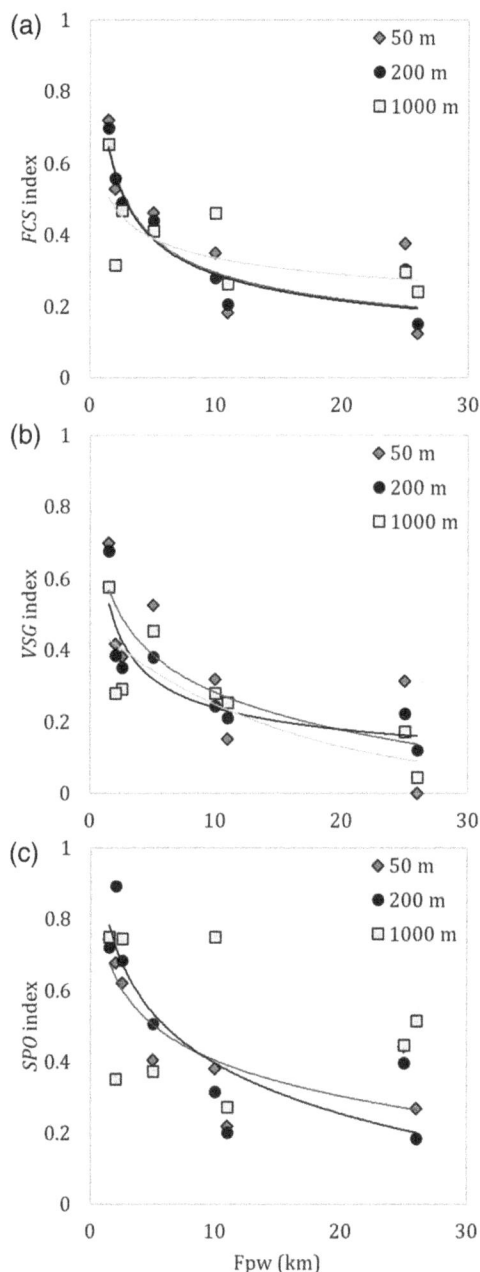

Figure 5. Power relationships between Fpw and (**a**) FSC, (**b**) VSG, and (**c**) SPO at each of the three window sizes.

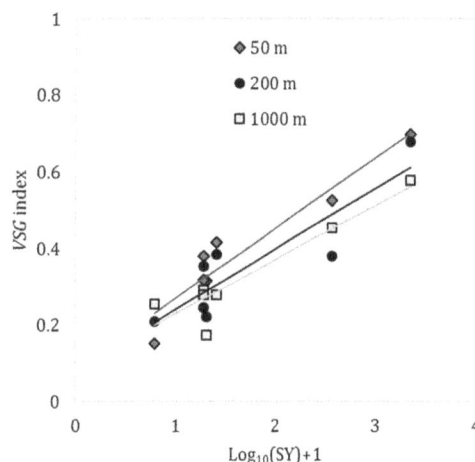

Figure 6. Linear relationships between log-transformed SY and VSG at each of the three window sizes with Gwydir removed.

gan and Nanson, 2004), which may have been overlooked in this index. In narrower, confined settings, where widths are less than 10 km, floodplain construction may be the result primarily of vertical processes (e.g. accretion/incision) leading to more prominent topographic features that exhibit a higher degree of spatial organisation and thus increased surface complexity (Nanson and Croke, 1992). Such complexity can lead to the concentration of flood energies in particular areas, promoting episodic catastrophic stripping (Nanson, 1986). The narrowest floodplain examined in this study was, on average, 1.5 km in width and the results presented in this study may not apply to narrower floodplains. In particular, there is known to be a loss of surface complexity when floodplains are contained between artificial levees or embankments (Florsheim and Mount, 2002; Gurnell and Petts, 2002), so floodplain surface complexity should not be considered to increase indefinitely with declining width in floodplains.

Contemporary sediment yield estimates were used in this study to investigate the influence of sediment yield on floodplain surface complexity. However, historical sediment yields are thought to be relatively more important in structuring floodplains (Panin et al., 1999). Substantial anthropogenic increases in sediment loads have been reported for the Gwydir floodplain (De Rose et al., 2003), and once this floodplain was removed as an outlier, variability in surface geometry was found to significantly increase with sediment yield. This result suggests that sediment yield may exert top-down control on the variability of floodplain surface geometry, although recent anthropogenic changes in sediment yields (Prosser et al., 2001), particularly increased erosion in the catchment due to land use changes, may have delayed "lag" effects on floodplain surfaces which have not yet been observed (sensu Thoms, 2006).

Valley slope was used in this study as a surrogate for stream energy, and this was not found to have any effect on overall floodplain surface complexity. More accurate measures of energy conditions such as specific stream power (Nanson and Croke, 1992) may reveal effects of energy conditions on floodplain surface complexity. It is also likely that variable flood energy conditions within each floodplain have an effect on localised surface complexity. For example, Fagan and Nanson (2004) found distinct differences in floodplain surface channel patterns among high, intermediate, and low energy areas of the semi-arid Cooper Creek in Australia.

They also found the energy of flood flows to be largely controlled by floodplain width.

Hydrology has been widely considered the main determinant of floodplain ecosystem patterns and processes (Junk et al., 1989; Hughes, 1990; Bayley, 1995; Whited et al., 2007). However, the research presented in this paper indicates that this may not be the case for floodplain surface complexity. None of the four hydrological variables measured here had a significant effect on floodplain surface complexity. This suggests that, although hydrology is largely important in driving floodplain ecosystem processes, floodplain width and sediment conditions appear to exert more control over the complexity of floodplain surfaces. This is important given that floodplain research and restoration is often focused on hydrology, particularly connectivity (e.g. Thoms, 2003; Thoms et al., 2005), whereas valley trough, sediment, and energy conditions may be more important in structuring and maintaining the physical template upon which hydrology acts as an ecosystem driver (Salo, 1990). Loss of floodplain surface complexity due to changes in sediment yield or calibre, or confinement between artificial levees, may be as ecologically important as changes to hydrology and should not be overlooked (Thoms, 2003). It is important to note, however, that some of the eight floodplains studied have experienced anthropogenic alterations to their hydrology. Thus, hydrological parameters based on contemporary data may not reflect the nature of the flow regime that was influential in establishing current surface conditions; lagged effects of altered hydrology on surface complexity may occur in the future (Thoms, 2006).

In terms of the origin and implications of floodplain surface complexity, this research focuses on top-down environmental drivers (sensu Thorp et al., 2008). Bottom-up feedbacks from the floodplain ecosystem are also likely to affect surface complexity. For example, vegetation establishment on deposited floodplain sediments is known to produce a positive feedback loop in which more sediment is trapped and semi-permanent morphological features such as islands develop (Nanson and Beach, 1977; Hupp and Osterkamp, 1996). Such feedbacks are likely to influence floodplain surface complexity, particularly in floodplains dominated by such features (Gurnell and Petts, 2002; Stanford et al., 2005). Bottom-up influences on floodplain surface complexity are difficult to quantify and were not examined in this study. Future research into the influence of vegetation type and density on floodplain surface complexity, particularly in relation to its hydraulic roughness, may provide valuable insights into bottom-up controls on floodplain surface complexity. Such data are also available through lidar (Straatsma and Baptist, 2008). Effects of floodplain surface complexity on biodiversity and productivity should also be examined in future research. The floodplain surface provides the primary geomorphic template upon which ecosystem processes are acted out (Salo, 1990) and it would be expected that increased surface complexity would promote the range of physical habitats re-quired to maintain floodplain biodiversity (Hamilton et al., 2007).

The inclusion of other floodplains, from different regions, in future studies of this nature would further determine whether the trends observed in this study extend beyond the floodplains investigated here. This study was limited to eight floodplains because of data availability. As high-resolution lidar data across many more floodplains are made available to researchers, other analyses such as multiple regression will be possible in studies such as this. Multiple regression would enable the interactive effects of environmental variables to be elucidated, whereas this study was limited to relatively simple linear regression because of the sample size of only eight floodplains.

5.3 The effect of scale

The different sampling scales used in this research indicate that the scale at which patterns in floodplain surfaces are most complex depends on the floodplain setting. In particular, wide, unconfined floodplains appear to have higher floodplain surface complexity when measured at larger sampling scales, whereas narrow, confined floodplains have so at smaller sampling scales. These results suggest that the scales of processes that maximise complexity, and potentially biodiversity and productivity (Tockner and Ward, 1999), in floodplains differ between different valley settings. This has implications for understanding and managing the complexity of floodplain ecosystems. Floodplain processes, which operate over certain temporal scales, elicit a response over relative spatial scales (Salo, 1990; Hughes, 1997). Consequently, managing processes at the appropriate scale to achieve desired outcomes is important (Parsons and Thoms, 2007). This has already been recognised for managing floodplain hydrology to maintain biodiversity (Amoros and Bornette, 2002), and these results indicate it is also important for managing the processes that maintain floodplain surface complexity.

Acknowledgements. The authors wish to thank Janet Hooke and two anonymous reviewers, whose comments on earlier versions greatly improved the manuscript. The authors wish to acknowledge support from the University of New England and the USGS Upper Midwest Environmental Sciences Center, without which this research would not have been possible. Any use of trade, product, or firm names is for descriptive purposes only and does not imply endorsement by the US Government. Data to support this article are available from the authors.

References

Amoros, C. and Bornette, G.: Connectivity and biocomplexity in waterbodies of riverine floodplains, Freshw. Biol., 47, 761–776, 2002.

Bar Massada, A. and Radeloff, V. C.: Two multi-scale contextual approaches for mapping spatial pattern, Landsc. Ecol., 25, 711–725, 2010.

Bayley, P. B.: Understanding large river: floodplain ecosystems, BioScience, 45, 153–158, 1995.

Cadenasso, M. L., Pickett, S. T. A., and Grove, J. M.: Dimensions of ecosystem complexity: heterogeneity, connectivity, and history, Ecol. Complex., 3, 1–12, 2006.

De Jager, N. R. and Rohweder, J. J.: Spatial patterns of aquatic habitat richness in the Upper Mississippi River floodplain, USA, Ecol. Ind., 13, 275–283, 2012.

De Jager, N. R., Thomsen, M., and Yin, Y.: Threshold effects of flood duration on the vegetation and soils of the Upper Mississippi River floodplain, USA, Forest Ecol. Manage., 270, 135–146, 2012.

De Rose, R. C., Prosser, I. P., Wiesse, M., and Hughes, A. O.: Patterns of erosion and sediment and nutrient transport in the Murray-Darling Basin, Technical Report 32/03, CSIRO Land and Water, Canberra, Australia, 2003.

Dollar, E. S. J., James, C. S., Rogers, K. H., and Thoms, M. C.: A framework for interdisciplinary understanding of rivers as ecosystems, Geomorphology, 89, 147–162, 2007.

Dorner, B., Lertzman, K., and Fall, J.: Landscape pattern in topographically complex landscapes: issues and techniques for analysis, Landscape Ecology, 17, 729–743, 2002.

Fagan, S. D. and Nanson, G. C.: The morphology and formation of floodplain-surface channels, Cooper Creek, Australia, Geomorphology, 60, 107–126, 2004.

Florsheim, J. L. and Mount, J. F.: Restoration of floodplain topography by sand-splay complex formation in response to intentional levee breaches, Lower Cosumnes River, California, Geomorphology, 44, 67–94, 2002.

Flotemersch, J. E., Leibowitz, S. G., Hill, R. A., Stoddard, J. L., Thoms, M. C., and Tharme, R. E.: A watershed integrity definition and assessment approach to support strategic management of watersheds, River Res. Appl., doi:10.1002/rra.2978, 2015.

Forman, R. T. T. and Godron, M.: Patches and structural components for a landscape ecology, BioScience, 31, 733–740, 1981.

Frost, N. J., Burrows, M. T., Johnson, M. P., Hanley, M. E., and Hawkins, S. J.: Measuring surface complexity in ecological studies, Limnol. Oceanogr.-Meth., 3, 203–210, 2005.

Gurnell, A. M. and Petts, G. E.: Island-dominated landscapes of large floodplain rivers, a European perspective, Freshw. Biol., 47, 581–600, 2002.

Hamilton, S. K., Kellndorfer, J., Lehner, B., and Tobler, M.: Remote sensing of floodplain geomorphology as a surrogate for biodiversity in a tropical river system (Madre de Dios, Peru), Geomorphology, 89, 23–38, 2007.

Holling, C. S.: Cross-scale morphology, geometry, and dynamics of ecosystems, Ecological Monographs, 62, 447–502, 1992.

Hughes, F. M. R.: The influence of flooding regimes on forest distribution and composition in the Tana River floodplain, Kenya, J. Appl. Ecol., 27, 475–491, 1990.

Hughes, F. M. R.: Floodplain biogeomorphology, Prog. Phys. Geogr., 21, 501–529, 1997.

Hupp, C. R. and Osterkamp, W. R.: Riparian vegetation and fluvial geomorphic processes, Geomorphology, 14, 277–295, 1996.

Hynes, H.: The stream and its valley, Verhandlungen des Internationalen Verein Limnologie, 19, 1–15, 1975.

Junk, W. J., Bayley, P. B., and Sparks, R. E.: The flood pulse concept in river-floodplain systems, in Proceedings of the International Large River Symposium, edited by: Dodge, D. P., 110–127, Canadian Special Publication of Fisheries and Aquatic Sciences No. 106, Canadian Government Publishing Centre, Ottawa, Ontario, 1989.

Kovalenko, K., Thomaz, S., and Warfe, D.: Habitat complexity: approaches and future directions, Hydrobiologia, 685, 1–17, 2012.

Levin, S. A.: Ecosystems and the biosphere as complex adaptive systems, Ecosystems, 1, 431–436, 1998.

Loreau, M., Mouquet, N., and Hold, R. D.: Meta-ecosystems: a theoretical framework for a spatial ecosystem ecology, Ecol. Lett., 6, 673–679, 2003.

McGarigal, K., Tagil, S., and Cushman, S.: Surface metrics: an alternative to patch metrics for the quantification of landscape structure, Landsc. Ecol., 24, 433–450, 2009.

Miller, A. J.: Valley morphology and boundary conditions influencing spatial patterns of flood flow, in Natural and Anthropogenic Influences in Fluvial Geomorphology: The Wolman Volume, edited by: Costa, J. E., Miller, A. J., Potter, K. W., and Wilcock, P. R., 57–82, American Geophysical Union, Washington, D. C., 1995.

Murray, O., Thoms, M., and Rayburg, S.: The diversity of inundated areas in semiarid flood plain ecosystems, in Sediment Dynamics and the Hydromorphology of Fluvial Systems, edited by: Rowan, J. S., Duck, R. W., and Werritty, A., 277–286, IAHS Press, Wallingford, UK, 2006.

Nanson, G. C.: Episodes of vertical accretion and catastrophic stripping: a model of disequilibrium flood-plain development, Geol. Soc. Am. Bull., 97, 1467–1475, 1986.

Nanson, G. C. and Beach Forest, H. F.: succession and sedimentation on a meandering-river floodplain, northeast British Columbia, Canada, J. Biogeogr., 4, 229–251, 1977.

Nanson, G. C. and Croke, J. C.: A genetic classification of floodplains, Geomorphology, 4, 459–486, 1992.

Norris, R. H., Linke, S., Prosser, I. P., Young, W. J., Liston, P., Bauer, N., Sloane, N., Dyer, F., and Thoms, M.: Very-broad-scale assessment of human impacts on river condition, Freshw. Biol., 52, 959–976, 2007.

O'Neill, R. V., Johnson, A. R., and King, A. W.: A hierarchical framework for the analysis of scale, Landsc. Ecol., 3, 193–205, 1989.

Panin, A. V., Sidorchuk, A. Y., and Chernov, A. V.: Historical background to floodplain morphology: examples from the East European Plain, in Floodplains: Interdisciplinary Approaches, edited by: Marriott, S. B. and Alexander, J., 217–229, Geological Society Special Publication No. 163, Geological Society, London, UK, 1999.

Papadimitriou, F.: Modelling indicators and indices of landscape complexity: an approach using G.I.S., Ecol. Ind., 2, 17–25, 2002.

Parsons, M. and Thoms, M. C.: Hierarchical patterns of physical–biological associations in river ecosystems, Geomorphology, 89, 127–146, 2007.

Phillips, J. D.: Sources of nonlinearity and complexity in geomorphic systems, Prog. Phys. Geogr., 27, 1–23, 2003.

Pinay, G., Black, V. J., Planty-Tabacchi, A. M., Gumiero, B., and Décamps, H.: Geomorphic control of denitrification in large river floodplain soils, Biogeochemistry, 50, 163–182, 2000.

Pollock, M. M., Naiman, R. J., and Hanley, T. A.: Plant species richness in riparian wetlands – a test of biodiversity theory, Ecology, 79, 94–105, 1998.

Prosser, I. P., Rutherfurd, I. D., Olley, J. M., Young, W. J., Wallbrink, P. J., and Moran, C. J.: Large-scale patterns of erosion and sediment transport in river networks, with examples from Australia, Mar. Freshw. Res., 52, 81–99, 2001.

Rossi, R. E., Mulla, D. J., Journel, A. G., and Eldon, H. F.: Geostatistical tools for modeling and interpreting ecological spatial dependence, Ecol. Monogr., 62, 277–314, 1992.

Salo, J.: External processes influencing origin and maintenance of inland water-land ecotones, in The Ecology and Management of Aquatic-Terrestrial Ecotones, edited by: Naiman, R. J. and Décamps, H., 37–64, Unesco, Paris, France, 1990.

Schumm, S. A.: The Fluvial System, John Wiley and Sons, New York, NY, 338 pp., 1977.

Scown, M. W., Thoms, M. C., and De Jager, N. R.: Floodplain complexity and surface metrics: Influences of scale and geomorphology, Geomorphology, 245, 102–116, 2015a.

Scown, M. W., Thoms, M. C., and De Jager, N. R.: Measuring floodplain spatial patterns using continuous surface metrics at multiple scales, Geomorphology, 245, 87–101, 2015b.

Scown, M. W., Thoms, M. C., and De Jager, N. R.: Measuring spatial pattern in floodplains: a step towards understanding the complexity of floodplain ecosystems, in River Science: Research and Management for the 21st Century, edited by: Greenwood, M. T., Thoms, M. C., and Wood, P. J., John Wiley and Sons, London, UK, 103–131, 2016.

Southwood, T. R. E.: Habitat, the templet for ecological strategies?, J. Animal Ecol., 46, 337–365, 1977.

Stanford, J. A., Lorang, M. S., and Hauer, F. R.: The shifting habitat mosaic of river ecosystems, Verhandlungen des Internationalen Verein Limnologie, 29, 123–136, 2005.

Straatsma, M. W. and Baptist, M. J.: Floodplain roughness parameterization using airborne laser scanning and spectral remote sensing, Remote Sens. Environ., 112, 1062–1080, 2008.

Thapa, R., Thoms, M. C., and Parsons, M.: An adaptive cycle hypothesis of semi-arid floodplain vegetation productivity in dry and wet resource states, Ecohydrology, doi:10.1002/eco.1609, 2015.

Thoms, M. C.: Floodplain–river ecosystems: lateral connections and the implications of human interference, Geomorphology, 56, 335–349, 2003.

Thoms, M. C.: Variability in riverine ecosystems, River Research and Applications, 22, 115–151, 2006.

Thoms, M. C., Parker, C. R., and Simons, M.: The dispersal and storage of trace metals in the Hawkesbury River valley, in River Management: The Australasian Experience, edited by: Brizga, S. and Finlayson, B., 197–219, John Wiley and Sons, Chichester, UK, 2000.

Thoms, M. C., Southwell, M., and McGinness, H. M.: Floodplain–river ecosystems: fragmentation and water resources development, Geomorphology, 71, 126–138, 2005.

Thorp, J. H., Thoms, M. C., and Delong, M. D.: The Riverine Ecosystem Synthesis: Towards Conceptual Cohesiveness in River Science, Elsevier, Amsterdam, 208 pp., 2008.

Tockner, K. and Ward, J. V.: Biodiversity along riparian corridors, Archiv für Hydrobiologie, Supplementband, Large Rivers, 11, 293–310, 1999.

Tokeshi, M. and Arakaki, S.: Habitat complexity in aquatic systems: fractals and beyond, Hydrobiologia, 685, 27–47, 2012.

Ward, J. V., Tockner, K., and Schiemer, F.: Biodiversity of floodplain river ecosystems: ecotones and connectivity, Regulated Rivers: Research & Management, 15, 125–139, 1999.

Warner, R. F.: Floodplain evolution in a New South Wales coastal valley, Australia: spatial process variations, Geomorphology, 4, 447–458, 1992.

Whited, D. C., Lorang, M. S., Harner, M. J., Hauer, F. R., Kimball, J. S., and Stanford, J. A.: Climate, hydrologic disturbance, and succession: drivers of floodplain pattern, Ecology, 88, 940–953, 2007.

Wiens, J. A.: Riverine landscapes: taking landscape ecology into the water, Freshw. Biol., 47, 501–515, 2002.

5

Development and verification of a real-time stochastic precipitation nowcasting system for urban hydrology in Belgium

L. Foresti[1]**, M. Reyniers**[1]**, A. Seed**[2]**, and L. Delobbe**[1]

[1]Royal Meteorological Institute of Belgium, Brussels, Belgium
[2]Bureau of Meteorology, Melbourne, Australia

Correspondence to: L. Foresti (loris.foresti@gmail.com)

Abstract. The Short-Term Ensemble Prediction System (STEPS) is implemented in real-time at the Royal Meteorological Institute (RMI) of Belgium. The main idea behind STEPS is to quantify the forecast uncertainty by adding stochastic perturbations to the deterministic Lagrangian extrapolation of radar images. The stochastic perturbations are designed to account for the unpredictable precipitation growth and decay processes and to reproduce the dynamic scaling of precipitation fields, i.e., the observation that large-scale rainfall structures are more persistent and predictable than small-scale convective cells. This paper presents the development, adaptation and verification of the STEPS system for Belgium (STEPS-BE). STEPS-BE provides in real-time 20-member ensemble precipitation nowcasts at 1 km and 5 min resolutions up to 2 h lead time using a 4 C-band radar composite as input. In the context of the PLURISK project, STEPS forecasts were generated to be used as input in sewer system hydraulic models for nowcasting urban inundations in the cities of Ghent and Leuven. Comprehensive forecast verification was performed in order to detect systematic biases over the given urban areas and to analyze the reliability of probabilistic forecasts for a set of case studies in 2013 and 2014. The forecast biases over the cities of Leuven and Ghent were found to be small, which is encouraging for future integration of STEPS nowcasts into the hydraulic models. Probabilistic forecasts of exceeding $0.5\,\mathrm{mm\,h^{-1}}$ are reliable up to 60–90 min lead time, while the ones of exceeding $5.0\,\mathrm{mm\,h^{-1}}$ are only reliable up to 30 min. The STEPS ensembles are slightly under-dispersive and represent only 75–90 % of the forecast errors.

1 Introduction

The use of radar measurements for urban hydrological applications has substantially increased during the last years (e.g., Berne et al., 2004; Einfalt et al., 2004; Bruni et al., 2015). Given the fast response time of urban catchments and sewer systems, radar-based very short-term precipitation forecasting (nowcasting) has the potential to extend the lead time of hydrological and hydraulic flow predictions.

Nowcasting concerns the accurate description of the current weather situation together with very short-term forecasts obtained by extrapolating the real-time observations. Quantitative precipitation nowcasting (QPN) is traditionally done by estimating the apparent movement of radar precipitation fields using optical flow or variational echo tracking techniques and extrapolating the last observed precipitation field into the future (e.g., Germann and Zawadzki, 2002; Bowler et al., 2004a). During recent years there has been significant progress in NWP modeling with radar data assimilation techniques (see a review in Sun et al., 2014), which reduces the useful lead time of extrapolation-based nowcasts compared with NWP forecasts. The development of seamless forecasting systems that optimally blend the extrapolation nowcast with the output of NWP models makes the definition of the nowcasting time range even fuzzier (see, e.g., Pierce et al., 2010).

Due to the lack of predictability of rainfall growth and decay processes at small spatial scales (Radhakrishna et al., 2012), it is very important to provide together with a forecast an estimation of its uncertainty. The established method to represent the forecast uncertainty in Numerical Weather Prediction (NWP) is to generate an ensemble of forecasts by perturbing the initial conditions of the model in the directions exhibiting the largest error growth, which amplify more the spread of the obtained ensemble. However, in the nowcasting range the computation of large NWP ensembles (50–100 members) that resolve features at the scales of 1 km and are updated every 5 min is still impossible to achieve. Consequently, the efforts in nowcasting research have recently focused on developing heuristic techniques for probabilistic precipitation nowcasting, which was the topic of the *Heuristic Probabilistic Forecasting Workshop* that was organized in Munich, Germany (Foresti et al., 2014).

Probabilistic QPN methods can be categorized into three main classes: analog, local Lagrangian and stochastic approaches. The analog-based approach derives the forecast probability density function (pdf) by retrieving a set of similar situations from an archive of precipitation events (Panziera et al., 2011; Foresti et al., 2015), the local Lagrangian approach derives the pdf by collecting the precipitation values in a neighborhood of a given grid point in Lagrangian frame of reference (Hohti et al., 2000; Germann and Zawadzki, 2004) and the stochastic approach exploits a random number generator to compute an ensemble of equally likely precipitation fields, for example by adding stochastic perturbations to a deterministic extrapolation nowcast (Pegram and Clothier, 2001a, b; Bowler et al., 2006; Metta et al., 2009; Berenguer et al., 2011; Seed et al., 2013; Atencia and Zawadzki, 2014; Dai et al., 2015). The stochastic approach is also extensively used to produce ensembles of precipitation fields that characterize the radar measurement uncertainty (e.g., Jordan et al., 2003; Germann et al., 2009) and for design storm studies (e.g., Willems, 2001a; Paschalis et al., 2013).

Uncertainty quantification is nowadays an integral part of both weather and hydrological forecasting (Pappenberger and Beven, 2006). Not surprisingly, an important part of hydro-meteorological research aims at understanding how to propagate the uncertainty of precipitation observations and forecasts into the hydrological models (e.g., Willems, 2001b; Cloke and Pappenberger, 2009; Collier, 2009; Zappa et al., 2010).

Several studies already analyzed the value of deterministic nowcasting systems for catchment hydrology (e.g., Berenguer et al., 2005) and for better control of urban drainage systems (e.g., Achleitner et al., 2009; Verworn et al., 2009; Thorndahl and Rasmussen, 2013). Since an important fraction of the uncertainty of hydrological predictions is due to the uncertainty of the input rainfall observations and forecasts, radar-based ensemble nowcasting systems are increasingly used as inputs for flood and sewer system modeling (e.g., Ehret et al., 2008; Silvestro and Rebora, 2012; Silvestro et al., 2013; Xuan et al., 2009, 2014). At longer forecast ranges, the NWP ensembles are also exploited for uncertainty propagation into hydrological models (see Roulin and Vannitsem, 2005; Thielen et al., 2009; Schellekens et al., 2011).

The Short-Term Ensemble Prediction System (STEPS) is a probabilistic nowcasting system developed at the Australian Bureau of Meteorology and the UK MetOffice (see the series of papers: Seed, 2003; Bowler et al., 2006; Seed et al., 2013). STEPS is operationally used at both weather services and provides short-term ensemble precipitation forecasts using both the extrapolation of radar images and the downscaled precipitation output of NWP models. The main idea behind STEPS is to represent the uncertainty due to the unpredictable precipitation growth and decay processes by adding stochastic perturbations to the deterministic extrapolation of radar images. The stochastic perturbations are designed to represent the scale-dependence of the predictability of precipitation and to reproduce the correct spatio-temporal correlation and growth of the forecast errors.

One of the first applications of STEPS in hydrology is presented in Pierce et al. (2005), who used the STEPS ensemble nowcasts to quantify the accuracy of flow predictions in a medium-sized catchment in the UK. The value of STEPS nowcasts for urban hydrology was extensively analyzed by Liguori and Rico-Ramirez (2012), Liguori et al. (2012), Liguori and Rico-Ramirez (2013) and Xuan et al. (2014). Liguori and Rico-Ramirez (2012) concluded that the performance of the radar-based extrapolation nowcast can be improved after 1 h lead time if blended with the output of a NWP model. They also found that, according to the Receiver Operating Characteristic (ROC) curve, the probabilistic nowcasts have more discrimination power than the deterministic ones. Liguori et al. (2012) integrated STEPS nowcasts as inputs into sewer system hydraulic models in an urban catchment in Yorkshire (UK). They concluded that the blending of radar and NWP forecasts has the potential to increase the lead time of flow predictions, but is strongly limited by the low accuracy of the NWP model in forecasting small-scale features. Liguori and Rico-Ramirez (2013) performed a detailed verification of the accuracy of flow predictions and concluded that the STEPS ensembles provide a similar performance as using a deterministic STEPS control forecast, but the ensembles lead to a slight underestimation of the flow predictions. Xuan et al. (2014) used ensemble STEPS nowcasts as inputs in a lumped hydrological model for a medium-sized catchment in the southwest of the UK. The hydrological model calibrated with rain gauges had lower RMSE than the one using radar data, but the ability of STEPS to account for the forecast uncertainty was useful in capturing some of the high flow peaks and extending the forecast lead time. However, the conclusions of the previous studies are strongly affected by the limited number of flood events analyzed. An extensive review of the usage of precipitation forecast systems for operational hydrological predictions in the UK from very short to long ranges (including STEPS) is provided in Lewis et al. (2015).

The goal of this paper is to present the development and verification of the STEPS system at the Royal Meteorological Institute of Belgium (RMI), referred to as STEPS-BE.

STEPS-BE provides in real-time 20-member ensemble precipitation nowcasts at 1 km and 5 min resolutions up to 2 h lead time on a 512×512 km domain using the Belgian 4 C-band radar composite as input. It was developed in the framework of the Belspo project PLURISK for better management of rainfall-induced risks in the urban environment. With respect to the original implementation of STEPS (Bowler et al., 2006), STEPS-BE includes two main improvements, which are designed to generate better STEPS nowcasts without NWP blending. The first one is related to the optical flow algorithm, which is extended with a kernel-based interpolation method to obtain smoother velocity fields. The second one concerns the generation of stochastic noise only within the advected radar composite. While the verification of STEPS nowcasts with NWP blending has already been extensive (Bowler et al., 2006; Seed et al., 2013), this paper analyzes the accuracy of STEPS ensemble nowcasts without NWP blending in the 0–2 h forecasting range.

Ensemble STEPS nowcasts are computed for a set of sewer overflow cases that affected the cities of Leuven and Ghent in 2013 and 2014. The accuracy of the ensemble mean forecast is verified using both continuous verification scores (multiplicative bias, RMSE) and categorical scores derived from the contingency table (probability of detection, false alarm ratio and Gilbert skill score). However, the most interesting part of this paper is the probabilistic and ensemble verification of STEPS nowcasts using both stratiform and convective rainfall events. Probabilistic nowcasts are verified using reliability diagrams and ROC curves. On the other hand, the dispersion of the nowcast ensembles is verified using rank histograms and by comparing the ensemble spread to the error of the ensemble mean.

The paper is structured as follows. Section 2 presents the radar data processing and case studies that are used to generate and verify the STEPS forecasts. Section 3 describes the STEPS nowcasting system and its extension and local implementation for Belgium (STEPS-BE). Section 4 illustrates the forecast verification results. Section 5 concludes the paper and discusses future perspectives.

2 Radar data and precipitation case studies

STEPS-BE integrates as input a composite image produced from the C-band radars of Wideumont (RMI, single-pol), Zaventem (Belgocontrol, single-pol), Jabbeke (RMI, dual-pol) and Avesnois (Meteo-France, dual-pol). The composite is produced on a 500 m resolution grid by combining single-radar pseudo Constant Altitude Plan Position Indicators (CAPPI) at a height of 1500 m a.s.l. The compositing algorithm takes the maximum reflectivity value from each radar at each grid point.

The radars have different hardware and scanning strategies, and are operated by different agencies (RMI, Belgocontrol and Meteo-France), which inevitably leads to differences in the data processing. The Wideumont and Zaventem radars eliminate the non-meteorological echoes using standard Doppler filtering. The Jabbeke radar includes an additional clutter filtering that uses a fuzzy logic algorithm based on the dual-polarization moments (essentially the co-polar correlation coefficient, the texture of the differential reflectivity and the texture of the specific differential phase shift). A static ground clutter map and a statistical filter are used by Meteo-France to remove the non-meteorological echoes of the Avensois radar. The French radar data processing chain is described in Tabary (2007) and in Figueras i Ventura and Tabary (2013).

Since the Zaventem radar is mainly used for aviation applications, its scanning strategy is optimized for the measurement of winds. Except for the lowest elevation scan, a dual PRF mode (1200/800 Hz) is used. The azimuths that are scanned with a high PRF (1200 Hz) only have a maximum range of 125 km and are more affected by the second trip echoes caused by convective cells located beyond the 125 km range.

All radars use the standard Marshall–Palmer relationship $Z = 200R^{1.6}$ to convert the measured reflectivity to rainfall rate. A composite image with more advanced radar-based quantitative precipitation estimation (QPE), that includes better ground clutter removal algorithms and also a correction for the bright band, was recently developed and the verification of the new product is ongoing.

STEPS forecasts were generated and verified for a set of sewer system overflow cases that affected the cities of Ghent and Leuven (see Table 1). The Ghent cases have a more stratiform character and occurred in late autumn and winter. On the other hand, the Leuven cases are more convective and occurred in summer months. A detailed climatology of convective storms in Belgium can be found in Goudenhoofdt and Delobbe (2009).

3 Short-Term Ensemble Prediction System (STEPS)

3.1 STEPS description

The Short-Term Ensemble Prediction System (STEPS) was jointly developed by the Australian Bureau of Meteorology (BOM) and the UK MetOffice (Bowler et al., 2006). STEPS forecasts are produced operationally at both weather services and are distributed to weather forecasters and a number of external users, in particular the hydrological services.

The key idea behind STEPS is to account for the unpredictable rainfall growth and decay processes by adding stochastic perturbations to the deterministic extrapolation of radar images (Seed, 2003). In order to be effective, the stochastic perturbations need to reproduce important statistical properties of both the precipitation fields and the forecast errors:

Table 1. List of precipitation events that caused sewer system floods in Ghent and Leuven.

City	Date	Event start (UTC)	Event end (UTC)	Duration	Predominant precipitation	Main wind direction
Ghent	10 Nov 2013	13:50	22:00	8:10 h	Stratiform	WNW → NNW
Ghent	3 Jan 2014	03:00	14:00	11 h	Stratiform	SW → WSW
Leuven	9–10 June 2014	06:30, 9th	15:30, 10th	33 h	Convective	SW
Leuven	19–20 July 2014	22:00, 19th	06:30, 20th	8:30 h	Convective	SSW

1. spatial scaling of precipitation fields,

2. dynamic scaling of precipitation fields,

3. spatial correlation of the forecast errors,

4. temporal correlation of the forecast errors.

The *spatial scaling* considers the precipitation field as arising from multiplicative cascade processes (Schertzer and Lovejoy, 1987; Seed, 2003). The presence of spatial scaling can be demonstrated by computing the 2-D Fourier power spectrum (PS) of a precipitation field. A 1-D PS can be obtained by radially averaging the 2-D PS. The precipitation field is said to be *scaling* if the 1-D PS draws a straight line on the log-log plot of the power against the spatial frequency (power law), which can be parametrized by one or two spectral exponents (see, e.g., Seed et al., 2013; Foresti and Seed, 2014). Within the multiplicative framework, a rainfall field is not represented as a collection of convective cells of a characteristic size but rather as a hierarchy of precipitation structures embedded in each other over a continuum of scales. STEPS considers the spatial scaling by decomposing the radar rainfall field into a multiplicative cascade using a fast Fourier transform (FFT) to isolate a set of eight spatial frequencies (Seed, 2003; Bowler et al., 2006; Seed et al., 2013). The top cascade levels (0, 1 and 2) represent the low spatial frequencies (large precipitation structures), while the bottom cascade levels (5, 6, 7) represent the high spatial frequencies (small precipitation structures). Another important behavior of rainfall fields is known as *dynamic scaling*, which is the empirical observation that the rate of temporal development of rainfall structures is a power law function of their spatial scale (Venugopal et al., 1999; Foresti and Seed, 2014). This means that large precipitation features are more persistent and hence predictable compared with small precipitation cells, which is closely related to the concept of scale-dependence of the predictability of precipitation (Germann and Zawadzki, 2002; Turner et al., 2004).

The stochastic perturbations should be able to reflect the properties of the forecast errors. Generating spatially and temporally correlated forecast errors is mandatory for hydrological applications, in particular when the correlation length of the errors is comparable or superior to the size and response time of the catchment. *Spatially correlated stochastic noise* can be constructed by applying a power law filter to a white noise field (Schertzer and Lovejoy, 1987). In practice it consists of three steps: computing the FFT of a white noise field, multiplying the obtained components in frequency domain by a given filter and applying the inverse FFT to return back to the spatial domain. The 1-D or 2-D power spectra of the rainfall field can be used as a filter to obtain noise fields that have the same scaling and spatial correlation of the rainfall field. The 1-D PS of the precipitation fields often appears to be a power law of the spatial frequency and explains why the procedure is also called power law filtering of white noise. In order to represent the anisotropies of the precipitation field, the 2-D PS can also be used as a filter. In the absence of a target precipitation field from which to derive the PS, the filter can be parametrized by using a climatological power law (see Seed et al., 2013). Finally, the *temporal correlations* are imposed by auto-regressive (AR) filtering. A hierarchy of AR processes defines the temporal evolution of the cascade levels. With the exception of forecast lead times beyond 2–3 h (Atencia and Zawadzki, 2014), an AR process of order 1 or 2 is already a good approximation to describe the temporal decorrelation of the forecast errors.

The practical implementation of STEPS to reproduce these important properties consists of the following steps (see Bowler et al., 2006; Foresti and Seed, 2014).

1. Estimation of the velocity field using optical flow on the last two radar rainfall images (Bowler et al., 2004a)

2. Decomposition of both rainfall fields into a multiplicative cascade using an FFT to isolate a set of eight spatial frequencies

3. Estimation of the rate of temporal evolution of rainfall features at each level of the cascade (Lagrangian auto-correlation)

4. Generation of a cascade of spatially correlated stochastic noise using as a filter the 1-D or 2-D power spectra of the last observed radar rainfall field. A Gaussian filter is used to isolate a given spatial frequency (see Foresti and Seed, 2014).

5. Stochastic perturbation of the rainfall cascade using the noise cascade (level by level)

6. Extrapolation of the cascade levels using a semi-Lagrangian advection scheme

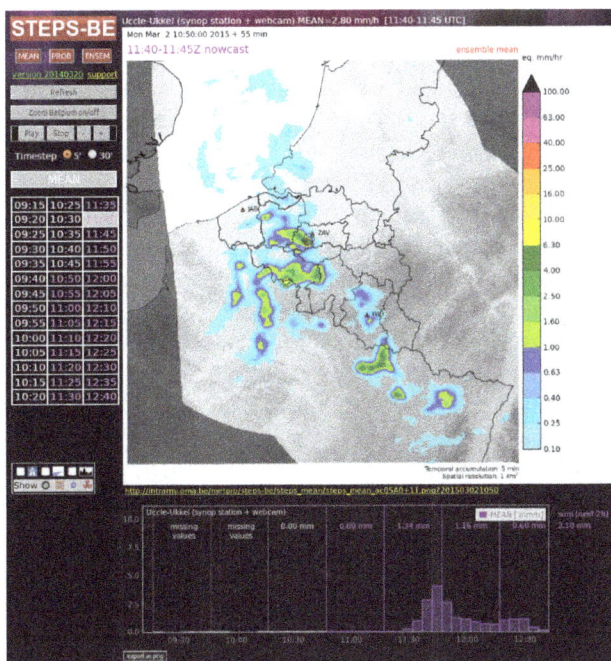

Figure 1. Web platform of STEPS-BE showing the ensemble mean forecast.

7. Application of the AR(1) or AR(2) model for the temporal update of the cascade levels at each forecast lead time using the Lagrangian auto-correlations estimated in step (3)

8. Recomposition of the cascade into a rainfall field

9. Probability matching of the forecast rainfall field with the original observed field (Ebert, 2001)

10. Computation of the forecast rainfall accumulations from the instant forecast rainfall rates. This procedure is known as advection correction and consists of advecting the instant rainfall rate forward over the 5 min period by discretizing the advection into smaller time steps.

3.2 STEPS implementation at RMI (STEPS-BE)

Bowler et al. (2006) introduced a general framework for blending a radar-based extrapolation nowcast with one or more outputs of downscaled NWP models (see also Pierce et al., 2010; Seed et al., 2013). Because of being designed for urban applications, the maximum lead time of STEPS-BE is restricted to 2 h. The operational NWP model of RMI (ALARO) runs only four times daily using a grid spacing of 4 km. Considering the model spin-up time and the absence of radar data assimilation, it is very unlikely that ALARO provides useful skill for blending its output with a radar-based extrapolation nowcast within the considered nowcasting range. It must also be remembered that the effective res-

olution of NWP models is much larger than the grid spacing. For instance, Grasso (2000) estimates the effective resolution to be at least 4 times the grid spacing, while Skamarock (2004) estimates it to be up to 7 times the grid spacing. ALARO would then only be able to resolve features that are greater than 20 km. For all these reasons, STEPS-BE only involves an extrapolation nowcast without NWP blending.

The STEPS-BE forecast domain is smaller than the extent of the 4 C-band radars composite (see Fig. 1). The radar field was upscaled from the original resolution of 500 m to 1 km and a sub-region of 512×512 grid points centered over Belgium was extracted. The forecast domain was extended by 32 pixels on each side to reduce the edge effects due to the FFT. This leads to an eight-level multiplicative cascade representing the following spatial scales (rounded to the nearest integer): $576-256-114, 256-114-51, 114-51-23, 51-23-11, 23-11-4, 11-4-2, 4-2-1$ and $2-1$ km. Italic characters mark the scales on which the Gaussian filter is centered in the frequency domain (see Foresti and Seed, 2014, for a more detailed explanation and visualization of the Gaussian FFT filter). The Gaussian filters of the largest and smallest spatial scales are truncated in order to preserve the power of the field. The top cascade level represents scales between the 576 and 256 km wavelengths and is not a perfect Gaussian filter. One can notice that the spatial scales are not exact multiples of 2. In fact, a multiplication factor of 2.246 was employed to match the enlarged STEPS-BE domain size.

STEPS-BE includes a couple of improvements compared with the original implementation of the BOM:

1. kernel interpolation of optical flow vectors;

2. generation of stochastic noise only within the advected radar mask.

The optical flow algorithm of Bowler et al. (2004a) estimates the velocity field by dividing the radar domain into a series of blocks within which the optical flow equation is solved. The minimization of the field divergence is only performed at the level of the block, which leaves sharp discontinuities in the velocity field between the blocks. In order to overcome this issue, a Gaussian kernel smoothing was applied to interpolate the velocity vectors located at the center of the blocks onto the fine radar grid. The bandwidth of the Gaussian kernel was chosen to be $\sigma = 24\,\text{km} = 0.4k$, where $k = 60$ grid points is the block size. This setting has the advantage of obtaining velocity fields that are less affected by the differential motion of small rainfall features and the presence of ground clutter. A too precise velocity field would provide increased predictability at very short lead times but worse forecasts at longer lead times due to excessive convergence and divergence of precipitation features during the advection. Smooth velocity fields could also be obtained by using a smaller block size and by compensating with a larger bandwidth of the smoothing kernel.

In STEPS-BE the 1-D power spectrum of the last observed rainfall field is used as a filter to generate the spatially correlated stochastic perturbations. The PS is parameterized using two spectral slopes to account for a scaling break that is often observed at the wavelength of 40 km (see Seed et al., 2013; Foresti and Seed, 2014). To simplify the computations, an auto-regressive model of order 1 (AR(1)) was employed for imposing the temporal correlations and to model the growth of forecast errors.

The original STEPS implementation (Bowler et al., 2006) was designed to blend the radar extrapolation nowcasts with the output of NWP models. However, the domain covered by the radars is smaller than the rectangular domain of the NWP model and small amounts of stochastic noise are generated by default also outside the radar composite. This setting was not adapted for radar-based nowcasts without NWP blending and needed some adaptation. In fact, when advecting the radar mask over several time steps, large areas with small amounts of stochastic rain appear outside the validity domain of the forecast and perturb the probability matching. In STEPS-BE the stochastic perturbations are only generated within the advected radar domain and set to zero elsewhere.

STEPS-BE can also account for the uncertainty in the estimation of the velocity field. The STEPS version that is implemented in the UK (Bowler et al., 2006) includes a detailed procedure to generate velocity perturbations that reproduce various statistical properties of the differences between the forecast velocity and the actual future diagnosed velocity (see details in Bowler et al., 2004b). In the BOM and RMI implementations a simpler procedure is applied. The diagnosed velocity field is multiplied by a single factor C that is drawn from the following distribution:

$$C = 10^{1.5N/10}, \qquad (1)$$

where N is a normally distributed random variable with zero mean and unit variance. In other words, the velocity field is accelerated or decelerated by a single random factor without affecting the direction of the vectors. In fact, the uncertainty on the diagnosed speed was observed to be higher than that of the direction of movement (Bowler et al., 2006).

The BOM and RMI versions of STEPS also include a stochastic model for the radar measurement error, a broken-line model to account for the unknown future evolution of the mean areal rainfall and the possibility to use time-lagged ensembles. However, a nowcasting model with too many components is harder to calibrate and complicates the interpretation of the forecast fields. Because of these reasons, STEPS-BE only exploits the basic stochastic model for the velocity field and for the evolution of rainfall fields (due to growth and decay processes).

The core of STEPS and the STEPS-BE extensions are implemented in C/C^{++} and the production of figures in python. Bash scripts were written to call multiple STEPS instances and compute the ensemble members in parallel over several processors. Once all the ensemble members are computed, a separate script collects the corresponding netCDF files and calculates the forecast probabilities. Most of the computational cost of STEPS consists of filtering the white noise field with FFT, advecting and updating the radar cascade with the AR model. The re-calculation of optical flow fields on each processor takes less than 10 % of the total computational time.

The python matplotlib library is used to read the netCDF files, export the PNG figures and the time series of observed and forecast rainfall at the location of major cities and weather stations. A single STEPS nowcast generates more than 600 figures, which takes a significant fraction of the total computational time. In order to optimize the timing, a bash script monitors continuously the directory with incoming radar composites and triggers STEPS-BE once a field with a new time stamp is found. All these implementation details ensure that the user/forecaster can have access to an ensemble STEPS nowcast in less than 5 min after receiving the radar composite image.

The visualization system of STEPS-BE is very similar to the one of INCA-BE, the local Belgian implementation of the Integrated Nowcasting through Comprehensive Analysis system (INCA, Haiden et al., 2011) developed at the Austrian weather office (ZAMG). Figure 1 illustrates the web interface with an example of an ensemble mean nowcast. The user can highlight the major cities, weather stations and click to visualize the time series of observed and forecast precipitation/probability, which appears at the bottom of the web page. The navigation through the observations and forecast lead times is facilitated by the scroll wheel of the mouse. On the other hand, by clicking on the image it is possible to easily scroll through the various ensemble members or probability levels for a given lead time. Scrolling the ensemble members at different lead times is very instructive and can make the user aware of the forecast uncertainty. In fact, at a lead time of 5 min the ensemble members agree very well on the intensity and location of precipitation. This means that the ensemble spread is small and the probabilistic forecast is sharp; i.e., most of the forecast probabilities are close to 1 or 0 (see an explanation in Appendix A). On the other hand, at 1 or 2 h lead time the ensemble members disagree on the location and intensity of rainfall, which enhances the ensemble spread and decreases the sharpness of the probabilistic forecast. The web page includes extensive documentation to guide the user and a set of case studies to help understanding the strengths and limitations of STEPS. The visualization system was implemented with great attention to take full advantage of the multi-dimensional information content of probabilistic and ensemble forecasts.

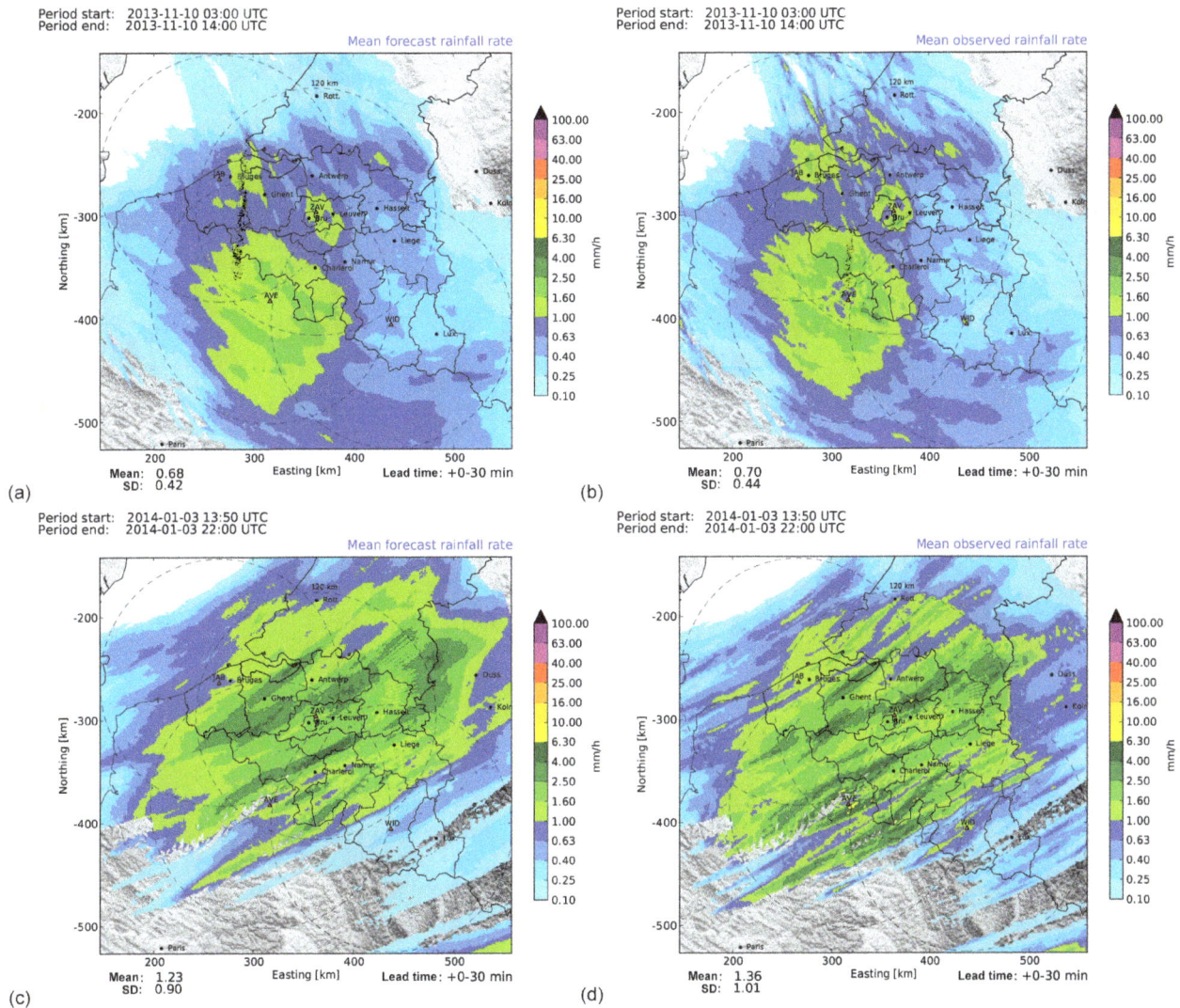

Figure 2. Average forecast and observed rainfall accumulations for the Ghent cases. (**a**) Forecast and (**b**) observed 0–30 min rainfall accumulations on 10 November 2013. (**c**) Forecast and (**d**) observed 0–30 min rainfall accumulations on 3 January 2014. The mean and standard deviation of the field within the 120 km range of the radars are shown on the bottom left corner. Field values are shown only if there are at least 10 samples for the computation of the mean. The red triangles denote the location of the Wideumont (WID, coordinates 438 km east/−405 km north), Zaventem (ZAV, 363/−296), Jabbeke (JAB, 266/−263) and Avesnois (AVE, 317/−382) radars. The 120 km range from the radar is displayed as a dashed circle. The mountain range of the Ardennes covers the three most southern districts of Belgium and Luxembourg (LUX).

4 Forecast verification

4.1 Verification set-up

This section presents the verification of STEPS-BE forecasts using a set of case studies (see Sect. 2). The accumulated radar observations were employed as reference for the verification. The rainfall rates are accumulated over the last 5 min by reversing the field vectors based on the observations and then performing the advection correction. The 30 min ensemble mean forecast was verified against the observed 30 min radar accumulations using both continuous and categorical verification scores. The deterministic verification procedure follows the one presented in Foresti and Seed (2015), which was designed to analyze the spatial distribution of the forecast errors. More details about the forecast verification setup and scores are given in Appendix A.

The continuous scores include the multiplicative bias and the root mean squared error (RMSE), while the categorical scores include the probability of detection (POD), false alarm ratio (FAR) and Gilbert skill score (GSS) derived from the contingency table for rainfall thresholds of 0.5 and $5.0 \, \mathrm{mm \, h^{-1}}$. The rainfall thresholds are given in equivalent intensity independently of the forecast rainfall accumulation.

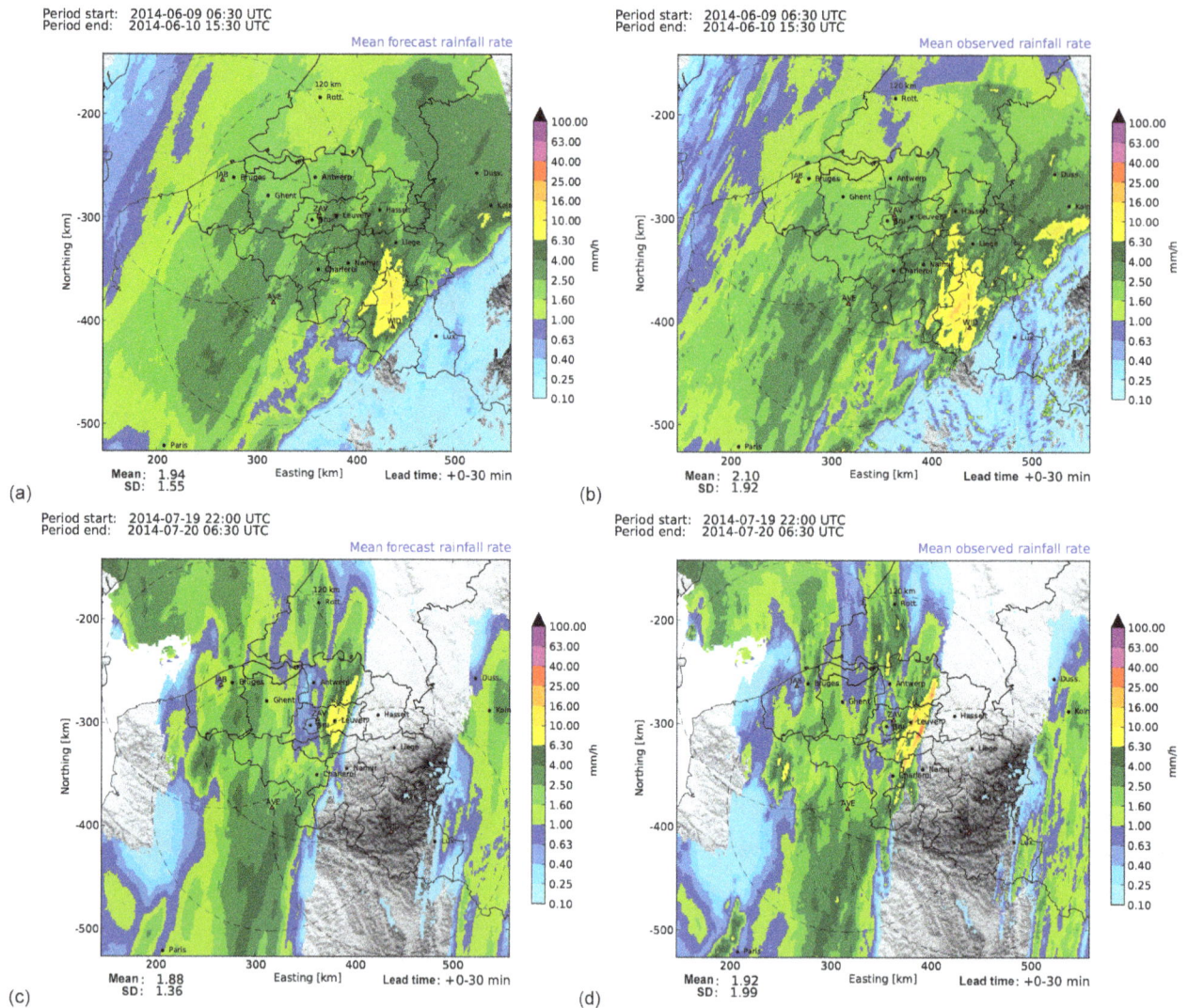

Figure 3. Average observed and forecast rainfall accumulations for the Leuven cases. (**a**) Forecast and (**b**) observed 0–30 min rainfall accumulations on 9–10 June 2014. (**c**) Forecast and (**d**) observed 0–30 min rainfall accumulations on 19–20 July 2014.

Thus, a threshold of $5.0\,\mathrm{mm\,h^{-1}}$ on a 30 min accumulation corresponds to 2.5 mm of rain. The multiplicative bias and the RMSE were evaluated only at the locations where the forecast or the verifying observations exceeded $0.1\,\mathrm{mm\,h^{-1}}$, which can be referred to as a *weakly conditional verification*. The probabilistic forecast of exceeding 0.1, 0.5 and $5.0\,\mathrm{mm\,h^{-1}}$ was verified using the reliability diagrams and ROC curves. Finally, the dispersion of the ensemble was analyzed by comparing the ensemble spread to the RMSE of the ensemble mean and by using rank histograms. The probabilistic and ensemble verification does not consider the spatial distribution of the errors and pools the data together in both space and time to derive the statistics.

4.2 Deterministic verification

Figures 2 and 3 show the average forecast and observed rainfall rates corresponding to the 0–30 min ensemble mean accumulation nowcast for the Ghent and Leuven cases, respectively. In other words, they represent the average forecast and observed rainfall rates over the duration of the precipitation event (for the 0–30 min lead time). The average was computed using the weak conditional principle explained above.

The average forecast and observed accumulations generally agree very well for the 0–30 min lead time forecast. The Ghent case on 10 November 2013 (Fig. 2a and b) is the only one with northwesterly flows and is characterized by the lowest average rainfall rates. The Avesnois radar demonstrates very well the range dependence of the average rainfall rates, which gradually decrease with increasing distance from the

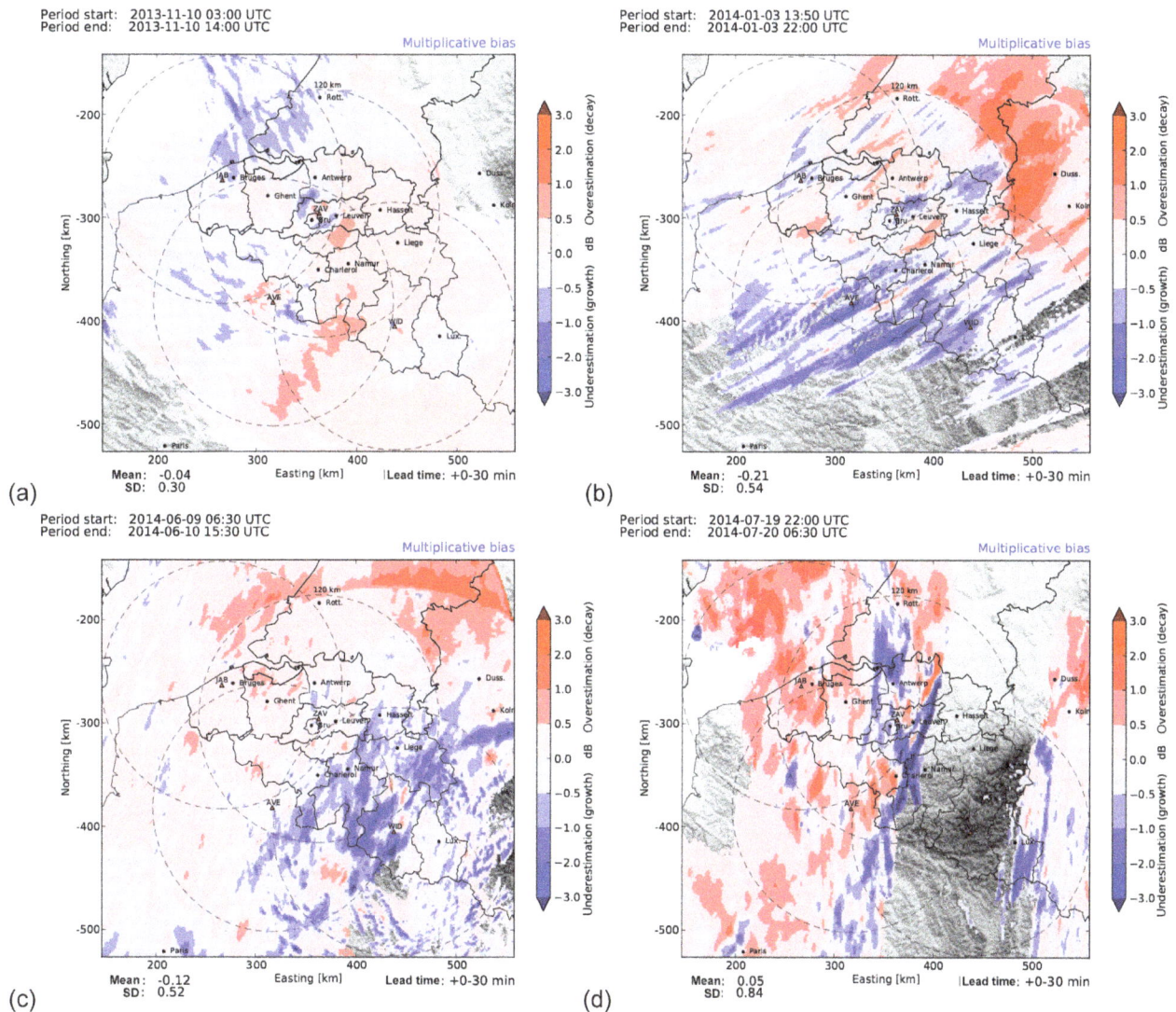

Figure 4. Average 0–30 min multiplicative forecast biases for the Ghent cases on (**a**) 10 November 2013 and on (**b**) 3 January 2014 and the Leuven cases on (**c**) 9–10 June 2014 and on (**d**) 19–20 July 2014. The interpretation of under- and over-estimations by STEPS as systematic rainfall growth and decay or simply as radar measurement biases is subject to interpretation as explained in the text.

radar. On the contrary, the smaller ring of high rainfall rates around the Zaventem radar is mostly due to the bright band (Fig. 2b).

The bright band effect influences the radar observations and hence the nowcasts based on their extrapolation. At longer lead times the larger rainfall estimates due to the bright band are extrapolated far from the location of the radar. The stochastic perturbations of STEPS can help to gradually dissolve the circular patterns introduced by the bright band effect. However, the bright band affects more the observations used for the verification, in particular when the rainfall is advected from upstream over the radar region. In such a case, the local larger rainfall estimates lead to a verification bias and the forecasts are wrongly supposed of rainfall underestimation. In spite of these issues, bright band effects might

not be so important for urban hydrological applications. In fact, except for one stratiform case presented in this paper, pluvial floods mainly happen in summer with convective precipitation events, during which the bright band is absent or negligible.

The Ghent case on 3 January 2014 has higher rainfall rates and the elongated structures of precipitation areas demonstrate well the southwesterly flow regime (Fig. 2c and d). For this case the measurements of the Zaventem radar are also affected by second trip echoes, which appear as a set of radially oriented rainfall structures northwest of the radar. These alternating patterns are explained by the dual PRF mode of Zaventem (see Sect. 2).

The Leuven cases on 9 June and 19–20 July 2014 have an important convective activity (Fig. 3a–d). The maximum av-

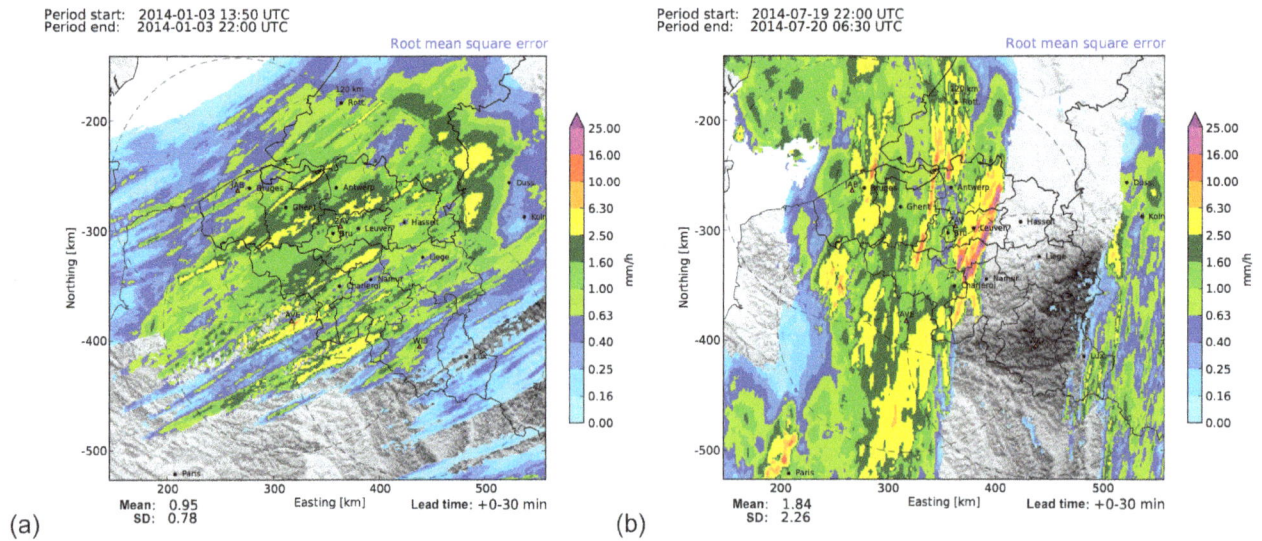

Figure 5. Average 0–30 min forecast RMSE for (**a**) the Ghent winter case on 3 January 2014 and (**b**) the Leuven summer case on 19–20 July 2014.

erage rainfall rates are located over the Ardennes mountain range and the city of Leuven, respectively. Since urban flash floods can be triggered by a single convective cell, the average rainfall rate over the duration of the event may not be as high in the considered city (e.g., Fig. 3b).

Figure 4 illustrates the multiplicative bias of the 0–30 min nowcast averaged over each of the four events. A detailed interpretation of such forecast biases using Australian radar data and their connection to orographic features is given in Foresti and Seed (2015), which point out that an important fraction of the forecast errors is caused by the biases of the verifying radar observations rather than systematic rainfall growth and decay processes due to orography. In Fig. 4a it is easy to notice the effect of bright band, which causes a series of systematic forecast biases around the Zaventem radar and perpendicularly oriented with respect to the prevailing flow direction (NW). Systematic rainfall underestimation occurs along the Belgian coast of the North Sea. One factor which contributes to this underestimation is the absence of visibility of the radar at longer ranges. The incoming precipitation is suddenly detected by the radar and therefore strongly underestimated by STEPS. The only situation where the range dependence of the rainfall estimation does not affect the forecast verification occurs when the velocity field is perfectly rotational and centered on the radar (assuming no beam blockage). All the other cases have to deal with the fact that the rainfall nowcast also extrapolates the biases of the radar observations! Contrary to expectation, on the upwind side of the Ardennes there is overestimation, which may depict a region of systematic rainfall decay. The bias over the city of Ghent is fortunately small and is included in the range from 0 to +0.5 dB (light overestimation, rainfall decay). Having small systematic biases over the cities of interest is very im-

portant for future integration of STEPS nowcasts as input in hydraulic models.

In Fig. 4b the systematic underestimation is also located upstream with respect to the prevailing winds (SW). The strong overestimations in Germany and the Netherlands are mostly due to the underestimation of rainfall by the verifying radar observations rather than caused by systematic rainfall decay. This is particularly visible after a range of 125 km from the Zaventem radar, which demonstrates again that discontinuities and biases in the radar observations lead to biases in the extrapolation forecast. Also in this case the bias over the city of Ghent is small but in the range from −0.5 to 0 dB (light underestimation, rainfall growth). Radar composite discontinuities are also visible in Fig. 4c but this time located at a range of 240 km north of the Wideumont radar when entering the area covered by the Jabbeke radar. This forecast bias is mainly explained by the negative calibration bias of the Jabbeke radar, which is known to slightly underestimate the rainfall rates with respect to the Wideumont radar. Strong underestimation occurs over the Ardennes due to the systematic initiation and growth of convection that cannot be predicted by STEPS (Fig. 4c). Fortunately the city of Leuven is located in a region with small biases in the range from −0.5 to +0.5 dB. Figure 4d is quite interesting since strong underestimations are located in front of the rain band (from Charleroi to Leuven and beyond) and overestimations at the rear of the rain band (west of the Jabbeke radar). The underestimations are due to systematic rainfall initiation in front of the rain band, while the overestimations are probably caused by a too slow extrapolation of rainfall, which tends to drag at the rear of the rain band. The two bands of underestimations south of Leuven are caused by two different thunderstorms. The first one passed over the city of Leuven and had

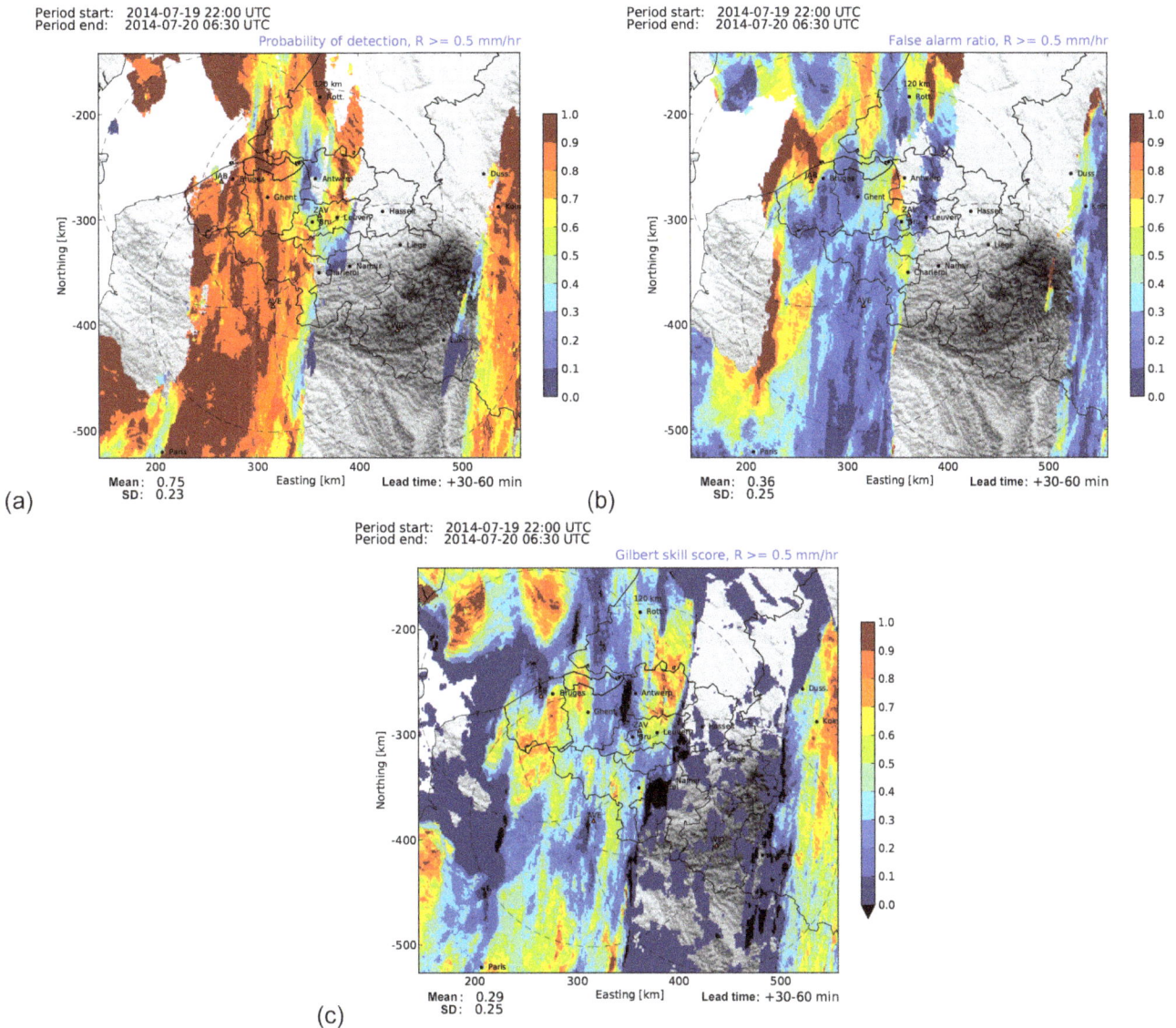

Figure 6. (a) POD, **(b)** FAR and **(c)** GSS of the 30–60 min ensemble mean forecast of exceeding $0.5\,\mathrm{mm\,h^{-1}}$ for the Leuven case on 19–20 July 2014.

a stronger westerly component with respect to the prevailing southerly flow. The second thunderstorm was weaker and had a stronger easterly component. When isolated convection does not follow the prevailing movement of the rainfall field, strong biases can appear in the nowcast during the first lead times.

Figure 5 shows the spatial distribution of the RMSE for the stratiform event on 3 January 2014 in Ghent and the convective event on 19–20 July 2014 in Leuven. If compared with Figs. 2d and 3d it is clear that the RMSE is strongly correlated with the regions having the highest mean rainfall accumulations (proportional effect). Thus, it is not surprising that the RMSE of the convective case (Fig. 5b) displays values exceeding $10\,\mathrm{mm\,h^{-1}}$ over the city of Leuven. The winter

case only shows RMSE values below $2\,\mathrm{mm\,h^{-1}}$ over the city of Ghent.

Figure 6 illustrates an example of categorical verification of the 30–60 min ensemble mean forecast for the Leuven case on 19–20 July 2014 relative to the rainfall threshold of $0.5\,\mathrm{mm\,h^{-1}}$. The probability of detection is high everywhere (mean of 0.75) except in the neighborhood of Antwerp and south of Leuven, where the initiation of thunderstorms could not be predicted by STEPS (Fig. 6a). The false alarm ratio is quite low (mean of 0.36) and the regions with high values are mainly located at the rear of the front where the rainfall is advected too slowly compared with the actual movement of the front (Fig. 6b). A high Gilbert skill score generally coincides with the regions with the highest rainfall accumulations and

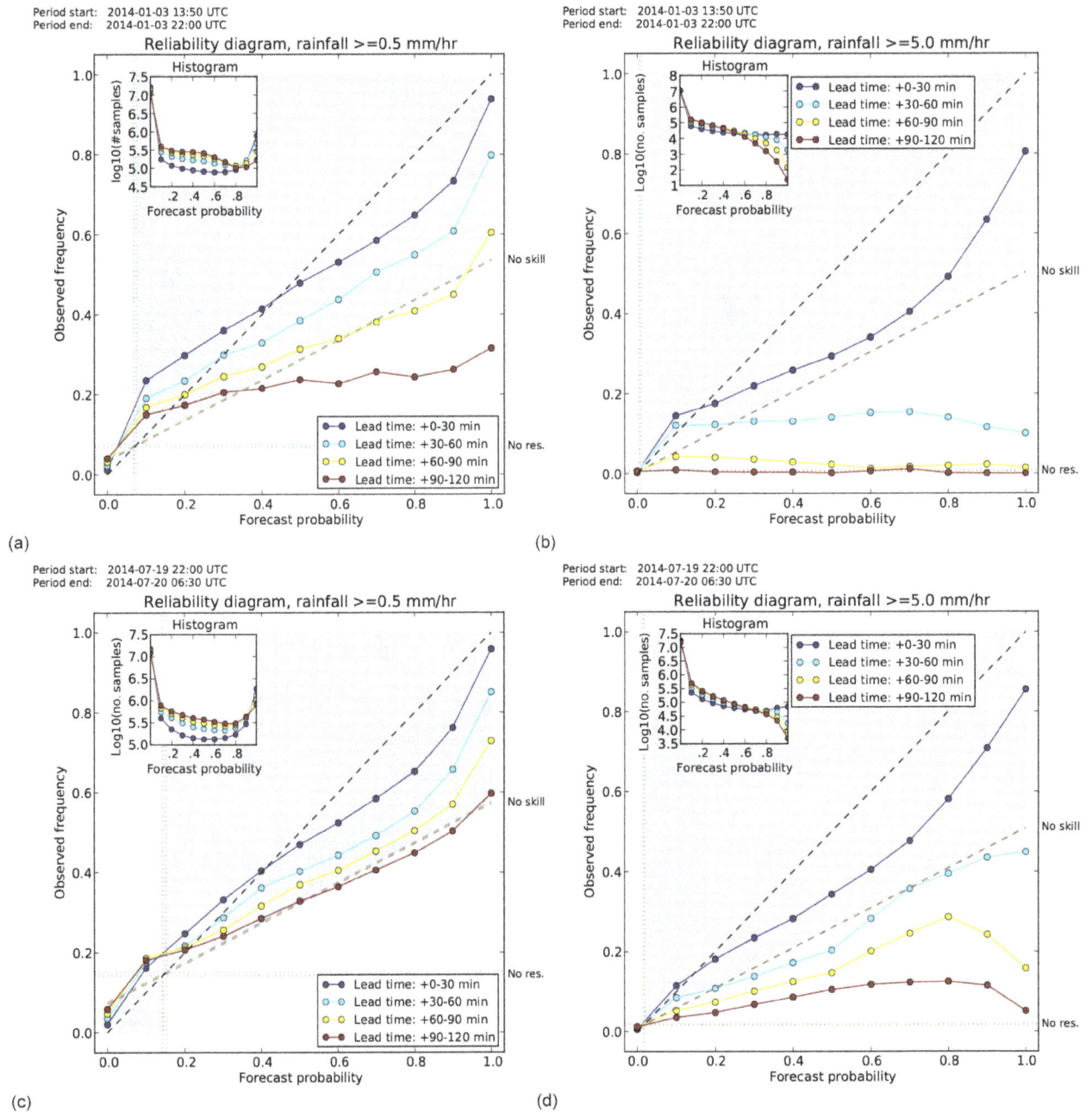

Figure 7. Reliability diagrams for the Ghent case on 3 January 2014 relative to the probabilistic forecast of exceeding (**a**) $0.5\,\mathrm{mm\,h^{-1}}$ and (**b**) $5.0\,\mathrm{mm\,h^{-1}}$. (**c, d**) Same as (**a, b**) but for the Leuven case on 19–20 July 2014.

becomes lower at the edges of the rain areas (Fig. 6c). This finding can be explained conceptually if one thinks about the verification of the future path of a single convective cell. The regions with the highest uncertainty are located along the edges of the predicted thunderstorm path and the highest skill is obtained in the center of the predicted path.

4.3 Probabilistic verification

Figure 7 shows the reliability diagrams relative to the probabilistic forecast of exceeding the 0.5 and $5.0\,\mathrm{mm\,h^{-1}}$ rainfall thresholds for the Ghent case on 3 January 2014 (Fig. 7a and b) and the Leuven case on 19–20 July 2014 (Fig. 7c and d). The reference probabilistic forecast is taken as the climatological frequency of exceeding a given rainfall threshold during that precipitation event (horizontal dashed line). Unexpectedly, the forecasts of the stratiform case in Ghent

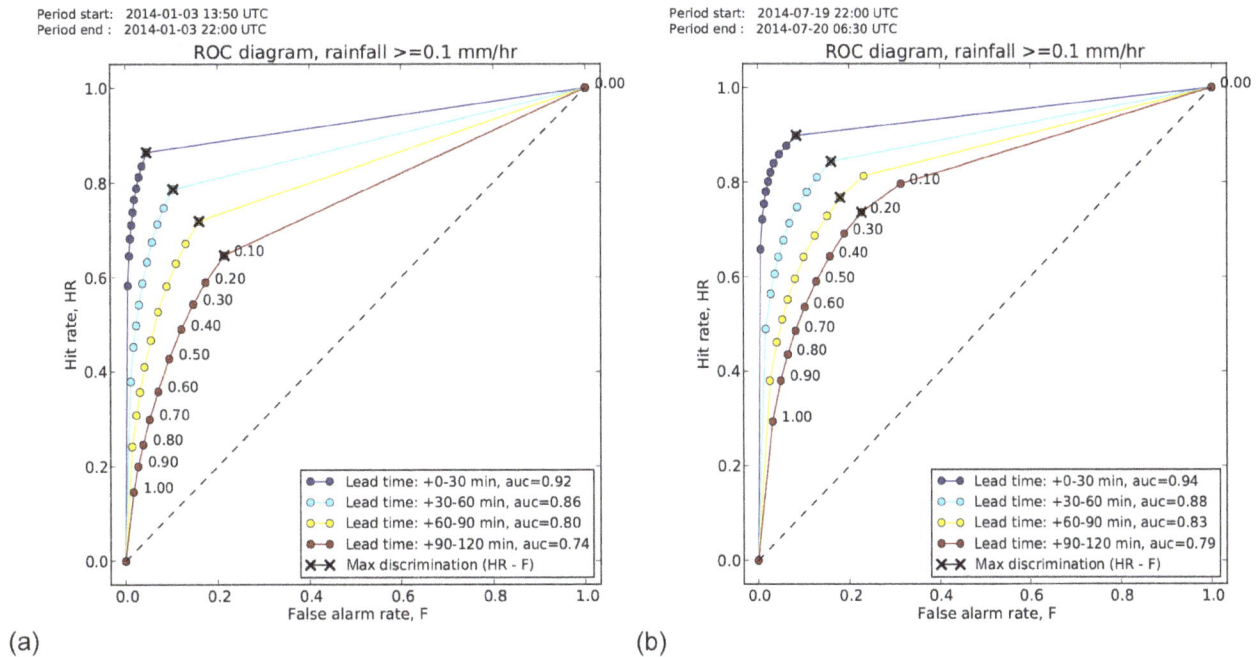

Figure 8. ROC curves relative to the probabilistic forecast of exceeding $0.1 \, \text{mm h}^{-1}$ for **(a)** the Ghent case on 3 January 2014 and **(b)** the Leuven case on 19–20 July 2014.

are less reliable than the ones of the convective case in Leuven for both rainfall thresholds. Probabilistic forecasts of exceeding $0.5 \, \text{mm h}^{-1}$ for the Ghent case have a good reliability and positive Brier skill score (BSS) up to 60 min lead time (Fig. 7a). The higher rainfall threshold of $5.0 \, \text{mm h}^{-1}$ is harder to predict and there is skill only up to 30 min lead time (Fig. 7b). The convective case in Leuven is more predictable and the probabilistic forecast of exceeding $0.5 \, \text{mm h}^{-1}$ exhibits skill up to 90 min lead time (Fig. 7c). It is interesting to note that forecast probabilities that are close to the climatological frequency (intersection of lines around the probability 0.15) can easily fall outside the skillful region (Fig. 7c). In fact, a small systematic forecast bias is likely to be worse than the event climatology at those frequencies. The rainfall threshold of $5.0 \, \text{mm h}^{-1}$ shows again a limit of predictability of 30 min (Fig. 7d). Despite having a negative BSS, the following lead times (Fig. 7d) have higher resolution than the stratiform case in Ghent (Fig. 7b).

Figure 8 illustrates the ROC curves relative to the probabilistic forecast of exceeding $0.1 \, \text{mm h}^{-1}$ for the Ghent case on 3 January 2014 (Fig. 8a) and the Leuven case on 19–20 July 2014 (Fig. 8b). All the ROC curves are very far from the diagonal line of no skill. The probability level that is marked with a cross is the one that maximizes the difference between the hit rate (HR) and the false alarm rate (F) (not to be confused with the false alarm ratio, which is conditioned on the forecasts). This point is located within the probabilities 0.1 and 0.2, which means that an optimal forecast of the probability of rain is achieved when only 10–20 % of the

ensemble members exceed the $0.1 \, \text{mm h}^{-1}$ threshold. A forecaster who is not scared of making false alarms would choose a lower probability level to increase the number of hits. On the contrary, an unconfident forecaster who would like to minimize the false alarms would choose a higher probability level, which has however the consequence of reducing the number of hits. As expected, the area under the ROC curves (AUC) decreases for increasing lead times. The discrimination skill for the convective event in Leuven is slightly higher than the one of the stratiform event in Ghent, which confirms the findings on the reliability diagrams (Fig. 7). This does not mean that small-scale features are easier to forecast than larger-scale features, which is known to be false (see Foresti and Seed, 2014). It means that the predictability of well-defined and organized convective systems is higher than the one of more moderate convection with shorter lifetime, at least for the cases analyzed in this paper.

4.4   Ensemble verification

Figure 9 compares the error of the ensemble mean (RMSE) and the ensemble spread for the Ghent case on 3 January 2014 and the Leuven case on 19–20 July 2014 (see interpretation of ensemble spread in Appendix A). In both cases the RMSE increases up to a lead time of 50–60 min and then starts a slow decrease, which can be counter-intuitive. However, it must be remembered that the ensemble mean forecast becomes smoother for increasing lead times, which reduces the double penalty error due to forecasting a thunderstorm at the wrong location. The ensemble spread also increases up to

Period start: 2014-01-03 13:50 UTC
Period end : 2014-01-03 22:00 UTC

Period start: 2014-07-19 22:00 UTC
Period end : 2014-07-20 06:30 UTC

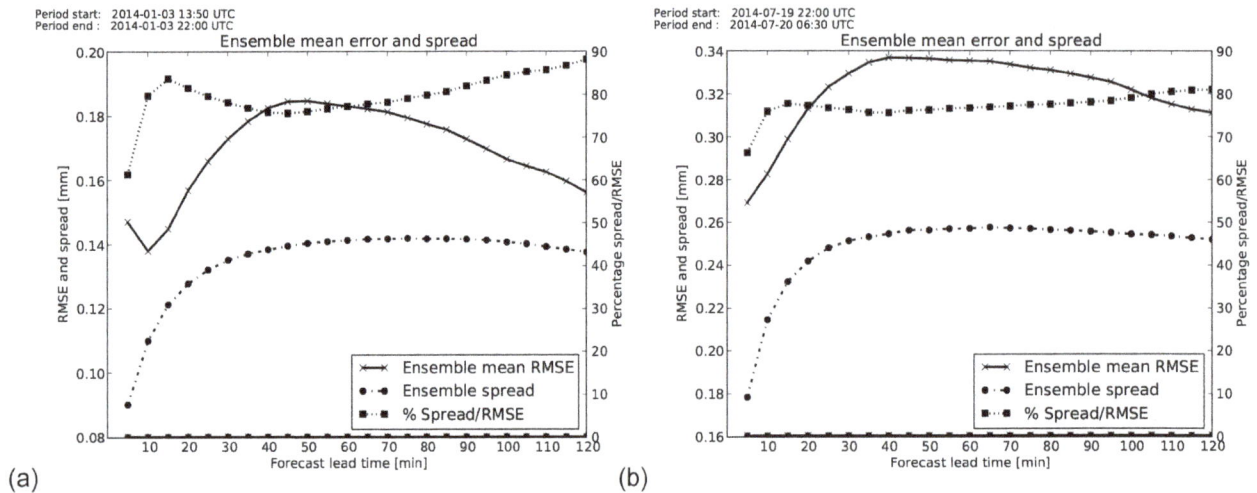

(a)

(b)

Figure 9. Comparison of ensemble spread and RMSE of the ensemble mean forecast at 5 min resolution for **(a)** the Ghent case on 3 January 2014 and **(b)** the Leuven case on 19–20 July 2014.

Period start: 2014-07-19 22:00 UTC
Period end : 2014-07-20 06:30 UTC

Period start: 2014-07-19 22:00 UTC
Period end : 2014-07-20 06:30 UTC

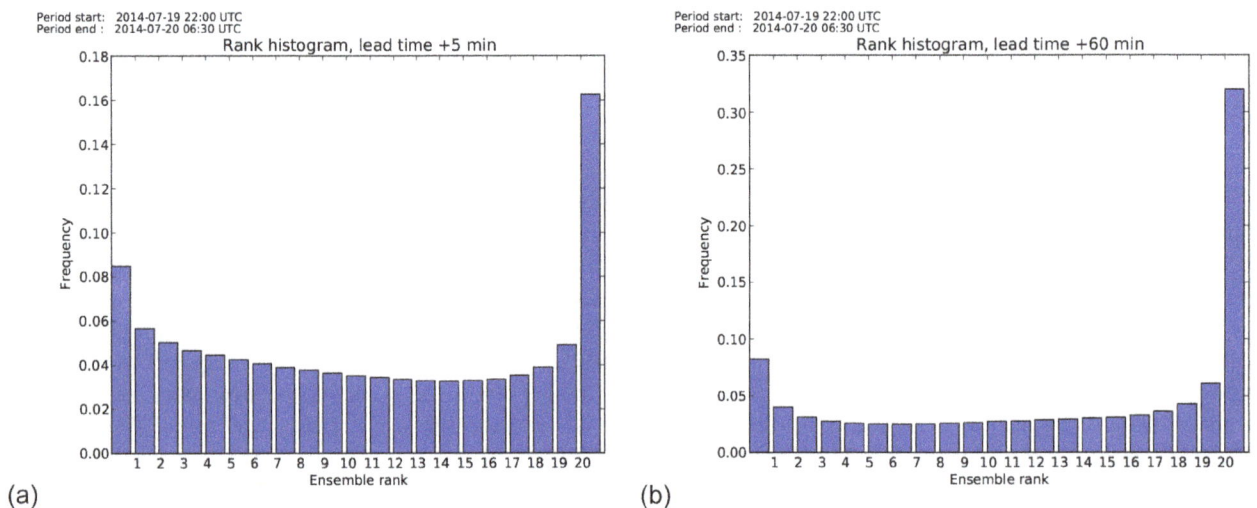

(a)

(b)

Figure 10. Rank histograms for the Leuven case on 19–20 July 2014 for a lead time of **(a)** 5 min and **(b)** 60 min.

50–60 min lead time and then slowly stabilizes. For both the Ghent and Leuven cases the ensemble spread is lower than the error of the ensemble mean at all lead times, which suggests that the ensemble forecasts are under-dispersive. The degree of under-dispersion is highest at a lead time of 5 min, with the spread values being equal to 60 % of the forecast error for the winter event in Ghent (Fig. 9a) and 70 % for the summer event in Leuven (Fig. 9b). Except for the 5 min lead time, the ensemble spread represents 75–90 % of the forecast error for the Ghent case (Fig. 9a) and 75–80 % for the Leuven case (Fig. 9b), which is a good result. It is not yet clear why the RMSE at a lead time of 5 min is higher than the one at 10 min for the winter case in Ghent (Fig. 9a).

The underestimation of the ensemble dispersion at the first lead time can be attributed to both the underestimation of the ensemble spread and the overestimation of the ensem-

ble mean RMSE, but with different degrees according to the different causes. High RMSEs at the start of the now-cast can be due to using a very smooth velocity field for the advection (see Sect. 3.2), which does not exploit sufficiently the very short-term predictability of small-scale precipitation features, but is optimized for predictions at longer lead times. Another explanation for this underestimation of ensemble dispersion could be due to the space–time variability of the Z–R relationship. Spatial and temporal changes in the drop size distribution (DSD) can lead to changes in the estimated rainfall rate that is used for the verification. Therefore, there could be a mismatch between the "fixed" DSD of the forecasts and the variable DSD underlying the verifying observations. Another possible source of mismatch could be due to the advection correction with optical flow when computing the rainfall accumulations. The forecast accumulations are computed by advecting the previous rainfall field forward. On the other hand, the observed accumulations are computed by reversing the optical flow vectors and advecting the rainfall field backwards (see Sect. 4.1). This choice increases the differences when comparing the +0–5 min forecast accumulations (advection of the 0 min image forward) with the +0–5 min observed accumulations a posteriori (advection of the +5 min image backwards). The ideal approach would be to derive the accumulation by advecting both the previous image forward and the last image backwards. An optimal accumulation could be computed by a weighted average of the two advected images by discretizing the 5 min interval. However, such an approach is not very pragmatic and would require additional computational time in order to obtain a marginal improvement in the forecasts.

Figure 10 illustrates the rank histograms for the Leuven case on 19–20 July 2014 for lead times of 5 and 60 min. The U-shape of the rank histograms demonstrates again a certain degree of ensemble under-dispersion. In particular, all the ensemble members for the 5 min lead time are inferior to the observations in $\sim$ 16 % of the cases (Fig. 10a), while for the 60 min lead time it happens in more than $\sim$ 30 % of the cases (Fig. 10b). On the other hand, the fraction of observations falling below the value of the lowest ensemble member is only 8 % for both lead times. Despite the fact that STEPS is designed to reproduce the space–time variability of rainfall, it underestimates a certain fraction of the observed rainfall extremes. This underestimation grows with increasing lead time and depicts an increasing smoothness of the STEPS ensembles, which is probably due to the advection of the radar rainfall cascade (see Sect. 3, step 6). In fact, the small-scale rainfall features represented by the bottom cascade levels suffer more from numerical diffusion during the Lagrangian extrapolation, which is observed as a gradual loss of variability in the forecast ensembles.

4.5 Verification summary of the events

Table 2 provides a comparison of the verification scores for each event. The average standard deviation of the multiplicative biases of the 30 min lead time forecast is in the range 0.3–0.8. Except for the event on 19–20 July 2014 the biases remain well below 1 dB for all lead times, which is a positive result. Of course, these are average values, and locally they can even exceed 3 dB (see Fig. 4).

On the other hand, the RMSE values mark more the distinction between the two winter cases in Ghent and the two summer cases in Leuven. For the winter cases the RMSE values increase from 0.38–0.95 at a lead time of 30 min to 0.78–1.48 at 120 min, while for the summer cases from 1.84–2.45 to 2.52–3.38 mm h^{-1}. Thus, the RMSE of a 30 min lead time nowcast of the two convective cases is higher than the RMSE of a 120 min nowcast of the two stratiform cases, as might be expected. It is interesting to mention that linear verification scores such as the RMSE strongly depend on the variance

of the data. Consequently, it would be difficult to compare the error of the STEPS ensemble mean nowcast with the one of a deterministic nowcast, for example computed by INCA-BE. In fact, the ensemble averaging process filters out the unpredictable precipitation features and is rewarded in terms of RMSE. Similar results were already observed in Foresti et al. (2015), who also pointed out the difficulty of comparing ensemble prediction systems having a different number of ensemble members.

The probability of detection relative to the 0.5 mm h^{-1} threshold decreases from 78–86 to 33–58 %, while the false alarm ratio increases from 10–17 to 46–65 %. The Gilbert skill score starts with values of 0.58–0.64 and 0.29–0.40 at the 30 and 60 min lead times, respectively, and decays to values of 0.08–0.20 at 120 min. Wang et al. (2009) reported a critical success index value of 0.45 for STEPS nowcasts of 0–60 accumulations relative to the 1 mm h^{-1} threshold. Considering that the GSS is the CSI corrected by random chance, this value is comparable with the ones of the 30–60 min accumulations obtained in this paper. The GSS values relative to the threshold of 5.0 mm h^{-1} are much lower. They oscillate between 0.15 and 0.44 for the first lead time and become very low and close to 0 afterwards. Thus, the predictability of rainfall structures exceeding 5.0 mm h^{-1} rarely exceeds 30 min according to the GSS.

The areas under the ROC curve values characterizing the potential discrimination power of the probabilistic forecast of exceeding 0.5 mm h^{-1} start at 0.92–0.95 at 30 min lead time and decrease to 0.69–0.79 at 120 min. For the probabilistic forecast of exceeding 5.0 mm h^{-1} they start at 0.88–0.90 and decrease to 0.62 for the convective cases and to 0.50 for the stratiform cases (no discrimination).

From all these results we can conclude that there is not much predictability beyond 2 h lead time by extrapolating the 4 C-band composite radar images in Belgium. Therefore, a maximum lead time of 2 h in STEPS-BE is a good choice. Extending this lead time requires blending the radar-extrapolation nowcast with the output of NWP models to increase the predictability of precipitation.

5 Conclusions

The Short-Term Ensemble Prediction System (STEPS) is a probabilistic nowcasting system based on the extrapolation of radar images developed at the Australian Bureau of Meteorology in collaboration with the UK MetOffice. The principle behind STEPS is to produce an ensemble forecast by perturbing a deterministic extrapolation nowcast with stochastic noise. The perturbations are designed to reproduce the spatial and temporal correlations of the forecast errors and the scale dependence of the predictability of precipitation.

This paper presented the local implementation, adaptation and verification of STEPS at the Royal Meteorological Institute of Belgium, referred to as STEPS-BE. STEPS-BE pro-

Table 2. Summary of the forecast verification scores of the next four 30 min accumulation forecasts for the precipitation events in Ghent and Leuven. The lead time shown is the end of the 30 min accumulation period (e.g., 60 min is relative to the 30–60 min accumulation). The bias values correspond to the standard deviation of the multiplicative bias, which is more interesting than its mean (often close to 0).

Event	Bias 30 min	Bias 60 min	Bias 90 min	Bias 120 min	RMSE 30 min	RMSE 60 min	RMSE 90 min	RMSE 120 min
	(dB)				(mm h^{-1})			
10/11/2013	0.30	0.49	0.61	0.70	0.38	0.59	0.71	0.78
03/01/2014	0.54	0.74	0.82	0.89	0.95	1.39	1.53	1.48
9–10/06/2014	0.52	0.63	0.66	0.69	2.45	3.26	3.40	3.38
19–20/07/2014	0.84	1.18	1.30	1.35	1.84	2.36	2.49	2.52

Event	POD 30 min	POD 60 min	POD 90 min	POD 120 min	FAR 30 min	FAR 60 min	FAR 90 min	FAR 120 min
	Forecast $>=0.5$ mm h^{-1}				Forecast $>=0.5$ mm h^{-1}			
10/11/2013	0.83	0.71	0.62	0.54	0.17	0.30	0.38	0.46
03/01/2014	0.80	0.63	0.49	0.33	0.10	0.25	0.45	0.65
9–10/06/2014	0.78	0.65	0.55	0.46	0.15	0.32	0.44	0.54
19–20/07/2014	0.86	0.75	0.66	0.58	0.17	0.36	0.50	0.61

Event	GSS 30 min	GSS 60 min	GSS 90 min	GSS 120 min	GSS 30 min	GSS 60 min	GSS 90 min	GSS 120 min
	Forecast $>=0.5$ mm h^{-1}				Forecast $>=5.0$ mm h^{-1}			
10/11/2013	0.58	0.38	0.27	0.20	0.15	0.02	0.0	0.0
03/01/2014	0.64	0.40	0.20	0.08	0.28	0.06	0.0	0.0
9–10/06/2014	0.59	0.38	0.26	0.17	0.44	0.20	0.09	0.04
19–20/07/2014	0.58	0.29	0.14	0.07	0.27	0.09	0.04	0.02

Event	AUC 30 min	AUC 60 min	AUC 90 min	AUC 120 min	AUC 30 min	AUC 60 min	AUC 90min	AUC 120 min
	Forecast $>=0.5$ mm h^{-1}				Forecast $>=5.0$ mm h^{-1}			
10/11/2013	0.95	0.89	0.84	0.79	0.88	0.67	0.56	0.50
03/01/2014	0.92	0.85	0.78	0.69	0.90	0.72	0.57	0.50
9–10/06/2014	0.93	0.86	0.81	0.76	0.89	0.77	0.68	0.62
19–20/07/2014	0.94	0.87	0.82	0.77	0.88	0.75	0.68	0.62

duces in real-time 20-member ensemble nowcasts at 1 km and 5 min resolutions up to 2 h lead time using the four C-band radar composite of Belgium. Compared with the original implementation, STEPS-BE includes a kernel-based interpolation of optical flow vectors to obtain smoother velocity fields and an improvement to generate stochastic noise only within the advected radar composite to respect the validity domain of the nowcasts.

The performance of STEPS-BE was verified using the radar observations as reference on four case studies that caused sewer system floods in the cities of Ghent and Leuven during the years 2013 and 2014. The ensemble mean forecast of the next four 30 min accumulations was verified using the multiplicative bias, the RMSE as well as some categorical scores derived from the contingency table: the probability of detection, false alarm ratio and Gilbert skill score (equitable skill score). The spatial distribution of multiplicative biases revealed regions of systematic over- and under-estimation by STEPS. The underestimations are often associated with the locations of convective initiation and thunderstorm growth, which cannot be predicted by STEPS. On the other hand, the regions of overestimation are mostly due to the underestimation of rainfall by the verifying observations rather than systematic rainfall decay (see Foresti and Seed, 2015, for a more detailed discussion). In order to disentangle the forecast and observation biases, detailed knowledge about the spatial distribution of the radar measurement errors for a given weather situation is needed. The multiplicative biases over the cities of Leuven and Ghent are very low (from −0.5 to +0.5 dB), which is a good starting point to integrate STEPS nowcasts as inputs into sewer system hydraulic models. The categorical forecast verification helped discovering the places with low probability of detection due to convective initiation at the front of the rain band and high false alarm ratio at the rear of the rain band, likely due to a too slow rainfall extrapolation by STEPS. Reliability diagrams demonstrated that probabilistic forecasts of exceeding $0.5\,mm\,h^{-1}$ have skill up to 60–90 min lead time. On the contrary, convective features exceeding $5.0\,mm\,h^{-1}$ are only predictable up to 30 min. In terms of reliability and discrimination ability, it was also observed that the forecasts of convective events have more skill than the ones of stratiform events. The STEPS ensembles are characterized by a certain degree of underestimation of the forecast uncertainty, with values of the ensemble spread close to 75–90 % of the forecast error.

The current contribution focused on the verification of STEPS-BE nowcasts using only four precipitation cases of different character. The deterministic and categorical verifications require many more cases to analyze the climatological distribution of the forecast errors, e.g., as done in Foresti and Seed (2015). On the other hand, the probabilistic and ensemble verification pools the data in both space and time and converges much faster to stable statistics.

From a research perspective, STEPS-BE could also be extended by including a stochastic model to account for the residual radar measurement errors, in particular to obtain more accurate estimations of the forecast uncertainty at short range. The STEPS framework also allows blending of the extrapolation nowcast with the output of NWP models, which is a necessary step to increase the predictability of precipitation for lead times beyond 2 h.

Table A1. Correspondence between the decibel scale and the power ratio.

dB	−6	−3	−1	−0.5	0	+0.5	+1	+3	+6
Power ratio (F/O)	0.251	0.501	0.794	0.891	1	1.122	1.259	1.995	3.981

Appendix A: Forecast verification scores

Forecast verification is an important aspect of a forecasting system. A forecast without an estimation of its accuracy is not very informative. For an in-depth description of forecast verification science and corresponding scores we refer to Jolliffe and Stephenson (2011) and the verification website maintained at the Bureau of Meteorology (http://www.cawcr.gov.au/projects/verification/).

The STEPS *ensemble mean forecast* was verified using the following scores.

- Multiplicative bias:

$$\text{bias} = \frac{1}{N} \sum_{i=1}^{N} 10\log_{10}\left(\frac{F_i + b}{O_i + b}\right), \tag{A1}$$

where F_i is the forecast rainfall at a given grid point, O_i is the observed rainfall at a given grid point, $b = 2\,\text{mm h}^{-1}$ is an offset to eliminate the division by zero and to reduce the contribution of the forecast errors at low rainfall intensities, and N is the number of samples. For the specific case of the verification of the spatial distribution of forecast biases, the summation is performed over time. Thus, N corresponds to the number of forecasts where either the forecast or the observed rainfall are greater than 0.1 mm h^{-1} at a given grid point (denoted as weak conditional verification). The bias is given in decibels (dB) in order to obtain a more symmetric distribution of the multiplicative errors centered at 0, which is not possible with the simple power ratio F/O. Table A1 summarizes the correspondence between the decibel scale and the power ratio. For example, a bias of +3 dB occurs when the forecast rainfall F is twice as much as the observed rainfall O.

- Root mean square error:

$$\text{RMSE} = \sqrt{\frac{1}{N} \sum_{i=1}^{N} (F_i - O_i)^2} \tag{A2}$$

- Contingency table of a dichotomous (yes/no) forecast, see Table A2, where the "hits" is the number of times that both the observation and the forecast exceed a given rainfall threshold (at a given grid point), the "false alarms" is the number of times that the forecast exceeds the threshold but the observation does not, the "misses" is the number of times that the forecast does not exceed

Table A2. Contingency table of a categorical forecast.

		Observed		
		Yes	No	Total
Forecast	Yes	Hits	False alarms	Forecast yes
	No	Misses	Correct negatives	Forecast no
	Total	Observed yes	Observed no	Total

the threshold but the observation does and the "correct negatives" is the number of times that both the observation and the forecast do not exceed the threshold.

- Different scores can be derived from the contingency table to characterize a particular feature or skill of the forecasting system:

  - Probability of detection (hit rate):

  $$\text{POD} = \frac{\text{hits}}{\text{hits} + \text{misses}} = \frac{\text{hits}}{\text{observed yes}}, \tag{A3}$$

  The "POD" characterizes the fraction of observed events that were correctly forecast and is also known as the hit rate (HR).

  - False alarm ratio:

  $$\text{FAR} = \frac{\text{false alarms}}{\text{hits} + \text{false alarms}} = \frac{\text{false alarms}}{\text{forecast yes}}, \tag{A4}$$

  The "FAR" characterizes the fraction of forecast events that were wrongly forecast.

  - False alarm rate:

  $$F = \frac{\text{false alarms}}{\text{false alarms} + \text{correct negatives}}$$
  $$= \frac{\text{false alarms}}{\text{observed no}}. \tag{A5}$$

  The false alarm rate F is conditioned on the observations, while the false alarm ratio FAR on the forecasts.

  - Gilbert skill score (equitable threat score):

  $$\text{GSS} = \tag{A6}$$
  $$\frac{\text{hits} - \text{hits}_{\text{random}}}{\text{hits} + \text{misses} + \text{false alarms} - \text{hits}_{\text{random}}},$$

  where

  $$\text{hits}_{\text{random}} = \frac{(\text{hits} + \text{misses})(\text{hits} + \text{false alarms})}{\text{total}}$$
  $$= \frac{(\text{observed yes})(\text{forecast yes})}{\text{total}} \tag{A7}$$

  is the number of hits obtained by random chance, which is calculated by multiplying the marginal sums of the observed and forecast events (such as computing the theoretical frequencies for the Chisquared test). The GSS characterizes the detection skill of the forecasting system with respect to random chance. In practice it corresponds to the critical success index (CSI) adjusted for the hits obtained by random chance.

The accuracy of probabilistic forecasts can be verified in various ways. In this paper we employ the reliability diagram and the Receiver Operating Characteristic (ROC) curve. The reliability diagram compares the forecast probability with the observed frequency. Reliability characterizes the agreement between the forecast probability and observed frequency. For a reliable forecasting system the two values should be the same, which happens for example when we observe rain 80 % of the time when it is forecast with 80 % probability (in average, diagonal line of Fig. 8). Unreliable forecasts exhibit departures from this optimum (bias). Resolution characterizes the ability of the forecasts to categorize the observed frequencies into distinct classes. The complete lack of resolution occurs when the forecast probabilities are completely unable to distinguish the observed frequencies, which generally corresponds to the climatological frequency of exceeding a given precipitation threshold (horizontal dashed line in Fig. 8). The Brier skill score (BSS) characterizes the relative accuracy of the probabilistic forecast compared to a reference system (see Jolliffe and Stephenson, 2011). Although the climatology or sample climatology of the event is often used as a reference, the BSS can also be computed against other reference forecasts, e.g., another probabilistic forecasting method or even a deterministic forecasting method treated as a probabilistic binary forecast. However, in such cases it is not possible to draw a unique horizontal line representing complete lack of skill in Fig. 8. The region where the probabilistic forecast has a positive BSS, i.e., it is better than the climatological frequency, is grayed out. In fact, the points located below the no skill line are closer to the climatological frequency and produce a negative BSS. Reliability diagrams usually contain the histogram of the forecast probabilities to analyze the sharpness of the forecasts (small inset in Fig. 8). Sharpness characterizes the ability to forecast probabilities that are different from the reference forecast. Sharp forecasting systems are "confident" about their predictions and give many probabilities around one and zero.

The ROC curve is used to analyze the discrimination power of a probabilistic forecast of exceeding a given rainfall threshold. It is constructed by plotting the hit rates and false alarm rates evaluated at increasing probability thresholds to make the binary decision whether it will rain or not. The ROC curve of a random probabilistic forecast system lies on the diagonal where the hit rate equals the false alarm rate (no skill): the forecast probabilities do not have discrimination power. When the false alarm rate is higher than the hit rate the forecast is worse than that obtained by random

chance (below the diagonal). A skilled forecasting system is observed when the hit rates are higher than the false alarm rates, which draws a characteristic curve. The area under the ROC curve (AUC) measures the discrimination power of the probabilistic forecasts, with a maximum value of 1 (100 % of hits and 0 % of false alarms) and a minimum value of 0.5 for a random forecasting system. Values below 0.5 denote a forecasting system that performs worse than random chance. The AUC is computed by integrating over all the trapezoids that can be drawn below the ROC curve. The AUC is not sensitive to the forecast bias and the reliability of the forecast could still be improved through calibration. For this reason the AUC is only a measure of potential skill.

The ensemble forecasts are verified to detect whether there is over- or under-dispersion. It is common practice to compare the "skill" (error) of the ensemble mean with the ensemble spread (Whitaker and Loughe, 1998; Foresti et al., 2015):

$$\text{spread} = \frac{1}{N} \sum_{i=1}^{N} \sqrt{\frac{1}{M-1} \sum_{m=1}^{M} \left(F_{im} - \overline{F_i} \right)^2}, \qquad \text{(A8)}$$

where M is the number of ensemble members (ensemble size), F_{im} is the forecast of a given ensemble member and $\overline{F_i}$ is the ensemble mean forecast (at a given grid point). Since we are not analyzing the spatial or temporal distribution of the ensemble spread, N corresponds to the total number of samples in space and time, which is the number of forecasts within a rainfall event multiplied by the number of grid points within a radar field. The weak conditional verification is also applied to the computation of the spread. The ensemble spread characterizes the variability of the ensemble members about the ensemble mean (standard deviation). For a good ensemble prediction system, the ensemble spread should be equal to the average variability of the observations about the ensemble mean, as measured by the RMSE of the ensemble mean (Eq. 2). If the spread is larger than the RMSE, the ensemble is overestimating the forecast uncertainty (over-dispersion), otherwise it is underestimating it (under-dispersion). It is interesting to mention that the ensemble mean RMSE and ensemble spread could also be computed starting from the logarithm of rainfall rates to account for the skewed distribution of precipitation (not used in this paper).

Another way to analyze the spread of ensemble forecasts is based on rank histograms (also known as a Talagrand diagram). First, the precipitation values of the ensemble members are ranked in increasing order. Then, the rank of the observation is evaluated by checking in which of the $M + 1$ bins it falls. By repeating the operation for a large number of cases and forecasts it is possible to construct a histogram. A good ensemble prediction system displays a flat histogram; i.e., the observations are indistinguishable from the forecasts and each ensemble member is an equi-probable realization of the future state of the atmosphere. A bell-shaped histogram with a peak in the middle is observed in the case of ensemble over-dispersion. On the contrary, a U-shape histogram with peaks at the edges is observed in the case of ensemble under-dispersion, which is more common (in particular for NWP ensembles). In this case the values of the observations often fall below or above the lowest or highest value of the ranked ensemble, which is not dispersive enough to capture the extremes.

Acknowledgements. This research was funded by Belgian Science Policy Office (BelSPO) project PLURISK: Forecasting and management of rainfall-induced risks in the urban environment (SD/RI/01A). We thank Clive Pierce for the detailed discussions about the STEPS implementation and the guidance for various code improvements. We also acknowledge Meteo-France for providing the Avesnois data.

References

Achleitner, S., Fach, S., Einfalt, T., and Rauch, W.: Nowcasting of rainfall and of combined sewage flow in urban drainage systems, Water Sci. Technol., 59, 1145–51, 2009.

Atencia, A. and Zawadzki, I.: A comparison of two techniques for generating nowcasting ensembles – Part I: Lagrangian ensemble technique, Mon. Weather Rev., 142, 4036–4052, 2014.

Berenguer, M., Corral, C., Sánchez-Diezma, R., and Sempere-Torres, D.: Hydrological validation of a radar-based nowcasting technique, J. Hydrometeorol., 6, 532–549, 2005.

Berenguer, M., Sempere-Torres, D., and Pegram, G. G. S.: SBMcast – an ensemble nowcasting technique to assess the uncertainty in rainfall forecasts by Lagrangian extrapolation, J. Hydrol., 404, 226–240, 2011.

Berne, A., Delrieu, G., Creutin, J.-D., and Obled, C.: Temporal and spatial resolution of rainfall measurements required for urban hydrology, J. Hydrol., 299, 166–179, 2004.

Bowler, N. E. H., Pierce, C. E., and Seed, A. W.: Development of a precipitation nowcasting algorithm based upon optical flow techniques, J. Hydrol., 288, 74–91, 2004a.

Bowler, N. E. H., Pierce, C. E., and Seed, A. W.: STEPS: a probabilistic precipitation forecasting scheme which merges an extrapolation nowcast with downscaled NWP, Forecast Research Technical Report No. 433, MetOffice, Exeter, UK, 2004b.

Bowler, N. E. H., Pierce, C. E., and Seed, A. W.: STEPS: A probabilistic precipitation forecasting scheme which merges an extrapolation nowcast with downscaled NWP, Q. J. Roy. Meteor. Soc., 132, 2127–2155, 2006.

Bruni, G., Reinoso, R., van de Giesen, N. C., Clemens, F. H. L. R., and ten Veldhuis, J. A. E.: On the sensitivity of urban hydrodynamic modelling to rainfall spatial and temporal resolution, Hydrol. Earth Syst. Sci., 19, 691–709, doi:10.5194/hess-19-691-2015, 2015.

Cloke, H. L. and Pappenberger, F.: Ensemble flood forecasting: a review, J. Hydrol., 375, 613–626, 2009.

Collier, C. G.: On the propagation of uncertainty in weather radar estimates of rainfall through hydrological models, Meteorol. Appl., 16, 35–40, 2009.

Dai, Q., Rico-Ramirez, M. A., Han, D., Islam, T., and Liguori, S.: Probabilistic radar rainfall nowcasts using empirical and theoretical uncertainty models, Hydrol. Process., 29, 66–79, 2015.

Ebert, E. E.: Ability of a poor man's ensemble to predict the probability and distribution of precipitation, Mon. Weather Rev., 129, 2461–2480, 2001.

Ehret, U., Götzinger, J., Bárdossy, A., and Pegram, G. G. S.: Radar-based flood forecasting in small catchments, exemplified by the Goldersbach catchment, Germany, International Journal of River Basin Management, 6, 323–329, 2008.

Einfalt, T., Arnbjerg-Nielsen, K., Golz, C., Jensen, N. E., Quirmbachd, M., Vaes, G., and Vieux, B.: Towards a roadmap for use of radar rainfall data in urban drainage, J. Hydrol., 299, 186–202, 2004.

Figueras i Ventura, J. and Tabary, P.: The new French operational polarimetric radar rainfall rate product, J. Appl. Meteorol. Clim., 52, 1817–1835, 2013.

Foresti, L. and Seed, A.: The effect of flow and orography on the spatial distribution of the very short-term predictability of rainfall from composite radar images, Hydrol. Earth Syst. Sci., 18, 4671–4686, doi:10.5194/hess-18-4671-2014, 2014.

Foresti, L. and Seed, A.: On the spatial distribution of rainfall nowcasting errors due to orographic forcing, Meteorol. Appl., 22, 60–74, 2015.

Foresti, L., Seed, A., and Zawadzki, I.: Report of the Heuristic Probabilistic Forecasting Workshop, Munich, Germany, 13 pp., 30–31 August 2014, available at: https://sites.google.com/site/lorisforesti/projects/nowcasting/ScientificReport_HeuristicProbForecastingWorkshop_Munich_2014_121214.pdf, last access: 3 July 2015, 2014.

Foresti, L., Panziera, L., Mandapaka, P. V., Germann, U., and Seed, A.: Retrieval of analogue radar images for ensemble nowcasting of orographic rainfall, Meteorol. Appl., 22, 141–155, 2015.

Germann, U. and Zawadzki, I.: Scale-dependence of the predictability of precipitation from continental radar images – Part I: Methodology, Mon. Weather Rev., 130, 2859–2873, 2002.

Germann, U. and Zawadzki, I.: Scale-dependence of the predictability of precipitation from continental radar images – Part II: Probability forecasts, J. Appl. Meteorol., 43, 74–89, 2004.

Germann, U., Berenguer, M., Sempere-Torres, D., and Zappa, M.: REAL – ensemble radar precipitation estimation for hydrology in a mountainous region, Q. J. Roy. Meteor. Soc., 135, 445–456, 2009.

Goudenhoofdt, E. and Delobbe, L.: Evaluation of radar-gauge merging methods for quantitative precipitation estimates, Hydrol. Earth Syst. Sci., 13, 195–203, doi:10.5194/hess-13-195-2009, 2009.

Grasso, L. D.: The differentiation between grid spacing and resolution and their application to numerical modeling, B. Am. Meteorol. Soc., 81, 579–580, 2000.

Haiden, T., Kann, A., Wittmann, C., Pistotnik, G., Bica, B., and Gruber, C.: The Integrated Nowcasting through Comprehensive Analysis (INCA) system and its validation over the eastern Alpine region, Weather Forecast., 26, 166–183, 2011.

Hohti, H., Koistinen, J., Nurmi, P., Saltikoff, E., and Holmlund, K.: Precipitation nowcasting using radar-derived atmospheric motion vectors, in: Proc. of the 1st European Conf. on Radar in Meteorology and Hydrology (ERAD), Bologna, Italy, 4–8 September 2000, Physics and Chemistry of the Earth, Part B: Hydrology, Oceans and Atmosphere, 25, 1323–1327, 2000.

Jolliffe, I. T. and Stephenson, D. B.: Forecast Verification: a Practitioner's Guide in Atmospheric Science, 2nd edn., John Wiley and Sons, Chichester, 2011.

Jordan, P., Seed, A. W., and Weinnman, P. E.: A stochastic model of radar measurement errors in rainfall accumulations at catchment scale, J. Hydrometeorol., 4, 841–855, 2003.

Lewis, H., Mittermaier, M., Mylne, K., Norman, K., Scaife, A., Neal, R., Pierce, C., Harrison, D., Jewell, S., Kendon, M., Saunders, R., Brunet, G., Golding, B., Kitchen, M., Davies, P., and Pilling, C.: From months to minutes – exploring the value of high-resolution rainfall observation and prediction during the UK winter storms of 2013/2014, Meteorol. Appl., 22, 90–104, 2015.

Liguori, S. and Rico-Ramirez, M. A.: Quantitative assessment of short-term rainfall forecasts from radar nowcasts and MM5 forecasts, Hydrol. Process., 26, 3842–3857, 2012.

Liguori, S. and Rico-Ramirez, M. A.: A practical approach to the assessment of probabilistic flow predictions, Hydrol. Process., 27, 18–32, 2013.

Liguori, S., Rico-Ramirez, M. A., Schellart, A., and Saul, A.: Using probabilistic radar rainfall nowcasts and NWP forecasts for flow prediction in urban catchments, Atmos. Res., 103, 80–95, 2012.

Metta, S., Rebora, N., Ferraris, L., von Hardernberg, J., and Provenzale, A.: PHAST: a phase-diffusion model for stochastic nowcasting, J. Hydrometeorol., 10, 1285–1297, 2009.

Panziera, L., Germann, U., Gabella, M., and Mandapaka, P. V.: NORA – nowcasting of orographic rainfall by means of analogues, Q. J. Roy. Meteor. Soc., 137, 2106–2123, 2011.

Pappenberger, F. and Beven, K. J.: Ignorance is bliss: or seven reasons not to use uncertainty analysis, Water Resour. Res., 42, W05302, doi:10.1029/2005WR004820, 2006.

Paschalis, A., Molnar, P., Fatichi, S., and Burlando, P.: A stochastic model for high-resolution space–time precipitation simulation, Water Resour. Res., 49, 8400–8417, 2013.

Pegram, G. G. S. and Clothier, A. N.: High resolution space–time modelling of rainfall: the "String of Beads" model, J. Hydrol., 241, 26–41, 2001a.

Pegram, G. G. S. and Clothier, A. N.: Downscaling rainfields in space and time, using the String of Beads model in time series mode, Hydrol. Earth Syst. Sci., 5, 175–186, doi:10.5194/hess-5-175-2001, 2001b.

Pierce, C., Bowler, N., Seed, A., Jones, A., Jones, D., and Moore, R.: Use of a stochastic precipitation nowcast scheme for fluvial flood forecasting and warning, Atmos. Sci. Lett., 6, 78–83, 2005.

Pierce, C., Hirsch, T., and Bennett, A. C.: Formulation and evaluation of a post-processing algorithm for generating seamless, high resolution ensemble precipitation forecasts, Forecasting R&D Technical Report 550, MetOffice, Exeter, UK, 2010.

Radhakrishna, B., Zawadzki, I., and Fabry, F.: Predictability of precipitation from continental radar images. Part V: growth and decay, J. Atmos. Sci., 69, 3336–3349, 2012.

Roulin, E. and Vannitsem, S.: Skill of medium-range hydrological ensemble predictions, J. Hydrometeorol., 6, 729–744, 2005.

Schellekens, J., Weerts, A. H., Moore, R. J., Pierce, C. E., and Hildon, S.: The use of MOGREPS ensemble rainfall forecasts in operational flood forecasting systems across England and Wales, Adv. Geosci., 29, 77–84, doi:10.5194/adgeo-29-77-2011, 2011.

Schertzer, D. and Lovejoy, S.: Physical modelling and analysis of rain and clouds by anisotropic scaling multiplicative processes, J. Geophys. Res., 92, 9696–9714, 1987.

Seed, A.: A dynamic and spatial scaling approach to advection forecasting, J. Appl. Meteorol., 42, 381–388, 2003.

Seed, A. W., Pierce, C. E., and Norman, K.: Formulation and evaluation of a scale decomposition-based stochastic precipitation nowcast scheme, Water Resour. Res., 49, 6624–6641, 2013.

Silvestro, F. and Rebora, N.: Operational verification of a framework for the probabilistic nowcasting of river discharge in small and medium size basins, Nat. Hazards Earth Syst. Sci., 12, 763–776, doi:10.5194/nhess-12-763-2012, 2012.

Silvestro, F., Rebora, N., and Cummings, G.: An attempt to deal with flash floods using a probabilistic hydrological nowcasting chain: a case study, Nat. Hazards Earth Syst. Sci. Discuss., 1, 7497–7515, doi:10.5194/nhessd-1-7497-2013, 2013.

Sun, J., Xue, M., Wilson, J. W., Zawadzki, I., Ballard, S. P., Onvlee-Hooimeyer, J., Joe, P., Barker, D. M., Li, P.-W., Golding, B., Xu, M., and Pinto, J.: Use of NWP for nowcasting convective precipitation: recent progress and challenges, B. Am. Meteor. Soc., 95, 409–426, 2014.

Tabary, P.: The new French operational radar rainfall product – Part I: Methodology, Weather Forecast., 22, 393–408, 2007.

Thielen, J., Bartholmes, J., Ramos, M.-H., and de Roo, A.: The European Flood Alert System – Part 1: Concept and development, Hydrol. Earth Syst. Sci., 13, 125–140, doi:10.5194/hess-13-125-2009, 2009.

Thorndahl, S. and Rasmussen, M. R.: Short-term forecasting of urban storm water runoff in real-time using extrapolated radar rainfall data, J. Hydroinform., 15, 897–912, 2013.

Turner, B. J., Zawadzki, I., and Germann, U.: Predictability of precipitation from continental radar images. Part III: Operational nowcasting implementation (MAPLE), J. Appl. Meteorol., 43, 231–248, 2004.

Venugopal, V., Foufoula-Georgiou, E., and Sapozhnikov, V.: Evidence of dynamic scaling in space–time rainfall, J. Geophys. Res., 104, 31599–31610, 1999.

Verworn, H. R., Rico-Ramirez, M. A., Krämer, S., Cluckie, I., and Reichel, F.: Radar-based flood forecasting for river catchments, Water Manage., 162, 159–168, 2009.

Wang, J. Keenan, T., Joe, P., Wilson, J., Lai, E. S. T., Liang, F., Wang, Y., Ebert, E. E., Ye, Q., Bally, J., Seed, A., Chen, M., Xue, J., and Conway, B.: Overview of the Beijing 2008 Olympics Project. Part I: Forecast Demonstration Project, WMO World Weather Research Programme, Report, 133 pp., 2009.

Whitaker, J. S. and Loughe, A. F.: The relationship between ensemble spread and ensemble mean skill, Mon. Weather Rev., 26, 3292–3302, 1998

Willems, P.: A spatial rainfall generator for small spatial scales, J. Hydrol., 252, 126–144, 2001a.

Willems, P.: Stochastic description of the rainfall input errors in lumped hydrological models, Stoch. Env. Res. Risk A., 15, 132–152, 2001b.

Skamarock, W. C.: Evaluating mesoscale NWP models using kinetic energy spectra, Mon. Weather Rev., 132, 3019–3032, 2004.

Xuan, Y., Cluckie, I. D., and Wang, Y.: Uncertainty analysis of hydrological ensemble forecasts in a distributed model utilising short-range rainfall prediction, Hydrol. Earth Syst. Sci., 13, 293–303, doi:10.5194/hess-13-293-2009, 2009.

Xuan, Y., Zhu, D., Triballi, P., and Cluckie, I.: Forecast uncertainty of a lumped hydrological model coupled with the STEPS radar rainfall nowcasts, in: Int. Symp. Weather Radar and Hydrol., Washington DC, US, 7–10 April 2014, 9 pp., 2014.

Zappa, M., Beven, K., Bruen, M., Cofino, A., Kok, K., Martin, E., Nurmi, P., Orfila, B., Roulin, E., Seed, A., Schroter, K., Szturc, J., Vehvilainen, B., Germann, U., and Rossa, A.: Propagation of uncertainty from observing systems and NWP into hydrological models: COST-731 Working Group 2, Atmos. Sci. Lett., 11, 83–91, 2010.

6

Time series of tritium, stable isotopes and chloride reveal short-term variations in groundwater contribution to a stream

C. Duvert[1], **M. K. Stewart**[2], **D. I. Cendón**[3,4], **and M. Raiber**[5]

[1]Queensland University of Technology, Brisbane, QLD 4001, Australia
[2]Aquifer Dynamics Ltd & GNS Science, P.O. Box 30368, Lower Hutt, 5040, New Zealand
[3]Australian Nuclear Science and Technology Organisation, Kirrawee DC, NSW 2232, Australia
[4]School of Biological, Earth & Environmental Sciences, University of New South Wales, Sydney, NSW 2052, Australia
[5]CSIRO Land & Water, Dutton Park, Brisbane, QLD 4102, Australia

Correspondence to: C. Duvert (clement.duvert@gmail.com)

Abstract. A major limitation to the assessment of catchment transit time (TT) stems from the use of stable isotopes or chloride as hydrological tracers, because these tracers are blind to older contributions. Yet, accurately capturing the TT of the old water fraction is essential, as is the assessment of its temporal variations under non-stationary catchment dynamics. In this study we used lumped convolution models to examine time series of tritium, stable isotopes and chloride in rainfall, streamwater and groundwater of a catchment located in subtropical Australia. Our objectives were to determine the different contributions to streamflow and their variations over time, and to understand the relationship between catchment TT and groundwater residence time. Stable isotopes and chloride provided consistent estimates of TT in the upstream part of the catchment. A young component to streamflow was identified that was partitioned into quickflow (mean TT $\approx$ 2 weeks) and discharge from the fractured igneous rocks forming the headwaters (mean TT $\approx$ 0.3 years). The use of tritium was beneficial for determining an older contribution to streamflow in the downstream area. The best fits between measured and modelled tritium activities were obtained for a mean TT of 16–25 years for this older groundwater component. This was significantly lower than the residence time calculated for groundwater in the alluvial aquifer feeding the stream downstream ($\approx$ 76–102 years), emphasising the fact that water exiting the catchment and water stored in it had distinctive age distributions. When simulations were run separately on each tritium streamwater sample, the TT of old water fraction varied substantially over time, with values

averaging 17 ± 6 years at low flow and 38 ± 15 years after major recharge events. This counterintuitive result was interpreted as the flushing out of deeper, older waters shortly after recharge by the resulting pressure wave propagation. Overall, this study shows the usefulness of collecting tritium data in streamwater to document short-term variations in the older component of the TT distribution. Our results also shed light on the complex relationships between stored water and water in transit, which are highly non-linear and remain poorly understood.

1 Introduction

Catchment transit time (TT) can be defined as the time water spends travelling through a catchment, from infiltrating precipitation until its exit through the stream network (McDonnell et al., 2010). Because this parameter integrates information on storage, flow pathways and source of water in a single value, it has been increasingly used as a generic indicator of catchment dynamics (McGuire and McDonnell, 2006). Accurate quantification of TT is of prime importance for water resource management issues, in particular for the assessment of catchment sensitivity to anthropogenic inputs such as fertilizers or herbicides (e.g. van der Velde et al., 2010; Benettin et al., 2013), and for the provision of additional constraints on catchment-scale hydrological models (e.g. Gusyev et al., 2013). TT is estimated by relating the signature of a tracer measured in a sample taken at the outlet of a catchment to the history of the tracer input in rainfall-derived recharge water. Interpretation of TT data is often problematic because a single sample typically contains water parcels with different recharge histories, different flowpaths to the stream and thus different ages. This is exacerbated when the catchment is underlain by heterogeneous aquifers, as dispersion and mixing of different water sources can lead to very broad spectra of ages (Weissmann et al., 2002). Rather than a single scalar value, samples are therefore characterised by a TT distribution (i.e. probability density function of the TTs contained in the sample). The residence time (RT) distribution is another useful indicator that refers to the distribution of ages of water resident within the system, rather than exiting it. RT distributions are generally used to characterise subsurface water or deeper groundwater that is stored in the catchment.

In the last 2 decades, a great deal of effort has been directed to the determination of catchment TTs in a variety of streams and rivers worldwide (e.g. Maloszewski et al., 1992; Burns et al., 1998; Soulsby et al., 2000; Rodgers et al., 2005; Dunn et al., 2010). Attempts have been made to correlate the TTs to catchment characteristics such as topography (McGuire et al., 2005; Mueller et al., 2013; Seeger and Weiler, 2014), geology (Katsuyama et al., 2010) or soil type (Tetzlaff et al., 2009, 2011; Timbe et al., 2014). Assessment of the relationship between groundwater RT and catchment TT has also been undertaken occasionally (Matsutani et al., 1993; Herrmann et al., 1999; Reddy et al., 2006). Because catchment storage is highly non-stationary, catchment TTs are known to vary over time (McDonnell et al., 2010), yet the importance of temporal dynamics in TT distributions has been overlooked until recently. One of the reasons is that this non-stationarity is not accounted for in the models commonly used in catchment TT research. In the last 5 years, an ever-growing number of studies has transferred its focus to assessing dynamic TT distributions (Hrachowitz et al., 2010, 2013; Roa-García and Weiler, 2010; Rinaldo et al., 2011; Cvetkovic et al., 2012; Heidbüchel et al., 2012, 2013; McMillan et al., 2012; Tetzlaff et al., 2014; Birkel et al., 2015; van der Velde et al., 2015; Benettin et al., 2015; Harman, 2015; Klaus et al., 2015a; Kirchner, 2015). Most of these studies agreed on the importance of considering storage dynamics, because the RT distribution of storage water and the TT distribution of water transiting at the outlet of the catchment are likely to be very different. Concurrently to these recent advances in catchment hydrology, groundwater scientists have also developed new theoretical bases for the incorporation of transient conditions in RT distribution functions (Massoudieh, 2013; Leray et al., 2014). Nonetheless, the determination of time-variant TT and RT distributions requires data-intensive computing, which still largely limits their use in applied studies (Seeger and Weiler, 2014).

A simple, yet still widely used alternative to more sophisticated models is the lumped-parameter modelling approach, which has been developed since the 1960s to interpret age tracer data (Vogel, 1967; Eriksson, 1971; Maloszewski and Zuber, 1982). Lumped models require minimal input information, and are based on the assumptions that the shape of the TT or RT distribution function is a priori known and that the system is at steady state. The relationship between input and output signatures is determined analytically using a convolution integral, i.e. the amount of overlap of the TT or RT distribution function as it is shifted over the input function. Some of the lumped models consider only the mechanical advection of water as driver of tracer transport (e.g. exponential model), while others also account for the effects of dispersion–diffusion processes (e.g. dispersion model). Nonparametric forms of RT distribution functions have recently been developed (Engdahl et al., 2013; Massoudieh et al., 2014b; McCallum et al., 2014), but again, these more recent approaches require a higher amount of input data, which makes the standard lumped-parameter approach a method of choice for the time being.

Commonly used to determine TT distributions using such models are the stable isotopes of water (δ^2H and δ^{18}O). Because they are constituents of the water molecule itself, ^{2}H and ^{18}O follow almost the same response function as the traced material, hence are generally referred to as "ideal" tracers. Another tracer that behaves relatively conservatively and has been often used in the literature is chloride. An important issue with using ^{2}H, ^{18}O and/or chloride as TT indicators is that detailed catchment-specific input functions are needed (ideally at a weekly sampling frequency for several years), and such data are rare globally. More importantly, Stewart et al. (2010, 2012) criticised the use of these tracers to assess catchment TTs, arguing that TT distributions are likely to be truncated when only ^{2}H and/or ^{18}O are used. In an earlier study, Stewart et al. (2007) reported differences of up to an order of magnitude between the TTs determined using stable isotopes as compared to those determined using tritium (^{3}H). Later works by Seeger and Weiler (2014) and Kirchner (2015) reinforced the point that "stable isotopes are effectively blind to the long tails of TT distributions" (Kirchner, 2015). The effects of older groundwater contributions to streamflow have largely been ignored until recently (Smerdon et al., 2012; Frisbee et al., 2013), and according to Stewart et al. (2012), new research efforts need to be focused on relating deeper groundwater flow processes to catchment response. Accounting for potential delayed contributions from deeper groundwater systems therefore requires the addition of a tracer, such as ^{3}H, that is capable of determining longer TTs.

^{3}H is a radioactive isotope of hydrogen with a half-life of 12.32 years. Like ^{2}H and ^{18}O it is part of the water molecule and can therefore be considered an "ideal" tracer. Fractionation effects are small and can be ignored relative to measurement uncertainties and to its radioactive decay (Michel, 2005). The bomb pulse ^{3}H peak that occurred in the 1960s was several orders of magnitude lower in the Southern Hemisphere than in the Northern Hemisphere (Freeze and Cherry, 1979; Clark and Fritz, 1997), and the ^{3}H concentrations of remnant bomb pulse water have now decayed well below that of modern rainfall (Morgenstern and Daughney, 2012). These characteristics allow the detection of relatively older groundwater (up to 200 years) and, importantly, the calculation of unique TT distributions from a single ^{3}H value, provided the measurement is accurate enough (Morgenstern et al., 2010; Stewart et al., 2010). Other age tracers such as chlorofluorocarbons and sulfur hexafluoride have shown potential for estimating groundwater RT (e.g. Cook and Solomon, 1997; Lamontagne et al., 2015); however, these tracers are less suitable for streamwater because of gas exchange with the atmosphere (Plummer et al., 2001).

Long-term evolution of ^{3}H activity within catchments has been reported in a number of studies, both for the determination of RT in groundwater systems (e.g. Zuber et al., 2005; Stewart and Thomas, 2008; Einsiedl et al., 2009; Manning et al., 2012; Blavoux et al., 2013) and for the assessment of TT in surface water studies (Matsutani et al., 1993; Stewart et al., 2007; Morgenstern et al., 2010; Stolp et al., 2010; Stewart, 2012; Gusyev et al., 2013; Kralik et al., 2014). Most of these studies had to assume stationarity of the observed system by deriving a unique estimate of TT or RT from ^{3}H time-series data, in order to circumvent the bomb pulse issue. Benefiting from the much lower ^{3}H atmospheric levels in the Southern Hemisphere, Morgenstern et al. (2010) were the first to use repeated streamwater ^{3}H data to assess the temporal variations in TT distributions. Using simple lumped parameter models calibrated to each ^{3}H sample, they established that catchment TT was highly variable and a function of discharge rate. Following the same approach, Cartwright and Morgenstern (2015) explored the seasonal variability of ^{3}H activities in streamwater and their spatial variations from headwater tributaries to a lowland stream. They showed that different flowpaths were likely to have been activated under varying flow conditions, resulting in a wide range of TTs. To the extent of our knowledge, shorter-term (i.e. less than monthly) variations in streamwater ^{3}H and their potential to document rapid fluctuations in the older groundwater component in streamflow have not been considered in the literature.

This study investigates the different contributions to streamflow in a subtropical headwater catchment subjected to highly seasonal rainfall, as well as their variations over time. The overarching goal is to advance our fundamental understanding of the temporal dynamics in groundwater contributions to streams, through the collection of time series of seasonal tracers, i.e. tracers subject to pronounced seasonal cycles (^{2}H, ^{18}O and chloride), and ^{3}H. We postulate that ^{3}H time-series data may provide insight into the non-linear processes of deeper groundwater contribution to rivers. Specifically, the questions to be addressed are the following.

i. Can simple lumped models provide reliable estimates of catchment TTs in catchments characterised by intermittent recharge and high evapotranspiration rates?

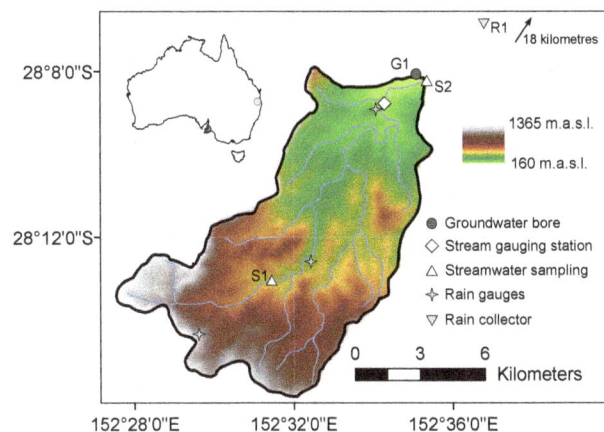

Figure 1. Upper Teviot Brook catchment and location of sampling sites. The stream gauging station corresponds to Teviot Brook at Croftby (145011A; operated by the Queensland Department of Natural Resources and Mines). The rainfall gauges correspond to Wilsons Peak Alert (040876), Carneys Creek The Ranch (040490) and Croftby Alert (040947), all run by the Bureau of Meteorology.

ii. Can short-term variations in older (5–100 years) groundwater contributions be captured by ^{3}H time-series data?

iii. How dissimilar are the RT of aquifers adjacent to streams (i.e. storage water) and the transit time of streamwater (i.e. exiting water)?

2 Study area

2.1 Physical setting

The upper Teviot Brook catchment is located south-west of Brisbane (south-eastern Queensland, Australia), with its headwaters in the Great Dividing Range (Fig. 1). It covers an area of 95 km^2, and elevations range between 160 and 1375 m a.s.l.. Climate in the region is humid subtropical with extremely variable rainfall: mean annual precipitation for the catchment is 970 mm (1994–2014 period), of which 76% falls from November to April. While Teviot Brook is a perennial stream, the distribution of discharge is uneven throughout the year: the mean annual discharge is 120 mm (1994–2014 period), with highest and lowest streamflow occurring in February (average 40 mm) and September (average 2 mm), respectively. The headwaters support undisturbed subtropical rainforest, while the valley supports open woodland and grassland.

The first sampling location (S1) is situated in a steep, narrow valley where the stream erodes into the fractured, silica-rich igneous rocks forming the headwaters. At this upstream location, boulders, gravel and sand constitute the streambed substrate as well as near-channel deposits. The second sampling location (S2) lies further downstream where the valley is flatter and forms a wide alluvial plain. At this downstream location the stream is incised into the alluvial deposits, which at G1 are composed of fine-grained material, i.e. mostly gravel and silty clay. Underlying the alluvial deposits is a sedimentary bedrock formation (Walloon Coal Measures) consisting of irregular beds of sandstone, siltstone, shale and coal, some of which contain significant volumes of groundwater. Duvert et al. (2015b, a) reported high Fe concentrations and low ^{3}H activities for some groundwaters of the sedimentary bedrock.

Hydraulic gradient analysis indicates that the alluvium mostly drains into the stream; hydrochemical and isotopic data also revealed a close connection between the alluvium and surface water in the Teviot Brook catchment (Duvert et al., 2015b). Borehole G1 is 13.9 m deep and it is screened from 12.3 m to its bottom, i.e. entirely within the alluvial stratum. The horizontal distance between G1 and S2 is 60 m.

2.2 Catchment hydrology

The monitoring period spans over 2 years, from mid-2012 to late 2014. Daily streamflow data were obtained from a gauging station operated by the Queensland Department of Natural Resources and Mines (Croftby station; 145011A) and located 2 km upstream of S2 (Fig. 1). Daily precipitation data were available at three rain gauges spread across the catchment and operated by the Australian Bureau of Meteorology. Average precipitation was calculated from the three records using the Thiessen method. Annual precipitation amounted to 1010 mm in 2012, 1190 mm in 2013 and 960 mm in 2014. The rainfall depths recorded in the headwaters were 100 to 250 mm higher than those in the floodplain. The maximum daily rainfall amount was 275 mm and occurred in late January 2013, with a weekly value of 470 mm for this same event (Fig. 2a). This intense episode of rainfall generated a daily peak flow of 137 $m^3 s^{-1}$ upstream of S2 (Fig. 3b), which corresponds to a 22-year return period event at that station – calculated by fitting long-term data to a Galton distribution. Earlier work has shown that this major event contributed significantly to recharge of the alluvial and bedrock aquifers in the headwaters (Duvert et al., 2015a, b). Another high flow event occurred in late March 2014, with a daily peak flow of 39 $m^3 s^{-1}$. Generally, examination of the hydrograph reveals that extended recession periods followed peak flows. Low flow conditions ($Q < 0.01 m^3 s^{-1}$) occurred towards the end of the dry season, i.e. approximately from November through to January (Fig. 2b). The stream did not dry up during the study period although very low flow ($Q < 0.001 m^3 s^{-1}$) occurred for 30 consecutive days in February–March 2014.

3 Methods

3.1 Sample collection and analysis

Bulk samples of precipitation were collected at R1 (Fig. 1) at fortnightly to monthly intervals using a Palmex RS1 rainfall collector, which allows virtually evaporation-free sampling (Gröning et al., 2012). Streamwater and groundwater samples were collected at S1 and S2 (stream sampling locations) and G1 (alluvial aquifer) following the same sampling design as the rainfall samples. Samples at G1 were taken after measuring the water table level and purging a minimum of three casing volumes with a stainless steel submersible pump (Hurricane XL, Proactive). All samples were filtered through 0.45 μm membrane filters, and care was taken to seal the bottles and vials tightly to avoid evaporation.

Stable isotopes and chemical elements were measured for all samples at R1, S1, S2, and G1. ^{3}H activity was determined at S2 for most samples, and at G1 for one sample. Chloride concentrations were measured using ion chromatography (ICS-2100, Dionex), while iron and silicon were measured using inductively coupled plasma optical emission spectrometry (Optima 8300, Perkin Elmer). Total alkalinity was measured by titrating water samples with hydrochloric acid to a pH endpoint of 4.5. Major ions were assessed for accuracy by evaluating the charge balance error, which was < 10 % for all samples and < 5 % for 93 % of the samples. Samples were also analysed for ^{18}O and ^{2}H, using a Los Gatos Research water isotope analyser (TIWA-45EP). All isotopic compositions in this study are expressed relative to the VSMOW standard (δ notation). Between-sample memory effects were minimised by pre-running all samples and subsequently re-measuring them with decreasing isotopic ratios, as recommended in Penna et al. (2012). Replicate analyses indicate that analytical error was ± 1.1‰ for δ^2H and ± 0.3‰ for δ^{18}O. All these analyses were conducted at the Queensland University of Technology (QUT) in Brisbane. In addition, ^{3}H was analysed at the Australian Nuclear Science and Technology Organisation (ANSTO) in Sydney. Samples were distilled and electrolytically enriched 68-fold prior to counting with a liquid scintillation counter for several weeks. The limit of quantification was 0.05 tritium units (TU) for all samples, and uncertainty was ± 0.06 TU. A sample collected in August 2013 was excluded from the data set since it was analysed twice and yielded inconsistent results.

3.2 Tracer-based calculation of transit and residence times

3.2.1 Using stable isotopes and chloride

Mean TTs were determined through adjustment of a TT distribution function to observations of fortnightly input and output signatures (here the term "signature" is meant to encompass either an ionic concentration or an isotopic com-

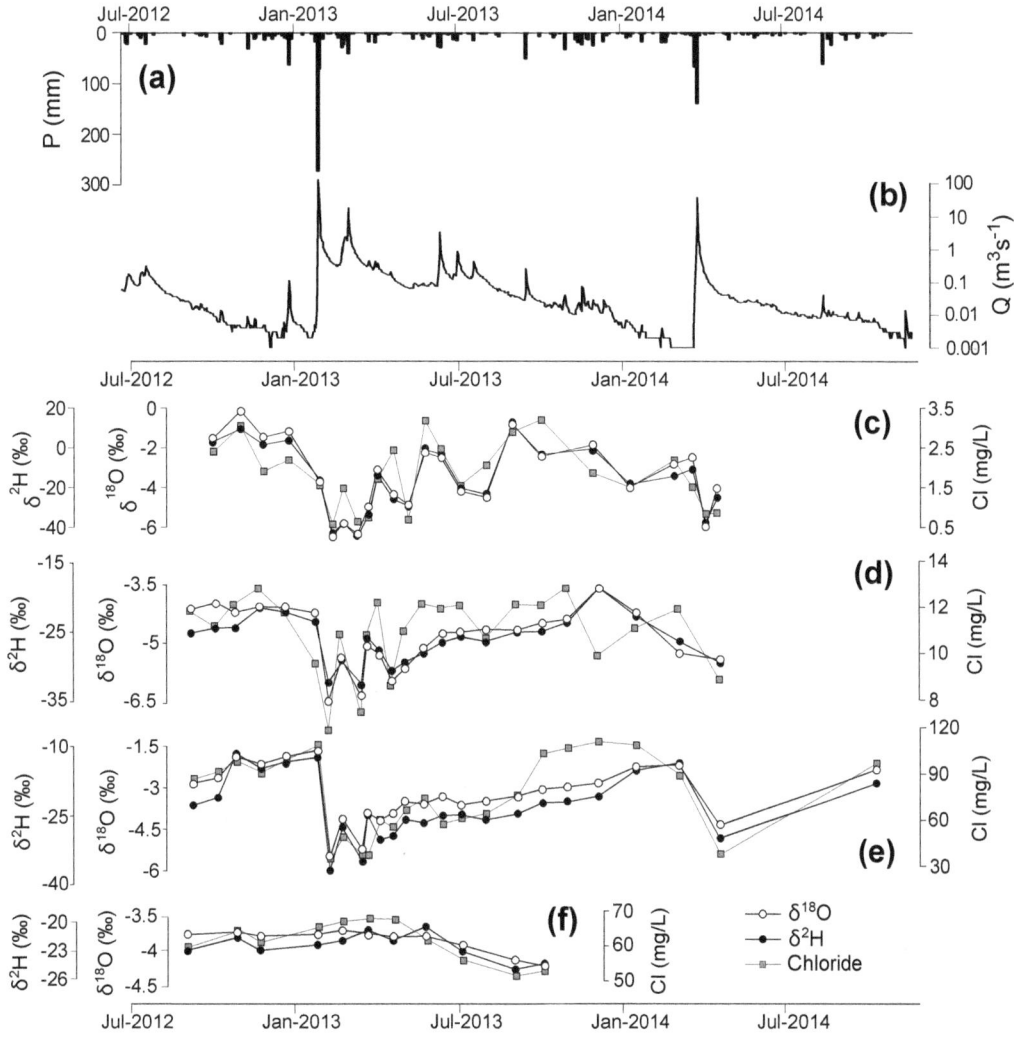

Figure 2. Time series of Thiessen-averaged precipitation (**a**), daily discharge at Croftby (DNRM station 145011A) (**b**), and δ^2H, δ^{18}O and chloride at R1 (rainfall) (**c**), S1 (**d**) and S2 (streamwater) (**e**), and G1 (groundwater) (**f**). Note that the y axes of δ^2H, δ^{18}O and chloride have different scales for each individual plot.

position). An input recharge function was initially computed from the measured input data that accounts for loss due to evapotranspiration (e.g. Bergmann et al., 1986; Stewart and Thomas, 2008):

$$C_r(t) = \frac{R(t)}{\overline{R}}(C_p(t) - \overline{C_r}) + \overline{C_r}, \qquad (1)$$

where $C_r(t)$ is the weighted input recharge signature at time t; $\overline{C_r}$ is the average recharge signature (taken at G1); $C_p(t)$ is the input rainfall signature; $R(t)$ is the fortnightly recharge as calculated by the difference between precipitation and evapotranspiration; and $\overline{R}$ is the average recharge amount.

The weighted input was then convoluted to the selected TT distribution function (g) to obtain output signatures (Mal-

oszewski and Zuber, 1982):

$$C_{out}(t) = [g \cdot C_r](t) = \int_0^\infty C_r(t - t_e)g(t_e)e^{(-\lambda t_e)}dt_e, \qquad (2)$$

where t_e is time of entry; $C_{out}(t)$ is the output signature; $C_r(t)$ is the weighted input signature; $g(t_e)$ is an appropriate TT distribution function; and $e^{(-\lambda t_e)}$ is the term that accounts for decay if a radioactive tracer is used ($\lambda = 0$ for stable isotopes and chloride). In this study we used both the exponential and dispersion models; the reader is referred to Maloszewski and Zuber (1982) and Stewart and McDonnell (1991) for a detailed overview of TT distribution functions.

In some instances, two models were combined to represent more complex systems on the basis of our understanding of the catchment behaviour (Fig. 3). This was to distinguish be-

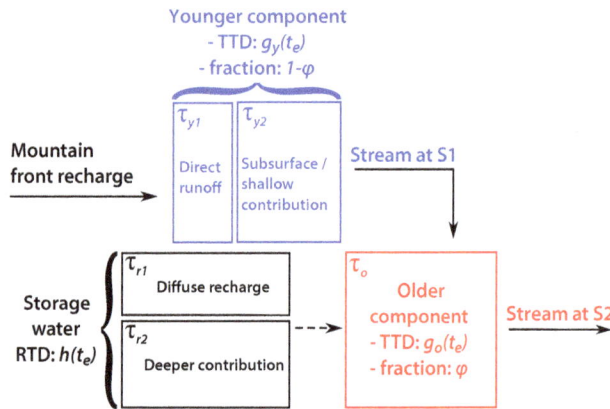

Figure 3. Conceptual diagram showing the flow components and their transit times to be characterised in this study.

tween a shallower and a deeper flow component with shorter and longer TT, respectively. Bimodal models were obtained by linearly combining two TT distributions:

$$C_{\text{out}}(t) = \phi \int_0^\infty C_r(t - t_e) g_o(t_e) e^{(-\lambda t_e)} dt_e$$

$$+ (1 - \phi) \int_0^\infty C_r(t - t_e) g_y(t_e) e^{(-\lambda t_e)} dt_e, \quad (3)$$

where ϕ is the fraction of the older component ($0 < \phi < 1$), and $g_o(t_e)$ and $g_y(t_e)$ are the TT distribution functions of the older and younger components, respectively (Fig. 3). Bimodal distributions combined either two dispersion models or one exponential and one dispersion model. The mean TTs, noted τ, were then derived from the fitted distributions by calculating their first moment:

$$\tau = \int_0^\infty t g(t) dt. \quad (4)$$

In the following the mean TT of the younger component is referred to as τ_y (subdivided into τ_{y1} and τ_{y2}), while the mean TT of the older component is referred to as τ_o, and the mean RT of storage groundwater is referred to as τ_r (subdivided into τ_{r1} and τ_{r2}) (Fig. 3).

For chloride, the measured input and output series were highly dissimilar due to the significant effect of evaporative enrichment in soils. To get around this issue, a correction factor was applied to the predictions obtained using Eqs. (2) and (3): $C_{\text{out}}(t)$ values were multiplied by $F = \frac{P}{(P-\text{ET})}$ (i.e. ratio between precipitation and recharge over the preceding 12 months). The reasoning behind the use of this correction factor was that all chloride ions find their way through the soil, whereas much of the rainfall is evaporated off.

To estimate the fraction of older water that contributed to streamflow, a simple two-component hydrograph separation

was carried out (Sklash and Farvolden, 1979) based on fortnightly data of each of the three seasonal tracers. This allowed one to obtain time-varying values of ϕ:

$$\phi(t) = \frac{\delta_{S1}(t) - \delta_{R1}(t)}{\delta_{G1} - \delta_{R1}(t)}, \quad (5)$$

where δ_{S1}, δ_{R1} and δ_{G1} are the tracer values of streamflow, rainfall and groundwater, respectively. The use of a chemical mass balance approach to partition streamflow was preferred over recursive digital filtering (Nathan and McMahon, 1990), because the former method is less likely to include delayed sources, such as bank return flow and/or interflow, in the older water component (Cartwright et al., 2014).

3.2.2 Using tritium

The occurrence of seasonal variations in rainfall ^{3}H concentrations has been widely documented (e.g. Stewart and Taylor, 1981; Tadros et al., 2014). These variations can be significant and have to be considered for achieving reliable estimates of TT distributions. Monthly ^{3}H precipitation data measured by ANSTO from bulk samples collected at Brisbane Aero were used to estimate the ^{3}H input function for the Teviot Brook catchment. Because Brisbane Aero is ca. 100 km north-east of Teviot Brook, the rainfall ^{3}H concentrations are likely to be significantly different between these two locations due to oceanic and altitudinal effects. According to Tadros et al. (2014), ^{3}H values for Toowoomba (i.e. located in the Great Dividing Range near Teviot Brook) were about 0.4 TU above those for Brisbane Aero for the period 2005–2011. Based on this work, an increment of +0.4 TU was applied to values measured at Brisbane Aero in order to obtain a first estimate of rainfall ^{3}H concentrations for Teviot Brook (input series A2 in Table 1). A second estimate was obtained by comparing the historical ^{3}H data between Toowoomba and Brisbane Aero for the period with overlap between the two stations, i.e. 1968–1982. All monthly values with precipitation > 100 mm, corresponding to rainfall likely contributing to recharge, were included in the analysis ($n = 31$). A scaling factor of 1.24 was derived from the correlation between the two stations ($R^2 = 0.80$). This factor was used to compute input series B2 (Table 1).

To account for losses due to evapotranspiration as rainfall infiltrates into the ground, a weighting procedure similar to the one reported by Stewart et al. (2007) was developed. Monthly ^{3}H recharge was estimated by subtracting monthly evapotranspiration from monthly precipitation, and weighting the ^{3}H rainfall concentrations by the resulting recharge. Instead of calculating single annual values, 6-month and 1 yr sliding windows were used to obtain monthly values as follows:

$$C_i = \frac{\sum_{i-t}^i C_j r_j}{\sum_{i-t}^i r_j}, \quad (6)$$

Table 1. Description of the different ^{3}H input series computed for the Teviot Brook catchment.

Input series	Description of input parameters
A1	A2 − 25 %
A2	Brisbane Aero ^{3}H values + 0.4 TU
A3	A2 + 25 %
B1	B2 − 90 % CI slope
B2	Brisbane Aero ^{3}H values × 1.24 TU
B3	B2 + 90 % CI slope

CI refers to the confidence interval on the Toowoomba vs. Brisbane Aero regression slope.

where C_i is the monthly ^{3}H recharge for the ith month, C_j and r_j are the monthly ^{3}H precipitation and monthly recharge rate for the jth month, and t is 6 or 12 depending on the span of the sliding interval used. To avoid edge effects, a Tukey filter (Tukey, 1968) with coefficient 0.6 was applied to the sliding windows.

Input (recharge) and output (streamwater) ^{3}H concentrations were then related using the same convolution integral as the one used for stable isotopes (Eqs. 2 and 3), with λ the ^{3}H decay constant such that $\lambda = 1.54 \times 10^{-4}$ day^{-1}. To account for the uncertainty in input parameters and to assess the sensitivity of TT distribution calculations to the input function, four additional input series were derived from A2 and B2 (Table 1), and all six input series were subsequently used in the calculations. Least square regressions were used, and root mean square errors (RMSE) were calculated to find the best data fit for each simulation using a trial and error process. All data processing and analyses were performed using Matlab version 8.4.0 (R2014b), with the Statistics toolbox version 9.1.

4 Results

4.1 Seasonal tracers in precipitation, streamwater and groundwater

4.1.1 Description

Stable isotope ratios and chloride signatures in precipitation were highly variable throughout the study period (Figs. 2c and 4). The δ^2H and δ^{18}O rainfall values ranged between −41 and +12‰ (average −12‰) and between −6.5 and −0.1‰ (average −3.1‰), respectively, while chloride concentrations ranged between 0.6 and 3.2 mg L^{-1} (average 1.8 mg L^{-1}). Generally, the most significant rainfall events had isotopically depleted signatures. As an example, there was a considerable drop in all tracers during the January 2013 event (e.g. for δ^2H: decrease from −16 to −41‰; Fig. 2c). The local meteoric water line derived from rainfall samples had an intercept of 15.8 and a slope of 8.4 (Duvert et al., 2015b), similar to that of Brisbane (Fig. 4a). The stable iso-

tope ratios measured in streamwater at S1 (Fig. 2d) and S2 (Fig. 2e) also covered a wide range of values, and followed similar temporal patterns to those for rainfall. However, the overall variations were less pronounced in streamwater, with evident dampening of input signals. Average values were lower for S1 (δ^2H = −25 and δ^{18}O = −4.9‰) than for S2 (δ^2H = −20 and δ^{18}O = −3.7‰), both locations having lower average values than rainfall. All S1 samples aligned close to the meteoric water line, whereas most S2 samples plotted along a linear trend to the right of the line (Fig. 4a). Chloride concentrations in streamwater ranged between 6.4 and 12.8 mg L^{-1} at S1, and between 35.1 and 111.1 mg L^{-1} at S2 (Figs. 2d and e, 4b). At S2, higher chloride values were consistent with higher δ^{18}O values and *vice versa*, whereas there was a weaker correlation between the two tracers at S1 (Fig. 4b). The fluctuations in stable isotopes and chloride in groundwater were considerably attenuated as compared to rain and streamwater (Figs. 2f and 4). The δ^2H, δ^{18}O and chloride values recorded at G1 tended to slightly decrease during the rainy season, although they stayed within the ranges −22 ± 3, −3.9 ± 0.4‰ and 60 ± 10 mg L^{-1}, respectively (Fig. 2f). Consistent displacement to the right of the meteoric line was observed for all G1 samples (Fig. 4a).

4.1.2 Interpretation

The large temporal variability observed in rainfall isotopic and chloride records (Fig. 2c) may be attributed to a combination of factors. First, there was an apparent seasonal cycle as values were higher in the dry season and tended to decrease during the wet season. These are well-known features for rainfall that can be related to the "amount effect" (Dansgaard, 1964) where raindrops during drier periods experience partial evaporation below the cloud base, typical in tropical to subtropical areas (Rozanski et al., 1993). Second, more abrupt depletions of ^{2}H and ^{18}O occurred during significant precipitation events (Fig. 2c), as has been reported in other parts of eastern Australia (Hughes and Crawford, 2013; King et al., 2015). In streamwater, isotopic ratios were generally lower for S1 and S2 than for rainfall, which most likely reflects the predominant contribution of depleted rainfall to recharge (Duvert et al., 2015b). Also, the position of S1 and S2 samples relative to the meteoric line (Fig. 4a) indicates that fractionation due to evaporation occurred at S2, because unlike those measured at S1, isotopic ratios measured at S2 followed a clear evaporation trend. Elevated chloride concentrations are further evidence of the occurrence of evaporative enrichment downstream, with values one order of magnitude higher at S2 than at S1 (Fig. 4b). These results are in line with field observations, showing that the streambed at S2 featured a gentler slope and that lateral inflows from evaporation-prone tributaries may have contributed to streamflow at this location. It can also be noted that the enrichment of chloride at S2 was much higher than that of stable isotopes (Fig. 4b). This is a common observa-

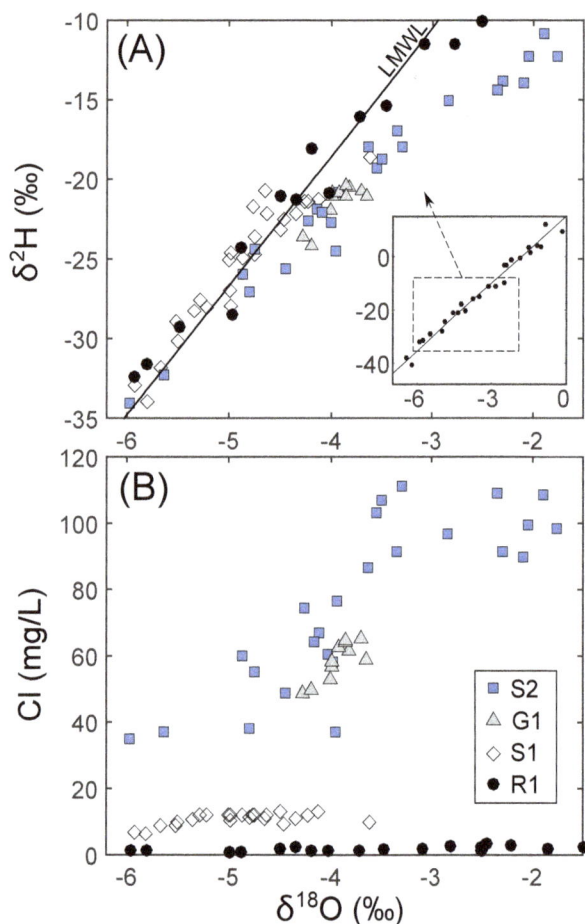

Figure 4. Relationships between (**a**) δ^2H and δ^{18}O and (**b**) chloride and δ^{18}O for rainfall, streamwater and groundwater of the Teviot Brook catchment. The local meteoric water line plotted in (**a**) follows the equation δ^2H $= 8.4 \cdot \delta^{18}$O$+15.8$ (Duvert et al., 2015b). The eight R1 samples with δ^{18}O values either $< -6.2\,$‰ or $> -1.5\,$‰ are not shown in (**b**); chloride concentrations for these samples were in the range 0.6–3 mg L^{-1}.

tion in Australian catchments, largely attributed to high rates of evapotranspiration that concentrate cyclic salts in the unsaturated zone, thereby increasing the salinity of subsurface water before it discharges into streams (e.g. Allison et al., 1990; Cartwright et al., 2004; Bennetts et al., 2006).

4.2 Tritium in precipitation, streamwater and groundwater

4.2.1 Description

The groundwater sample collected at G1 in October 2012 yielded a ^{3}H activity of 1.07 ± 0.06 TU. Additional data was obtained from Please et al. (1997), who collected a sample at the same location in 1994. This earlier sample had an activity of 1.80 ± 0.20 TU. The 20 samples of streamwater collected

Table 2. Kendall's τ and Pearson's r correlation coefficients between ^{3}H and other variables at S2.

Variable	r	τ
Mean daily discharge (m^3 s^{-1})	0.47	0.06
δ^2H (‰)	−0.27	−0.06
δ^{18}O (‰)	−0.23	0.02
Cl (mg L^{-1})	−0.12	0.03
Si (mg L^{-1})	0.35	0.11
Alkalinity (mg L^{-1})	−0.32	−0.13
Fe (mg L^{-1})	0.25	0.11
Antecedent P in the last 15 days (mm)	0.32	−0.01
Last day with $P > 2$ mm (–)	0.11	0.03

No value was statistically significant at $p < 0.05$ for both tests.

at S2 showed variable ^{3}H activities ranging between 1.16 ± 0.06 and 1.43 ± 0.06 TU (Fig. 5).

In order to estimate a ^{3}H input signal for the Teviot Brook catchment, several precipitation time series were calculated from Brisbane Aero monthly ^{3}H data set, as detailed in Table 1. Recharge time series were then derived from these precipitation time series using Eq. (6). An example of the calculated monthly precipitation and recharge time series for the 2003–2014 period is presented in Fig. 6 for scenario A2. While the ^{3}H activity in rainfall ranged between 1.1 and 6.4 TU for A2, most of the rainfall events contributing to recharge (i.e. for which monthly precipitation prevailed over monthly evapotranspiration; red circles in Fig. 6) remained in the narrower range 1.5–2.5 TU.

4.2.2 Interpretation

The ^{3}H activity in rainfall showed considerable month-to-month variability. Winter (dry season) values generally were higher than summer (wet season) values, consistent with results from Tadros et al. (2014). Among the 20 ^{3}H values obtained at S2, higher values tended to coincide with higher flow conditions, although it was not systematic (Fig. 5). For instance, the sample collected in January 2013 under low flow conditions yielded 1.35 ± 0.06 TU; by contrast, the sample collected in April 2014 during the falling limb of a major runoff event yielded 1.19 ± 0.06 TU, i.e. among the lowest values on record. Kendall's rank correlation and Pearson's coefficients were calculated between the ^{3}H measurements in streamwater and other hydrological, hydrochemical and isotopic variables (Table 2). ^{3}H activity was not significantly correlated with any of the other variables. Unlike in Morgenstern et al. (2010) and Cartwright and Morgenstern (2015), there was no strong linear relationship between flow rate and ^{3}H activity in the stream. The lack of strong correlation between ^{3}H and variables such as antecedent wetness conditions and the number of days since the last high flow event occurred, implies that more complex mechanisms governed the short-term fluctuations of ^{3}H in streamwater.

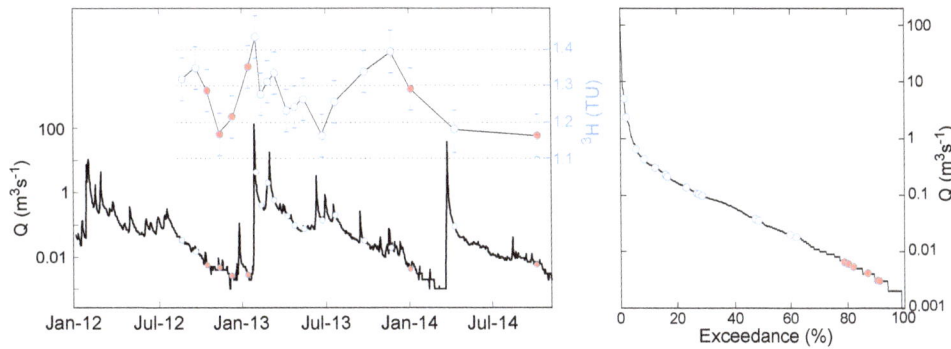

Figure 5. Time series of ^{3}H activity at S2 and daily discharge data (left). Flow duration curve at S2 (right). The six red circles correspond to samples used to fit the low baseflow model (see Fig. 9). The whiskers correspond to measurement uncertainty (± 0.06 TU for all samples).

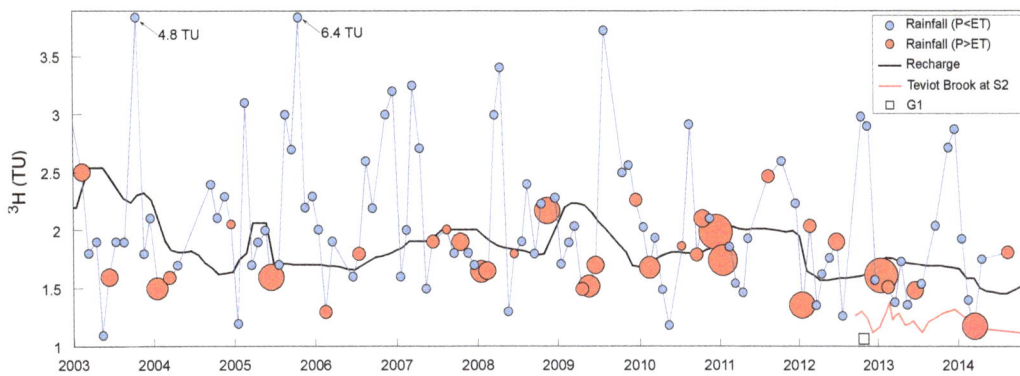

Figure 6. Temporal evolution of input ^{3}H in precipitation (circles) and recharge (black line) for the Teviot Brook catchment considering the A2 scenario. The plotted circles correspond to rainfall collected at Brisbane Aero and adjusted to Teviot Brook according to A2. The recharge time series was obtained using Eq. (6) and a 12-month sliding window. The marker size for rainfall contributing to recharge (red circles) reflects the recharge rate.

4.3 Residence time estimate for storage water

The sample collected at G1 in October 2012 (^{3}H $= 1.07 \pm 0.06$ TU) suggests that alluvial groundwater contains a substantial modern component, because its ^{3}H concentration was only slightly below that of modern rainfall. An earlier ^{3}H value reported by Please et al. (1997) was re-interpreted and combined with our more recent measurement to provide additional constraints on the RT at G1. Two steady-state models were adjusted to the data points. The first model to be tested was a unimodal dispersion model while the second one was a bimodal exponential–dispersion model. For the bimodal model, the mean RT of younger components τ_{r1} was constrained to 1 year, and the fraction of younger water was constrained to 57 % as these parameters provided best fits on average.

Results for both models are presented in Table 3 and the two fits using A2 as an input function are shown in Fig. 7. As expected, mean RTs varied as a function of the input function chosen: values were generally lowest with A1 and B1 and highest with B3. Both models provided reasonably good fits, although for all simulations the bimodal distribu-

tion described more accurately the measured data (median RMSE 0.04 vs. 0.20 TU; Table 3). Unimodal distributions had τ_r ranging between 40 (using A3 as input series) and 62 years (using B2 as input series), with a standard deviation of 7 years among all simulations. The older water fraction of bimodal models had τ_{r2} between 76 (using A1 as input series) and 102 years (using B3 as input series), with a standard deviation of 9 years.

4.4 Transit time estimates using seasonal tracers

Lumped parameter models were adjusted to the stable isotope and chloride time series at S1. Due to the limited number of fortnightly data, all values were included in the analysis, i.e. samples collected under both low baseflow and higher flow conditions. Two models were tested and compared for this purpose, a unimodal exponential model and a bimodal exponential–dispersion model (Table 4; Fig. 8).

While both models provided reasonably low RMSE, unimodal models were less successful in capturing the high-frequency variations observed in output measurements (e.g. lowest values in late January and late February 2013; blue

Table 3. Results of model simulations of residence time for G1 using ^{3}H.

| Input series | \multicolumn{3}{c}{Unimodal DM} | | | \multicolumn{4}{c}{Bimodal EM–DM} | | | |
	τ_r (years)	D_P	RMSE (TU)	τ_{r1} (years)	τ_{r2} (years)	D_P	RMSE (TU)
A1	46.9	0.70	±0.19	1	75.8	0.29	±0.02
A2	48.2	0.71	±0.18	1	82.9	0.30	±0.01
A3	39.8	0.71	±0.18	1	89.0	0.28	±0.03
B1	48.5	0.69	±0.22	1	86.8	0.30	±0.06
B2	61.6	0.70	±0.20	1	95.0	0.29	±0.05
B3	54.6	0.69	±0.21	1	102.5	0.29	±0.05

DM stands for dispersion model; EM–DM stands for exponential–dispersion model; D_P stands for dispersion parameter. For the EM–DM, τ_{r1} was constrained to 1 year, and the fraction of younger water was constrained to 57 %.

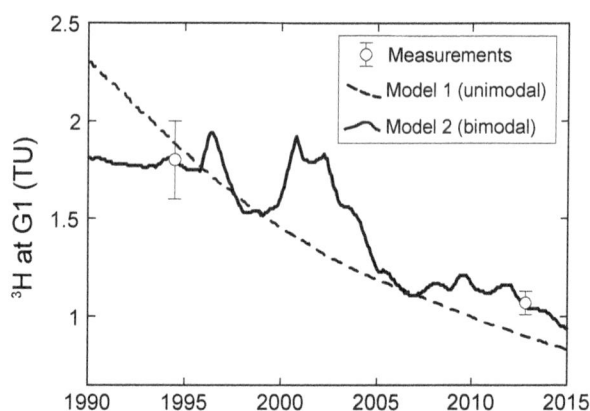

Figure 7. Fits of two models at G1 using A2 as input ^{3}H series. The unimodal model is a dispersion model with first moment 48.2 years and dispersion parameter 0.71. The bimodal model is an exponential–dispersion model: a younger component (exponential distribution; fraction 57 %) with first moment 1 year and an older component (dispersion distribution; fraction 43%) with first moment 82.9 years and dispersion parameter 0.30. The 1994 measurement is from Please et al. (1997).

lines in Fig. 8). All three tracers yielded comparable exponential TT distribution functions, with τ_y ranging between 65 and 70 days (Table 4). The bimodal models provided slightly more satisfactory fits for all tracers (black lines in Fig. 8), with lower RMSE overall. Bimodal TT distribution functions derived from data at S1 had a younger fraction (27 %) with τ_{y1} between 14 and 16 days, and an older fraction (73 %) with τ_{y2} between 113 and 146 days (Table 4) depending on which tracer was used.

Calibration was also carried out on the tracer time series collected at S2 and following the same procedure (Table 4). When considering a unimodal exponential distribution, all three tracers yielded comparable TT distribution functions, with τ_y ranging between 71 and 85 days, which was slightly longer than the mean TTs calculated at S1. When considering a bimodal exponential–dispersion distribution, the younger fraction had τ_{y1} of 23 to 24 days, while the older fraction had τ_{y2} of 99 to 109 days (Table 4).

4.5 Transit time estimates using tritium

4.5.1 Model adjustment to low baseflow samples

A lumped parameter model was fitted to the six ^{3}H samples that were taken under low baseflow conditions, i.e. $Q < 0.01 \, \mathrm{m^3 \, s^{-1}}$. The model chosen for this purpose was a bimodal exponential–dispersion model; the fitting procedure was as follows:

– The dispersion parameter of the older component was loosely constrained to around 0.3 in order to mimic the shape of the TT distribution identified at G1 (Sect. 4.3). The old water fraction ϕ was constrained to 82 %, i.e. the average value obtained for the six baseflow samples using tracer-based hydrograph separation following Eq. (5).

– Initial simulations were run using the six input series with no further model constraint. For the six scenarios, τ_y consistently converged to 0.33 ± 0.08 years.

– All models were then re-run while adding the additional constraint as noted above, so that the only parameter to be determined by fitting was τ_o.

Figure 9 provides an example of the adjustment using A2 as input ^{3}H function. Reasonably good fits were obtained for all simulations ($0.14 \, \mathrm{TU} < \mathrm{RMSE} < 0.16 \, \mathrm{TU}$), with τ_o between 15.8 and 24.5 years, average 20.1 ± 3.9 years (Table 5).

4.5.2 Model adjustment to single tritium values

Unlike for rainfall ^{3}H values where high temporal variability was observed, the derived time series for recharge was relatively constant over the last decade (Fig. 6). This characteristic in principle allows reliable assessment of catchment TTs with single ^{3}H measurements, providing the ^{3}H remaining in the hydrosphere is too small to cause ambiguous ages, as it is in the Southern Hemisphere (Morgenstern et al., 2010; Stewart et al., 2010). All 20 samples collected at S2 were fitted separately using the same lumped model for each point, so that the only parameter to be determined by fitting was

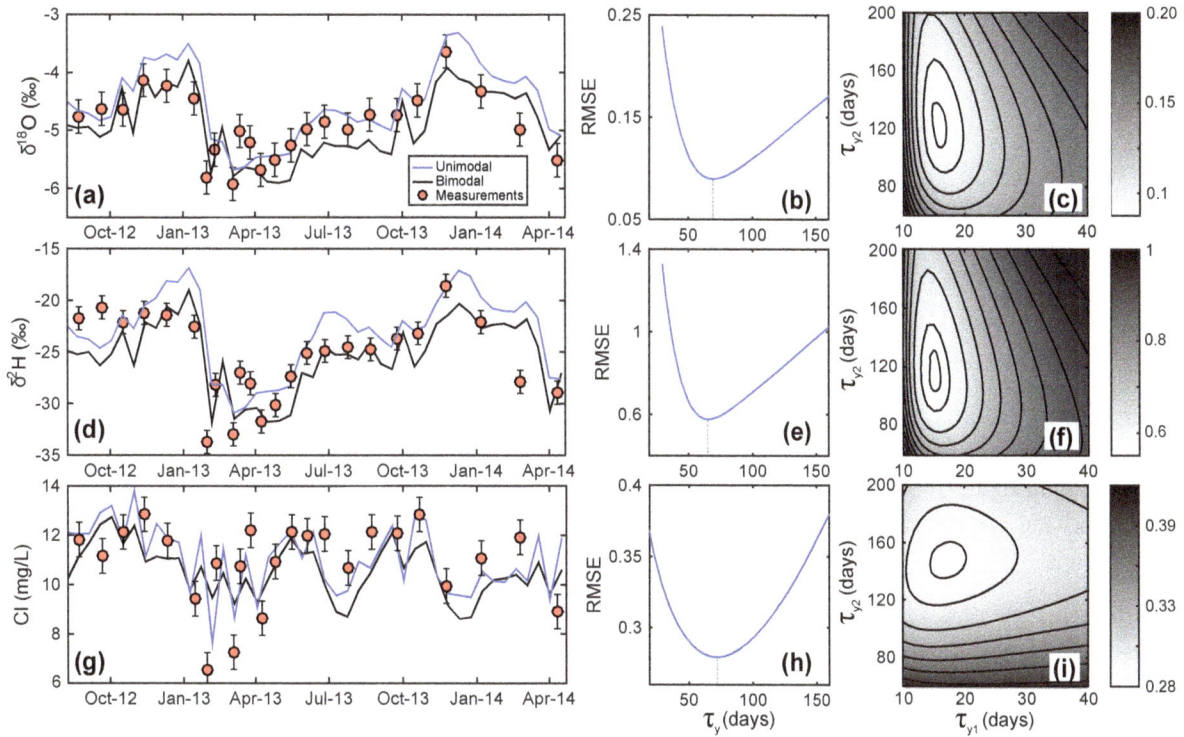

Figure 8. Exponential (blue) and exponential–dispersion (black) models calibrated to the $\delta^{18}O$ (**a**), δ^2H (**d**) and chloride (**g**) time series at S1. Whiskers correspond to the measurement uncertainty as given in the Methods section. Root mean square errors (RMSEs) of the exponential model as a function of τ_y for the three tracers (**b, e, h**). RMSE of the exponential–dispersion model (27 % younger component; dispersion parameter 0.3) as a function of mean transit times of the younger (τ_{y1}) and older (τ_{y2}) fractions for the three tracers (**c, f** and **i**). Lighter colours are for lower RMSE, and the smallest contours correspond to the range of acceptable fit, arbitrarily defined as the values for which the RMSE are lower than the lowest RMSE obtained with the exponential models. Results for these simulations are reported in Table 4.

Table 4. Results of model simulations of transit time for S1 and S2 using δ^2H, $\delta^{18}O$ and chloride.

Sampling location	Tracer	Unimodal EM		Bimodal EM–DM		
		τ_y (days)	RMSE	τ_{y1} (days)	τ_{y2} (days)	RMSE
S1	$\delta^{18}O$	69	$\pm0.09\,\permil$	15	121	$\pm0.08\,\permil$
	δ^2H	65	$\pm0.58\,\permil$	15	113	$\pm0.52\,\permil$
	Chloride	70	$\pm0.28\,\mathrm{mg\,L^{-1}}$	16	146	$\pm0.26\,\mathrm{mg\,L^{-1}}$
S2	$\delta^{18}O$	85	$\pm0.16\,\permil$	23	109	$\pm0.16\,\permil$
	δ^2H	71	$\pm0.75\,\permil$	24	99	$\pm0.72\,\permil$
	Chloride	76	$\pm4.89\,\mathrm{mg\,L^{-1}}$	24	106	$\pm4.68\,\mathrm{mg\,L^{-1}}$

EM stands for exponential model; EM–DM stands for exponential–dispersion model. For the EM–DM, the dispersion parameter of the second mode was 0.3 and the fraction of younger water was 27 %.

the TT of the old water fraction (τ_o). The model parameters were chosen according to the best fit obtained for baseflow samples (i.e. mean TT of young component τ_y 0.33 years, dispersion parameter of old component 0.3; Sect. 4.5.1). In addition, for each sample the fraction of old water ϕ was constrained to the value obtained using tracer-based hydrograph separation according to Eq. (5). Conceptually, this approach appeared more meaningful than another option that would have consisted in constraining τ_o and subsequently determining the old water fractions ϕ, because there was no

indication that τ_o remained constant over time. Simulations were carried out for all three hydrograph separation tracers and all six input series, and the sensitivity of simulations to both the ^3H measurement uncertainty ($\pm0.06\,\mathrm{TU}$) and the error related to the hydrograph separation procedure were also calculated.

Time series of τ_o were derived for each input function, and Fig. 10 shows the results obtained with A2 as an input series. The old water fraction ϕ varied between 0.39 and 1, and while there was a good agreement between the three

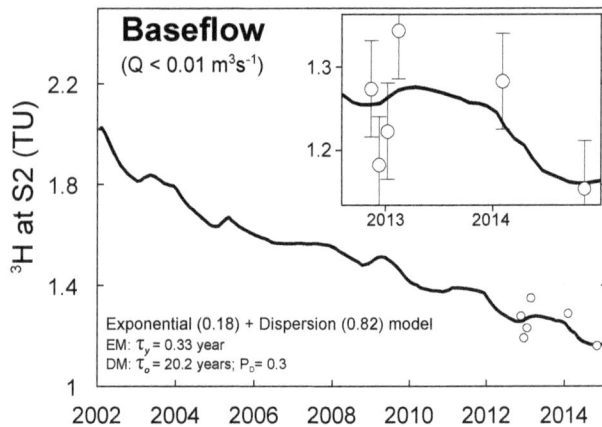

Figure 9. Bimodal model fitted to the ^{3}H activities at S2 under low baseflow conditions (i.e. daily $Q < 0.01$ m^3 s^{-1}). A2 was used as input ^{3}H series for this case. Results using other input series are listed in Table 5.

Table 5. Results of model simulations of transit time for S2 under low baseflow conditions (i.e. daily $Q < 0.01$ m^3 s^{-1}), using ^{3}H and an exponential–dispersion model.

Input series	τ_o (years)	RMSE (TU)
A1	15.8	±0.15
A2	20.2	±0.15
A3	24.5	±0.15
B1	15.8	±0.14
B2	19.8	±0.16
B3	24.4	±0.16

The mean TT of younger components (τ_y) was constrained to 0.33 years, the dispersion parameter of older components was constrained to 0.3, and the ratio of older water was constrained to 82 %.

tracers, hydrograph separation based on chloride generally yielded lower variations in ϕ over time (Fig. 10a). Generally, the older component was lowest during high flow conditions and greatest during recession periods. The simulated τ_o values varied considerably over time, and variations exceeded the uncertainties related to measurement uncertainties, chemical mass balance calculation errors and input estimates (Fig. 10b–d). ^{18}O was the least accurate in evaluating the variations in τ_o (wider range for the red shaded area in Fig. 10c), while chloride was the most accurate despite less pronounced τ_o variations (narrower range for the red shaded area in Fig. 10d). Yet, all three tracers provided comparable results, with a consistent shift in values either upwards or downwards. As a general rule, there was a negative correlation between ϕ and τ_o. When using A2 as input function, τ_o fluctuated between 11.9 and 58.0 years (^{2}H; Fig. 10b), 11.6 and 63.2 years (^{18}O; Fig. 10c) and 11.5 and 42.1 years (chloride; Fig. 10d). For clarity purposes the τ_o values reported in the text do not consider errors related to measurement uncertainty. Values were highest after the major recharge events

that occurred in January and February 2013, with τ_o between 26.8 and 63.2 years in late February, and in April 2014, with τ_o between 28.3 and 55.1 years. They were lowest during periods undergoing sustained low flow such as in September 2012 (τ_o between 11.6 years for ^{18}O and 13.1 years for ^{2}H) and in September 2013 (τ_o between 11.5 years for chloride and 11.9 years for ^{2}H). Of note is the timing of the highest τ_o value in late February 2013, i.e. 1 month after the major recharge episode.

5 Discussion

5.1 Conceptual framework

According to our conceptual understanding of the upper Teviot Brook catchment, we have partitioned streamflow into two major components (Fig. 3). The first end-member represents the contribution of younger waters from rapid recharge through the highly fractured igneous rocks forming the mountain front, as outlined in previous studies (Duvert et al., 2015b, a). This younger component was further divided into (i) quick flow and (ii) relatively delayed contribution of waters seeping from the rock fractures (Fig. 3). We assume that the TTs of the younger end-member can be accurately described through analysis of the seasonal tracers' signal dampening. Waters originating from this component typically had low total dissolved solid (TDS) concentrations, although high Si concentrations at high flow.

The second end-member we postulate contains older waters derived from the aquifer stores located in the lowland section of the study area (Fig. 3). Specifically, these are waters discharging from both the alluvial aquifer and the underlying sedimentary bedrock aquifer. Although a distinction between the two groundwater stores would be ideal, the lack of clear differentiation between both water types led us to consider one single "older water" component. We assume that the TTs of the older end-member may be accurately described through ^{3}H data analysis. The ^{3}H activities in both aquifers were generally lower than those in surface water; the sedimentary bedrock aquifer had on average lower ^{3}H values than the alluvial aquifer, and waters from both aquifers had varying but generally high TDS concentrations (Duvert et al., 2015b). Furthermore, higher Fe concentrations were observed in the sedimentary bedrock waters shortly after recharge (Duvert et al., 2015a).

In the next sections of the discussion, a stepwise approach is followed to evaluate the accuracy of the conceptual model outlined above. In particular, the younger and older components in streamflow are assessed and discussed in Sect. 5.2 and 5.3, respectively. Section 5.4 considers the relationships between the older streamflow component and groundwater stored in the catchment. The variations over time of the TTs of the older component τ_o are then quantified and elucidated (Sect. 5.5). Lastly, Sect. 5.6 addresses the limitations of the

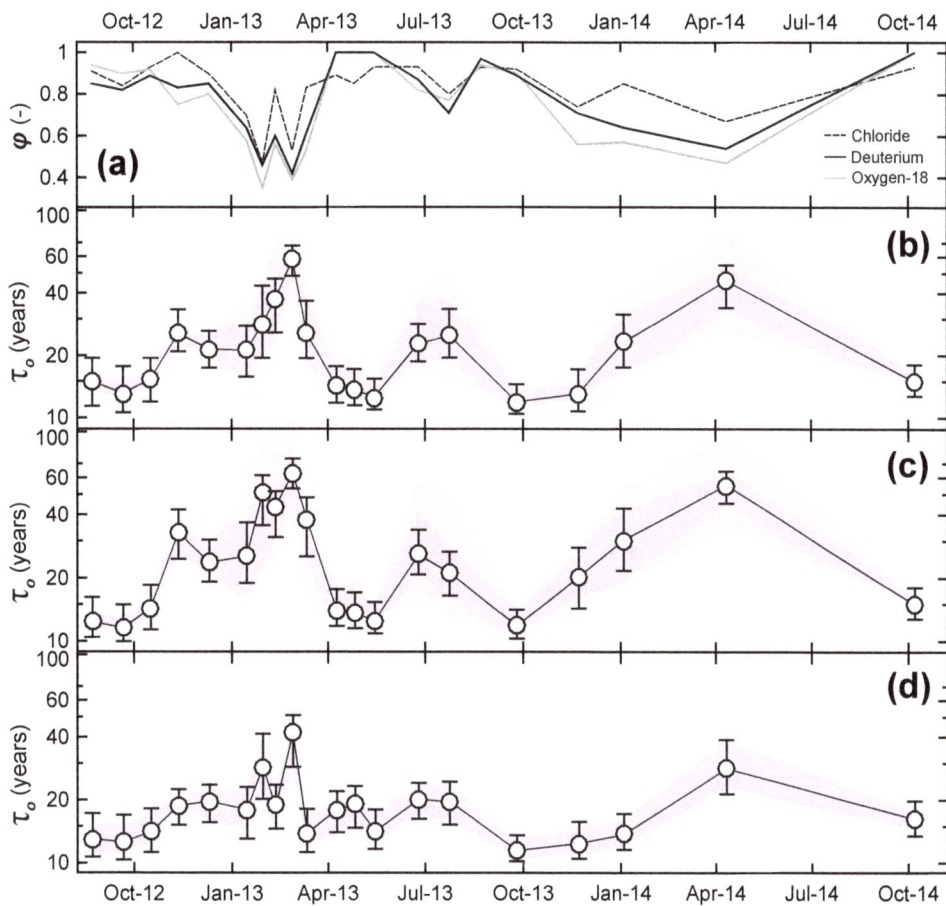

Figure 10. Variations in the older component fraction ϕ according to the three seasonal tracers (using Eq. (5)) (**a**). Variations in the TT of older fraction τ_o at S2 based on hydrograph separation using [2]H (**b**), [18]O (**c**) and chloride (**d**). Values in (**b–d**) were obtained through the adjustment of exponential–dispersion models to each ^{3}H sample separately, and using A2 as input series and a 12-month sliding window. Whiskers represent the error range due to the measurement uncertainty on each sample (i.e. ± 0.06 TU). The blue shaded areas represent the range of values due to uncertainties in the estimation of recharge input (i.e. for the six ^{3}H input time series), while the red shaded areas represent the range of error related to the calculation of ϕ, which was estimated according to the method described in Genereux (1998) and propagated to the calculation of τ_o.

current methodology and raises new questions for future research.

5.2 Identification of a younger component in streamflow

The younger end-member was defined by adjusting lumped models to the seasonal tracer time series (Sect. 4.4; Fig. 8). Among all the TT distributions described in the literature, the exponential model was selected because it considers all possible flowpaths to the stream – the shortest flowpath having a TT equal to zero and the longest having a TT equal to infinity (e.g. Stewart et al., 2010). Importantly, this distribution assumes heavy weighting of short flowpaths, which in our case may accurately replicate the prompt response of streamflow to rainfall inputs in the headwaters.

At S1, the bimodal distribution provided the most accurate simulations (Table 4), which lends support to the occurrence

of two end-members contributing to streamflow at this upstream location. The first (exponential) component may reflect quick flow and subsurface waters feeding the stream (τ_{y1} between 14 and 16 days), while the second (dispersion) component may be attributed to the contribution of waters discharging from the highly fractured igneous rocks (τ_{y2} between 113 and 146 days; Fig. 8). Results at S2 were also slightly more accurate when using a bimodal distribution, suggesting a dual contribution to streamflow at S2 as well. More importantly, the fits for S2 were not as accurate as those for S1, regardless of the distribution and tracer used (Table 4). This reflects the likely importance of other concurrent processes in the downstream section of the catchment. Among them, evaporation may be a major limitation to applying steady-state lumped models at S2. It has been reported that [18]O is generally more sensitive to the effects of evaporation than [2]H (Klaus and McDonnell, 2013; Klaus et al.,

2015b). However, in this study there were no significant differences between TT distributions derived from the two stable isotopes. Calibration of the models on chloride measurements did not yield as accurate results as those for stable isotopes at S1 and to a higher extent at S2, which may be attributed to the higher effects of evaporative enrichment on chloride. Based on flux tracking methods, Hrachowitz et al. (2013) showed that processes such as evaporation can result in considerable biases in TT distribution estimates when using chloride as a tracer.

It is increasingly recognised that stable isotopes cannot provide realistic estimates of longer TT waters, regardless of the lumped model used (Stewart et al., 2012; Seeger and Weiler, 2014; Kirchner, 2015). In this study, it is very likely that older water (i.e. > 5 years) contributed to streamflow at S2 (see Sect. 5.3) but also possibly at S1, and only using stable isotopes and chloride does not allow detection of such contribution. Therefore the ages defined above should be regarded as partial TTs that reflect the short-term and/or intermediate portions of the overall TT distribution for the system, i.e. τ_y rather than τ (Seeger and Weiler, 2014).

5.3 Identification of an older component in streamflow

The transfer function that provided the most accurate estimates of TT for the baseflow samples at S2 was an exponential–dispersion model (Sect. 4.5.1). While other distributions could have been tested, there is a large body of literature that has reported good agreement between exponential, exponential-piston flow and dispersion models calibrated to ^{3}H data (e.g. Maloszewski et al., 1992; Herrmann et al., 1999; Stewart et al., 2007; Cartwright and Morgenstern, 2015). The good fits obtained using this bimodal function (Fig. 9; Table 5) confirm that two major water sources contributed to streamflow at S2. It can be argued that the exponential component captured all young contributions from upstream, i.e. quick flow + soil water + discharge from fractured igneous rocks, as identified in Sect. 5.2 ($\tau_y =$ 0.33 years), while the dispersion component encompassed the delayed groundwater flowpaths (τ_o between 15.8 and 24.5 years). This older contribution to streamflow may originate from the alluvial aquifer, potentially supplemented by seepage from the bedrock storage, as discussed in Sect. 5.1.

A number of studies were carried out in the last 4 decades that also used ^{3}H to assess TTs of the baseflow component to streams. For catchment areas in the range 10–200 km^2, TT estimates were between 3 and 157 years ($n = 39$; median 12 years; data presented in Stewart et al. (2010) supplemented with later papers by Morgenstern et al. (2010), Kralik et al. (2014) and Cartwright and Morgenstern (2015)). While our results compare relatively well to the literature, estimates can vary greatly even within single catchments (e.g. Morgenstern et al., 2010). Also, all reported studies were conducted in temperate regions, this work being the first one carried out in a subtropical setting.

5.4 Storage water and its relationships with the older streamflow component

Simulations of groundwater RT using ^{3}H as a tracer are generally insensitive to the type of lumped parameter model chosen, given that ambient ^{3}H levels are now almost at pre-bomb levels (e.g. Stewart and Thomas, 2008). At G1, better fits were obtained for bimodal functions (Fig. 7; Table 3). This may be interpreted as the probable partitioning of groundwater into one contribution of younger waters by diffuse recharge or flood-derived recharge ($\tau_{r1} \approx 1$ year) coupled with a second contribution of older waters, potentially seeping from the underlying sedimentary bedrock aquifer ($\tau_{r2} \approx 80$ to 100 years).

While the older component to streamflow as identified in Sect. 5.3 was characterised by relatively old waters with TT in the range 15.8–24.5 years, this contribution could not be directly related to the RT of storage waters (i.e. $\tau_o \neq \tau_r$). Despite the exclusive use of samples taken under low baseflow conditions to determine τ_o, the obtained values were significantly lower than the estimates of τ_{r2} for the alluvial aquifer (average 20.1 ± 3.9 vs. 88.7 ± 9.3 years, respectively). This confirms that water stored in the catchment (resident water) and water exiting the catchment (transit water) are fundamentally different and do not necessarily follow the same variations, as recognised in recent work (e.g. Hrachowitz et al., 2013; van der Velde et al., 2015). Results from a dynamic model of chloride transport revealed that water in transit was generally younger than storage water (Benettin et al., 2015). Differences between RTs and TTs also indicate that the assumption of complete mixing was not met for the Teviot Brook catchment. This corroborates the findings from van der Velde et al. (2015), who established that complete mixing scenarios resulted in incorrect TT estimates for a catchment subjected to high seasonal rainfall variability. For instance, shallow flowpaths may be activated or deactivated under varying storage. Among the few studies that investigated the relations between catchment TT and groundwater RT based on ^{3}H measurements, Matsutani et al. (1993) reported that streamwater was formed by a mixture of longer RT groundwater (19 years) and shorter RT soil water (< 1 year). Overall, more work is needed to better define the two distributions and to assess how they relate to each other under non-stationary storage conditions.

5.5 Drivers of the variability in the older component transit time

When fitting models to each ^{3}H value in streamwater, τ_o was found to vary substantially over time (Fig. 10). In order to better apprehend the factors influencing the variations in τ_o, the obtained values were compared to other hydrological and hydrochemical variables, particularly the antecedent wetness conditions, dissolved Fe concentrations and the old water discharge rate (Fig. 11). Under sustained dry conditions ($P_{15} < 5$ mm), there was no consistent relationship between τ_o and the amount of precipitation during the 15 days prior to sampling, with τ_o ranging between 14.9 and 23.1 years ($n = 3$; Fig. 11a). For higher values of P_{15} (i.e. $P_{15} \geq 10$ mm), there was a positive correlation between the two variables ($n = 17$, R^2 for power law fit = 0.47, p-value = 0.002). The TT of the old water fraction was lowest for P_{15} between 10 and 50 mm (τ_o 11.9 to 25.5 years), and it increased when antecedent precipitation increased (τ_o 25.6 to 58.0 years for $P_{15} > 100$ mm). Generally, values averaged 17.0 ± 5.6 years at low flow and 38.3 ± 14.7 years after major high flow events. This was in accordance with results from Fig. 10, and suggestive of the predominant contribution of older alluvial and/or bedrock waters shortly after recharge episodes.

There was also a positive relationship between τ_o and Fe concentrations at S2 ($n = 20$, R^2 for power law fit = 0.48, p-value = 0.001), with all the values > 0.2 mg L^{-1} corresponding to $\tau_o > 30$ years (Fig. 11b). In contrast, no significant relationship was observed at S1, as Fe values at this station ranged between < 0.01 and 0.96 mg L^{-1}. Duvert et al. (2015a) reported increasing Fe concentrations after a major recharge event for some groundwaters of the sedimentary bedrock. The increase in streamflow Fe might therefore be a result of enhanced discharge of these waters into the drainage network, which is coherent with older τ_o values. However, other chemical parameters distinctive of the bedrock groundwaters did not produce a characteristic signature in streamflow during high flow conditions. Or else, high Fe concentrations may be simply due to higher weathering rates at higher flows, although this hypothesis disregards the high value measured for the April 2014 sample (Fe = 4.15 mg L^{-1}) despite relatively low discharge ($Q = 0.095$ m^3 s^{-1}).

As discussed previously, a modification in storage due to a change in recharge dynamics may have activated different groundwater flowpaths and hence water parcels with different RTs (Heidbüchel et al., 2013; van der Velde et al., 2015; Cartwright and Morgenstern, 2015). When the rate of recharge was highest, flushing out of waters located in the deeper, older bedrock aquifer may have been triggered by the resulting pressure wave propagation. By contrast, the relatively younger τ_o observed during lower flow conditions may be attributed to waters that originate from shallower parts of the alluvium and/or from subsurface layers. This is reflected in the relationship between τ_o and Q_o, i.e. the portion of streamflow provided by the older component ($Q_o = Q \cdot \phi$; Fig. 11c). In this figure the groundwater end-member corresponds to τ_r (using the highest recorded Q_o through the study period), while the baseflow end-member corresponds to the τ_o value calculated using the six baseflow samples. The two end-members were linearly connected in an area that represents the extent of possible fluctuations of τ_o, from lower old water contributions to higher old water contributions. The individual τ_o values broadly followed this mixing

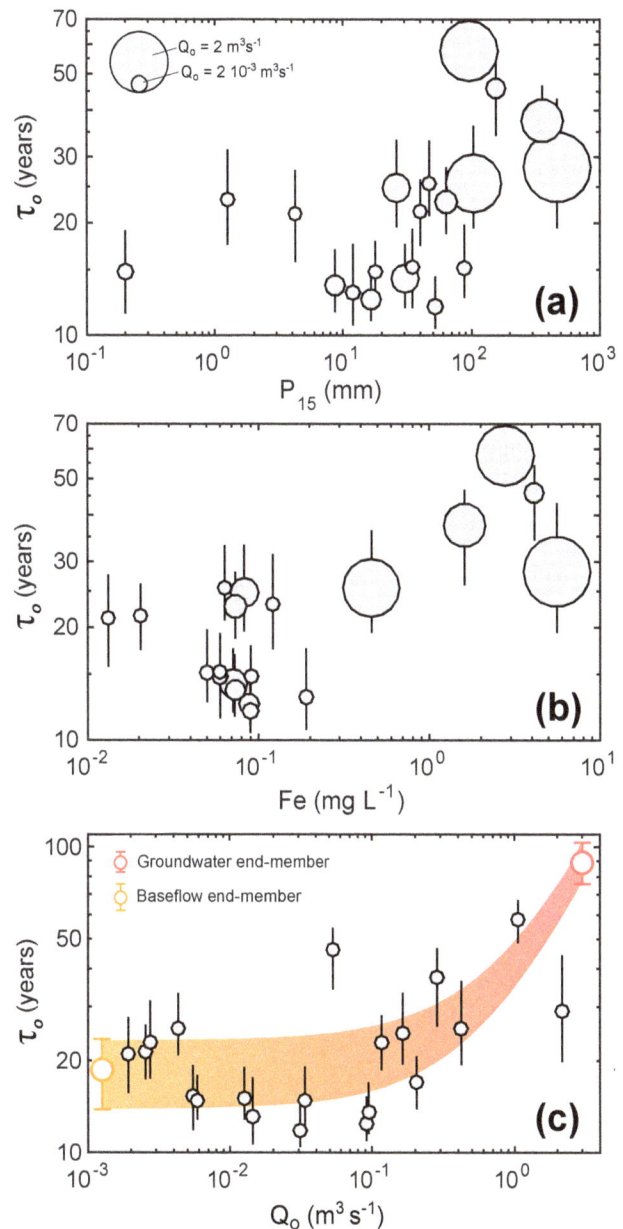

Figure 11. Relationship between the transit time of old water fraction (τ_o) and antecedent precipitation P_{15}, i.e. precipitation depth over the catchment during the 15 days prior to sampling (**a**). Relationship between τ_o and dissolved Fe concentrations (**b**). Relationship between τ_o and Q_o ($Q_o = Q \cdot \phi$) (**c**). Values were obtained using A2 as input series and ^{2}H as a hydrograph separation tracer. Whiskers correspond to simulations using upper and lower measurement uncertainty errors. The size of markers in (**a**) and (**b**) provides an indication on the value of Q_o during sampling. In (**c**), the groundwater (red) end-member corresponds to the RT calculated at G1, while the baseflow (orange) end-member corresponds to the TT of the old water fraction calculated at S2 using the six baseflow samples. The shaded area in (**c**) represents simple linear mixing between the two end-members.

trend (Fig. 11c), which lends support to the assumptions that (i) the TT of the older end-member may not be characterised by a single value but rather by a range of possible ages that fluctuate depending on flow conditions, and (ii) during and shortly after higher flows, a near steady state was reached in which the TT of the old water fraction increased and approached the RT of stored water (i.e. $\tau_o \to \tau_r$). Overall, the large scattering observed in Fig. 11 suggests that many processes led to the variations in τ_o, and that these processes were largely non-linear.

Importantly, the finding that TTs of the old water component increased with increasing flow has not been reported before. Our results are in stark contrast with the previous observation by Morgenstern et al. (2010) and Cartwright and Morgenstern (2015) that ^{3}H-derived TTs were higher at low flow conditions and lower at high flow conditions. However, these two studies did not account for a younger component to streamflow (i.e. ϕ was effectively constrained to 1 for all samples), which may explain the disagreement with our results. Hrachowitz et al. (2015) reported an increase in storage water RT at the start of the wet season in an agricultural catchment in French Brittany, which they related to changes in storage dynamics (i.e. more recent water bypassing storage at higher flow). The authors did not comment on potential changes in streamwater TT during the same period, however.

We also recognise that the results reported here might be due to partially incorrect interpretation of the obtained data set: underestimation of the old water fraction ϕ during high flow events might be responsible for the apparent positive correlation between Q_o and τ_o, although this is unlikely because the three seasonal tracers yielded very similar flow partitions. Another potential bias in our calculations is the possible lack of representation of the discharge from the fractured igneous rocks in the headwaters, which might contribute significantly to the young component during high flow events. Such enhanced contribution might result in slightly longer τ_y, hence shorter τ_o. Because no ^{3}H measurement was conducted at S1, this hypothesis could not be tested further (see Sect. 5.2). More generally, our work emphasises the current lack of understanding of the role and dynamics of deeper groundwater contributions to streams, and suggests that more multi-tracer data is needed to better assess the TTs of the old water fraction. Our findings also indicate that the so-called "old water fraction" (also referred to as "pre-event water" or "baseflow component" in tracer studies; e.g. Klaus and McDonnell, 2013; Stewart, 2015) should not be regarded as one single, time-invariant entity, but rather as a complex component made up of a wide range of flowpaths that can be hydrologically disconnected – and subsequently reactivated – as recharge and flow conditions evolve.

5.6   Limitations of this study and way forward

Several assumptions have been put forward in this study that need to be carefully acknowledged. Firstly, there are limita-

tions related to the use of seasonal tracers (i.e. stable isotopes and chloride):

1. The lumped convolution approach used for the assessment of TTs of the younger contribution to streamflow relied on assumptions of stationarity. Such assumptions are very likely not satisfied in headwater catchments, particularly those characterised by high responsiveness and high seasonal variability in their climate drivers (Rinaldo et al., 2011; McDonnell and Beven, 2014). Unfortunately, the data set obtained as part of this study did not enable characterisation of time-varying TT distribution functions, since this approach would require longer tracer records (e.g. Hrachowitz et al., 2013; Birkel et al., 2015) and/or higher sampling frequencies (e.g. Birkel et al., 2012; Benettin et al., 2013, 2015). Nonetheless, Seeger and Weiler (2014) recently noted that in the current state of research, the calculation of time-invariant TT distributions from lumped models still represents a useful alternative to more complex, computer-intensive modelling methods.

2. Using tracers that are notoriously sensitive to evapotranspiration in environments where this process commonly occurs can be problematic. Hrachowitz et al. (2013) established that evaporation can severely affect the calculations of TTs when chloride is used as an input–output tracer. Although evapotranspiration was considered in our recharge calculations (Eq. (1)), a detailed analysis of catchment internal processes would be needed to verify whether evapotranspiration modifies the storage water RTs and subsequent catchment TTs. Using data from a catchment subjected to high rainfall seasonal variability, van der Velde et al. (2015) showed that younger water was more likely to contribute to evapotranspiration, which tended to result in longer catchment TTs.

3. The partitioning of streamflow relied on the assumption that two main components contributed to streamwater, although this may not be the case at S2 because soil water may explain the higher chloride concentration and more enriched $\delta^{18}O$ observed at this location (Klaus and McDonnell, 2013; Fig. 4). However, we hypothesise that the occurrence of this third end-member would not significantly affect the calculation of τ_o, because the TT of soil water is likely to be considerably shorter than that of the older streamflow component (e.g. Matsutani et al., 1993; Muñoz-Villers and McDonnell, 2012).

Secondly, there are a number of limitations related to the use of ^{3}H:

1. The most significant uncertainties were those related to the computed ^{3}H input functions. These may be reduced by regularly collecting rainfall ^{3}H on site. The accuracy of ^{3}H measurements was another source of uncertainty,

and further improving analytical precision of ^{3}H activity in water samples may allow more rigorous assessment of short-term TT variations (e.g. Morgenstern and Daughney, 2012).

2. Changes in ^{3}H concentrations due to phase changes such as evaporation are commonly ignored, however, high evaporation environments such as that of the lower Teviot Brook catchment might significantly affect ^{3}H activity in streamwater. Future research is needed to examine more thoroughly the potential interferences on ^{3}H due to evaporation (Koster et al., 1989).

3. While stationarity may be a reasonable assumption for groundwater, inter-annual variations in recharge can affect RTs substantially (Manning et al., 2012). Further work aimed at providing additional constraints on RT variability is therefore required, by routinely collecting age tracer data in groundwater. Massoudieh et al. (2014a) showed that using multiple years of tracer records can allow more realistic quantification of the uncertainty on RT distributions. Also uncertain in our work is the spatial representativeness of waters collected at G1.

4. Despite yielding longer TTs than seasonal tracers, the use of ^{3}H did not preclude the potential omission of any older contribution (i.e. > 100 years) to the stream. Frisbee et al. (2013) argued that even studies based on ^{3}H measurements might miss a significant part of the TT distributions rather than just their tail. In our case, the likelihood of waters with much longer RTs seeping from the sedimentary bedrock could not be verified using ^{3}H only. Other tracers that can capture older water footprints, such as terrigenic helium-4 (Smerdon et al., 2012) or carbon-14 (Bourke et al., 2014), would need to be tested for that purpose.

5. Another issue that has been raised recently is the potential aggregation biases affecting the calculation of TT distributions in complex systems (Kirchner, 2015). Based on the use of seasonal tracers, the author demonstrated that mean TTs are likely to be underestimated in heterogeneous catchments, i.e. those composed of sub-catchments with contrasting TT distributions. A similar benchmark study should be undertaken for ^{3}H in order to verify whether TTs derived from ^{3}H measurements in heterogeneous catchments are also biased.

6 Conclusions

Based on time-series observations of seasonal tracers (stable isotopes and chloride) and ^{3}H in a subtropical mountainous catchment, we assessed the different contributions to streamflow as well as the variations in catchment TT and groundwater RT. Calibrating lumped parameter models to seasonal tracer data provided consistent estimates of TTs in the upstream part of the catchment, where evaporation was not a major process. In the downstream location, lumped models reproduced the tracers' output signals less accurately, partly because evapotranspiration complicated the input–output relationships, but also because of the increased hydrological complexity at this scale (i.e. interactions with deeper storage waters).

In this context, the use of ^{3}H time series was highly beneficial for (i) determining an older groundwater contribution to streamflow in the downstream area, and (ii) providing insight into the temporal variations of this old water fraction. The old water fraction TT was significantly younger than the RT of groundwater stored in the catchment, which outlines the necessary distinction between transit and storage waters in catchment process conceptualisation. When simulations were run separately on each ^{3}H streamwater sample, the TT of old water fraction was found to vary substantially over time, with values averaging 17 ± 6 years at low flow and 38 ± 15 years after major recharge events – other parameters being held constant. These variations were interpreted as the activation of longer, deeper flowpaths carrying older waters when the rate of recharge was highest.

Overall, this study suggests that collecting high-resolution ^{3}H data in streamwater can be valuable to document short-term variations in the TT of old water fraction. If confirmed by further studies and corroborated by the use of other dating tracers, the occurrence of fluctuations in older contributions to streamflow may have important implications for water resource management and particularly contamination issues, because these fluctuations may control the timescales of retention and release of contaminants. It is therefore essential to collect longer-term experimental data that will contribute to identifying older groundwater contributions and to quantifying them with more confidence.

Acknowledgements. Funding for the tritium analyses was provided by the Australian Institute of Nuclear Science and Engineering (ALNGRA14026). Continuous financial support from the School of Earth, Environmental & Biological Sciences and M. E. Cox (QUT) are greatly appreciated. We would like to thank A. Bonfanti, M. Citati, G. Destefano, J. López and C. Ranchoux for their assistance with fieldwork. R. Chisari (ANSTO) and J. Brady (QUT) carried out most laboratory analyses. The Brisbane Aero tritium rainfall data set was kindly provided by S. Hollins (ANSTO). Insightful discussion with S. Lamontagne (CSIRO) during the course of this study is gratefully acknowledged. D. Owen (QUT) is thanked for assistance with English. Comments by two anonymous reviewers and the Editor L. Pfister helped us improve the manuscript substantially. C. Duvert is supported by an Endeavour Scholarship (Australian Government).

References

Allison, G. B., Cook, P. G., Barnett, S. R., Walker, G. R., Jolly, I. D., and Hughes, M. W.: Land clearance and river salinisation in the western Murray Basin, J. Hydrol., 119, 1–20, doi:10.1016/0022-1694(90)90030-2, 1990.

Benettin, P., van der Velde, Y., van der Zee, S. E. A. T. M., Rinaldo, A., and Botter, G.: Chloride circulation in a lowland catchment and the formulation of transport by travel time distributions, Water Resour. Res., 49, 4619–4632, doi:10.1002/wrcr.20309, 2013.

Benettin, P., Kirchner, J. W., Rinaldo, A., and Botter, G.: Modeling chloride transport using travel time distributions at Plynlimon, Wales, Water Resour. Res., 51, 3259–3276, doi:10.1002/2014WR016600, 2015.

Bennetts, D. A., Webb, J. A., Stone, D. J. M., and Hill, D. M.: Understanding the salinisation process for groundwater in an area of south-eastern Australia, using hydrochemical and isotopic evidence, J. Hydrol., 323, 178–192, doi:10.1016/j.jhydrol.2005.08.023, 2006.

Bergmann, H., Sackl, B., Maloszewski, P., and Stichler, W.: Hydrological investigation in a small catchment area using isotope data series, in: Fifth International Symposium on Underground Water Tracing, IAHS Publication no. 215, 255–272, Institute of Geology and Mineral Exploration, Athens (Greece), 1986.

Birkel, C., Soulsby, C., Tetzlaff, D., Dunn, S., and Spezia, L.: High-frequency storm event isotope sampling reveals time-variant transit time distributions and influence of diurnal cycles, Hydrol. Process., 26, 308–316, doi:10.1002/hyp.8210, 2012.

Birkel, C., Soulsby, C., and Tetzlaff, D.: Conceptual modelling to assess how the interplay of hydrological connectivity, catchment storage and tracer dynamics controls nonstationary water age estimates, Hydrol. Process., 29, 2956–2969, doi:10.1002/hyp.10414, 2015.

Blavoux, B., Lachassagne, P., Henriot, A., Ladouche, B., Marc, V., Beley, J.-J., Nicoud, G., and Olive, P.: A fifty-year chronicle of tritium data for characterising the functioning of the Evian and Thonon (France) glacial aquifers, J. Hydrol., 494, 116–133, doi:10.1016/j.jhydrol.2013.04.029, 2013.

Bourke, S. A., Harrington, G. A., Cook, P. G., Post, V. E., and Dogramaci, S.: Carbon-14 in streams as a tracer of discharging groundwater, J. Hydrol., 519, 117–130, doi:10.1016/j.jhydrol.2014.06.056, 2014.

Burns, D. A., Murdoch, P. S., Lawrence, G. B., and Michel, R. L.: Effect of groundwater springs on NO3- concentrations during summer in Catskill Mountain streams, Water Resour. Res., 34, 1987–1996, doi:10.1029/98WR01282, 1998.

Cartwright, I. and Morgenstern, U.: Transit times from rainfall to baseflow in headwater catchments estimated using tritium: the Ovens River, Australia, Hydrol. Earth Syst. Sci., 19, 3771–3785, doi:10.5194/hess-19-3771-2015, 2015.

Cartwright, I., Weaver, T. R., Fulton, S., Nichol, C., Reid, M., and Cheng, X.: Hydrogeochemical and isotopic constraints on the origins of dryland salinity, Murray Basin, Victoria, Australia, Appl. Geochem., 19, 1233–1254, doi:10.1016/j.apgeochem.2003.12.006, 2004.

Cartwright, I., Gilfedder, B., and Hofmann, H.: Contrasts between estimates of baseflow help discern multiple sources of water contributing to rivers, Hydrol. Earth Syst. Sci., 18, 15–30, doi:10.5194/hess-18-15-2014, 2014.

Clark, I. D. and Fritz, P.: Environmental Isotopes in Hydrogeology, Lewis, New York, USA, 174–179, 1997.

Cook, P. and Solomon, D.: Recent advances in dating young groundwater: chlorofluorocarbons, $^3H^3He$ and ^{85}Kr, J. Hydrol., 191, 245–265, doi:10.1016/S0022-1694(96)03051-X, 1997.

Cvetkovic, V., Carstens, C., Selroos, J.-O., and Destouni, G.: Water and solute transport along hydrological pathways, Water Resour. Res., 48, W06537, doi:10.1029/2011WR011367, 2012.

Dansgaard, W.: Stable isotopes in precipitation, Tellus, 16, 436–468, doi:10.1111/j.2153-3490.1964.tb00181.x, 1964.

Dunn, S. M., Birkel, C., Tetzlaff, D., and Soulsby, C.: Transit time distributions of a conceptual model: their characteristics and sensitivities, Hydrol. Process., 24, 1719–1729, doi:10.1002/hyp.7560, 2010.

Duvert, C., Cendón, D. I., Raiber, M., Seidel, J.-L., and Cox, M. E.: Seasonal and spatial variations in rare earth elements to identify inter-aquifer linkages and recharge processes in an Australian catchment, Chem. Geol., 396, 83–97, doi:10.1016/j.chemgeo.2014.12.022, 2015a.

Duvert, C., Raiber, M., Owen, D. D. R., Cendón, D. I., Batiot-Guilhe, C., and Cox, M. E.: Hydrochemical processes in a shallow coal seam gas aquifer and its overlying stream–alluvial system: implications for recharge and inter-aquifer connectivity, Appl. Geochem., 61, 146–159, doi:10.1016/j.apgeochem.2015.05.021, 2015b.

Einsiedl, F., Maloszewski, P., and Stichler, W.: Multiple isotope approach to the determination of the natural attenuation potential of a high-alpine karst system, J. Hydrol., 365, 113–121, doi:10.1016/j.jhydrol.2008.11.042, 2009.

Engdahl, N. B., Ginn, T. R., and Fogg, G. E.: Using groundwater age distributions to estimate the effective parameters of Fickian and non-Fickian models of solute transport, Adv. Water Resour., 54, 11–21, doi:10.1016/j.advwatres.2012.12.008, 2013.

Eriksson, E.: Compartment Models and Reservoir Theory, Ann. Rev. Ecol. Syst., 2, 67–84, doi:10.1146/annurev.es.02.110171.000435, 1971.

Freeze, R. A. and Cherry, J. A.: Groundwater, Prentice-Hall, Englewood Cliffs, USA, 136–137, 1979.

Frisbee, M. D., Wilson, J. L., Gomez-Velez, J. D., Phillips, F. M., and Campbell, A. R.: Are we missing the tail (and the tale) of residence time distributions in watersheds?, Geophys. Res. Lett., 40, 4633–4637, doi:10.1002/grl.50895, 2013.

Genereux, D.: Quantifying uncertainty in tracer-based hydrograph separations, Water Resour. Res., 34, 915–919, doi:10.1029/98WR00010, 1998.

Gröning, M., Lutz, H. O., Roller-Lutz, Z., Kralik, M., Gourcy, L., and Pöltenstein, L.: A simple rain collector preventing water re-evaporation dedicated for $\delta^{18}O$ and δ^2H analysis of cumulative precipitation samples, J. Hydrol., 448, 195–200, doi:10.1016/j.jhydrol.2012.04.041, 2012.

Gusyev, M. A., Toews, M., Morgenstern, U., Stewart, M., White, P., Daughney, C., and Hadfield, J.: Calibration of a transient transport model to tritium data in streams and simulation of groundwater ages in the western Lake Taupo catchment, New Zealand, Hydrol. Earth Syst. Sci., 17, 1217–1227, doi:10.5194/hess-17-1217-2013, 2013.

Harman, C. J.: Time-variable transit time distributions and transport: Theory and application to storage-dependent transport of chloride in a watershed, Water Resour. Res., 51, 1–30, doi:10.1002/2014WR015707, 2015.

Heidbüchel, I., Troch, P. A., Lyon, S. W., and Weiler, M.: The master transit time distribution of variable flow systems, Water Resour. Res., 48, W06520, doi:10.1029/2011WR011293, 2012.

Heidbüchel, I., Troch, P. A., and Lyon, S. W.: Separating physical and meteorological controls of variable transit times in zero-order catchments, Water Resour. Res., 49, 7644–7657, doi:10.1002/2012WR013149, 2013.

Herrmann, A., Bahls, S., Stichler, W., Gallart, F., and Latron, J.: Isotope hydrological study of mean transit times and related hydrogeological conditions in Pyrenean experimental basins (Vallcebre, Catalonia), in: Integrated methods in catchment hydrology – tracer, remote sensing, and new hydrometric techniques. Proceedings of IUGG 99 Symposium HS4, IAHS Publication no. 258, 101–110, International Association of Hydrological Sciences, Birmingham (UK), 1999.

Hrachowitz, M., Soulsby, C., Tetzlaff, D., Malcolm, I. A., and Schoups, G.: Gamma distribution models for transit time estimation in catchments: Physical interpretation of parameters and implications for time-variant transit time assessment, Water Resour. Res., 46, W10536, doi:10.1029/2010WR009148, 2010.

Hrachowitz, M., Savenije, H., Bogaard, T. A., Tetzlaff, D., and Soulsby, C.: What can flux tracking teach us about water age distribution patterns and their temporal dynamics?, Hydrol. Earth Syst. Sci., 17, 533–564, doi:10.5194/hess-17-533-2013, 2013.

Hrachowitz, M., Fovet, O., Ruiz, L., and Savenije, H. H. G.: Transit time distributions, legacy contamination and variability in biogeochemical 1/fα scaling: how are hydrological response dynamics linked to water quality at the catchment scale?, Hydrol. Process. 29, 5241–5256, doi:10.1002/hyp.10546, 2015.

Hughes, C. E. and Crawford, J.: Spatial and temporal variation in precipitation isotopes in the Sydney Basin, Australia, J. Hydrol., 489, 42–55, doi:10.1016/j.jhydrol.2013.02.036, 2013.

Katsuyama, M., Tani, M., and Nishimoto, S.: Connection between streamwater mean residence time and bedrock groundwater recharge/discharge dynamics in weathered granite catchments, Hydrol. Process., 24, 2287–2299, doi:10.1002/hyp.7741, 2010.

King, A. C., Raiber, M., Cendón, D. I., Cox, M. E., and Hollins, S. E.: Identifying flood recharge and inter-aquifer connectivity using multiple isotopes in subtropical Australia, Hydrol. Earth Syst. Sci., 19, 2315–2335, doi:10.5194/hess-19-2315-2015, 2015.

Kirchner, J. W.: Aggregation in environmental systems: seasonal tracer cycles quantify young water fractions, but not mean transit times, in spatially heterogeneous catchments, Hydrol. Earth Syst. Sci. Discuss., 12, 3059–3103, doi:10.5194/hessd-12-3059-2015, 2015.

Klaus, J. and McDonnell, J. J.: Hydrograph separation using stable isotopes: Review and evaluation, J. Hydrol., 505, 47–64, doi:10.1016/j.jhydrol.2013.09.006, 2013.

Klaus, J., Chun, K. P., McGuire, K. J., and McDonnell, J. J.: Temporal dynamics of catchment transit times from stable isotope data, Water Resour. Res., 51, 4208–4223, doi:10.1002/2014WR016247, 2015a.

Klaus, J., McDonnell, J. J., Jackson, C. R., Du, E., and Griffiths, N. A.: Where does streamwater come from in low-relief forested

watersheds? A dual-isotope approach, Hydrol. Earth Syst. Sci., 19, 125–135, doi:10.5194/hess-19-125-2015, 2015b.

Koster, R. D., Broecker, W. S., Jouzel, J., Suozzo, R. J., Russell, G. L., Rind, D., and White, J. W. C.: The global geochemistry of bomb-produced tritium: General circulation model compared to available observations and traditional interpretations, J. Geophys. Res.-Atmos., 94, 18305–18326, doi:10.1029/JD094iD15p18305, 1989.

Kralik, M., Humer, F., Fank, J., Harum, T., Klammler, G., Gooddy, D., Sültenfuß, J., Gerber, C., and Purtschert, R.: Using $^{18}O/^2H$, $^3H/^3He$, ^{85}Kr and CFCs to determine mean residence times and water origin in the Grazer and Leibnitzer Feld groundwater bodies (Austria), Appl. Geochem., 50, 150–163, doi:10.1016/j.apgeochem.2014.04.001, 2014.

Lamontagne, S., Taylor, A. R., Batlle-Aguilar, J., Suckow, A., Cook, P. G., Smith, S. D., Morgenstern, U., and Stewart, M. K.: River infiltration to a subtropical alluvial aquifer inferred using multiple environmental tracers, Water Resour. Res., 51, 4532–4549, doi:10.1002/2014WR015663, 2015.

Leray, S., de Dreuzy, J.-R., Aquilina, L., Vergnaud-Ayraud, V., Labasque, T., Bour, O., and Borgne, T. L.: Temporal evolution of age data under transient pumping conditions, J. Hydrol., 511, 555–566, doi:10.1016/j.jhydrol.2014.01.064, 2014.

Maloszewski, P. and Zuber, A.: Determining the turnover time of groundwater systems with the aid of environmental tracers: 1. Models and their applicability, J. Hydrol., 57, 207–231, doi:10.1016/0022-1694(82)90147-0, 1982.

Maloszewski, P., Rauert, W., Trimborn, P., Herrmann, A., and Rau, R.: Isotope hydrological study of mean transit times in an alpine basin (Wimbachtal, Germany), J. Hydrol., 140, 343–360, doi:10.1016/0022-1694(92)90247-S, 1992.

Manning, A. H., Clark, J. F., Diaz, S. H., Rademacher, L. K., Earman, S., and Plummer, L. N.: Evolution of groundwater age in a mountain watershed over a period of thirteen years, J. Hydrol., 460–461, 13–28, doi:10.1016/j.jhydrol.2012.06.030, 2012.

Massoudieh, A.: Inference of long-term groundwater flow transience using environmental tracers: A theoretical approach, Water Resour. Res., 49, 8039–8052, doi:10.1002/2013WR014548, 2013.

Massoudieh, A., Leray, S., and de Dreuzy, J.-R.: Assessment of the value of groundwater age time-series for characterizing complex steady-state flow systems using a Bayesian approach, Appl. Geochem., 50, 240–251, doi:10.1016/j.apgeochem.2013.10.006, 2014a.

Massoudieh, A., Visser, A., Sharifi, S., and Broers, H. P.: A Bayesian modeling approach for estimation of a shape-free groundwater age distribution using multiple tracers, Appl. Geochem., 50, 252–264, doi:10.1016/j.apgeochem.2013.10.004, 2014b.

Matsutani, J., Tanaka, T., and Tsujimura, M.: Residence times of soil water, ground, and discharge waters in a mountainous headwater basin, central Japan, traced by tritium, in: Tracers in Hydrology, edited by Peters, N. E., Hoehn, E., Leibundgut, C., Tase, N., and Walling, D. E., IAHS Publication no. 215, 57–63, International Association for Hydrological Science, Wallingford (UK), 1993.

McCallum, J. L., Engdahl, N. B., Ginn, T. R., and Cook, P. G.: Nonparametric estimation of groundwater residence time distributions: What can environmental tracer data tell us about groundwater residence time?, Water Resour. Res., 50, 2022–2038, doi:10.1002/2013WR014974, 2014.

McDonnell, J. J. and Beven, K.: Debates – The future of hydrological sciences: A (common) path forward? A call to action aimed at understanding velocities, celerities and residence time distributions of the headwater hydrograph, Water Resour. Res., 50, 5342–5350, doi:10.1002/2013WR015141, 2014.

McDonnell, J. J., McGuire, K., Aggarwal, P., Beven, K. J., Biondi, D., Destouni, G., Dunn, S., James, A., Kirchner, J., Kraft, P., Lyon, S., Maloszewski, P., Newman, B., Pfister, L., Rinaldo, A., Rodhe, A., Sayama, T., Seibert, J., Solomon, K., Soulsby, C., Stewart, M., Tetzlaff, D., Tobin, C., Troch, P., Weiler, M., Western, A., Wörman, A., and Wrede, S.: How old is streamwater? Open questions in catchment transit time conceptualization, modelling and analysis, Hydrol. Process., 24, 1745–1754, doi:10.1002/hyp.7796, 2010.

McGuire, K. J. and McDonnell, J. J.: A review and evaluation of catchment transit time modeling, J. Hydrol., 330, 543–563, doi:10.1016/j.jhydrol.2006.04.020, 2006.

McGuire, K. J., McDonnell, J. J., Weiler, M., Kendall, C., McGlynn, B. L., Welker, J. M., and Seibert, J.: The role of topography on catchment-scale water residence time, Water Resour. Res., 41, W05002, doi:10.1029/2004WR003657, 2005.

McMillan, H., Tetzlaff, D., Clark, M., and Soulsby, C.: Do time-variable tracers aid the evaluation of hydrological model structure? A multimodel approach, Water Resour. Res., 48, W05501, doi:10.1029/2011WR011688, 2012.

Michel, R. L.: Tritium in the hydrologic cycle, in: Isotopes in the water cycle: past, present, and future of a developing science, edited by: Aggarwal, P. K., Gat, J. R., and Froehlich, K. F. O., 53–66, Springer, Dordrecht The Netherlands, doi:10.1007/1-4020-3023-1, 2005.

Morgenstern, U. and Daughney, C. J.: Groundwater age for identification of baseline groundwater quality and impacts of land-use intensification – The National Groundwater Monitoring Programme of New Zealand, J. Hydrol., 456–457, 79–93, doi:10.1016/j.jhydrol.2012.06.010, 2012.

Morgenstern, U., Stewart, M. K., and Stenger, R.: Dating of streamwater using tritium in a post nuclear bomb pulse world: continuous variation of mean transit time with streamflow, Hydrol. Earth Syst. Sci., 14, 2289–2301, doi:10.5194/hess-14-2289-2010, 2010.

Mueller, M. H., Weingartner, R., and Alewell, C.: Importance of vegetation, topography and flow paths for water transit times of base flow in alpine headwater catchments, Hydrol. Earth Syst. Sci., 17, 1661–1679, doi:10.5194/hess-17-1661-2013, 2013.

Muñoz-Villers, L. E. and McDonnell, J. J.: Runoff generation in a steep, tropical montane cloud forest catchment on permeable volcanic substrate, Water Resour. Res., 48, W09528, doi:10.1029/2011WR011316, 2012.

Nathan, R. J. and McMahon, T. A.: Evaluation of automated techniques for base flow and recession analyses, Water Resour. Res., 26, 1465–1473, doi:10.1029/WR026i007p01465, 1990.

Penna, D., Stenni, B., Šanda, M., Wrede, S., Bogaard, T. A., Michelini, M., Fischer, B. M. C., Gobbi, A., Mantese, N., Zuecco, G., Borga, M., Bonazza, M., Sobotková, M., Čejková, B., and Wassenaar, L. I.: Technical Note: Evaluation of between-sample memory effects in the analysis of δ^2H and δ^{18}O of water samples measured by laser spectroscopes, Hydrol. Earth Syst. Sci., 16, 3925–3933, doi:10.5194/hess-16-3925-2012, 2012.

Please, P. M., Bauld, J., and Watkins, K. L.: A groundwater quality assessment of the alluvial aquifers in the Logan-Albert catchment, SE Queensland, Tech. Rep. 1996/048, Australian Geological Survey Organisation, Canberra (Australia), 1997.

Plummer, L., Busenberg, E., Böhlke, J., Nelms, D., Michel, R., and Schlosser, P.: Groundwater residence times in Shenandoah National Park, Blue Ridge Mountains, Virginia, USA: a multi-tracer approach, Chem. Geol., 179, 93–111, doi:10.1016/S0009-2541(01)00317-5, 2001.

Reddy, M. M., Schuster, P., Kendall, C., and Reddy, M. B.: Characterization of surface and ground water δ^{18}O seasonal variation and its use for estimating groundwater residence times, Hydrol. Process., 20, 1753–1772, doi:10.1002/hyp.5953, 2006.

Rinaldo, A., Beven, K. J., Bertuzzo, E., Nicotina, L., Davies, J., Fiori, A., Russo, D., and Botter, G.: Catchment travel time distributions and water flow in soils, Water Resour. Res., 47, W07537, doi:10.1029/2011WR010478, 2011.

Roa-García, M. C. and Weiler, M.: Integrated response and transit time distributions of watersheds by combining hydrograph separation and long-term transit time modeling, Hydrol. Earth Syst. Sci., 14, 1537–1549, doi:10.5194/hess-14-1537-2010, 2010.

Rodgers, P., Soulsby, C., Waldron, S., and Tetzlaff, D.: Using stable isotope tracers to assess hydrological flow paths, residence times and landscape influences in a nested mesoscale catchment, Hydrol. Earth Syst. Sci., 9, 139–155, doi:10.5194/hess-9-139-2005, 2005.

Rozanski, K., Araguás-Araguás, L., and Gonfiantini, R.: Isotopic Patterns in Modern Global Precipitation, in: Climate Change in Continental Isotopic Records, edited by Swart, P. K., Lohman, K. C., McKenzie, J., and Savin, S., 1–36, American Geophysical Union, Washington D.C., USA, doi:10.1029/GM078p0001, 1993.

Seeger, S. and Weiler, M.: Reevaluation of transit time distributions, mean transit times and their relation to catchment topography, Hydrol. Earth Syst. Sci., 18, 4751–4771, doi:10.5194/hess-18-4751-2014, 2014.

Sklash, M. G. and Farvolden, R. N.: Role of groundwater in storm runoff, J. Hydrol., 43, 45–65, doi:10.1016/0022-1694(79)90164-1, 1979.

Smerdon, B. D., Gardner, W. P., Harrington, G. A., and Tickell, S. J.: Identifying the contribution of regional groundwater to the baseflow of a tropical river (Daly River, Australia), J. Hydrol., 464–465, 107–115, doi:10.1016/j.jhydrol.2012.06.058, 2012.

Soulsby, C., Malcolm, R., Helliwell, R., Ferrier, R. C., and Jenkins, A.: Isotope hydrology of the Allt a' Mharcaidh catchment, Cairngorms, Scotland: implications for hydrological pathways and residence times, Hydrol. Process., 14, 747–762, 2000.

Stewart, M. K.: A 40-year record of carbon-14 and tritium in the Christchurch groundwater system, New Zealand: Dating of young samples with carbon-14, J. Hydrol., 430–431, 50–68, doi:10.1016/j.jhydrol.2012.01.046, 2012.

Stewart, M. K.: Promising new baseflow separation and recession analysis methods applied to streamflow at Glendhu Catchment, New Zealand, Hydrol. Earth Syst. Sci., 19, 2587–2603, doi:10.5194/hess-19-2587-2015, 2015.

Stewart, M. K. and McDonnell, J. J.: Modeling base flow soil water residence times from deuterium concentrations, Water Resour. Res., 27, 2681–2693, doi:10.1029/91WR01569, 1991.

Stewart, M. K. and Taylor, C. B.: Environmental isotopes in New Zealand hydrology; 1. Introduction. The role of oxygen-18, deuterium, and tritium in hydrology, New Zeal. J. Sci., 24, 295–311, 1981.

Stewart, M. K. and Thomas, J. T.: A conceptual model of flow to the Waikoropupu Springs, NW Nelson, New Zealand, based on hydrometric and tracer (^{18}O, Cl, ^{3}H and CFC) evidence, Hydrol. Earth Syst. Sci., 12, 1–19, doi:10.5194/hess-12-1-2008, 2008.

Stewart, M. K., Mehlhorn, J., and Elliott, S.: Hydrometric and natural tracer (oxygen-18, silica, tritium and sulphur hexafluoride) evidence for a dominant groundwater contribution to Pukemanga Stream, New Zealand, Hydrol. Process., 21, 3340–3356, doi:10.1002/hyp.6557, 2007.

Stewart, M. K., Morgenstern, U., and McDonnell, J. J.: Truncation of stream residence time: how the use of stable isotopes has skewed our concept of streamwater age and origin, Hydrol. Process., 24, 1646–1659, doi:10.1002/hyp.7576, 2010.

Stewart, M. K., Morgenstern, U., McDonnell, J. J., and Pfister, L.: The 'hidden streamflow' challenge in catchment hydrology: a call to action for stream water transit time analysis, Hydrol. Process., 26, 2061–2066, doi:10.1002/hyp.9262, 2012.

Stolp, B. J., Solomon, D. K., Suckow, A., Vitvar, T., Rank, D., Aggarwal, P. K., and Han, L. F.: Age dating base flow at springs and gaining streams using helium-3 and tritium: Fischa-Dagnitz system, southern Vienna Basin, Austria, Water Resour. Res., 46, W07503, doi:10.1029/2009WR008006, 2010.

Tadros, C. V., Hughes, C. E., Crawford, J., Hollins, S. E., and Chisari, R.: Tritium in Australian precipitation: A 50 year record, J. Hydrol., 513, 262–273, doi:10.1016/j.jhydrol.2014.03.031, 2014.

Tetzlaff, D., Seibert, J., and Soulsby, C.: Inter-catchment comparison to assess the influence of topography and soils on catchment transit times in a geomorphic province; the Cairngorm mountains, Scotland, Hydrol. Process., 23, 1874–1886, doi:10.1002/hyp.7318, 2009.

Tetzlaff, D., Soulsby, C., Hrachowitz, M., and Speed, M.: Relative influence of upland and lowland headwaters on the isotope hydrology and transit times of larger catchments, J. Hydrol., 400, 438–447, doi:10.1016/j.jhydrol.2011.01.053, 2011.

Tetzlaff, D., Birkel, C., Dick, J., Geris, J., and Soulsby, C.: Storage dynamics in hydropedological units control hillslope connectivity, runoff generation, and the evolution of catchment transit time distributions, Water Resour. Res., 50, 969–985, doi:10.1002/2013WR014147, 2014.

Timbe, E., Windhorst, D., Crespo, P., Frede, H.-G., Feyen, J., and Breuer, L.: Understanding uncertainties when inferring mean transit times of water trough tracer-based lumped-parameter models in Andean tropical montane cloud forest catchments, Hydrol. Earth Syst. Sci., 18, 1503–1523, doi:10.5194/hess-18-1503-2014, 2014.

Tukey, J.: An introduction to the calculations of numerical spectrum analysis, in: Spectral Analysis of Time Series, edited by Harris, B., 25–46, Wiley, New York, USA, 1968.

van der Velde, Y., de Rooij, G. H., Rozemeijer, J. C., van Geer, F. C., and Broers, H. P.: Nitrate response of a lowland catchment: On the relation between stream concentration and travel time distribution dynamics, Water Resour. Res., 46, W11534, doi:10.1029/2010WR009105, 2010.

van der Velde, Y., Heidbüchel, I., Lyon, S. W., Nyberg, L., Rodhe, A., Bishop, K., and Troch, P. A.: Consequences of mixing assumptions for time-variable travel time distributions, Hydrol. Process., 29, 3460–3474, doi:10.1002/hyp.10372, 2015.

Vogel, J. C.: Investigation of groundwater flow with radiocarbon, in: Isotopes in hydrology, 355–369, International Atomic Energy Agency, Vienna (Austria), 1967.

Weissmann, G. S., Zhang, Y., LaBolle, E. M., and Fogg, G. E.: Dispersion of groundwater age in an alluvial aquifer system, Water Resour. Res., 38, 1198, doi:10.1029/2001WR000907, 2002.

Zuber, A., Witczak, S., Rozanski, K., Sliwka, I., Opoka, M., Mochalski, P., Kuc, T., Karlikowska, J., Kania, J., Jackowicz-Korczynski, M., and Dulinski, M.: Groundwater dating with ^{3}H and SF$_6$ in relation to mixing patterns, transport modelling and hydrochemistry, Hydrol. Process., 19, 2247–2275, doi:10.1002/hyp.5669, 2005.

7

Estimating spatially distributed soil water content at small watershed scales based on decomposition of temporal anomaly and time stability analysis

W. Hu[2,3] and B. C. Si[2,1]

[1]College of Hydraulic and Architectural Engineering, Northwest A & F University, Yangling 712100, China
[2]University of Saskatchewan, Department of Soil Science, Saskatoon, SK S7N 5A8, Canada
[3]New Zealand Institute for Plant & Food Research Limited, Private Bag 4704, 8140 Christchurch, New Zealand

Correspondence to: B. Si (bing.si@usask.ca)

Abstract. Soil water content (SWC) is crucial to rainfall-runoff response at the watershed scale. A model was used to decompose the spatiotemporal SWC into a time-stable pattern (i.e., temporal mean), a space-invariant temporal anomaly, and a space-variant temporal anomaly. The space-variant temporal anomaly was further decomposed using the empirical orthogonal function (EOF) for estimating spatially distributed SWC. This model was compared to a previous model that decomposes the spatiotemporal SWC into a spatial mean and a spatial anomaly, with the latter being further decomposed using the EOF. These two models are termed the temporal anomaly (TA) model and spatial anomaly (SA) model, respectively. We aimed to test the hypothesis that underlying (i.e., time-invariant) spatial patterns exist in the space-variant temporal anomaly at the small watershed scale, and to examine the advantages of the TA model over the SA model in terms of the estimation of spatially distributed SWC. For this purpose, a data set of near surface (0–0.2 m) and root zone (0–1.0 m) SWC, at a small watershed scale in the Canadian Prairies, was analyzed. Results showed that underlying spatial patterns exist in the space-variant temporal anomaly because of the permanent controls of *static* factors such as depth to the CaCO$_3$ layer and organic carbon content. Combined with time stability analysis, the TA model improved the estimation of spatially distributed SWC over the SA model, especially for dry conditions. Further application of these two models demonstrated that the TA model outperformed the SA model at a hillslope in the Chinese Loess Plateau, but the performance of these two models in

the GENCAI network ($\sim 250\,\text{km}^2$) in Italy was equivalent. The TA model can be used to construct a high-resolution distribution of SWC at small watershed scales from coarse-resolution remotely sensed SWC products.

1 Introduction

Soil water content (SWC) of surface soils exerts a major influence on a series of hydrological processes such as runoff and infiltration (Famiglietti et al., 1998; Vereecken et al., 2007; She et al., 2013a). Soil water content in the root zone is, in many cases, linked to vegetative growth (Wang et al., 2012; Ward et al., 2012; Jia and Shao, 2013). Obtaining accurate information on the spatiotemporal SWC is crucial for improving hydrological prediction and soil water management (Venkatesh et al., 2011; Champagne et al., 2012; She et al., 2013b; Zhao et al., 2010). While remote sensing has advanced SWC measurements of surface soils (<5 cm in depth) at basin (2500–25 000 km^2) and continental scales (Robinson et al., 2008), characterization of spatially distributed SWC at small watershed (0.1–80 km^2) scales still poses a challenge. A method is needed for estimating spatially distributed SWC in the near surface and root zone at watershed scales.

Time stability of SWC, which refers to similar spatial patterns of SWC across different measurement times (Vachaud et al., 1985; Brocca et al., 2009), has been used for estimating spatially distributed SWC (Starr, 2005; Perry and Niemann, 2007; Blöschl et al., 2009). This method is conceptually ap-

Figure 1 diagram:

- Mean of time-stable pattern $M_{\hat{t}\hat{n}}$ + Residuals $V_{\hat{t}n}$
- Spatiotemporal SWC S_{tn}
  - Time-stable pattern (Spatial forcing: Soil & Topography) $M_{\hat{t}n}$
  - +
  - Space-dependent dynamics (Interactions: Spatial & Temporal forcing) R_{tn}
  - +
  - Space-invariant dynamics (Temporal forcing: Meteorology & Vegetation) $A_{t\hat{n}}$
- Underlying patterns EOFs × Time coefficients ECs
- Spatial anomaly Z_{tn} + Mean of time-stable pattern $M_{\hat{t}\hat{n}}$
- Underlying patterns EOFs × Time coefficients ECs + Spatial mean $S_{t\hat{n}}$
- Temporal anomaly A_{tn}

SDD model: $S_{tn} = M_{\hat{t}n} + A_{t\hat{n}} + R_{tn}$, where $R_{tn} = \sum EOF^{xg} \times (EC^{xg})^T$

SA model (Perry and Niemann, 2007): $S_{tn} = S_{t\hat{n}} + Z_{tn}$, where $Z_{tn} = \sum EOF^{xg} \times (EC^{xg})^T$

Mittelbach and Seneviratne (2012): $S_{tn} = M_{\hat{t}n} + A_{tn}$

Vanderlinden et al. (2012): $S_{tn} = M_{\hat{t}n} + V_{\hat{t}n} + A_{t\hat{n}} + R_{tn}$

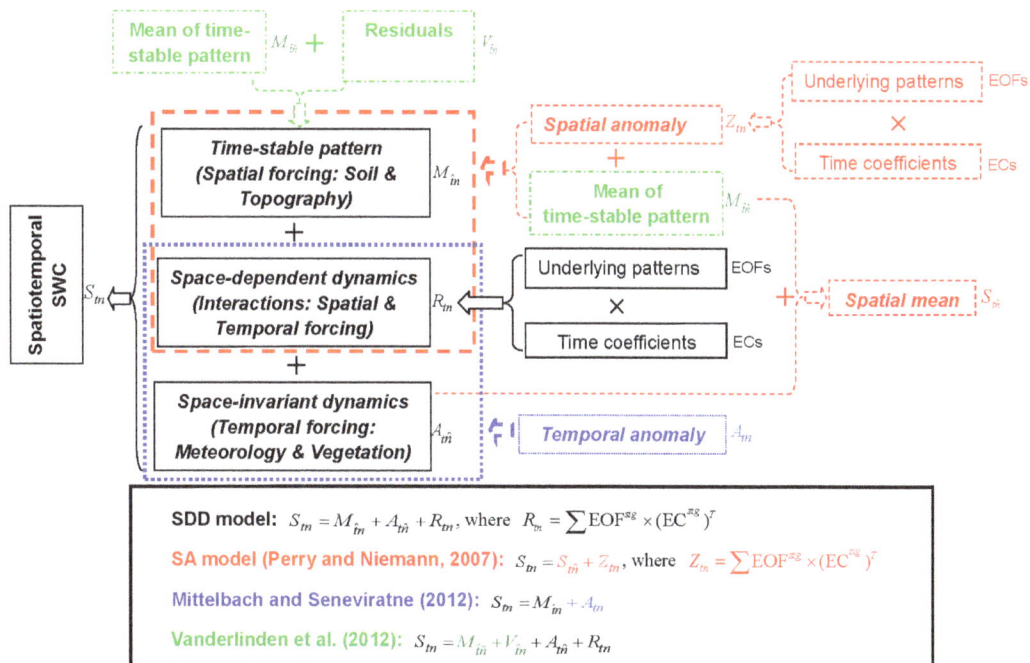

Figure 1. Decomposition of spatiotemporal soil water content (SWC) in different models.

pealing, but assumes completely time-stable spatial patterns of SWC.

The time-stable pattern does not explain all of the spatial variances in SWC, indicating the existence of time-variant components (Starr, 2005). In order to identify underlying patterns of SWC that have time-variant components, the spatiotemporal SWC was decomposed into a spatial mean and a spatial anomaly. The spatial anomaly of the SWC was further decomposed into the sum of the product of time-invariant spatial patterns (EOFs) and temporally varying, but spatially constant coefficients (ECs) using the empirical orthogonal function (EOF) (Fig. 1) (Jawson and Niemann, 2007; Perry and Niemann, 2007, 2008; Joshi and Mohanty, 2010; Korres et al., 2010; Busch et al., 2012). Spatially distributed SWC estimates based on the decomposition of spatial anomaly outperformed those based on time-stable patterns (Perry and Niemann, 2007).

Recently, the spatiotemporal SWC was also decomposed into a temporal mean and a temporal anomaly (Mittelbach and Seneviratne, 2012) (Fig. 1). Previous studies indicated that the contribution of the temporal anomaly to the total spatial variance was notable (Mittelbach and Seneviratne, 2012; Brocca et al., 2014; Rötzer et al., 2015). These studies, however, only focused on surface soils at large scales (>250 km²). Vanderlinden et al. (2012) suggested that the temporal mean may be further decomposed into its spatial mean and residuals, and the temporal anomaly may be further decomposed into space-invariant term (i.e., spatial mean of temporal anomaly) and space-variant term (i.e., spatial residuals of temporal anomaly) (Fig. 1). Note that the spatial variance

in the temporal anomaly (Mittelbach and Seneviratne, 2012) equals that of the space-variant term of the temporal anomaly (Vanderlinden et al., 2012). The further decomposition of the temporal anomaly may be physically meaningful, because the space-invariant and space-variant terms in the temporal anomaly may be forced differently. However, the models of Mittelbach and Seneviratne (2012) and Vanderlinden et al. (2012) have not been used for estimating spatially distributed SWC. If the space-variant terms are ignored during the estimation of spatially distributed SWC, their models are equivalent to that based on time-stable patterns. Therefore, estimation of spatially distributed SWC may be improved by incorporating the space-variant term of the temporal anomaly if underlying (i.e., time-invariant) spatial patterns exist in the temporal anomaly.

To our knowledge, the importance of the space-variant term of the temporal anomaly and its physical meaning at small watershed scales is not well-known. Based on previous studies (Perry and Niemann, 2007; Mittelbach and Seneviratne, 2012; Vanderlinden et al., 2012), we assume soil water dynamics at watershed scales can be decomposed into three components (Fig. 1): (1) time-stable pattern (i.e., temporal mean, spatial forcing): the *static* factors such as soil and topography control the pattern; (2) space-invariant temporal anomaly (temporal forcing): the *dynamic* factors such as meteorological variables and vegetation change with time, and therefore modify SWC in time, regardless of spatial locations; and (3) space-variant temporal anomaly (interactions between spatial forcing and temporal forcing): this term represents interactions between static and dynamic factors. For example, SWC recharge introduced by a rainfall may be modified by topography through runoff processes; SWC loss triggered by evapotranspiration may be regulated by topography through solar radiation exposure.

The static factors may be persistent in the space-variant temporal anomaly, and their impacts on the space-variant temporal anomaly likely change with time. Thus, we hypothesize that some underlying (i.e., time-invariant) spatial patterns exist in the space-variant temporal anomaly, and their impacts can be modulated by a time coefficient, both of which can be obtained by the EOF method (Fig. 1). If the hypothesis is true, the estimation of spatially distributed SWC utilizing the EOF decomposition may outperform the one suggested by Perry and Niemann (2007). This is because: (1) the spatial anomaly, which was decomposed using the EOF in Perry and Niemann (2007), lumped the time-stable pattern and space-variant temporal anomaly together (Fig. 1); (2) the underlying spatial patterns in the spatial anomaly may not fully capture both time-stable patterns and patterns in the space-variant temporal anomaly due to the possible nonlinear relations between these two terms.

Therefore, the objectives were (1) to test the hypothesis that underlying spatial patterns exist in the space-variant temporal anomaly at small watershed scales and (2) to examine whether the decomposition of the space-variant temporal anomaly using the EOF has any advantages over the decomposition of the spatial anomaly (Perry and Niemann, 2007) for estimating spatially distributed SWC. Two steps were included in the estimation of spatially distributed SWC. First, the spatial mean SWC was upscaled from the SWC measurement at the most time-stable location using time stability analysis. Following this, the spatially distributed SWC was downscaled from the estimated spatial mean SWC. For the purpose of this study, spatiotemporal SWC data sets at depths of near surface (0–0.2 m) and root zone (0–1.0 m) from a Canadian Prairie landscape were used. Spatiotemporal SWC of samples taken 0–0.06 m from a hillslope (100 m) in the Chinese Loess Plateau and 0–0.15 m from the GENCAI network (~ 250 km^2) in Italy were also used to further demonstrate conditions under which the decomposition of the spatial anomaly was beneficial to the estimation of spatially distributed SWC.

2   Materials and methods

2.1   Study area and data collection

This study was mainly conducted in the Canadian Prairie pothole region (hereafter abbreviated as Canadian site) at St. Denis National Wildlife Area (52°12′ N, 106°50′ W) with an area of 3.6 km^2. This area has a humid continental climate (Peel et al., 2007), and had a mean annual air temperature of 1.9 °C and a mean annual precipitation of 402 mm during the study period (Fig. 2). A variety of depressions,

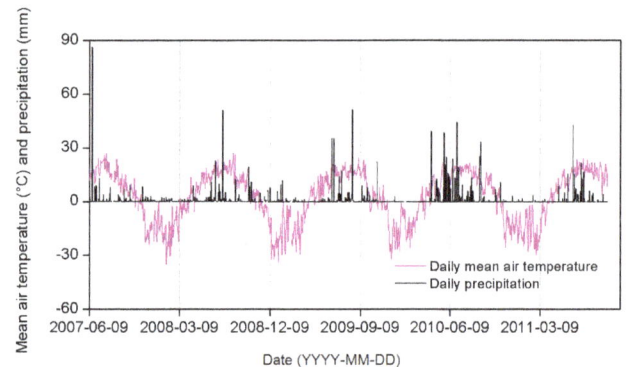

Figure 2. Daily mean air temperature and precipitation during the study period.

knolls, and knobs result in a sequence of undulating slopes (Biswas and Si, 2011). The elevation varies from 554.8 to 557.5 m. The soils are dominated by clay loam textured Mollisols (Soil Survey Staff, 2010) and covered by mixed grass, i.e., smooth brome grass (*Bromus inermis*) and alfalfa (*Medicago sativa* L.). The near-surface soil porosity ranges from 38 % (knolls) to 70 % (depressions). Calcium carbonates (CaCO$_3$) derived mostly from fragments of limestone rocks are common in the Canadian Prairies. The CaCO$_3$ is dissolved by the slightly acidic rainwater moving through the upper horizons and deposited to lower horizons. The heterogeneous amount of infiltrated water resulted in a varying depth of CaCO$_3$ layer ranging from almost 0 m in the knolls to 2.1 m in the depressions. A 576 m long sampling transect with 128 sampling locations spaced at 4.5 m intervals was established over several rounded knolls and depressions. At each location, a time domain reflectometry probe was used to measure SWC of the near-surface soil (0–0.2 m), and a neutron probe was used to collect SWC measurements at 0.2 m intervals between a depth of 0.2 and 1.0 m. The SWC was measured on a volumetric basis and expressed as a percentage (%) volume of water per unit soil volume. The SWC of the root zone was calculated by averaging the SWC of 0–0.2, 0.2–0.4, 0.4–0.6, 0.6–0.8, and 0.8–1.0 m. Soil water content was measured on 23 dates from 17 July 2007 to 29 September 2011. The SWC data set was collected in all seasons except winter, and accurately portrays the variations in soil water conditions in the study area. In addition to the SWC data set, the soil, vegetative, and topographical properties were obtained at each sampling location. These properties included soil particle components (clay, silt, and sand contents), bulk density, soil organic carbon (SOC) content for the surface layer, A horizon depth, C horizon depth, depth to the CaCO$_3$ layer, leaf area index, elevation, cos (aspect), slope, curvature, gradient, upslope length, solar radiation, specific contributing area, convergence index, wetness index, and flow connectivity. Detailed information on the measurements can be found in Biswas et al. (2012). The data sets from the Canadian site were used to demonstrate the following two aspects in detail: (1) different components of spatiotemporal SWC and their contributing factors, and (2) the advantages of the new decomposition method over the method suggested by Perry and Niemann (2007) in terms of the estimation of spatially distributed SWC.

To further test the applicability of the new method, we compared its performance at two other sites, covering both the hillslope and the large watershed scale. Along a hillslope of 100 m in length in the Chinese Loess Plateau, SWC of 0–0.06 m was measured 136 times from 25 June 2007 to 30 August 2008 by a Delta-T Devices Theta probe (ML2x) at 51 locations (Hu et al., 2011). The hillslope was covered by *Stipa bungeana* Trin. and *Medicago sativa* L. in sandy loam and silt loam soils. In the GENCAI network ($\sim 250\,\mathrm{km^2}$) in Italy, SWC of 0–0.15 m was measured by a TDR probe at 46 locations, 34 times from February to December in 2009 (Brocca et al., 2012, 2013). The GENCAI area was dominated by grassland with a flat topography, in silty clay soils.

2.2 Statistical models for decomposing soil water content

Spatiotemporal SWC at small watershed scales was decomposed into three components: time-stable pattern, space-invariant temporal anomaly, and space-variant temporal anomaly. This model was compared to the one that decomposed SWC into spatial mean and spatial anomaly (Perry and Niemann, 2007). Both the space-variant temporal anomaly and spatial anomaly were decomposed using the EOF method. The two models are termed the temporal anomaly (TA) model and the spatial anomaly (SA) model. Figure 1 displays the differences between the two models. Each component will be explained in detail later. The explanation of nomenclatures is listed in Table A1. Because we focus on estimating spatial distribution of SWC at any given time, only spatial variances of SWC were taken into account. Therefore, the variance or covariance denotes the quantity in space without specifications.

2.2.1 The SA model

Perry and Niemann (2007) expressed SWC at location n and time t (S_{tn}) as (Fig. 1):

$$S_{tn} = S_{t\hat{n}} + Z_{tn}, \tag{1}$$

where $S_{t\hat{n}}$ is the spatial mean SWC at time t (temporal forcing) and Z_{tn} is the spatial anomaly of SWC (lumped spatial forcing and interactions). The subscript $\hat{n}$ ($\hat{t}$) indicates a space (time) averaged quantity.

According to Perry and Niemann (2007), $S_{t\hat{n}}$ can be estimated by remote sensing, water balance models, and in situ soil water measurement at a representative (or time-stable) location. The in situ soil water measurement method was selected because the representative location can be easily de-

termined with prior SWC data sets. By measuring SWC only at the most time-stable location (s) and future time t (S_{ts}), $S_{t\hat{n}}$ can be estimated using (Grayson and Western, 1998)

$$S_{t\hat{n}} = \frac{S_{ts}}{1 + \delta_{\hat{t}s}}, \tag{2}$$

where the s was identified using the time stability index of mean absolute bias error (Hu et al., 2010, 2012). The $\delta_{\hat{t}s}$ is the temporal mean relative difference of SWC at the s, which was calculated with prior measurements.

Spatial anomaly (Z_{tn}) can be reconstructed by the sum of the product of time-invariant spatial structures (EOFs) and temporally varying coefficients (ECs) using the EOF method (Perry and Niemann, 2007; Joshi and Mohanty, 2010; Vanderlinden et al., 2012). The ECs correspond to the eigenvectors of the matrix of spatial covariance of the Z_{tn}, and the EOFs are obtained by projecting the Z_{tn} onto the matrix ECs as $EOFs = Z_{tn}$ ECs. The number of EOF (or EC) series equals the number of sampling dates. Each EOF series corresponds to one value at each location, and each EC series has one value at each measurement time. Each EOF is chosen to be orthogonal to other EOFs, and the lower-order EOFs account for as much variance as possible. The sum of variances of all EOFs equals the sum of variances of Z_{tn} from all measurement times.

Usually, a substantial amount of variance can be explained by a small number of EOFs. Johnson and Wichern (2002) suggested the eigenvalue confidence limits method for selecting the number of EOFs. Once the number of significant EOFs at a confidence level of 95 % is selected, Z_{tn} can be estimated as the sum of the product of significant EOFs and associated ECs as

$$Z_{tn} = \sum EOF^{sig} \times \left(EC^{sig}\right)^T, \tag{3}$$

where EOF^{sig} represents the significant EOFs of the Z_{tn} obtained during model development, EC^{sig} is the associated temporally varying coefficient, and the superscript T represents matrix transpose. Following Perry and Niemann (2007), the associated significant EC at time t (EC_t), is estimated by the cosine relationship between EC and $S_{t\hat{n}}$ developed using prior measurements:

$$EC_t = a + b\cos\left(\frac{2\pi}{c}S_{t\hat{n}} - d\right), \tag{4}$$

where $a, b, c,$ and d are the fitted parameters using prior measurements and $S_{t\hat{n}}$ is estimated from Eq. (2). By using the continuous function, EC_t can be estimated at any $S_{t\hat{n}}$ values, which allows for the estimation of spatially distributed SWC at any soil water conditions.

2.2.2 The TA model

Mittelbach and Seneviratne (2012) decomposed the S_{tn} into a time-stable pattern (i.e., temporal mean) and a temporal anomaly component (Fig. 1):

$$S_{tn} = M_{\hat{\imath}n} + A_{tn}, \qquad (5)$$

where $M_{\hat{\imath}n}$ is the time-stable pattern (spatial forcing) controlled by static factors such as soil properties and topography; A_{tn} refers to the temporal anomaly (lumped temporal forcing and interactions). The variance of SWC ($\sigma_{\hat{n}}^2 (S_{tn})$) is the sum of variance of the $M_{\hat{\imath}n}$ ($\sigma_{\hat{n}}^2 (M_{\hat{\imath}n})$), variance of the A_{tn} ($\sigma_{\hat{n}}^2 (A_{tn})$), and two times of covariance between $M_{\hat{\imath}n}$ and A_{tn} ($2\,\mathrm{cov}\,(M_{\hat{\imath}n}, A_{tn})$), which can be expressed as:

$$\sigma_{\hat{n}}^2 (S_{tn}) = \sigma_{\hat{n}}^2 \left(M_{\hat{\imath}n}\right) + 2\mathrm{cov}\left(M_{\hat{\imath}n}, A_{tn}\right) + \sigma_{\hat{n}}^2 (A_{tn}). \qquad (6)$$

Because the A_{tn} in Mittelbach and Seneviratne (2012) is a lumped term, it can be further decomposed into space-invariant temporal anomaly ($A_{t\hat{n}}$, i.e., temporal forcing) and space-variant temporal anomaly (R_{tn}, i.e., interactions) (Vanderlinden et al., 2012). At a watershed scale, the $A_{t\hat{n}}$ is controlled by temporally varying factors such as meteorological variables and vegetation. Positive and negative $A_{t\hat{n}}$ correspond to relatively wet and dry periods, respectively. The R_{tn} refers to the redistribution of $A_{t\hat{n}}$ among different locations due to the interactions between spatial forcing and temporal forcing. For example, soil and topography regulate how much rainfall enters soil and how much water runs off or runs on at a location. This, in turn, dictates vegetation growth in a water-limited environment. Therefore, S_{tn} can also be expressed as (Fig. 1)

$$S_{tn} = M_{\hat{\imath}n} + A_{t\hat{n}} + R_{tn}. \qquad (7)$$

The temporal trends of $A_{t\hat{n}}$ in Eq. (7) and $S_{t\hat{n}}$ in Eq. (1) are the same as both represent temporal forcing. Because the $A_{t\hat{n}}$ is space-invariant and orthogonal to the $M_{\hat{\imath}n}$ and R_{tn} in a space, $\sigma_{\hat{n}}^2 (S_{tn})$ in Eq. (6) can also be written as

$$\sigma_{\hat{n}}^2 (S_{tn}) = \sigma_{\hat{n}}^2 \left(M_{\hat{\imath}n}\right) + 2\mathrm{cov}\left(M_{\hat{\imath}n}, R_{tn}\right) + \sigma_{\hat{n}}^2 (R_{tn}), \qquad (8)$$

where $\mathrm{cov}\,(M_{\hat{\imath}n}, R_{tn})$ is the covariance between the $M_{\hat{\imath}n}$ and R_{tn}, and $\sigma_{\hat{n}}^2 (R_{tn})$ is the variance of the R_{tn}. Apparently, $2\,\mathrm{cov}\,(M_{\hat{\imath}n}, R_{tn})$ equals $2\,\mathrm{cov}\,(M_{\hat{\imath}n}, A_{tn})$, and $\sigma_{\hat{n}}^2 (R_{tn})$ equals $\sigma_{\hat{n}}^2 (A_{tn})$. The percent (%) of $\sigma_{\hat{n}}^2 (M_{\hat{\imath}n})$, $2\,\mathrm{cov}\,(M_{\hat{\imath}n}, R_{tn})$, and $\sigma_{\hat{n}}^2 (R_{tn})$ out of the $\sigma_{\hat{n}}^2 (S_{tn})$ are calculated. The cov $(M_{\hat{\imath}n}, R_{tn})$ can be negative at some conditions, for example, when the depressions correspond to greater $M_{\hat{\imath}n}$ and more negative R_{tn} values in the discharge periods. This resulted in percentage of $\sigma_{\hat{n}}^2 (M_{\hat{\imath}n})$ and $\sigma_{\hat{n}}^2 (R_{tn}) > 100\,\%$ and percentage of $2\,\mathrm{cov}\,(M_{\hat{\imath}n}, R_{tn}) < 0\,\%$ (Mittelbach and Seneviratne, 2012; Brocca et al., 2014; Rötzer et al., 2015). If R_{tn} is zero at any time or location, there are no interactions between spatial forcing and temporal forcing, $\sigma_{\hat{n}}^2 (S_{tn})$ and the spatial trends

of SWC are consistent over time. Therefore, R_{tn} is directly responsible for temporal change in the spatial variability of SWC.

If some underlying spatial patterns exist in R_{tn}, R_{tn} can be reconstructed by the sum of the product of time-invariant spatial structures (EOFs) and time-dependent coefficients (ECs) using the EOF method. Note that the number of EOF (or EC) series also equals the number of sampling dates.

For estimation of spatially distributed SWC, R_{tn} is estimated by the same method as Z_{tn} using Eq. (3). The $M_{\hat{\imath}n}$ is estimated with prior measurements by

$$M_{\hat{\imath}n} = \frac{1}{m} \sum_{j=1}^{m} S_{tn}, \qquad (9)$$

where m is the number of previous measurement times, and $A_{t\hat{n}}$ is estimated by:

$$A_{t\hat{n}} = S_{t\hat{n}} - M_{\hat{\imath}\hat{n}}, \qquad (10)$$

where $M_{\hat{\imath}\hat{n}}$ is the spatial mean of $M_{\hat{\imath}n}$, and $S_{t\hat{n}}$ is estimated from SWC measurements at the most time-stable location using Eq. (2).

The Pearson correlation coefficient (R) is used to explore the linear relationships between various spatial components in the two models (i.e., EOF1 of the Z_{tn} in the SA model, $M_{\hat{\imath}n}$, and EOF1 of the R_{tn} in the TA model) and environmental factors (i.e., soil, vegetative, and topographical properties). The multiple stepwise regressions are conducted to determine the percentage of variations in the spatial components which the controlling factors explain.

2.3 Validation and performance parameter

The TA model is more complicated than the SA model. In order to evaluate the two models for parsimony, AICc values are calculated (Burnham and Anderson, 2002) as

$$\mathrm{AICc} = 2k + n\ln(\mathrm{RSS}/n) + 2k(k+1)/(n-k-1), \qquad (11)$$

where k is the number of parameters, n is the sample size, and RSS is the residual sum of squares.

Both cross-validation and split sample validation are used to estimate SWC distribution with both models. For the cross-validation, an iterative removal of 1 of the 23 dates is made for model development, and the SWC along the transect corresponding to the removed date is estimated iteratively. For the split sample validation, SWC from 14 dates of the first 2 years (from 17 July 2007 to 27 May 2009) is used for model development, and the SWC distribution of 9 dates in the second 2 years (from 21 July 2009 to 29 September 2011) is estimated.

The Nash–Sutcliffe coefficient of efficiency (NSCE) is used to evaluate the quality of estimation of spatially distributed SWC, which is expressed as

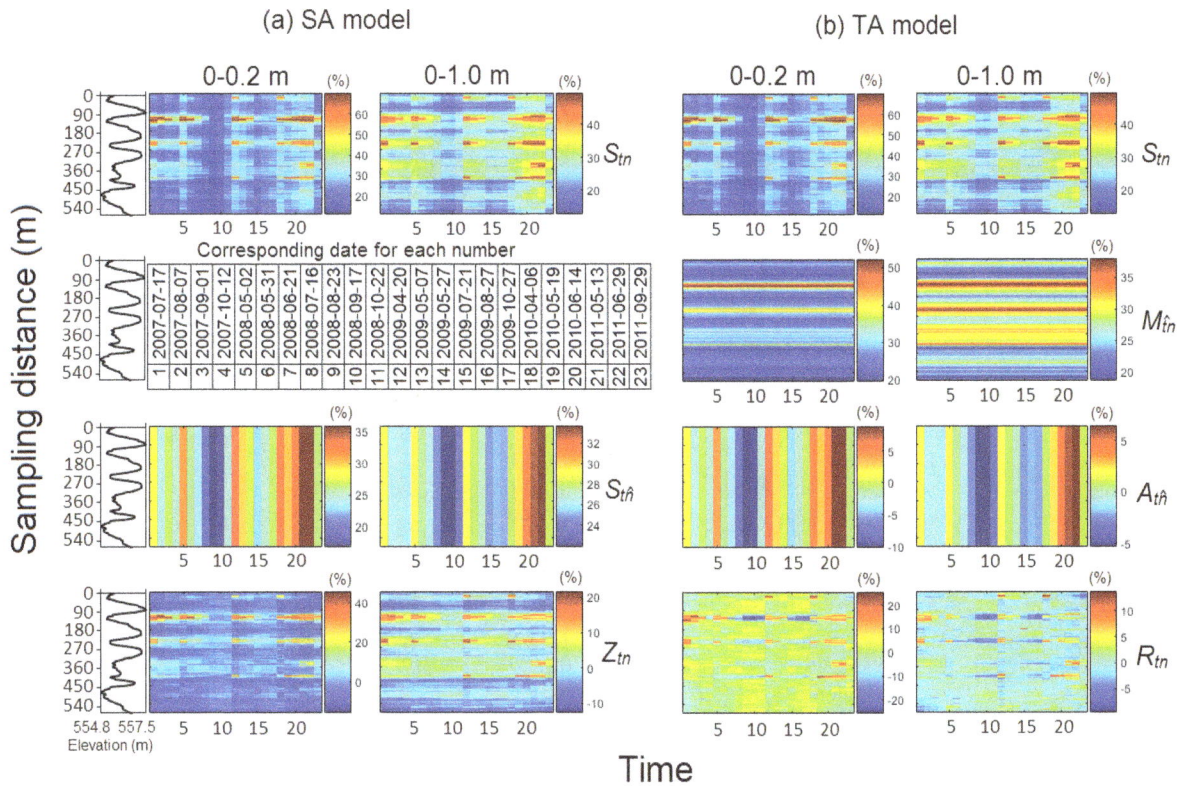

Figure 3. Components of soil water content in (**a**) the SA model (spatial mean soil water content $S_{t\hat{n}}$ and spatial anomaly Z_{tn}) and in (**b**) the TA model (time-stable pattern $M_{\hat{t}n}$, space-invariant temporal anomaly $A_{t\hat{n}}$, and space-variant temporal anomaly R_{tn}) for 0–0.2 and 0–1.0 m. Also shown is the elevation.

$$\text{NSCE} = 1 - \frac{\sigma_{\varepsilon}^2}{\sigma_{\text{measure}}^2}, \tag{12}$$

where $\sigma_{\text{measure}}^2$ is the variance of measured SWC, and σ_{ε}^2 is the mean squared estimation error. A larger NSCE value implies a better quality of estimation. A paired samples T test is used to test whether the NSCE values between the TA model and the SA model are statistically significant at $P < 0.05$.

Many factors may affect the relative performance of spatially distributed SWC estimation between the TA model and the SA model. First, the degree of outperformance of the TA model over the SA model may depend on the amount of R_{tn} variance considered in the TA model. On one hand, the two models are identical if variance of R_{tn} is close to zero or there are negligible interactions between the spatial and temporal components (Fig. 1). On the other hand, if no underlying spatial patterns exist in the R_{tn} or the underlying spatial patterns accounted for little variance of the R_{tn}, the outperformance will also be very limited. Therefore, the greater the variance of R_{tn} considered in the TA model, the more likely the TA model can outperform the SA model. Second, the way of EOF decomposition may also affect the relative performance. In the SA model, EOF decomposition is performed on lumped time-stable patterns ($M_{\hat{t}n}$) and space-variant temporal anomaly (R_{tn}). In the TA model, however, EOF decom-

position is made only on the R_{tn}. In theory, the two models will be identical if the $M_{\hat{t}n}$ and the first underlying spatial pattern (i.e., EOF1) of the R_{tn} were perfectly correlated. If a nonlinear relationship exists between them, lumping the $M_{\hat{t}n}$ and R_{tn} together, as in the SA model, would weaken the model performance as compared to the TA model. From this aspect, the greater deviation from a linear relationship between the $M_{\hat{t}n}$ and EOF1 of the R_{tn}, may lead to a greater outperformance of the TA model over the SA model. Finally, the performances of both models rely on the estimation accuracy of the EC_t which depends on both goodness of fit of the cosine function (i.e., Eq. 4) and estimation accuracy of the $S_{t\hat{n}}$. Because the same $S_{t\hat{n}}$ values are used for the two models, the relative performance of the two models is related to the goodness of fit of Eq. (4).

3   Results

3.1   Components of SWC and their controls

3.1.1   Spatial mean ($S_{t\hat{n}}$) and spatial anomaly (Z_{tn})

The values of spatial mean ($S_{t\hat{n}}$) in the SA model varied with the seasons (Fig. 3a). In the spring, such as 2 May 2008 and 20 April 2009, snowmelt infiltration resulted in rela-

Table 1. Pearson correlation coefficients between time-stable pattern $M_{\hat{t}n}$, EOF1 of space-variant temporal anomaly R_{tn} and various properties.

	0–0.2 m		0–1.0 m	
	$M_{\hat{t}n}$	EOF1	$M_{\hat{t}n}$	EOF1
Sand content	−0.52**	−0.36**	−0.66**	−0.26**
Silt content	0.29**	0.14	0.40**	0.06
Clay content	0.43**	0.38**	0.51**	0.33**
Organic carbon	0.78**	0.83**	0.73**	0.76**
Wetness index	0.64**	0.59**	0.68**	0.56**
Depth to $CaCO_3$ layer	0.77**	0.84**	0.65**	0.88**
A horizon depth	0.51**	0.62**	0.44**	0.65**
C horizon depth	0.66**	0.69**	0.58**	0.76**
Bulk density	−0.58**	−0.67**	−0.46**	−0.62**
Elevation	−0.24**	−0.28**	−0.24**	−0.32**
Specific contributing area	0.20*	0.24**	0.24**	0.23**
Convergence index	−0.58**	−0.56**	−0.55**	−0.58**
Curvature	−0.10	−0.08	−0.19*	−0.16
Cos (aspect)	0.05	0.04	0.08	0.05
Gradient	−0.12	−0.09	−0.21*	−0.02
Slope	−0.51**	−0.48**	−0.56**	−0.44**
Upslope length	0.19*	0.21*	0.21*	0.25**
Solar radiation	−0.07	0.03	−0.11	0.08
Flow connectivity	0.45**	0.43**	0.49**	0.49**
Leaf area index	−0.07	0.06	−0.10	−0.14
Variance explained[1]	74.5 %	81.6 %	75.6 %	81.0 %

[1] Percent of variance explained by the controlling factors obtained by the multiple stepwise regressions. * Significant at $P < 0.05$; ** significant at $P < 0.01$.

tively great $S_{\hat{t}n}$ values. In the summer, however, even 1 month after large rainfall events (such as on 19 July 2008 and 21 June 2009), the high evapotranspiration by fast-growing vegetation resulted in small $S_{\hat{t}n}$ values. The values of $S_{\hat{t}n}$ also varied between inter-annual meteorological conditions. In 2008, there was less precipitation and higher air temperature than in 2010 (Fig. 2). As a result, $S_{\hat{t}n}$ was relatively smaller in 2008 than in 2010.

The spatial patterns of spatial anomaly (Z_{tn}) were similar to those of the original SWC patterns (Fig. 3a). The values of Z_{tn} in wet periods (e.g., 13 May 2011) were much greater than in dry periods (e.g., 23 August 2008) in depressions (e.g., at a distance of 123 and 250 m); at other locations, however, the spatial anomaly was slightly less in wet periods than in dry periods for both soil layers. Moreover, the spatial anomaly in depressions during the wet periods was much greater in the near surface than in the root zone.

When SWCs of all 23 dates were used for model development, only EOF1 was statistically significant (Fig. 4a), which accounted for 84.3 % (0–0.2 m) and 86.5 % (0–1.0 m) of the variances in the Z_{tn}. Correlation analysis indicated that the spatial pattern of EOF1 in the Z_{tn} was identical to the time-stable patterns ($M_{\hat{t}n}$) in the TA model ($R = 1.0$). The controls of EOF1 was therefore the same as those of $M_{\hat{t}n}$, and will be discussed later. The relationship between associated EC1

and $S_{\hat{t}n}$ can be fitted well by the cosine function ($R^2 = 0.73$ at both the near surface and root zone) (Fig. 4b).

3.1.2 Time-stable pattern ($M_{\hat{t}n}$), space-invariant temporal anomaly ($A_{t\hat{n}}$), and space-variant temporal anomaly (R_{tn})

Figure 3b displays the three components in the TA model. The first component $M_{\hat{t}n}$ fluctuated along the transect, with high values in depressions and low values on knolls; the $M_{\hat{t}n}$ also had greater spatial variability in the near surface (variance = 36.7 %²) than in the root zone (variance = 19.5 %²). For both soil layers, SOC, depth to the $CaCO_3$ layer, sand content, and wetness index are the dominant factors of $M_{\hat{t}n}$; they together explained 74.5 % (near surface) and 75.6 % (root zone) of the variances in the $M_{\hat{t}n}$ (Table 1). In addition, the temporal trend of $A_{t\hat{n}}$ was the same as that of $S_{\hat{t}n}$ in the SA model (Fig. 3a) as both represent temporal forcing.

The R_{tn} varied among landscape positions (Fig. 3b). At a sampling distance of 123 m (in a depression), R_{tn} was negative in dry periods such as 23 August 2008 and positive in wet periods such as 13 May 2011. This was true for all depressions for both the near surface and the root zone. Therefore, topographically lower positions usually corresponded to more positive R_{tn} during the wet periods and more neg-

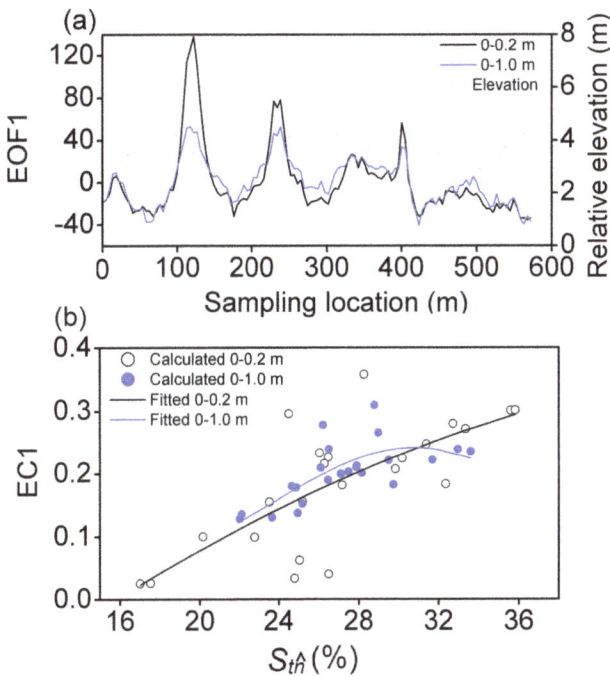

Figure 4. (a) The EOF1 of the spatial anomaly Z_{tn} and **(b)** relationships of associated EC1 versus spatial mean soil water content Z_{tn} fitted by the cosine function (Eq. 4).

ative R_{tn} during the dry periods. Furthermore, the absolute values of R_{tn} were generally greater in the near surface than the root zone, indicating a greater space-variant temporal anomaly for shallower depths.

The SWC variances and associated components (Eq. 8) also varied with time (Fig. 5). Often, wetter conditions corresponded to greater $\sigma_{\hat{n}}^2 (S_{tn})$, as further indicated by moderate correlation between $\sigma_{\hat{n}}^2 (S_{tn})$ and $S_{t\hat{n}}$ (R^2 of 0.51 and 0.38 for the near surface and the root zone, respectively). This was in agreement with others (Gómez-Plaza et al., 2001; Martínez-Fernández and Ceballos, 2003; Hu et al., 2011). Furthermore, there were greater $\sigma_{\hat{n}}^2 (S_{tn})$ values at the near surface than in the root zone, indicating greater variability of SWC in the near surface.

The time-invariant $\sigma_{\hat{n}}^2 (M_{\hat{i}n})$ accounted for the $\sigma_{\hat{n}}^2 (S_{tn})$ with percentages ranging from 25 to 795 % for the near surface and from 40 to 174 % for the root zone (Fig. 5). The $\sigma_{\hat{n}}^2 (M_{\hat{i}n})$ exceeded the $\sigma_{\hat{n}}^2 (S_{tn})$ mainly under dry conditions, such as July–October in 2008 and 2009. This excess was offset by the $\sigma_{\hat{n}}^2 (R_{tn})$ and $2 \operatorname{cov} (M_{\hat{i}n}, R_{tn})$, with the latter accounting for the $\sigma_{\hat{n}}^2 (S_{tn})$ negatively with mean absolute percentages of 210 % for the near surface and 17 % for the root zone. In the dry period, the absolute percentage of $2 \operatorname{cov} (M_{\hat{i}n}, R_{tn})$ was up to 1327 % for the near surface and 122 % for the root zone. These values are comparable to those in Mittelbach and Seneviratne (2012) and Brocca et al. (2014).

The $\sigma_{\hat{n}}^2 (R_{tn})$ accounted for less percentage of the $\sigma_{\hat{n}}^2 (S_{tn})$ than other components did (Fig. 5). The percentages of $\sigma_{\hat{n}}^2 (R_{tn})$ ranged from 11 to 632 % (arithmetic average of 118 %) for the near surface and from 6 to 48 % (arithmetic average of 19 %) for the root zone; the percentage of $\sigma_{\hat{n}}^2 (R_{tn})$ tended to be greater in drier periods. This indicates that the space-variant temporal anomaly cannot be ignored, particularly in dry conditions. Furthermore, the percentage of $\sigma_{\hat{n}}^2 (R_{tn})$ was greater in the near surface than in the root zone, confirming stronger temporal dynamics of soil water at the near surface. Compared with larger-scale studies (Mittelbach and Seneviratne, 2012; Brocca et al., 2014), the percentage of $\sigma_{\hat{n}}^2 (R_{tn})$ out of the $\sigma_{\hat{n}}^2 (S_{tn})$ at the near surface was greater, with a mean percentage of 118 %, versus 9–68 % in the other, larger-scale studies. This indicates that interactions between spatial and temporal forcing were stronger, resulting in relatively more intensive temporal dynamics of soil water in our study area than at larger scales.

Three significant EOFs of R_{tn} for both soil layers were identified when SWC of all 23 dates were used for model development. The first three EOFs explained 61.1, 13.4, and 8.1 %, respectively, of the total R_{tn} variance for the near surface, and 44.3, 20.2, and 12.4 %, respectively, of the total R_{tn} variance in the root zone. Therefore, our hypothesis that underlying spatial patterns exist in the R_{tn} was supported. Due to the negligible contribution of EOF2 and EOF3 to the estimation of spatially distributed SWC, only EOF1 is shown in Fig. 6a. The associated EC1 changed with soil water conditions ($S_{t\hat{n}}$) (Fig. 6b). When SWC was close to average levels, the EC1 was close to 0, resulting in negligible R_{tn}. This was in accordance with Mittelbach and Seneviratne (2012) and Brocca et al. (2014), who showed that the spatial variance of the temporal anomaly was the smallest when water contents were close to average levels. The cosine function (Eq. 4) explained a large amount of the variances in EC1 for both soil layers ($R^2 = 0.76$ at the near surface and 0.88 in the root zone).

The contribution of EOF1 to the space-variant temporal anomaly can be examined through the product of the EOF1 and the associated EC1. The EC1 values tended to be positive during wet periods and negative during dry periods (Fig. 6b); more positive EOF1 values were usually observed at locations with greater $M_{\hat{i}n}$ values (Figs. 3b and 6a). Therefore, the product of EOF1 and EC1 led to greater temporal SWC dynamics at wetter locations of both layers in both the wet and dry periods.

Depth to the $CaCO_3$ layer and SOC had significant, positive correlations with EOF1 for both soil layers (R ranging from 0.76 to 0.88; Table 1). They jointly accounted for 81.6 % (near surface) and 81.0 % (root zone) of the variances in EOF1. This implies that locations with a greater depth to the $CaCO_3$ layer and SOC, which correspond to wetter locations such as depressions, usually have greater temporal SWC dynamics during both wet and dry periods.

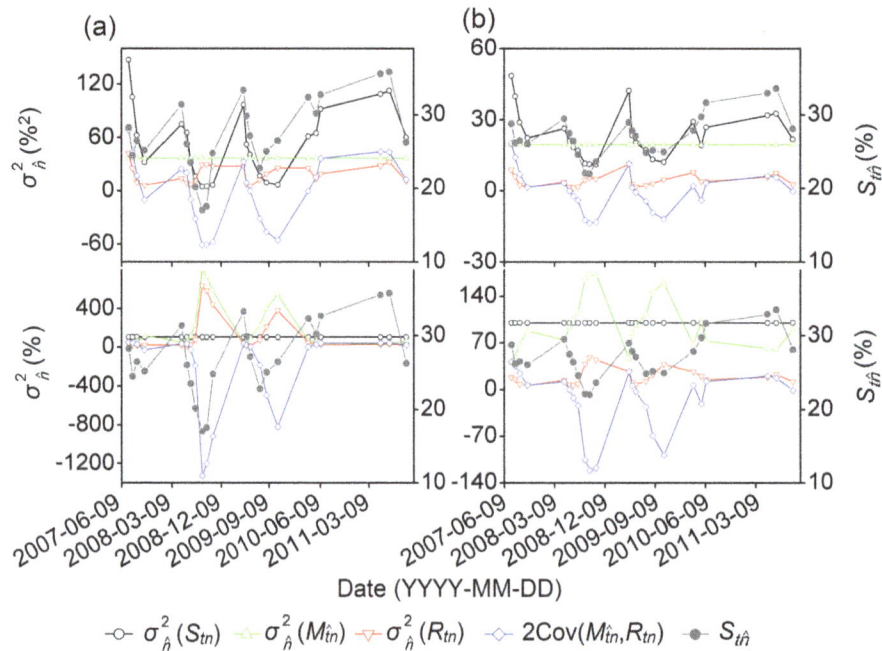

Figure 5. Spatial variances of different components in Eq. (8) expressed in $\%^2$ (upper panel) and as percentage (lower panel) for **(a)** 0–0.2 and **(b)** 0–1.0 m. Spatial mean soil water content $S_{t\hat{n}}$ on each measurement day is also shown.

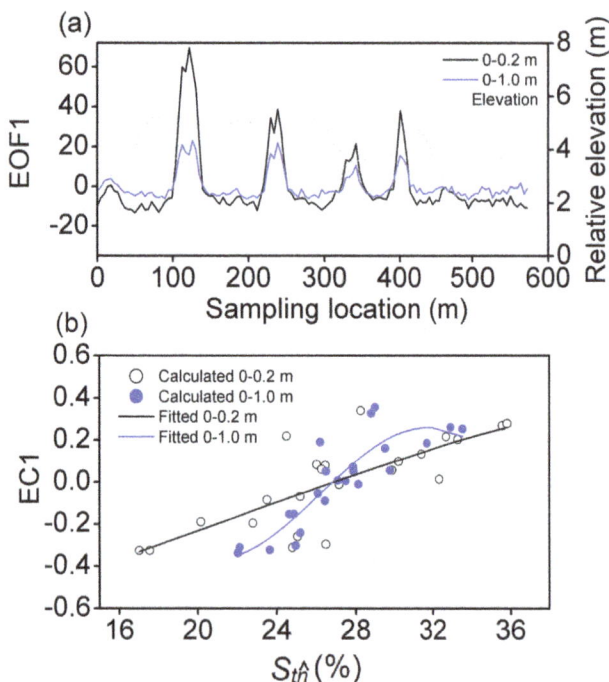

Figure 6. (a) The EOF1 of the space-variant temporal anomaly R_{tn} and **(b)** relationships of associated EC1 versus spatial mean soil water content $S_{t\hat{n}}$ fitted by the cosine function (Eq. 4).

3.2 Estimation of spatially distributed SWC

When all 23 data sets were used and only EOF1 was considered, the TA model had an AICc value of 4093 for the near surface and 562 for the root zone, while the corresponding values for the SA model were 6370 and 3460. This indicated that even when penalty for complexity was given, the TA model was better than the SA model. The two models in terms of spatially distributed SWC estimation are compared below.

3.2.1 The TA model

The R_{tn} terms and associated EOFs differed slightly with each validation. The number of significant EOFs varied between one (accounting for 60 % of the total cases) and three for both soil layers. A paired samples T test indicated that more EOFs did not result in a significant increase of NSCE in the estimation of spatially distributed SWC for both validation methods. This is also supported by the increasing AICc values with the increasing number of parameters resulting from more EOFs (data not shown). This indicates that higher-order EOFs, even if they are statistically significant, are negligible for SWC prediction. Therefore, SWC distribution was estimated with EOF1 only.

Estimated SWCs generally approximated those measured at different soil water conditions during the cross-validation (Fig. 7). However, on 27 October 2009, there were unsatisfactory overestimates at the 100–140 and 220–225 m locations near the surface (Fig. 7a). Unsatisfactory NSCE values

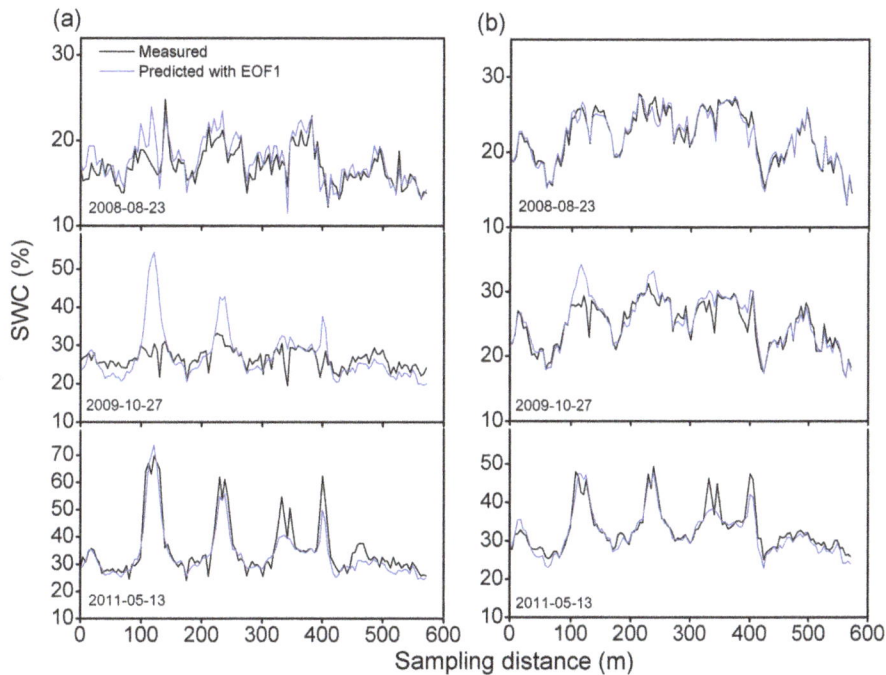

Figure 7. Estimated soil water content (SWC) versus measured SWC for three dates at different soil water conditions (23 August 2008, 27 October 2009, and 13 May 2011 are associated with relatively dry, medium, and wet days, respectively) using the TA model for **(a)** 0–0.2 and **(b)** 0–1.0 m.

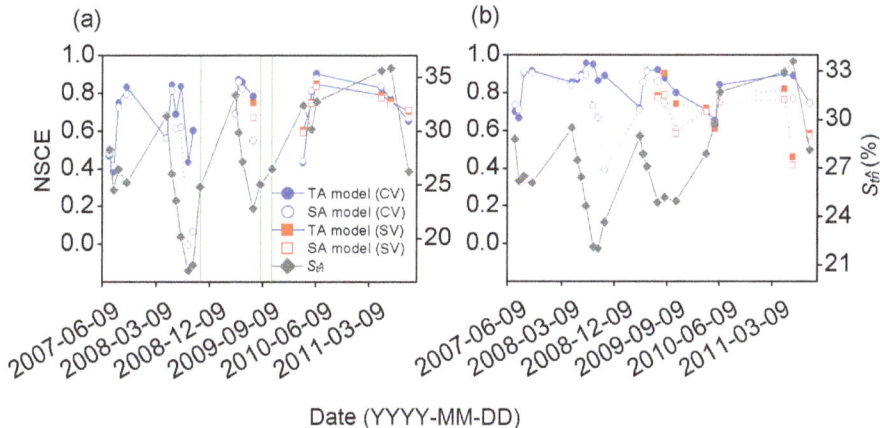

Figure 8. The Nash–Sutcliffe coefficient of efficiency (NSCE) of soil water content estimation using the TA and SA models for **(a)** 0–0.2 and **(b)** 0–1.0 m for both cross-validation (CV) and split sample validation (SV). At 0–0.2 m, three dates (22 October 2008, 27 August 2009, and 27 October 2009) as indicated by green lines present negative NSCE values (−4.05, −1.83, and −3.81, respectively, for the CV on the three dates; −2.63 and −5.12, respectively, for the SV on the latter two dates). Spatial mean soil water content $S_{t\hat{n}}$ on each measurement day is also shown.

of −4.05, −1.83, and −3.81 were obtained in the near surface in only three of the 23 dates, which were all in the fall (22 October 2008, 27 August 2009, and 27 October 2009). The poor performance obtained with the TA model on those dates (Fig. 8a) was a result of overestimation in depressions, which is shown for example on 27 October 2009 (Fig. 7a). These dates also corresponded to a high percentage of $\sigma_{\hat{n}}^2 (R_{tn})$ to the $\sigma_{\hat{n}}^2 (S_{tn})$ (203–439 %). For 23 August and

17 September in 2008, which were in dry periods, the percentage of $\sigma_{\hat{n}}^2 (R_{tn})$ at the near surface was also high (580 and 630 %). Because a fair amount of $\sigma_{\hat{n}}^2 (R_{tn})$ was accounted for with the TA model, the TA model performed satisfactorily (NSCE of 0.43 and 0.60). For the remaining 20 dates, the resulting NSCE value ranged from 0.38 to 0.90 in the near surface and from 0.65 to 0.96 in the root zone (Fig. 8). This

suggests that the TA model was generally satisfactory, with better performance in the root zone than in the near surface.

During the split sample validation, the TA model resulted in SWC estimations with NSCE values ranging from 0.61 to 0.85 near the surface and from 0.32 to 0.92 in the root zone, with exception of 2 days (27 August 2009 and 27 October 2009 with NSCE values of -2.63 and -5.12, respectively) at 0–0.2 m (Fig. 8). This suggested that the TA model performed well in estimating spatially distributed SWC patterns except on 27 August 2009 and 27 October 2009 at 0–0.2 m. The estimation in the root zone was also generally better than in the near surface.

3.2.2 Comparison with the SA model

One significant EOF of Z_{tn} was identified for both soil layers, irrespective of the validation method. The SA model with only EOF1 produced reasonable SWC estimations for both validations in all dates in the root zone and in every date except five dates (23 August 2008, 17 September 2008, 22 October 2008, 27 August 2009, and 27 October 2009) in the near surface (Fig. 8). Similarly, when more EOFs were included, NSCE values did not increase significantly (data not shown) and consequently, estimation of spatially distributed SWC was not improved. This was because EOF2 and EOF3 together explained a very limited ($< 10\%$) amount of variability of Z_{tn} and thus had low predictive power in terms of variance.

The difference in NSCE values between the TA and SA models for both validations are presented in Fig. 9. Generally, the difference decreased as $A_{t\hat{n}}$ increased, and then slightly increased with a further increase in $A_{t\hat{n}}$. A paired samples T test indicated that the NSCE values of the TA model were significantly ($P < 0.05$) greater than those of the SA model for both soil layers, irrespective of validation methods. This indicates that the TA model outperformed the SA model, particularly in dry conditions. This was because when the soil was dry, there was a high percentage of $\sigma_{\hat{n}}^2$ (R_{tn}), and thus strong variability in the space-variant temporal anomaly.

3.3 Further application at other two sites with different scales

3.3.1 A hillslope in the Chinese Loess Plateau

On average, the $\sigma_{\hat{n}}^2$ ($M_{\hat{t}n}$), $\sigma_{\hat{n}}^2$ (R_{tn}), and 2 cov ($M_{\hat{t}n}$, R_{tn}) accounted for 53, 74, and -27% out of the $\sigma_{\hat{n}}^2$ (S_{tn}), indicating that both time-stable pattern and temporal anomalies were the main contributors to the $\sigma_{\hat{n}}^2$ (S_{tn}). The EOF analysis showed that only the EOF1 was statistically significant for both the R_{tn} and Z_{tn}, and the EOF1 explained 23 and 47% of the total variances of R_{tn} and Z_{tn}, respectively. This illustrated that underlying spatial patterns exist in the R_{tn} on the hillslope. Cross-validation was used to estimate the

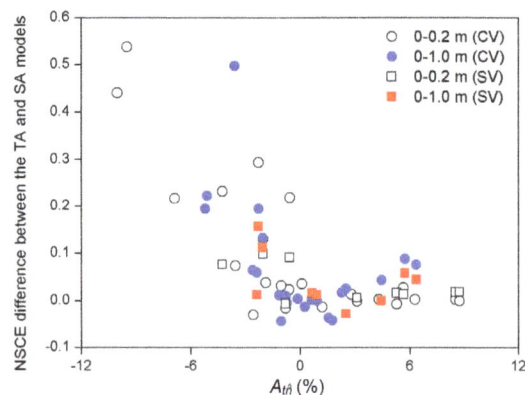

Figure 9. Nash–Sutcliffe coefficient of efficiency (NSCE) difference between the TA and SA models in terms of soil water content estimation using both cross-validation (CV) and split sample validation (SV) as a function of space-invariant temporal anomaly $A_{t\hat{n}}$ for (**a**) 0–0.2 and (**b**) 0–1.0 m.

spatially distributed SWC along the hillslope. The results showed that the NSCE varied from -4.25 to 0.83 (TA model) and from -4.30 to 0.81 (SA model), with a mean value of 0.25 and 0.19, respectively (Fig. 10a). A paired samples T test showed that the NSCE values for the TA model were significantly ($P < 0.05$) greater than those for the SA model, indicating that the TA model outperformed the SA model. As Fig. 10a shows, the outperformance was greater when SWC deviated from intermediate conditions, especially for dry conditions, which was similar to the Canadian site.

3.3.2 The GENCAI network in Italy

The $\sigma_{\hat{n}}^2$ ($M_{\hat{t}n}$), $\sigma_{\hat{n}}^2$ (R_{tn}), and 2 cov ($M_{\hat{t}n}$, R_{tn}) accounted for 38, 68, and -7% out of the $\sigma_{\hat{n}}^2$ (S_{tn}) (Brocca et al., 2014), indicating the dominant role of temporal anomalies in SWC variability. The first three EOFs of the R_{tn} explained 19, 16, and 8% of the total $\sigma_{\hat{n}}^2$ (R_{tn}), and no EOFs were statistically significant, indicating that no underlying spatial patterns exist in the R_{tn}. The EOF1 of the Z_{tn} was significant and accounted for 37% of the variances in the Z_{tn}. Although the EOF1 of the R_{tn} was not significant, it was considered in the TA model for estimating spatially distributed SWC. The cross-validation indicates that the NSCE varied from -0.79 to 0.50 (TA model) and from -0.87 to 0.56 (SA model), with mean values of 0.09 and 0.08, respectively (Fig. 10b). The SWC estimation based on these two models was not satisfactory except for a few days. As Fig. 10b shows, the differences in NSCE values between the two models were scattered around 0. A paired samples T test showed that the NSCE values between the TA model and the SA model were not significant ($P < 0.05$), indicating no differences in estimating spatially distributed SWC between these two models.

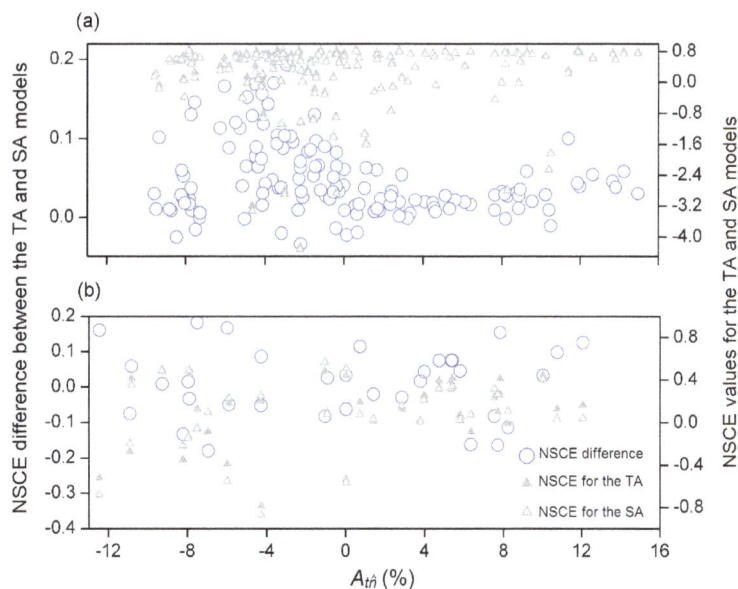

Figure 10. Nash–Sutcliffe coefficient of efficiency (NSCE) difference between the TA and SA models in terms of soil water content estimation using cross-validation as a function of space-invariant temporal anomaly $A_{t\hat{n}}$ for **(a)** 0–0.06 m of the Chinese Loess Plateau hillslope and **(b)** 0–0.15 m of the GENCAI network in Italy. The NSCE values for both models are also shown.

4　Discussion

4.1　Controls of the $M_{\hat{i}n}$ and R_{tn}

The R_{tn} played an important role in the temporal change in spatial patterns of the SWC. The underlying spatial patterns and physical meaning in the R_{tn} were examined in our study for the first time. Although three significant EOFs of the R_{tn} existed in some cases, only EOF1 rather than higher-order EOFs of the R_{tn} should be considered for the spatially distributed SWC estimation. Among many factors influencing the EOF1 of the R_{tn}, depth to the $CaCO_3$ layer followed by the SOC, were the most important factors. Depressions have deeper $CaCO_3$ layers than knolls, and the shallow $CaCO_3$ layer on knolls limited water infiltration during rainfall or snowmelt, resulting in less water recharge on knolls than in depressions. The depth to $CaCO_3$ layer and SOC were negatively correlated with elevation ($R = -0.54$, $P < 0.01$). Therefore, the influence of depth to $CaCO_3$ layer and SOC partially reflected the role of topography in driving snowmelt runoff along slopes in the spring, which contributes to increasing water recharge in depressions. As already demonstrated, topographically lower positions corresponded to more negative R_{tn} during the dry periods. This implies that depressions lost more water during discharge. This is because depressions usually corresponded to vegetation with a larger leaf area index, which would result in higher evapotranspiration and more water loss during discharge periods.

As Table 1 shows, both the depth to the $CaCO_3$ layer and SOC controlled the $M_{\hat{i}n}$. This was because deeper $CaCO_3$

layers and higher SOC were observed in depressions where soils were usually wetter in most of the year because of the snowmelt runoff in the spring and rainfall runoff in the summer and autumn (van der Kamp et al., 2003). Therefore, the roles of soil and topography were two-fold: On one hand, they were highly correlated with the time-stable patterns and thus the time stability of SWC (Gómez-Plaza et al., 2000; Mohanty and Skaggs, 2001; Grant et al., 2004); on the other hand, soil and topography, interplaying with temporal forcing, triggered local-specific soil water change and destroyed time stability of SWC. Their roles in protecting time stability persisted, but their roles in destroying time stability varied with time. Greater $\sigma_{\hat{n}}^2(R_{tn})$ implies greater contribution of these factors in soil water dynamics, resulting in less time stability of SWC.

4.2　Model performance for spatially distributed SWC estimation

The outperformance of the TA model for estimating spatial SWC at the Canadian site and Chinese site can be partly explained by the high percentages (average of 19–118%) of the $\sigma_{\hat{n}}^2(R_{tn})$ out of the total variance. When SWC is close to average levels, R_{tn} is also close to zero, resulting in negligible percentage of $\sigma_{\hat{n}}^2(R_{tn})$. In this case, the soil water patterns are stable in time, the SA model performs well, and there will be little difference between these two models. As is well known, the spatial patterns in soil water content are inherently time unstable. For example, when evapotranspiration becomes the dominant process at the small watershed scale, more water will be lost in depressions due to the denser vegetation than on knolls (Millar, 1971; Biswas et al., 2012), effectively diminishing the spatial patterns and increasing temporal instability. In this case, the $\sigma_{\hat{n}}^2$ (R_{tn}) accounts for more percentage of the total variance (e.g., high up to 632 %) and the TA model may outperform the SA model. This explained why the outperformance of the TA model was more obvious in the dry conditions. For the GENCAI network in Italy, although the $\sigma_{\hat{n}}^2$ (R_{tn}) accounted for 68 % of the total variance, the performance of the TA model was identical to the SA model. This was because there were no underlying spatial patterns in the R_{tn}. Similarly, because the first underlying spatial pattern (i.e., EOF1) explained greater percentages of the $\sigma_{\hat{n}}^2$ (R_{tn}) at the Canadian site (44–61 %) than the Chinese site (23 %), the outperformance of the TA model over the SA model was more obvious at the former site (Figs. 9 and 10a). Therefore, the TA model is advantageous only if the percentage of $\sigma_{\hat{n}}^2$ (R_{tn}) out of the total variance is substantial and underlying spatial patterns exist in the R_{tn}.

The existence of underlying spatial patterns in the R_{tn} is related to the controlling factors, which may be scale specific. At small scales, static factors such as the depth to the $CaCO_3$ layer and SOC at the Canadian site may affect not only the time-stable patterns but also the R_{tn}. The persistent influence of static factors on the R_{tn} resulted in significant underlying spatial patterns in the R_{tn}. Thus, the TA model outperformed the SA model at the small scales. At large scales such as the basin scale or greater, time-stable patterns may be controlled by, in addition to soil and topography (Mittelbach and Seneviratne, 2012), the climate gradient (Sherratt and Wheater, 1984); at those scales, R_{tn} is more likely to be controlled by the meteorological anomaly (i.e., spatially random variation) (Walsh and Mostek, 1980), and the effects of soil and topography may be reduced. Consequently, spatial patterns in the R_{tn} may be weakened and the TA model may have no advantages over the SA model such as for the Italian site.

The $M_{\hat{i}n}$ and the underlying spatial patterns (EOF1) in the R_{tn} were controlled by the same spatial forcing (e.g., depth to $CaCO_3$ layer and SOC) at the Canadian site (Table 1), and they were correlated with an R^2 of 0.83 for the near surface and 0.42 for the root zone. Although the relationships between $M_{\hat{i}n}$ and R_{tn} were strong, they were not strictly linear, suggesting that $M_{\hat{i}n}$ and R_{tn} were affected differently by these factors. Therefore, the nonlinear relationship between $M_{\hat{i}n}$ and R_{tn} partially contributed to the outperformance of the TA model over the SA model.

The relationship between the $S_{t\hat{n}}$ and EC1 was better fitted by the cosine function in the TA model than the SA model (Figs. 4b and 6b), with R^2 of 0.76 versus 0.73 in the near surface and 0.88 versus 0.73 in the root zone. The reduced scatter in the $S_{t\hat{n}}$ and EC1 relationship for the TA model may also partly explain the outperformance of the TA model over the SA model.

Therefore, the outperformance of the TA model over the SA model depends on counterbalance among the variance

of R_{tn} explained in the TA model, the linear correlation between the $M_{\hat{i}n}$ and EOF1 of the R_{tn}, and the goodness of fit for the $S_{t\hat{n}}$ and EC1 relationship. For example, the variance of EOF1 in the R_{tn} for the near surface (i.e., 264 %²) was much greater than that for the root zone (i.e., 43 %²). However, $M_{\hat{i}n}$ and underlying spatial patterns (EOF1) in the R_{tn} in the root zone deviated more from a linear relationship, and the reduced scatter in the $S_{t\hat{n}}$ and EC1 relationship in the TA model was more obviously in the root zone than in the near surface. As a result, the outperformance of the TA model was comparable between the near surface and root zone at the Canadian site (Fig. 9).

In the real world, the relations between the $M_{\hat{i}n}$ and underlying spatial patterns in the R_{tn} may rarely be perfectly linear. Therefore, when underlying spatial patterns exist in the R_{tn} and the R_{tn} has substantial variances, the TA model is preferable to the SA model for the estimation of spatially distributed SWC. On the other hand, when underlying spatial patterns do not exist in the R_{tn} or the R_{tn} has negligible variances, the SA model may be selected although these two models yield the same quality of SWC estimation. This is because the TA model needs one more spatial parameter (i.e., $M_{\hat{i}n}$) than the SA model.

Previous studies on SWC decomposition mainly focus on near-surface layers (Jawson and Niemann, 2007; Perry and Niemann, 2007, 2008; Joshi and Mohanty, 2010; Korres et al., 2010; Busch et al., 2012). This study decomposed spatiotemporal SWC using the TA model for both the near surface and the root zone. The results showed that the estimation of spatially distributed SWC at small watershed scales was improved by the TA method that considers the R_{tn}. The $\sigma_{\hat{n}}^2$ ($M_{\hat{i}n}$) was greater than the $\sigma_{\hat{n}}^2$ (R_{tn}) (Fig. 5), indicating that time stability was more important than time instability for SWC estimation. For the three dates in the fall (i.e., 22 October 2008, 27 August 2009, and 27 October 2009), strong evapotranspiration and deep drainage in depressions resulted in a much lower SWC at the near surface than in the spring. This resulted in reduced time stability of SWC patterns and poor performance of both models in terms of SWC evaluation (Fig. 8a). Because of the stronger time stability of SWC in deeper soil layers (Biswas and Si, 2011), SWC evaluation was more accurate for soil layers extending from the surface to greater depth. This is particularly important because SWC data for deeper soil layers in a watershed is more difficult to collect than that of surface soil.

5   Conclusions

The TA model was used to decompose spatiotemporal SWC into time-stable patterns $M_{\hat{i}n}$, space-invariant temporal anomaly $A_{t\hat{n}}$, and space-variant temporal anomaly R_{tn}. This study indicated that underlying spatial patterns may exist in the R_{tn} at small scales (e.g., small watersheds and hillslope) but may not exist at large scales such as the GENCAI network ($\sim 250\,km^2$) in Italy. This was because the R_{tn} at small scales was driven by *static* factors such as depth to the $CaCO_3$ layer and SOC at the Canadian site, while the R_{tn} at large scales may be dominated by *dynamic* factors such as meteorological anomaly. Compared to the SA model, estimation of spatially distributed SWC was improved with the TA model at small watershed scales. This was because the TA model considered a fair amount of spatial variance in the R_{tn}, which was ignored in the SA model. Furthermore, the improved performance was observed mainly when there was less or more soil water than the average level, especially in drier conditions due to the high $\sigma_{\hat{n}}^2$ (R_{tn}) value.

This study showed that outperformance of the TA model over the SA model is possible when $\sigma_{\hat{n}}^2$ (R_{tn}) accounts for substantial variance of SWC, and significant spatial patterns (or EOFs) exist in the R_{tn}. Further application of the TA model for the estimation of spatially distributed SWC at different scales and hydrological backgrounds is recommended. If the TA model parameters (i.e., $M_{\hat{tn}}$, EOF1 of the R_{tn}, and relationship between EC and $S_{t\hat{n}}$) are obtained from historical in situ SWC data sets, a detailed spatially distributed SWC of near-surface soil at watershed scales can be constructed from remotely sensed SWC. Note that both models rely on in situ SWC measurements for model parameters. Therefore, future research should be conducted to estimate spatially distributed SWC in un-gauged watersheds based on the estimation of the model parameters using pedotransfer functions. The codes for decomposing SWC with the SA and TA models and related EOF analysis were written in Matlab and are freely available from the authors upon request.

Appendix A

Table A1. Notations.

$M_{\hat{t}\hat{n}}$	spatial mean of $M_{\hat{t}n}$
R_{tn}	space-variant temporal anomaly of SWC at location n and time t
$A_{t\hat{n}}$	space-invariant temporal anomaly of SWC at time t
Z_{tn}	spatial anomaly of SWC at location n and time t
$S_{t\hat{n}}$	spatial mean SWC at time t
$\sigma_{\hat{n}}^2$	spatial variance
A_{tn}	temporal anomaly of SWC at location n and time t
$\delta_{\hat{t}n}$	temporal mean relative difference of SWC at location n
cov	spatial covariance
S_{tn}	SWC at location n and time t
$M_{\hat{t}n}$	time-stable pattern of SWC
ECs	temporally varying coefficients of R_{tn} (or Z_{tn})
EOFs	time-invariant spatial structures of R_{tn} (or Z_{tn})
NSCE	Nash–Sutcliffe coefficient of efficiency
R	Pearson correlation coefficient
SWC	soil water content

Acknowledgements. This project was funded by the National Science Foundation of China (K305021308) and the Natural Sciences and Engineering Research Council (NSERC) of Canada. We thank Asim Biswas, Henry Wai Chau, Trent Pernitsky, and Eric Neil for their help in data collection. We thank the anonymous reviewers and the Editor for their constructive comments.

References

Biswas, A. and Si, B. C.: Scales and locations of time stability of soil water storage in a hummocky landscape, J. Hydrol., 408, 100–112, doi:10.1016/j.jhydrol.2011.07.027, 2011.

Biswas, A., Chau, H. W., Bedard-Haughn, A., and Si, B. C.: Factors controlling soil water storage in the Hummocky landscape of the Prairie Pothole region of North America, Can. J. Soil Sci., 92, 649–663, doi:10.4141/CJSS2011-045, 2012.

Blöschl, G., Komma, J., and Hasenauer, S.: Hydrological downscaling of soil moisture, Final report to the H-SAF (Hydrology Satellite Application Facility) via the Austrian Central Institute for Meteorology and Geodynamics (ZAMG), Vienna University of Technology, Vienna, Austria, 2009.

Brocca, L., Melone, F., Moramarco, T., and Morbidelli, R.: Soil moisture temporal stability over experimental areas in Central Italy, Geoderma, 148, 364–374, doi:10.1016/j.geoderma.2008.11.004, 2009.

Brocca, L., Tullo, T., Melone, F., Moramarco, T., and Morbidelli, R.: Catchment scale soil moisture spatial-temporal variability, J. Hydrol., 422–423, 63–75, doi:10.1016/j.jhydrol.2011.12.039, 2012.

Brocca, L., Zucco, G., Moramarco, T., and Morbidelli, R.: Developing and testing a long-term soil moisture dataset at the catchment scale, J. Hydrol., 490, 144–151, doi:10.1016/j.jhydrol.2013.03.029, 2013.

Brocca, L., Zucco, G., Mittelbach, H., Moramarco, T., and Seneviratne, S. I.: Absolute versus temporal anomaly and percent of saturation soil moisture spatial variability for six networks worldwide, Water Resour. Res., 50, 5560–5576, doi:10.1002/2014WR015684, 2014.

Burnham, K. P. and Anderson, D. R.: Model selection and multimodel inference: A practical information-theoretic approach, 2nd Edn., Springer-Verlag, New York, 2002.

Busch, F. A., Niemann, J. D., and Coleman, M.: Evaluation of an empirical orthogonal function-based method to downscale soil moisture patterns based on topographical attributes, Hydrol. Process., 26, 2696–2709, doi:10.1002/hyp.8363, 2012.

Champagne, C., Berg, A. A., McNairn, H., Drewitt, G., and Huffman, T.: Evaluation of soil moisture extremes for agricultural productivity in the Canadian prairies, Agr. Forest Meteorol., 165, 1–11, doi:10.1016/j.agrformet.2012.06.003, 2012.

Famiglietti, J. S., Rudnicki, J. W., and Rodell, M.: Variability in surface moisture content along a hillslope transect: Rattlesnake Hill, Texas, J. Hydrol., 210, 259–281, doi:10.1016/S0022-1694(98)00187-5, 1998.

Gómez-Plaza, A., Alvarez-Rogel, J., Albaladejo, J., and Castillo, V. M.: Spatial patterns and temporal stability of soil moisture across a range of scales in a semi-arid environment, Hydrol. Process., 14, 1261–1277, doi:10.1002/(SICI)1099-1085(200005)14:7<1261::AID-HYP40>3.0.CO;2-D, 2000.

Gómez-Plaza, A., Martínez-Mena, M., Albaladejo, J., and Castillo, V. M.: Factors regulating spatial distribution of soil water content in small semiarid catchments, J. Hydrol., 253, 211–226, doi:10.1016/S0022-1694(01)00483-8, 2001.

Grant, L., Seyfried, M., and McNamara, J.: Spatial variation and temporal stability of soil water in a snow-dominated, mountain catchment, Hydrol. Process., 18, 3493–3511, doi:10.1002/hyp.5789, 2004.

Grayson, R. B. and Western, A. W.: Towards areal estimation of soil water content from point measurements: Time and space stability of mean response, J. Hydrol., 207, 68–82, doi:10.1016/S0022-1694(98)00096-1, 1998.

Hu, W., Shao, M. A., and Reichardt, K.: Using a new criterion to identify sites for mean soil water storage evaluation, Soil Sci. Soc. Am. J., 74, 762–773, doi:10.2136/sssaj2009.0235, 2010.

Hu, W., Shao, M. A., Han, F. P., and Reichardt, K.: Spatiotemporal variability behavior of land surface soil water content in shrub- and grass-land, Geoderma, 162, 260–272, doi:10.1016/j.geoderma.2011.02.008, 2011.

Hu, W., Tallon, L. K., and Si, B. C.: Evaluation of time stability indices for soil water storage upscaling, J. Hydrol., 475, 229–241, doi:10.1016/j.jhydrol.2012.09.050, 2012.

Jawson, S. D. and Niemann, J. D.: Spatial patterns from EOF analysis of soil moisture at a large scale and their dependence on soil, land-use, and topographic properties, Adv. Water Resour., 30, 366–381, doi:10.1016/j.advwatres.2006.05.006, 2007.

Jia, Y. H. and Shao, M. A.: Temporal stability of soil water storage under four types of revegetation on the northern Loess Plateau of China, Agr. Water Manage., 117, 33–42, doi:10.1016/j.agwat.2012.10.013, 2013.

Johnson, R. A. and Wichern, D. W.: Applied multivariate statistical analysis, Prentice Hall, Upper Saddle River, New Jersey, 2002.

Joshi, C. and Mohanty, B. P.: Physical controls of near-surface soil moisture across varying spatial scales in an agricultural landscape during SMEX02, Water Resour. Res., 46, W12503, doi:10.1029/2010WR009152, 2010.

Korres, W., Koyama, C. N., Fiener, P., and Schneider, K.: Analysis of surface soil moisture patterns in agricultural landscapes using Empirical Orthogonal Functions, Hydrol. Earth Syst. Sci., 14, 751–764, doi:10.5194/hess-14-751-2010, 2010.

Martínez-Fernández, J. and Ceballos, A.: Temporal stability of soil moisture in a large-field experiment in Spain, Soil Sci. Soc. Am. J., 67, 1647–1656, 2003.

Millar, J. B.: Shoreline-area ratios as a factor in rate of water loss from small sloughs, J. Hydrol., 14, 259–284, doi:10.1016/0022-1694(71)90038-2, 1971.

Mittelbach, H. and Seneviratne, I.: A new perspective on the spatiotemporal variability of soil moisture: Temporal dynamics versus time-invariant contributions, Hydrol. Earth Syst. Sci., 16, 2169–2179, doi:10.5194/hess-16-2169-2012, 2012.

Mohanty, B. P. and Skaggs, T. H.: Spatio-temporal evolution and time–stable characteristics of soil moisture within remote sensing footprints with varying soil slope and vegetation, Adv. Water Resour., 24, 1051–1067, doi:10.1016/S0309-1708(01)00034-3, 2001.

Peel, M. C., Finlayson, B. L., and McMahon, T. A.: Updated world map of the Köppen–Geiger climate classification, Hydrol. Earth Syst. Sci., 11, 1633–1644, doi:10.5194/hess-11-1633-2007, 2007.

Perry, M. A. and Niemann, J. D.: Analysis and estimation of soil moisture at the catchment scale using EOFs, J. Hydrol., 334, 388–404, doi:10.1016/j.jhydrol.2006.10.014, 2007.

Perry, M. A. and Niemann, J. D.: Generation of soil moisture patterns at the catchment scale by EOF interpolation, Hydrol. Earth Syst. Sci., 12, 39–53, doi:10.5194/hess-12-39-2008, 2008.

Robinson, D. A., Campbell, C. S., Hopmans, J. W., Hornbuckle, B. K., Jones, S. B., Knight, R., Ogden, F., Selker, J., and Wendroth, O.: Soil moisture measurement for ecological and hydrological watershed-scale observatories: A review, Vadose Zone J., 7, 358–389, doi:10.2136/vzj2007.0143, 2008.

Rötzer, K., Montzka, C., and Vereecken, H.: Spatio-temporal variability of global soil moisture products, J. Hydrol., 522, 187–202, doi:10.1016/j.jhydrol.2014.12.038, 2015.

She, D. L., Liu, D. D., Peng, S. Z., and Shao, M. A.: Multiscale influences of soil properties on soil water content distribution in a watershed on the Chinese Loess Plateau, Soil Sci., 178, 530–539, doi:10.1016/j.jhydrol.2014.08.034, 2013a.

She, D. L., Xia, Y. Q., Shao, M. A., Peng, S. Z., and Yu, S. E.: Transpiration and canopy conductance of Caragana Korshinskii trees in response to soil moisture in sand land of China, Agrofor. Syst., 87, 667–678, doi:10.1007/s10457-012-9587-4, 2013b.

Sherratt, D. J. and Wheater, H. S.: The use of surface-resistance soil-moisture relationships in soil-water budget models, Agr. Forest Meteorol., 31, 143–157, doi:10.1016/0168-1923(84)90016-9, 1984.

Soil Survey Staff: Soil Taxonomy, 11th Edn., USDA National Resources Conservation Services, Washington, D.C., 2010.

Starr, G. C.: Assessing temporal stability and spatial variability of soil water patterns with implications for precision water management, Agr. Water Manage., 72, 223–243, doi:10.1016/j.agwat.2004.09.020, 2005.

Vachaud, G., De Silans, A. P., Balabanis, P., and Vauclin, M.: Temporal stability of spatially measured soil water probability density function, Soil Sci. Soc. Am. J., 49, 822–828, 1985.

van der Kamp, G., Hayashi, M., and Gallen, D.: Comparing the hydrology of grassed and cultivated catchments in the semi-arid Canadian prairies, Hydrol. Process., 17, 559–575, doi:10.1002/hyp.1157, 2003.

Vanderlinden, K., Vereecken, H., Hardelauf, H., Herbst, M., Martinez, G., Cosh, M. H., and Pachepsky, Y. A.: Temporal stability of soil water contents: A review of data and analyses, Vadose Zone J., 11, 4, doi:10.2136/vzj2011.0178, 2012.

Venkatesh, B., Nandagiri, L., Purandara, B. K., and Reddy, V. B.: Modelling soil moisture under different land covers in a sub-humid environment of Western Ghats, India, J. Earth Syst. Sci., 120, 387–398, 2011.

Vereecken, H., Kamai, T., Harter, T., Kasteel, R., Hopmans, J., and Vanderborght, J.: Explaining soil moisture variability as a function of mean soil moisture: A stochastic unsaturated flow perspective, Geophys. Res. Lett., 34, L22402, doi:10.1029/2007GL031813, 2007.

Walsh, J. E. and Mostek, A.: A quantitative-analysis of meteorological anomaly patterns over the United-States, 1900–1977, Mon. Weather Rev., 108, 615–630, doi:10.1175/1520-0493(1980)108<0615:AQAOMA>2.0.CO;2, 1980.

Wang, Y. Q., Shao, M. A., Liu, Z. P., and Warrington, D. N.: Regional spatial pattern of deep soil water content and its influencing factors, Hydrolog. Sci. J., 57, 265–281, doi:10.1080/02626667.2011.644243, 2012.

Ward, P. R., Flower, K. C., Cordingley, N., Weeks, C., and Micin, S. F.: Soil water balance with cover crops and conservation agriculture in a Mediterranean climate, Field Crop. Res., 132, 33–39, doi:10.1016/j.fcr.2011.10.017, 2012.

Zhao, Y., Peth, S., Wang, X. Y., Lin, H., and Horn, R.: Controls of surface soil moisture spatial patterns and their temporal stability in a semi-arid steppe, Hydrol. Process., 24, 2507–2519, doi:10.1002/hyp.7665, 2010.

8

Socio-hydrological modelling: a review asking "why, what and how?"

P. Blair and W. Buytaert

Grantham Institute and Department of Civil and Environmental Engineering, Skempton Building, Imperial College London, SW7 2AZ, UK

Correspondence to: P. Blair (peter.blair14@imperial.ac.uk)

Abstract. Interactions between humans and the environment are occurring on a scale that has never previously been seen; the scale of human interaction with the water cycle, along with the coupling present between social and hydrological systems, means that decisions that impact water also impact people. Models are often used to assist in decision-making regarding hydrological systems, and so in order for effective decisions to be made regarding water resource management, these interactions and feedbacks should be accounted for in models used to analyse systems in which water and humans interact. This paper reviews literature surrounding aspects of socio-hydrological modelling. It begins with background information regarding the current state of socio-hydrology as a discipline, before covering reasons for modelling and potential applications. Some important concepts that underlie socio-hydrological modelling efforts are then discussed, including ways of viewing socio-hydrological systems, space and time in modelling, complexity, data and model conceptualisation. Several modelling approaches are described, the stages in their development detailed and their applicability to socio-hydrological cases discussed. Gaps in research are then highlighted to guide directions for future research. The review of literature suggests that the nature of socio-hydrological study, being interdisciplinary, focusing on complex interactions between human and natural systems, and dealing with long horizons, is such that modelling will always present a challenge; it is, however, the task of the modeller to use the wide range of tools afforded to them to overcome these challenges as much as possible. The focus in socio-hydrology is on understanding the human–water system in a holistic sense, which differs from the problem solving focus of other water management fields, and as such

models in socio-hydrology should be developed with a view to gaining new insight into these dynamics. There is an essential choice that socio-hydrological modellers face in deciding between representing individual system processes or viewing the system from a more abstracted level and modelling it as such; using these different approaches has implications for model development, applicability and the insight that they are capable of giving, and so the decision regarding how to model the system requires thorough consideration of, among other things, the nature of understanding that is sought.

1 Introduction

Land-use changes and water resource management efforts have altered hydrological regimes throughout history (Savenije et al., 2014), but the increase in the scale of human interference has led to an intensification in the effects that our interventions have upon the hydrology of landscapes around the world, as well as having significant impacts on societal development, via our co-evolution with water (Liu et al., 2014). Indeed the scale of human intervention that has taken place in meeting the requirements of a population that has expanded from 200 million to 7 billion over the last 2000 years has required such control that in many locations water now flows as man dictates, rather than as nature had previously determined (Postel, 2011). The pace and scale of change that anthropogenic activities are bringing to natural systems are such that hydroclimatic shifts may be brought about in the relatively short term (Destouni et al., 2012), as well as leading to a coupling between human and hydrological systems (Wagener et al., 2010); this coupling

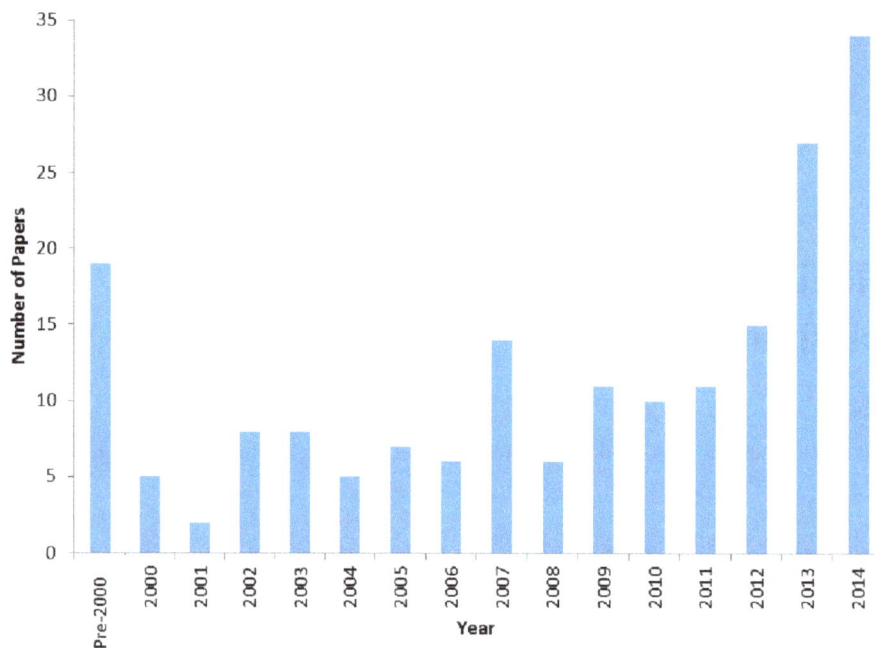

Figure 1. Distribution of years in which papers included in this review were published.

means that both positive and negative social impacts may be brought about via decisions that impact the hydrological system. The growing awareness of the impacts humans are having on a global scale and associated stewardship practices (Steffen et al., 2007) will, therefore, have impacts beyond the ecological and hydrological spheres.

A number of terms have been coined in order to develop the way in which the relationship between mankind and nature, and in particular water, are thought about: "Hydrosociology" (Falkenmark, 1979; Sivakumar, 2012), the "Hydro-social" (Swyngedouw, 2009) and "Hydrocosmological" (Boelens, 2013) cycles and "Ecohydrosolidarity" (Falkenmark, 2009), to name a few. The concept of "The Anthropocene" (Crutzen and Stoermer, 2000; Crutzen, 2002) to describe a new geological epoch in which we now exist, where mankind represents "a global geological force" (Steffen et al., 2007), rivalling the force of nature in the scale of impact on the earth system (Steffen et al., 2011), has been in circulation for some time, and the fact that man and water are linked through a "system of mutual interaction" (Falkenmark, 1977) has been recognised for many years. However, due to factors such as the implicit complexity and uncertainty involved in coupled human and natural systems, the feedbacks and interrelations between society and water are not commonly modelled when forecasting and developing policy. The relatively new field of "Socio-hydrology" (Sivapalan et al., 2012), however, seeks to change this by aiming to understand "the dynamics and co-evolution of coupled human-water systems".

This paper seeks to draw together relevant information and concepts pertaining to the modelling of socio-hydrological systems; it is structured as dealing with the questions of "why?", "what?" and "how?". The "why?" section deals with why socio-hydrological study would be conducted, the different contexts in which socio-hydrological models would be applied, and the possible applications that socio-hydrological models could have; the "what?" section first looks at the distinguishing features of socio-hydrology, as well as the characteristics it shares with other disciplines (and so the lessons that may be learned), before covering different concepts that need to be understood when developing socio-hydrological models; the "how?" section critically examines the application of different modelling techniques to the study of socio-hydrological systems. This structure is used so that the "why?" and "what?" being investigated can introduce readers to literature and concepts of importance to socio-hydrology, and the "how?" section can inform readers of the specific advantages and disadvantages of using different techniques when conducting socio-hydrological modelling. This paper is not intended to be a comprehensive review of all socio-hydrological modelling studies, since there are at this stage few socio-hydrological models in published literature; rather, this paper should be seen as an amalgamation of knowledge surrounding socio-hydrological modelling, such that understanding why and how it could be undertaken is easily accessible. Recently, there have been two excellent papers which have reviewed important aspects of socio-hydrology, which are mentioned here. Troy et al. (2015a) cover the current state of socio-hydrology and give an excellent outline of the different research methodologies that can be used in socio-hydrology (of which modelling is one); the role of the socio-hydrological researcher is also covered particularly well in this paper. Sivapalan and Blöschl (2015) give an in-depth analysis of: co-evolutionary processes (particularly in a mathematical sense); the differences between human and natural systems and the implications of these for modelling; and the overall socio-hydrological modelling process, common across modelling techniques and the different modelling archetypes that might be produced (i.e. stylised versus comprehensive models).

As can be seen in Fig. 1, the number of articles being published which relate to socio-hydrological modelling has increased dramatically over recent years, demonstrating interest in the subject (2015 is not included as this year was not complete at the time of writing, so its inclusion could cause confusion).

1.1 Some background to socio-hydrology

The subject of socio-hydrology, first conceived by Sivapalan et al. (2012), seeks to understand the "dynamics and co-evolution of coupled human-water systems", including the impacts and dynamics of changing social norms and values, system behaviours such as tipping points and feedback mechanisms, some of which may be emergent (unexpected), caused by non-linear interactions between processes occurring on different spatio-temporal scales. Such dynamics include "pendulum swings" that have been observed in areas such as the Murray–Darling Basin, where extensive agricultural development was followed by a realisation of the impacts this was having and subsequent implementation of environmental protection policies (Kandasamy et al., 2014; van Emmerik et al., 2014), the co-evolution of landscapes with irrigation practices and community dynamics (Parveen et al., 2015), as well as instances of catastrophe in which hydrological extremes not been catastrophic in themselves; rather, social processes that result in vulnerability have made extreme events catastrophic (Lane, 2014). There are also cases where social systems have not interacted with water in the way that was anticipated: examples include the virtual water efficiency and peak-water paradoxes discussed by Sivapalan et al. (2014), and yet others where the perception, rather than the actuality, that people have of a natural system determines the way it is shaped (Molle, 2007). Studying these systems requires not only an interdisciplinary approach, but also an appreciation of two potentially opposing ontological and epistemological views: the Newtonian view, whereby reductionism of seemingly complex systems leads to elicitation of fundamental processes, and the Darwinian view, in which patterns are sought, but complexity of system processes is maintained (Harte, 2002). Taking a dualistic worldview encompassing both of these perspectives, as well as the manner in which man and water are related (Falkenmark, 1979), allows for an appreciation of impacts that actions will have due

to physical laws, as well as other impacts that will be brought about due to adaptations from either natural or human systems.

In understanding socio-hydrology as a subject, it may be useful to also briefly understand the history of the terminology within hydrological thinking, and how this has led to the current understanding. Study of the hydrological cycle began to "serve particular political ends" (Linton and Budds, 2013), whereby maximum utility was sought through modification of the cycle, and was viewed initially as fairly separate from human interactions: after several decades this led to a focus on water resource development in the 1970s, language clearly indicative of a utility-based approach. However, a change in rhetoric occurred in the 1980s, when water resource management (WRM) became the focus, and from this followed integrated water resource management (IWRM) and adaptive water management (AWM) (Savenije et al., 2014), the shift from "development" to "management" showing a change in the framing of water, while the concepts of integrated analysis and adaptivity show a more holistic mindset being taken. The introduction of the hydrosocial cycle (Swyngedouw, 2009) shows another clear development in thought, which aimed to "avoid the pitfalls of reductionist…water resource management analysis" (Mollinga, 2014) for the purpose of better water management. "A science, but one that is shaped by economic and policy frameworks" (Lane, 2014), socio-hydrology also represents another advancement in hydrological study, which requires further rethinking of how hydrological science is undertaken.

It is also important to consider how modelling has progressed in the water sciences, particularly in reference to the inclusion of socio-economic aspects. Subjects such as integrated assessment modelling consider socio-economic decisions and impacts alongside biophysical subsystems (generally in a one-way fashion) and can be applied to water resource management problems (for more detail, see Letcher et al., 2007). Hydro-economic modelling includes the capacity to model many aspects of the human–water system via ascribing economic values to water, which reflect the need to allocate water as a scarce resource, and which change across space and time according to the availability and demand (more detail in Harou et al., 2009). Global water resource models have also seen fascinating development; initially considering human impacts on global resources as a boundary condition (considering demand and supply as essentially separate), they increasingly integrate these two aspects and consider the impacts of water availability on demand (Wanders and Wada, 2015; Wada et al., 2013; Haddeland et al., 2014). It is equally important to remember the points of departure between these subjects and socio-hydrology, with socio-hydrology focusing particularly on bi-directional interactions and feedbacks between humans and water, and involving particularly long timescales considering changing values and norms, where the previously mentioned disciplines tend either to treat one or the other system as a boundary condition, or to consider one-way interactions, and generally focus on slightly shorter timescales.

The importance of socio-hydrology has been recognised since its introduction: The International Association of Hydrological Sciences (IAHS) has designated the title of their "Scientific Decade" (2013–2022) as 'Panta Rhei (Everything flows)' (Montanari et al., 2013), in which the aim 'is to reach an improved interpretation of the processes governing the water cycle by focusing on their changing dynamics in connection with rapidly changing human systems' (Montanari et al., 2013). In the IAHS's assessment of hydrology at present (Montanari et al., 2013), it is recognised that current hydrological models are largely conditioned for analysis of pristine catchments and that societal interaction is generally included in separately developed models, so that interactions between the two are not well handled: socio-hydrological study is posited as a step towards deeper integration that has long been called for (Falkenmark, 1979). The recent series of "Debates" papers in *Water Resour. Res.* (Di Baldassarre et al., 2015b; Sivapalan, 2015; Gober and Wheater, 2015; Loucks, 2015; Troy et al., 2015b) shows a real, continued commitment to the development of socio-hyrology as a subject; the unified conclusion of these papers is that the inclusion of the interaction between society and water is necessary in modelling, though the authors varied in their views on how this should be conducted, the sphere within which socio-hydrology should operate, and the value that socio-hydrological models may have. The continued commitment necessary to the subject is highlighted via the statement that "if we who have some expertise in hydrologic modelling do not some other discipline will [include nonhydrologic components in hydrologic models]" (Loucks, 2015).

2 Why?

Regarding why socio-hydrology is necessary, continuing on from the recognised significance of socio-hydrology, understanding of water (perceived or otherwise), as well as intervention following this understanding, has led to large changes in landscapes, which have then altered the hydrological processes that were initially being studied (Savenije et al., 2014), and as such the goals of study in hydrology are subject to regular modification and refinement. The development of socio-hydrology has come from this iterative process. Troy et al. (2015b) point out that, as a subject still in its infancy, socio-hydrology is still learning the questions to ask. However, Sivapalan et al. (2014) sets out the main goals of socio-hydrological study.

– Analysis of patterns and dynamics on various spatiotemporal scales for discernment of underlying features of biophysical and human systems, and interactions thereof.

– Explanation and interpretation of socio-hydrological system responses, such that possible future system movements may be forecast (current water management approaches often result in unsustainable management practices due to current inabilities in prediction).

– Furthering the understanding of water in a cultural, social, economic and political sense, while also accounting for its biophysical characteristics and recognising its necessity for existence.

It is hoped that the achievement of these goals will lead to more sustainable water management and may, for example, lead to the ability to distinguish between human and natural influences on hydrological systems, which has thus far been difficult (Karoly, 2014). Achievement of these goals will involve study in several spheres, including in historical, comparative and process contexts (Sivapalan et al., 2012), as well as 'across gradients of climate, socio-economic status, ecological degradation and human management' (Sivapalan et al., 2014). In accomplishing all of this, studies in socio-hydrology should strive to begin in the correct manner; as Lane (2014) states, "a socio-hydrological world will need a strong commitment to combined social-hydrological investigations that frame the way that prediction is undertaken, rather than leaving consideration of social and economic considerations as concerns to be bolted on to the end of a hydrological study".

Socio-hydrology can learn many lessons from other, similarly interdisciplinary subjects. Ecohydrology is one such subject, whereby the interaction between ecology and hydrology is explicitly included. Rodriguez-Iturbe (2000) gives a number of the questions that ecohydrology attempts to answer, which may be very similar to the questions that socio-hydrology attempts to answer:

– "Is there emergence of global properties out of these [eco-hydrological] dynamics?"

– "Does it tend to any equilibrium values?"

– "Is there a spontaneous emergence…associated with the temporal dynamics?"

– "Can we reproduce some of the observed…patterns?"

– "Is there a hidden order in the space–time evolution which models could help to uncover?"

– "Does the system evolve naturally, for example, without being explicitly directed to do so?"

Ecohydrology could also necessarily be a constituent part of socio-hydrological models, since anthropogenic influences such as land cover change have ecological impacts, which will themselves create feedbacks with social and hydrological systems.

Another aspect of the question of "why socio-hydrology?" is that, in a world where the decisions that mankind makes have such influence, those who make those decisions should be well-informed as to the impacts their decisions may have. As such, those working in water resources should be well-versed in socio-hydrological interaction, seeking to be "T-shaped professionals" (McClain et al., 2012) (technical skills being vertical, coupled with "horizontal" integrated resource management skills), and as such training should certainly reflect this, perhaps learning from the way that ecohydrology is now trained to hydrologists. Beyond being "T-shaped", socio-hydrologists should also seek to collaborate and cooperate with social scientists and sociologists. Socio-hydrology will require study into subjects that many with backgrounds in hydrology or engineering will have little experience in, for instance modelling how social norms change and how these norms cascade into changing behaviours. Learning from and working with those who are experts in these subjects is the best way to move the subject forward.

Regarding why modelling would be conducted in socio-hydrology, there could be significant demand for socio-hydrological system models in several circumstances; however, there are three main spheres in which such modelling could be used (Kelly et al., 2013):

- system understanding

- forecasting and prediction

- policy and decision-making.

The purpose of this section is to give an idea of why socio-hydrological modelling may be conducted, as the techniques used should be steered by what is required of their outputs. This is linked to, though separated from, current and future applications, since the applications will likely require study in all three of the mentioned spheres in the solution of complex problems. In this section, the significance of modelling in each of these areas will be introduced, the limitations that current techniques have investigated, and so the developments that socio-hydrological modelling could bring determined. The three typologies of socio-hydrological study that Sivapalan et al. (2012) present (historical, comparative and process) could all be used in the different spheres. There are of course significant difficulties in socio-hydrological modelling, which should not be forgotten, in particular due to the fact that "characteristics of human variables make them particularly difficult to handle in models" (Carey et al., 2014), as well as issues brought about by emergence, as models developed on current understanding may not be able to predict behaviours that have not previously been observed, or they may indeed predict emergent properties that do not materialise in real-world systems.

2.1   System understanding

"Perhaps a way to combat environmental problems is to understand the interrelations between ourselves and nature" (Norgaard, 1995). Understanding the mechanisms behind system behaviour can lead to a more complete picture of how a system will respond to perturbations, and so guide action to derive the best outcomes. For example, understanding the mechanisms that bring about droughts, which can have exceptionally severe impacts, can allow for better preparation as well as mitigative actions (Wanders and Wada, 2015). Creating models to investigate system behaviour can lead to understanding in many areas; for example, Levin et al. (2012) give the examples of socio-ecological models leading to understanding of how individual actions create system-level behaviours, as well as how system-level influences can change individual behaviours.

IWRM has been the method used to investigate human–water interactions in recent years, but the isolation in which social and hydrological systems are generally treated in this framework leads to limitations in assimilating "the more informative co-evolving dynamics and interactions over long periods" (Elshafei et al., 2014) that are present. This isolation has also led to the understanding of mechanisms behind human–water feedback loops currently being poor, and so integration has become a priority (Montanari et al., 2013).

If models of the coupled human–water system could be developed, this could give great insight into the interactions that occur, the most important processes, parameters and patterns, and therefore how systems might be controlled (Kandasamy et al., 2014). Historical, comparative and process-based studies would all be useful in this regard, as understanding how systems have evolved (or indeed co-evolved Norgaard, 1981) through time, comparing how different locations have responded to change and investigating the linkages between different parameters are all valuable in the creation of overall system understanding. Improved system understanding would also lead to an improvement in the ability for interpretation of long-term impacts of events that have occurred (Kandasamy et al., 2014). It is important to note that, while this study focuses on modelling, system understanding cannot be brought about solely through modelling, and other, more qualitative studies are of value, particularly in the case of historical investigations (e.g. Paalvast and van der Velde, 2014).

2.1.1   Understanding socio-hydrology

Within the goal of system understanding, there should also be a sub-goal of understanding socio-hydrology, and indeed meta-understanding within this. As a subject in which relevance and applicability are gained from the understanding that it generates, but one which is currently in its infancy, there is space for the evaluation of what knowledge exists in socio-hydrology. While the end-goal for socio-hydrology may be to provide better predictions of system behaviour (though this may not be viewed as the goal by all) via better understanding of fundamental human–water processes, this should be informed by an understanding of how well we really understand these processes.

2.1.2 Insights into data

Another sub-goal of system understanding, which will develop alongside understanding, is gaining insight into the data that are required to investigate and describe these systems. When socio-hydrological models are developed, they will require data for their validation; however, these data will not necessarily be available and will not necessarily be conventional in their form (Troy et al., 2015b). As such, new data collection efforts will be required which use new and potentially unconventional techniques to collect new and potentially unconventional data. On the other side of this coin, the nature of data that are collected will surely influence models that are developed within socio-hydrology, and indeed theories on socio-hydrological processes. This brings forth the iterative data–theory–model development process, in which these aspects of knowledge interact to move each other forward (Troy et al., 2015b). The role of data in socio-hydrology is discussed further in Sect. 3.5.

2.2 Forecasting and prediction

Once a system is understood, it may be possible to use models to predict what will happen in the future. Predictive and forecasting models estimate future values of parameters based on the current state of a system and its known (or rather supposed) behaviours. Such models generally require the use of past data in calibration and validation. Being able to forecast future outcomes in socio-hydrological systems would be of great value, as it would aid in developing foresight as to the long-term implications of current decisions, as well as allowing a view to what adaptive actions may be necessary in the future. Wanders and Wada (2015) state that "Better scenarios of future human water demand could lead to more skilful projection for the 21st century", which could be facilitated by "comprehensive future socio-economic and land use projections that are consistent with each other", as well as the inclusion of human water use and reservoirs, which now have "substantial impacts on global hydrology and water resources", as well as "modelling of interacting processes such as human-nature interactions and feedback"; socio-hydrological modelling may be able to contribute in all of these areas.

An example area of study in prediction/forecasting is resilience: prediction of regime transitions is very important in this sphere (Dakos et al., 2015), and while IWRM does explore the relationship between people and water, it does so in a largely scenario-based fashion, which leaves its predictive capacity for co-evolution behind that of socio-hydrology

(Sivapalan et al., 2012), and so in study of such areas a co-evolutionary approach may be more appropriate.

However, there are significant issues in the usage of models for prediction, including the accumulation of enough data for calibration (Kelly et al., 2013). Issues of uncertainty are very important when models are used for forecasting and prediction, as the act of predicting the future will always involve uncertainty. This is a particular issue when social, economic and political systems are included, as they are far more difficult to predict than physically based systems. The necessity of including changing norms and values in socio-hydrology exacerbates this uncertainty, since the timescale and manner in which societies change their norms are highly unpredictable and often surprising. Wagener et al. (2010) also state that "to make predictions in a changing environment, one in which the system structure may no longer be invariant or in which the system might exhibit previously unobserved behaviour due to the exceedance of new thresholds, past observations can no longer serve as a sufficient guide to the future". However, it must surely be that guidance for the future must necessarily be based on past observations, and as such it could be that interpretations of results based on the past should change.

2.3 Policy and decision-making

Decision-making and policy formation are ultimately where model outputs can be put into practice to make a real difference. Models may be used to differentiate between policy alternatives, or optimise management strategies, as well as to frame policy issues, and can be very useful in all of these cases. However, there are real problems in modelling and implementing policy in areas such as in the management of water resources (Liebman, 1976): it is commonly stated that planning involves "wicked" problems, plagued by issues of problem formulation, innumerable potential solutions, issue uniqueness and the difficulties involved in testing of solutions (it being very difficult to accurately test policies without implementing them, and then where solutions are implemented, extricating the impact that a particular policy has had is difficult, given the number of variables typically involved in policy problems) (Rittel and Webber, 1973). Models necessarily incorporate the perceptions of developers, which can certainly vary, and so models developed to investigate the same issue can also be very different, and suggest varying solutions (Liebman, 1976). Appropriate timescales should be used in modelling efforts, as unless policy horizons are very short, neglecting slow dynamics in socio-ecological systems has been said to produce inadequate results (Crépin, 2007). There are also the issues of policies having time lags before impacts (this is compounded by discounting the value of future benefits), uncertainty in their long-term impacts at time of uptake, root causes of problems being obscured by complex dynamics and the fact that large-scale, top-down policy solutions tend not to produce the best results due to the tendency of water systems to be "resistant to fundamental change" (Gober and Wheater, 2014). While the difficulties in managing complex systems (such as human–water systems) are clear, they can, however, be good to manage, as multiple drivers mean that there are multiple targets for policy efforts that may make at least a small difference (Underdal, 2010).

Past water resource policy has been built around optimisation efforts, which have been criticised for having "a very tenuous meaning for complex human-water systems decision making" (Reed and Kasprzyk, 2009), since they assume "perfect problem formulations, perfect information and evaluation models that fully capture all states/consequences of the future" (Reed and Kasprzyk, 2009), meaning that they result in the usage of "optimal" policies that are not necessarily optimal for many of the possible future system states. Another tension in finding optimal or pareto-optimal solutions in complex systems exists where optimising for a given criterion yields solutions which, via the multiple feedbacks that exist, can impact the rest of the system in very different ways (impacts on the rest of the system may go unnoticed if a single criterion is focused on). Techniques such as multi-criteria/multi-objective methods (Hurford et al., 2014; Kain et al., 2007) attempt to improve upon this, producing pareto-efficient outcomes, but still rarely account explicitly for human–water feedbacks.

Good evidence is required for the formation of good policy (Ratna Reddy and Syme, 2014), and so providing this evidence to influence, and improve policy and best management practices should be an aim of socio-hydrology (Pataki et al., 2011), in particular socio-hydrological modelling. Changes in land use are brought about by socio-economic drivers, including policy, but these changes in land use can have knock-on effects that can impact upon hydrology (Ratna Reddy and Syme, 2014), and so land productivity, water availability and livelihoods to such an extent that policy may be altered in the future. Socio-hydrology should at least attempt to take account of these future policy decisions, and the interface between science and policy to improve long-term predictive capacity (Gober and Wheater, 2014). There is a call for a shift in the way that water resources are managed, towards an ecosystem-based approach, which will require a "better understanding of the dynamics and links between water resource management actions, ecological side-effects, and associated long-term ramifications for sustainability" (Mirchi et al., 2014). SES analysis has already been used in furthering perceptions on the best governance structures, and has found that polycentric governance can lead to increased robustness (Marshall and Stafford Smith, 2013), and it may well be that socio-hydrology leads to a similar view of SHSs.

In order for outputs from policy-making models to be relevant they must be useable by stakeholders and decision-makers, not only experts (Kain et al., 2007). Participatory modelling encourages this through the involvement of stakeholders in model formulation, and often improves "buy-in" of stakeholders, and helps in their making sensible decisions

(Kain et al., 2007), as well as an increase in uptake in policy (Sandker et al., 2010). This technique could be well used in socio-hydrological modelling. Gober and Wheater (2015) take the scope of socio-hydrology further, suggesting a need to include a "knowledge exchange" (Gober and Wheater, 2015) component in socio-hydrological study, whereby the communication of results to policy makers and their subsequent decision-making mechanisms are included to fully encompass socio-hydrological interactions. However, Loucks (2015) points out that the prediction of future policy decisions will be one of the most challenging aspects of socio-hydrology.

2.4 Current and future applications

This section follows from the areas of demand for socio-hydrological to give a few examples (not an exhaustive list) of potential, non-location-specific examples of how socio-hydrological modelling could be used. These applications will incorporate system understanding, forecasting and prediction and policy formation, and where these spheres of study are involved they will be highlighted. SES models have been applied to fisheries, rangelands, wildlife management, bioeconomics, ecological economics, resilience and complex systems (Schlüter, 2012), and have resulted in great steps forward. Application of socio-hydrological modelling in the following areas could too result in progress in understanding, forecasting, decision-making and the much-needed modernisation of governance structures (Falkenmark, 2011) in different scenarios. This section should provide insight as to the situations where socio-hydrological modelling may be used in the future, and so guide the discussion of suitable modelling structures.

2.4.1 Understanding system resilience and vulnerability

Resilience can be defined as the ability of a system to persist in a given state subject to perturbations (Folke et al., 2010; Berkes, 2007), and so this "determines the persistence of relationships within a system" and can be used to measure the "ability of these systems to absorb changes of state variables, driving variables, and parameters" (Holling, 1973). Reduced resilience can lead to regime shift, "a relatively sharp change in dynamic state of a system" (Reyer et al., 2015), which can certainly have negative social consequences. SES literature has studied resilience in a great number of ways, and has found it is often the case that natural events do not cause catastrophe on their own; rather, catastrophe is caused by the interactions between extreme natural events and a vulnerable social system (Lane, 2014). Design principles to develop resilience have been developed in many spheres (for instance, design principles for management institutions seeking resilience; Anderies et al., 2004), though in a general sense Berkes (2007) terms four clusters of factors which can build resilience:

- learning to live with change and uncertainty;

- nurturing various types of ecological, social and political diversity;

- increasing the range of knowledge for learning and problem solving; and

- creating opportunities for self-organisation.

Exposure to natural events can lead to emergent resilience consequences in some cases, as in the case where a policy regime may be altered to increase resilience due to the occurrence of a catastrophe, for example London after 1953 (Lumbroso and Vinet, 2011), or Vietnamese agriculture (Adger, 1999), where the same event could perhaps have caused a loss in resilience were a different social structure in place (Garmestani, 2013).

In all systems, the ability to adapt to circumstances is critical in creating resilience (though resilience can also breed adaptivity (Folke, 2006)); in the sphere of water resources, the adaptive capacity that a society has towards hydrological extremes determines its vulnerability to extremes to a great extent, and so management of water resources in the context of vulnerability reduction should involve an assessment of hydrological risk coupled with societal vulnerability (Pandey et al., 2011). An example scenario where socio-hydrological modelling may be used is in determining resilience/vulnerability to drought, the importance of which is highlighted by AghaKouchak et al. (2015) in their discussion of recognising the anthropogenic facets of drought; sometimes minor droughts can lead to major crop losses, whereas major droughts can sometimes result in minimal consequences, which would indicate differing socio-economic vulnerabilities between cases which "may either counteract or amplify the climate signal" (Simelton et al., 2009). Studies such as that carried out by Fraser et al. (2013), which uses a hydrological model to predict drought severity and frequency coupled with a socio-economic model to determine vulnerable areas, and Fabre et al. (2015), which looks at the stresses in different basins over time caused by hydrological and anthropogenic issues, have already integrated socio-economic and hydrological data to perform vulnerability assessments. Socio-hydrological modelling could make an impact in investigating how the hydrological and socio-economic systems interact (the mentioned studies involve integration of disciplines, though not feedbacks between systems) to cause long-term impacts, and so determine vulnerabilities over the longer term. The most appropriate form of governance in socio-hydrological systems could also be investigated further, as differing governance strategies lead to differing resilience characteristics (Schlüter and Pahl-Wostl, 2007): Fernald et al. (2015) has investigated community-based irrigation systems (Acequias) and found that they pro-

duce great system resilience to drought, due to the "complex self-maintaining interactions between culture and nature" and "hydrologic and human system connections". There is also a question of scale in resilience questions surrounding water resources, which socio-hydrology could be used to investigate: individual resilience may be developed through individuals' use of measures of self-interest (for example digging wells in the case of drought vulnerability), though this may cumulatively result in a long-term decrease in vulnerability (Srinivasan, 2013).

An area that socio-hydrological modelling would be able to contribute in is determining dynamics that are likely to occur in systems: this is highly relevant to resilience study, as system dynamics and characteristics that socio-hydrological models may highlight, such as regime shift, tipping points, bistable states and feedback loops, all feature in resilience science. The long-term view that socio-hydrology should take will be useful in this, as it is often long-term changes in slow drivers that drive systems towards tipping points (Biggs et al., 2009). Modelling of systems also helps to determine indicators of vulnerability that can be monitored in real situations. Areas where desertification has/may take place would be ideal case-studies, since desertification may be viewed as "a transition between stable states in a bistable ecosystem" (D'Odorico et al., 2013), where feedbacks between natural and social systems bring about abrupt changes. Socio-hydrology may be able to forecast indicators of possible regime shifts, utilising SES techniques such as identification of critical slowing down (CSD) (Dakos et al., 2015), a slowing of returning to "normal" after a perturbation which can point to a loss of system resilience, as well as changes in variance, skewness and autocorrelation, which may all be signs of altered system resilience (Biggs et al., 2009), to determine the most effective methods of combating this problem.

In studying many aspects of resilience, historical socio-hydrology may be used to examine past instances where vulnerability/resilience has occurred unexpectedly and comparative studies could be conducted to determine how different catchments in similar situations have become either vulnerable or resilient; combinations of these studies could lead to understanding of why different social structure, governance regimes, or policy frameworks result in certain levels of resilience. Modelling of system dynamics for the purposes of system understanding, prediction and policy development are all clearly of relevance when applied to this topic, since in these the coupling is key in determination of the capacity for coping with change (Schlüter and Pahl-Wostl, 2007).

2.4.2 Understanding risk in socio-hydrological systems

Risk is a hugely important area of hydrological study in the wider context: assessing the likelihood and possible consequences of floods and droughts constitutes an area of great importance, and models to determine flood/drought risk help to determine policy regarding large infrastructure decisions, as well as inform insurance markets on the pricing of risk. However, the relationship between humans and hydrological risk is by no means a simple one, due to the differing perceptions of risk as well as the social and cultural links that humans have with water (Linton and Budds, 2013), and so providing adequate evidence for those who require it is a great challenge.

The way in which risk is perceived determines the actions that people take towards it, and this can create potentially unexpected effects. One such impact is known as the "levee effect" (White, 1945), whereby areas protected by levees are perceived as being immune from flooding (though in extreme events floods exceed levees, and the impacts can be catastrophic when they do), and so are often heavily developed, leading people to demand further flood protection and creating a positive feedback cycle. Flood insurance is also not required in the USA if property is "protected" by levees designed to protect against 100-year events (Ludy and Kondolf, 2012), leading to exposure of residents to extreme events. Socio-hydrological thinking is slowly being applied to flood risk management, as is seen in work such as that of Falter et al. (2015), which recognises that "A flood loss event is the outcome of complex interactions along the flood risk chain, from the flood-triggering rainfall event through the processes in the catchment and river system, the behaviour of flood defences, the spatial patterns of inundation processes, the superposition of inundation areas with exposure and flood damaging mechanisms", and that determining flood risk involves "not only the flood hazard, e.g. discharge and inundation extent, but also the vulnerability and adaptive capacity of the flood-prone regions." Socio-hydrology could, however, further investigate the link between human perceptions of risk, the actions they take, the hydrological implications that this has, and therefore the impact this has on future risk to determine emergent risk in socio-hydrological systems.

The impact that humans have on drought is another area where socio-hydrology could be used; work on the impact that human water use has upon drought has been done (e.g. Wanders and Wada, 2015), where it was found that human impacts "increased drought deficit volumes up to 100 % compared to pristine conditions", and suggested that "human influences should be included in projections of future drought characteristics, considering their large impact on the changing drought conditions". Socio-hydrology could perhaps take this further and investigate the interaction between humans and drought, determining different responses to past drought and assessing how these responses may influence the probability of future issues and changes in resilience of social systems.

2.4.3 Transboundary water management

Across the world, 276 river basins straddle international boundaries (Dinar, 2014); the issue of transboundary water management is a clear case where social and hydrolog-

ical systems interact to create a diverse range of impacts that have great social consequences, but which are very hard to predict. These issues draw together wholly socially constructed boundaries with wholly natural hydrological systems when analysed. The social implications of transboundary water management have been studied and shown to lead to varying international power structures (Zeitoun and Allan, 2008) (e.g. "hydro-hegemony" Zeitoun and Warner, 2006), as well as incidences of both cooperation and conflict (in various guises) (Zeitoun and Mirumachi, 2008) dependent on circumstance. The virtual water trade (Hoekstra and Hung, 2002) also highlights an important issue of transboundary water management: the import and export of goods almost always involves some "virtual water" transfer since those goods will have required water in their production. This alters the spatial scale appropriate for transboundary water management (Zeitoun, 2013), and investigating policy issues related to this would be very interesting from a socio-hydrological perspective (Sivapalan et al., 2012).

Socio-hydrological modelling could be used to predict the implications that transboundary policies may have for hydrological systems, and so social impacts for all those involved. However, the prediction of future transboundary issues is highly uncertain and subject to a great many factors removed entirely from the hydrological systems that they may impact, and so presents a significant challenge.

2.4.4 Land-use management

The final example situation where socio-hydrological modelling may be applicable is in land-use management. Changes in land use can clearly have wide-ranging impacts on land productivity, livelihoods, health, hydrology, and ecosystem services, which all interact to create changes in perception, which can feed back to result in actions being taken that impact on land management. Fish et al. (2010) posits the idea of further integrating agricultural and water management: "Given the simultaneously human and non-human complexion of land-water systems it is perhaps not surprising that collaboration across the social and natural sciences is regarded as a necessary, and underpinning, facet of integrated land-water policy". Modelling in socio-hydrology may contribute in this sphere through the development of models which explore the feedbacks mentioned above, and which can determine the long-term impacts of interaction between human and natural systems in this context.

3 What?

The question of "what?" in this paper can be viewed in several different ways: What are the characteristics of socio-hydrological systems? What is to be modelled? What are the issues that socio-hydrological systems will present to modellers?

3.1 Socio-hydrology and other subjects

The question of what is different and new about socio-hydrology, and indeed what is not, is useful to investigate in order to then determine how knowledge of modelling in other, related subjects can or cannot be transferred and used in socio-hydrology. Here, the subject of socio-ecology (as a similar synthesis subject) is introduced, before the similarities and differences between socio-hydrology and other subjects are summarised.

3.1.1 Socio-ecology

The study of socio-ecological systems (SESs) and coupled human and natural systems (CHANS), involves many aspects similar to that of socio-hydrology: feedbacks (Runyan et al., 2012), non-linear dynamics (Garmestani, 2013), co-evolution (Hadfield and Seaton, 1999), adaptation (Lorenzoni et al., 2000), resilience (Folke et al., 2010), vulnerability (Simelton et al., 2009), issues of complexity (Liu et al., 2007a), governance (Janssen and Ostrom, 2006), policy (Ostrom, 2009) and modelling (Kelly et al., 2013; An, 2012) are all involved in thinking around, and analysis of, SESs. As such, there is much that socio-hydrology can learn from this fairly established (Crook, 1970) discipline, and so in this paper a proportion of the literature presented comes from the field of socio-ecology due to its relevance. Learning from the approaches taken in socio-ecological studies would be prudent for future socio-hydrologists, and so much can be learnt from the manner in which characteristics such as feedback loops, thresholds, time-lags, emergence and heterogeneity, many of which are included in a great number of socio-ecological studies (Liu et al., 2007a) are dealt with. Many key concepts are also applicable to both subject areas, including the organisational, temporal and spatial (potentially boundary-crossing) coupling of systems bringing about behaviour "not belonging to either human or natural systems separately, but emerging from the interactions between them" (Liu et al., 2007b), and the required nesting of systems on various spatio-temporal scales within one another.

Socio-hydrology may, in some ways, be thought of as a sub-discipline of socio-ecology (Troy et al., 2015a); indeed, some studies that have been carried out under the banner of socio-ecology could perhaps be termed socio-hydrological studies (e.g. Roberts et al., 2002; Schlüter and Pahl-Wostl, 2007; Marshall and Stafford Smith, 2013; Molle, 2007), and Welsh et al. (2013) term rivers "complicated socio-ecological systems that provide resources for a range of water needs". There are however, important differences between socio-ecology and socio-hydrology which should be kept in mind when transferring thinking between the two disciplines, for example infrastructure developments such as dams introduce system intervention on a scale rarely seen outside this sphere (Elshafei et al., 2014), and the speed at which some hydro-

logical processes occur at means that processes on vastly different temporal scales must be accounted for (Blöschl and Sivapalan, 1995). There are also unique challenges in hydrological data collection; for example, impracticably long timescales are often being required to capture hydrological extremes and regime changes (Elshafei et al., 2014). Water also flows and is recycled via the hydrological cycle, and so the way that it is modelled is very different to subjects modelled in socio-ecology.

In a study comparable to this, though related to socio-ecological systems, Schlüter (2012) gives research issues in socio-ecological modelling; these issues are also likely to be pertinent in socio-hydrological modelling:

– Implications of complex social and ecological structure for the management of SESs

– The need to address the uncertainty of ecological and social dynamics in decision making

– The role of coevolutionary processes for the management of SESs

– Understanding the macroscale effects of microscale drivers of human behaviour.

Along with studying similarly defined systems and the usage of similar techniques, socio-ecology has suffered problems that could also potentially afflict socio-hydrology. For example, different contributors have often approached problems posed in socio-ecological systems with a bias towards their own field of study, and prior to great efforts to ensure good disciplinary integration social scientists may have "neglected environmental context" (Liu et al., 2007b) and ecologists "focused on pristine environments in which humans are external" (Liu et al., 2007b). Even after a coherent SES framework was introduced (Liu et al., 2007b), some perceived it to be "lacking on the ecological side" (Epstein and Vogt, 2013), and as such missing certain "ecological rules". Since socio-hydrology has largely emerged via scholars with water resources backgrounds, inclusion of knowledge from the social sciences, and collaboration with those in this field, should therefore be high on the agenda of those working in socio-hydrology to avoid similar issues. Another issue that both socio-ecologists and socio-hydrologists face is the tension between simplicity and complexity: the complexity inherent in both types of coupled system renders the development of universal solutions to issues almost impossible, whereas decision-makers prefer solutions to be simple (Ostrom, 2007), and while the inclusion of complexities and interrelations in models is necessary, including a great deal of complexity can result in opacity for those not involved in model development, leading to a variety of issues. The complexity, feedbacks, uncertainties, and presence of natural variabilities in socio-ecological systems also introduce issues in learning from systems due to the obfuscation of system signals (Bohensky, 2014), and similar issues will also be prevalent in socio-hydrological systems.

3.1.2 Similarities between socio-hydrology and other subjects

- Complex systems and co-evolution: studies in socio-ecology and eco-hydrology have had complex and co-evolutionary systems techniques applied to them, and so socio-hydrology may learn from this. While this is one of the ways in which socio-hydrology is similar to socio-ecology and eco-hydrology, it is also one of the ways in which socio-hydrology separates itself from IWRM. The specific aspects of complex/co-evolutionary dynamics that may be learnt from include the following.

  - Non-linear dynamics: socio-hydrology will involve investigating non-linear dynamics, possibly including regime shift, tipping points and time lags, all of which have been investigated in socio-ecology.

  - Feedbacks: the two-way interactions between humans and water will bring about feedbacks between the two, which have important consequences. Discerning impacts and causations in systems with feedbacks, and learning to manage such systems have been covered in socio-ecology and eco-hydrology.

- Uncertainties: while some aspects of the uncertainty present in socio-hydrology are not found in other subjects (see Unique Aspects of Socio-hydrology), some aspects are common with socio-ecology and eco-hydrology. In particular, propogative uncertainties present due to feedbacks and interactions, and the nature of uncertainties brought about by the inclusion of social systems are shared.

- Inter-scale analysis: both socio-ecology and eco-hydrology involve processes which occur on different spatio-temporal scales, so methods for this integration can be found in these subjects.

- Incorporation of trans-/inter-disciplinary processes: socio-ecological models have needed to incorporate social and ecological processes, and so while the particular methods used to incorporate social and hydrological processes may be different, lessons may certainly be learnt in integrating social and biophysical processes.

- Disciplinary bias: researchers in socio-ecology generally came from either ecology or the social sciences, and so studies could occasionally be biased towards either of these. Critiquing and correcting these biases is something that socio-hydrologists can certainly learn from.

3.1.3 Unique aspects of socio-hydrology

- Nature of water combined with nature of social system: while socio-ecology has incorporated social and ecological systems, and eco-hydrology has incorporated hydrological and ecological systems, the integration of hydrological and social systems brings a unique challenge.

  - Nature of water: water is a unique subject to model in many ways. It obeys physical rules, but has cultural and religious significance beyond most other parts of the physical world. It flows, is recycled via the water cycle, and is required for a multitude of human and natural functions. Hydrological events of interest are also often extremes.

  - Nature of social system: aspects of social systems, such as decision-making mechanisms and organisational structures, require models to deal with more than biophysical processes.

  - Particular human–water interactions: there will be particular processes which occur on the interface between humans and people which are neither wholly social nor wholly physical processes. These will require special attention when being modelled, and will necessitate the use of new forms of data.

- The role of changing norms: one of the focuses of socio-hydrological study is the impact of changing social values. Norms change on long timescales and are highly unpredictable, and so will present great difficulties in modelling.

- Scale: socio-hydrological systems will involve inter-scale modelling, but the breadth of spatial and temporal scales necessary for modelling will present unique problems.

- Uncertainties: socio-hydrological systems will involve uncertainties beyond those dealt with in socio-ecology and traditional water sciences. The level of unknown (and indeed unknown unknown) is great, and brings about particular challenges (see later section on uncertainty)

3.2 Concepts

Another aspect to the question of "what?" in this paper is the topic of what concepts are involved when developing socio-hydrological models. These concepts underpin the theory behind socio-hydrology, and as such modelling of SHSs; only when they are properly understood is it possible to develop useful, applicable models. The following sections detail different concepts applicable to socio-hydrological modelling.

3.3   Human–water system representations

People interact with water in complex ways which extend between the physical, social, cultural and spiritual (Boelens, 2013). How the human–water system is perceived is a vital component of socio-hydrological modelling, since this perception will feed into the system conceptualisation (Sivapalan et al., 2003), which will then feed into the model, and as such its outputs. In the past, linear, one-way relationships have often been used, which observations have suggested "give a misleading representation of how social-ecological systems work" (Levin et al., 2012). This unidirectional approach may have been more appropriate in the past when anthropogenic influences were smaller, but since the interactions between hydrology and society have changed recently (as has been described previously), "new connections and, in particular, more significant feedbacks which need to be understood, assessed, modelled and predicted by adopting an interdisciplinary approach" (Montanari et al., 2013), and so the view of systems in models should appreciate this. Views and knowledge of the human–water system have changed over time, and these changes themselves have had a great impact on the systems due to the changes in areas of study and policy that perception and knowledge can bring about (Hadfield and Seaton, 1999).

The concept of the hydrosocial cycle has been a step forward in the way that the relationship between humans and water is thought about, as it incorporates both "material and sociocultural relations to water" (Wilson, 2014). This links well with the view of Archer (1995), who pictured society as a "heterogeneous set of evolving structures that are continuously reworked by human action, leading to cyclic change of these structures and their emergent properties" (Mollinga, 2014). Socio-hydrology uses this hydrosocial representation, and also incorporates human influences on hydrology, whereby "aquatic features are shaped by intertwining human and non-human interaction" to form a bi-directional view of the human–water system (Di Baldassarre et al., 2013a). Technology could also be included in these representations, as was the case in a study by Mollinga (2014), where irrigation was considered in both social and technical terms.

Socio-hydrological human–water system representations should be considered in a case-specific manner, due to the fact that the relationship is very different in different climates. To give an extreme example, the way in which humans and water interact is atypical in a location such as Abu Dhabi, where water is scarce, desalination and water recycling provide much of the freshwater, and as such energy plays a key role (McDonnell, 2013). In this case, energy should certainly be included in socio-hydrological problem formulations since it plays such a key role in the relationship (McDonnell, 2013).

Figure 2 shows an example of a conceptualised socio-hydrological system (Elshafei et al., 2014), which gives insight into the view that the author has of the system. It shows

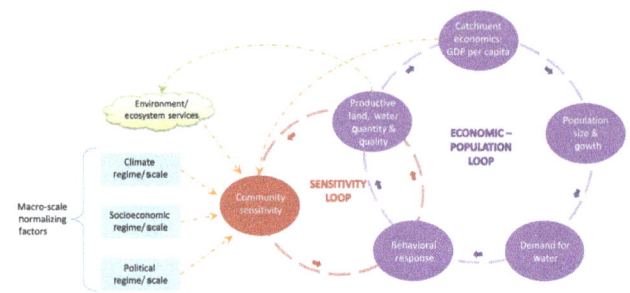

Figure 2. Elshafei et al. (2014), reproduced with permission under the CC Attribution License 3.0. A conceptual representation of a socio-hydrological system (Elshafei et al., 2014).

the linkage perceived between the social and hydrological systems, and the "order" in which the author feels interactions occur. In this system conceptualisation it is perceived that there are two feedback loops which interact to form system behaviour. One is a reinforcing loop, whereby increases in land productivity lead to economic gain, increased population, a higher demand for water and as such changes in management decisions, likely to be intensification of land use (and vice versa); the other loop is termed the "sensitivity loop" (Elshafei et al., 2014), whereby land intensification may impact upon ecosystem services, which, when the climate and socio-economic and political systems are taken into account may increase sensitivity to environmentally detrimental effects, and cause behavioural change. This second loop acts against the former and forms dynamic system behaviour. Others may have different views on the system, for example there may be more (or less) complexity involved in the system, as well as different interconnections between variables, and this would lead to a different conceptual diagram.

When forming a system representation, the topics of complex and co-evolutionary systems should be kept in mind so that these concepts may be applied where appropriate. These concepts are introduced in the following sections.

3.3.1   Complex systems

Complex systems have been studied in many spheres, from economics (Foster, 2005), physics, biology, engineering, mathematics, computer science, and indeed in inter-/transdisciplinary studies involving these areas of study (Chu et al., 2003), or other systems involving interconnected entities within heterogeneous systems (An, 2012). By way of a definition of complex systems, Ladyman et al. (2013) give their view on the necessary and sufficient conditions for a system to be considered complex.

- An "ensemble of many elements": there must be different elements within the system in order for interactions to occur, and patterns to emerge.

- "Interactions": elements within a system must be able to exchange or communicate.

- "Disorder": the distinguishing feature between simple and complex systems is the apparent disorder created by interactions between elements.

- "Robust order": elements must interact in the same way in order for patterns to develop.

- "Memory": robust order leads to memory within a system.

Complex systems representations rely on mechanistic relationships between variables, meaning that the dynamic relationships between different system components do not change over time (Norgaard, 1981), as opposed to evolutionary relationships, whereby responses between components change over time due to natural selection (Norgaard, 1981). Magliocca (2009) investigates the interactions between humans and their landscapes, and determines that emergent behaviours in these systems are due to the "induced coupling" between them, and so should be modelled and managed using complex-systems-appropriate techniques. Resilience has also been studied with regard to complex systems, and the interactions in complex systems have been said to lead to resilience (Garmestani, 2013). Complex systems are an excellent framework within which to study socio-hydrological systems, since they allow for the discernment of the origin of complex behaviours, such as cross-scale interactions, nonlinearity and emergence (Falkenmark and Folke, 2002), due to their structure being decomposable and formed of subsystems that may themselves be analysed.

3.3.2 Co-evolutionary systems

A related, though subtly different view of the human–water relationship is that of a co-evolutionary system. Sivapalan and Blöschl (2015) provide an excellent analysis of the application of the co-evolutionary framework to socio-hydrology, and so for an in-depth view of how to model co-evolutionary systems, the reader is directed here. In this paper an outline of what co-evolutionary systems are is given, before analysing whether this is applicable to socio-hydrology and reviewing applications of the co-evolutionary framework in human–water circumstances.

The strict meaning of a co-evolutionary system is occasionally "diluted" (Winder et al., 2005) in discussions of CHANS and socio-hydrology, though a looser usage of the term is certainly of relevance. In a strict application of the term co-evolutionary, two or more evolutionary systems are linked such that the evolution of each system influences that of the other (Winder et al., 2005); an evolutionary system is one in which entities exists, include responses that may vary with time (as opposed to mechanistic systems, in which responses are time-invariant), involving the mechanisms of

"variation, inheritance and selection" (Hodgson, 2003). Jeffrey and McIntosh (2006) give a guide in identification of co-evolutionary systems.

- Identify evolutionary (sub)systems and entities.

- Provide a characterisation of variation in each system.

- Identify mechanisms that generate, winnow and provide continuity for variation in each system.

- Describe one or more potential sequences of reciprocal change that result in an evolutionary change in one or more systems.

- Identify possible reciprocal interactions between systems.

- Identify effects of reciprocal interactions.

Whether or not the biophysical, hydrological system is viewed as evolutionary in nature determines whether socio-hydrological dynamics may be termed co-evolutionary, since Winder et al. (2005) state that "Linking an evolutionary system to a non-evolutionary system does not produce co-evolutionary dynamics. It produces simple evolutionary dynamics coupled to a mechanistic environment", which would imply that socio-hydrological systems are not co-evolutionary in nature, perhaps rather being complex systems, or systems of "cultural ecodynamics" (Winder et al., 2005). Norgaard (1981, 1984) allows for a looser definition of a co-evolutionary relationship, whereby two systems interact and impact one another such that they impact one another's developmental trajectory. Norgaard (1981, 1984) gives paddy rice agriculture as an example of a co-evolutionary system: in this example, changes in agricultural practice (investment in irrigation systems for example) led to higher land productivity and to societal development; the usage of paddy-based techniques then required the development of social constructs (water-management institutions and property rights) to sustain such farming methods, which served to socially perpetuate paddy farming and to alter ecosystems further in ways that made the gap between land productivity between farming techniques greater, and so led to yet greater societal and ecosystem change. Western monoculture may also be viewed in the same light, with social systems such as insurance markets, government bodies and agro-technological and agrochemical industries developed to be perfectly suited to current agriculture (Norgaard, 1984), but these constructs having been borne out of requirements by monocultures previously, and also serving to perpetuate monoculture and make its usage more attractive. The crucial difference between the two views is that Winder et al. (2005) do not consider biophysical systems, such as hydrological or agricultural systems, evolutionary in their nature (Kallis, 2007), since the biophysical mechanisms behind interactions in these systems are governed by Newtonian, rather than Darwinian, mechanisms.

Even if the strict definition of a co-evolutionary system does not apply to socio-hydrology, the co-evolutionary framework may be used as an epistemological tool (Jeffrey and McIntosh, 2006), a way to develop understanding, and so the subtle difference between complex and co-evolutionary systems should be kept in mind when developing socio-hydrological models, if for no other reason than it may remind developers that non-stationary responses may exist (whether this implies co-evolution or not), largely in terms of social response to hydrological change. The usage of a co-evolutionary framework also allows the usage of the teleological principle (i.e. an end outcome has a finite cause), which allows, for example, for policy implications to be drawn (Winder et al., 2005).

There are already examples where a co-evolutionary perspective has been taken on an issue that may be termed socio-hydrological/-ecological; these examples and how useful the co-evolutionary analogy is are examined here. Kallis (2010) uses a co-evolutionary perspective to look at how water resources have been developed in the past: Athens in Greece is used as an example, where expansions in water supply led to increases in demands, which required further expansion. However, this cycle is not seen as predetermined and unstoppable; rather, it is dependent on environmental conditions, governance regimes, technology and geo-politics, all of which are impacted by, and evolve with, the changes in water supply and demand, as well as each other. The relationship between the biophysical environment and technology is particularly interesting: the environment is non-stationary as water supply expands, as innovation and policy, driven by necessity to overcome environmental constraints, result in environmental changes, both expected and unforeseen, which then result in socioeconomic changes and new environmental challenges to be solved. The evolutionary perspective used in looking at innovation overcoming temporary environmental constraints, but also creating new issues in the future, is very useful in understanding how human–water systems develop. A study by Lorenzoni et al. (2000); Lorenzoni (2000) takes a co-evolutionary approach to climate change impact assessment and determines that using indicators of sustainability in a bi-directional manner (both as inputs to and outputs from climate scenarios) is possible, and that a co-evolutionary view of the human–climate system, involving adaptation as well as mitigation measures, results in a "more sophisticated and dynamic account of the potential feedbacks" (Lorenzoni et al., 2000). The dynamics that are implied using co-evolutionary frameworks are also interesting, as shown in studies by Liu et al. (2014), whereby the co-evolution of humans and water in a river basin system brings about long stable periods of system equilibrium, punctuated by shifts due to internal or external factors, which indicates a "resonance rather than a cause-effect relationship" (Falkenmark, 2003) between the systems.

The usage of a co-evolutionary framework could be beneficial in governance and modelling of socio-hydrological systems, and the previously mentioned IAHS paper (Montanari et al., 2013) states that the co-evolution of humans and water "needs to be recognized and modelled with a suitable approach, in order to predict their reaction to change". The co-evolution of societal norms with environmental state may be particularly interesting in this respect. The "lock-in" that is created by technological and policy changes in co-evolutionary systems, which can limit reversibility of decisions in terms of how resources are allocated (Van den Bergh and Gowdy, 2000), also means that improving the predictive approach taken should be a matter of priority, decisions taken now may result in co-evolutionary pathways being taken that cannot be altered later (Thompson et al., 2013). The implication of a potential lack of knowledge of long-term path dependencies for current policy decisions should be that, rather than seeking optimal policies in the short term, current decisions should be made that allow development in the long term and maintain the potential for system evolution in many directions (Rammel and van den Bergh, 2003).

3.3.3 Complex adaptive systems

In understanding the concept of sustainability, Jeffrey and McIntosh (2006) explains that the dynamic behaviour seen in natural systems, "is distinct from (simple or complex) dynamic or (merely) evolutionary change", and is instead a complex mixture of mechanistic and evolutionary behaviours. However, as was previously explained, the strict use of the term "co-evolutionary" is perhaps not applicable in socio-ecological systems, and so perhaps a better term to be used would be "complex adaptive systems" (Levin et al., 2012). Complex adaptive systems are a subset of complex systems in which systems or system components exhibit adaptivity (not necessarily all elements or subsystems); Lansing (2003) gives a good introduction. The important distinction between complex systems and complex *adaptive* systems is that, in complex systems, if a system reaches a previously seen state, this indicates a cycle, and so the system will return to this state at another point. Due to the adaptivity and time-variant responses, this is not the case in complex adaptive systems.

The complex adaptive systems paradigm has already been used in a socio-hydrological context, being used to investigate Balinese water temples that are used in irrigation (Lansing et al., 2009; Lansing and Kremer, 1993; Falvo, 2000). Policy implications of complex adaptive systems have also been investigated by Levin et al. (2012) and Rammel et al. (2007), and are summarised as the following.

- Nonlinearity – should be included in models such that surprises are not so surprising. Time-variant responses also mean that adaptive, changing management practices should be used, as opposed to stationary practices.

- Scale issues – processes occur on different spatial scales and timescales, and so analysis of policy impacts should be conducted on appropriate, and if possible on multiple, scales.

- Heterogeneity – heterogeneity in complex systems results in the application of homogeneous policies often being sub-optimal.

- Risk and uncertainty – Knightian (irreducible) uncertainty exists in complex adaptive systems.

- Emergence – surprising results should not be seen as surprising, due to the complex, changing responses within systems.

- Nested hierarchies – impacts of decisions can be seen on multiple system levels due to the hierarchies within complex adaptive systems.

As can be seen, these policy issues are very similar to those mentioned in previous sections relating to management of socio-hydrological and socio-ecological systems, which is not surprising.

Ultimately, in the modelling of socio-hydrological systems, it is not necessary to state whether the system is being treated as a complex system, a co-evolutionary system or a complex adaptive system; rather, it is the implications that the lens through which the system is seen has, via the representation of the system in model equations, that are most important. There are clearly dynamics that both do and do not vary in time in socio-hydrological systems, and so these should all be treated appropriately. Perhaps the most important outcome of the human–water system representation should be a mindset to be applied in socio-hydrological modelling, whereby mechanistic system components are used in harmony with evolutionary and adaptive components to best represent the system.

3.4 Space and time in socio-hydrological modelling

In several previous sections, the issues of scale that socio-ecological and socio-hydrological systems can face were presented and their significance stressed. As such, a section looking at space and time in socio-hydrology is warranted. Hydrology involves "feedbacks that operate at multiple spatiotemporal scales" (Ehret et al., 2014), and when coupled with human activities, which are also complex on spatial and temporal scales (Ren et al., 2002), this picture becomes yet more complicated, though these cross-scale interactions are the "essence of the human-water relationship" (Liu et al., 2014). As a method of enquiry, modelling allows for investigations to be conducted on spatiotemporal scales that are not feasible using other methods, such as experiments and observations (though the advent of global satellite observations is changing the role that observations have and the relationship between observations and modelling to one of modelling downscaling observations and converting raw observations into actionable information) (Reyer et al., 2015) (see Fig. 3),

and so is a useful tool in investigating socio-hydrology. However, ensuring the correct scale for modelling and policy implementation is of great importance, as both of these factors can have great impacts on the end results (Manson, 2008).

In terms of space, the interactions that occur between natural and constructed scales are superimposed with interactions occurring between local, regional and global spatial scales. Basins and watersheds are seen as "natural" (Blomquist and Schlager, 2005) scales for analysis, since these are the spatial units in which water flows (though there are of course watersheds of different scales and watersheds within basins, and so watershed-scale analysis does not answer the question of spatial scale on its own); however, these often do not match with the scales on which human activities occur, and indeed human intervention has, in some cases, rendered the meaning of a "basin" less relevant due to water transfers (Bourblanc and Blanchon, 2013). The importance of regional and global scales has been recognised, with Falkenmark (2011) stating that "the meso-scale focus on river basins will no longer suffice". Another issue of spatial scale is that of the extents at which issues are created and experienced (Zeitoun, 2013): some issues, for instance point-source pollution, are created locally and experienced more widely, whereas issues of climate are created globally, but problems are experienced more locally in the form of droughts and floods. This dissonance between cause and effect can only be combated with policy on the correct scale. Creating models involves scale decisions, often involving trade-offs between practicalities of computing power and coarseness of representation (Evans and Kelley, 2004), which can impact the quality of model output. The previous points all indicate there being no single spatial scale appropriate for socio-hydrological analysis; instead, each problem should be considered individually, with the relevant processes and their scales identified and modelling scales determined accordingly. This could result in potentially heterogeneous spatial scales within a model.

The interactions between slow and fast processes create the temporal dynamics seen in socio-ecological systems (Crépin, 2007); slow, often unnoticed, processes can be driven which lead to regime shift on a much shorter timescale (Hughes et al., 2013), and in modelling efforts these slow processes must be incorporated with faster processes. Different locations will evolve in a socio-hydrological sense at different paces, due to hydrogeological (Perdigão and Blöschl, 2014) and social factors, and so socio-hydrological models should be developed with this in mind. Also, different policy options are appropriate on different timescales, with efforts such as rationing and source-switching appropriate in the short term, as opposed to infrastructure decisions and water rights changes being more appropriate in the long term (Srinivasan et al., 2013). All of these factors mean that a variety of timescales, and interactions between these, should be included in models, and analyses on different timescales should not be seen as incompatible (Ertsen et al., 2014).

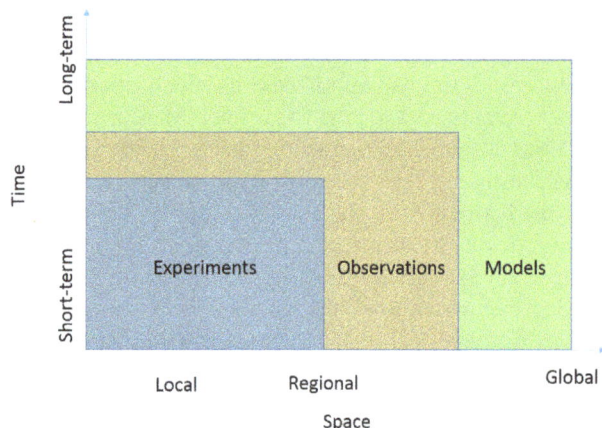

Figure 3. Temporal and spatial scales at which different research approaches are appropriate (adapted with permission from Reyer et al. (2015), used under the CC Attribution License 3.0).

3.5   Data

One of the cornerstones of study in hydrological sciences is data. However, there are significant problems in obtaining the data required in a socio-hydrological sense. Some of the issues present in this area are the following.

- Timescales: an issue in accruing data for long-term hydrological studies is that "detailed hydrologic data has a finite history" (Troy et al., 2015b). Good data from historical case studies are difficult to obtain, and so shorter-term studies sometimes have to suffice. The focus on long-term analysis that socio-hydrology takes exacerbates this problem, particularly since historical case studies are of great use during the system-understanding phase that the subject is currently in.

- Availability: where data are widely available, it may be possible for minimal analysis to be carried out, and for data-centric studies to be carried out (Showqi et al., 2013), but when the boundaries of the system of interest are expanded to include the social side of the system, data requirements naturally increase, and modellers are exposed to data scarcity in multiple disciplines (Cotter et al., 2014). Hydrological modelling often suffers from data unavailability (Srinivasan et al., 2015), but significant work has been carried out in recent years on prediction in ungauged basins (Hrachowitz et al., 2013; Wagener and Montanari, 2011) to reduce this, and so perhaps the potential multi-disciplinary data scarcity issues in socio-hydrology could borrow and adapt some techniques. Papers discussing solutions for a lack of data in a socio-hydrological context are also already appearing (Zlinszky and Timár, 2013). Data scarcity can heavily influence the modelling technique used (Odongo et al., 2014): lumped conceptual models tend to have "more

modest…data requirements" (Sivapalan et al., 2003), whereas distributed, physically based models tend to have "large data and computer requirements" (Sivapalan et al., 2003). A smaller amount of data may be necessary in some socio-hydrological studies, since the collection of a significant quantity of extra data (when compared to hydrological studies) also incurs an extra cost, both in terms of financial cost and time (Pataki et al., 2011).

- Interdisciplinary integration: the integration of different data types from different fields is complex (Cotter et al., 2014); socio-hydrology will have to cope with this, since some aspects of socio-hydrological study are necessarily quantitative and some qualitative. Since the subject of socio-hydrology has come largely from those with a hydrology background, integrating qualitative data sources with more quantitative sources that hydrologists are commonly more comfortable with could pose some issues (Troy et al., 2015b). However, the necessary interdisciplinary nature of socio-hydrology also means that communication between model developers from different subject areas should be enhanced (Cotter et al., 2014), so that everyone may gain.

- New data: in order to capture some of the complex socio-hydrological interactions, socio-hydrology should seek to go beyond merely summing together hydrological and social data, and instead investigate the use of new, different data types. Saying that this should be done is easy, but carrying it out in practice may be much more difficult, since the nature of these data and how they would be collected are presently unknown. To this end, Di Baldassarre et al. (2015b) point out that the use of stylised models can help to guide researchers towards the data that are needed, setting off an iterative process of model–data–theory development. With regard to unconventional data, Troy et al. (2015a) have propounded the use of proxy data in socio-hydrology where data do not exist, and Zlinszky and Timár (2013) have investigated the potential for an unconventional data source for socio-hydrology: historical maps.

3.6   Complexity

The expansion of system boundaries to include both social and hydrological systems introduces more complexity than when each system is considered separately. The increased complexity of the system leads to a greater degree of emergence present in the system, though this does not necessarily mean more complex behaviours (Kumar, 2011). The level of complexity required in a model of a more complex system will probably itself be more complex (though not necessarily, as Levin et al. (2012) said, "the art of modelling is to incorporate the essential details, and no more") than that of a simpler system, since model quality should be judged by the ability to match the emergent properties of the behaviour a system (Kumar, 2011). Manson (2001) introduces the different types of complexity:

- Algorithmic complexity: this may be split into two varieties of complexity. One is the computational effort required to solve a problem, and the other is complexity of the simplest algorithm capable of reproducing system behaviour.

  - While the first side of algorithmic complexity is important in socio-hydrological modelling, since mathematical problems should be kept as simple as is practicable, the second facet of algorithmic complexity is most applicable to socio-hydrological modelling, as modellers should be seeking to develop the simplest possible models that can replicate the behaviour of socio-hydrological systems.

- Deterministic complexity: the notion that every outcome has a root cause that may be determined, however detached they may seemingly be, is at the heart of deterministic complexity. Feedbacks, sensitivities to changes in parameters and tipping points are all part of deterministic complexity.

  - The study of complex systems using mechanistic equations implies that there are deterministic relationships within a system; since socio-hydrological modelling will use such techniques, deterministic complexity is of interest. Using deterministic principles, modellers may seek to determine the overall impacts that alterations to a system may have.

- Aggregate complexity: this is concerned with the interactions within a system causing overall system changes. The relationships within a system lead to the emergent behaviours that are of such interest, and determining the strengths of various correlations and how different interactions lead to system level behaviours gives an idea of the aggregate complexity of a system.

  - Aggregate complexity is of great interest to modellers of socio-hydrological systems. Determining how macro-scale impacts are created via interactions between system variables is a central challenge in the subject, and so determining the aggregate complexity of socio-hydrological systems may be an interesting area of study.

The increased complexity of the system, and the previously mentioned issues of possible data scarcity from multiple disciplines, could lead to issues. Including more complexity in models does not necessarily make them more accurate, particularly in the case of uncertain or poor resolution input data (Orth et al., 2015); this should be kept in mind when developing socio-hydrological models, and in some cases simple

models may outperform more complex models. Keeping in mind the various forms of complexity when developing models, socio-hydrologists should have an idea of how models should be developed and what they may be capable of telling us.

3.7 Model resolution

As well as being structured in different ways, there are different ways in which models can be used to obtain results via different resolutions. Methods include analytical resolution, Monte Carlo simulations, scenario-based techniques and optimisation (Kelly et al., 2013). Analytical resolutions, while they give a very good analysis of systems in which they are applied, will generally be inapplicable in socio-hydrological applications, due to the lack of certain mathematical formulations and deterministic relationships between variables which are required for analytical solutions. Monte Carlo analyses involve running a model multiple times using various input parameters and initial conditions. This is a good method for investigating the impacts that uncertainties can have (an important aspect in socio-hydrology), though the large number of model runs required can lead to large computational requirements. Optimisation techniques are useful when decisions are to be made; using computer programs to determine the "best" decision can aid in policy-making, however, optimisation techniques should be used with care: the impacts that uncertainties can have, as well as issues of subjectivity and model imperfections can (and have) lead to sub-optimal decisions being made. Techniques such as multi-objective optimisation (Hurford et al., 2014) seek to make more clear the trade-offs involved in determining "optimal" strategies.

3.8 Uncertainty

Uncertainty is an issue to be kept at the forefront of a modeller's mind before a modelling technique is chosen, while models are being developed and once they produce results. There are implications that uncertainty has in all modelling applications, and so it is important to cope appropriately with them, as well as to communicate their existence (Welsh et al., 2013). Some of the modelling techniques, for instance Bayesian networks, deal with uncertainty in an explicit fashion, while other techniques may require sensitivity analyses or scenario-based methods to deal with uncertainty. In any case, the method by which uncertainty is dealt with is an important consideration in determining an appropriate modelling technique.

Uncertainty in socio-hydrology could certainly be the subject of a paper on its own, and so while this paper outlines some of the aspects of uncertainty which have particular significance for modelling, some aspects are not covered in full detail. For more detailed coverage of uncertainty in a socio-hydrological context, the reader is directed towards Di Baldassarre et al. (2015a) and Merz et al. (2015).

3.8.1 Uncertainty in hydrological models

Hydrological models on their own are subject to great uncertainties, which arise for an array of reasons and from different places, including external sources (for instance uncertainties in precipitation or human agency, internal sources (model structure and parameterisation), as well as data issues and problem uniqueness (Welsh et al., 2013). In the current changing world, many of the assumptions on which hydrological models have been built, for instance non-stationarity (Milly et al., 2008), have been challenged, and new uncertainties are arising (Peel and Blöschl, 2011). However, the extensive investigations into dealing with uncertainty (particularly the recent focus on prediction in ungauged basins Wagener and Montanari, 2011) can only be of benefit to studies which widen system boundaries. The trade-offs between model complexity and "empirical risk" (Arkesteijn and Pande, 2013) in modelling, ways to deal with large numbers of parameters and limited data (Welsh et al., 2013), as well as statistical techniques to cope with uncertainties (Wang and Huang, 2014) have all been well investigated, and knowledge from these areas can certainly be applied to future studies.

3.8.2 Uncertainty in coupled socio-hydrological models

Interactive and compound uncertainties are an issue in many subjects, and indeed already in water science (particularly the policy domain). Techniques already exist in water resource management for taking action under such uncertainties, for instance the method used by Wang and Huang (2014), whereby upper and lower bounds are found for an objective function that is to be minimised/maximised to help identify the "best" decision, and to identify those that may suffer due to various uncertainties. This approach extends that taken in sensitivity analyses, and is a step forward, since sensitivity analyses usually examine "the effects of changes in a single parameter... assuming no changes in all other parameters" (Wang and Huang, 2014), which can fail to detect the impact of combined uncertainties in systems with a great deal of interconnections and feedbacks. The amplifications that feedback loops can induce in dynamic systems mean that the impact of uncertainties, particularly initial condition uncertainties, can be great (Kumar, 2011).

There are aspects to socio-hydrology which induce issues regarding uncertainties which are beyond mere propagation of deterministic uncertainty. The nature of the hydrological input brings about "aleatory" uncertainty (Di Baldassarre et al., 2015a), in which random variability brings uncertainty; this variability can be coped with in modelling to a certain extent by using probabilistic or stochastic methods; however, some of the effects that it brings about, for instance surprise

(Merz et al., 2015), have much more serious implications. The random nature of the times at which extreme hydrological events occur, and the often event-based response that humans take, means that very different trajectories can be predicted in socio-hydrological systems, depending on when events occur. Merz et al. (2015) argue that surprise should be accounted for more fully in flood risk assessment, and that thorough analyses should be carried out in which the possibility of surprise and the vulnerability of a system to surprising events are accounted for.

Another aspect of uncertainty that socio-hydrology needs to consider is that which Di Baldassarre et al. (2015a) term epistemic uncertainty. At present, understanding of the nature of human–water system dynamics is relatively poor, and this lack of knowledge means that significant uncertainty exists around whether representations of these dynamics are correct. Di Baldassarre et al. (2015a) characterise epistemic uncertainty as arising from three sources: known unknowns, unknown unknowns and wrong assumptions. These three sources of uncertainty lead to the present approach to modelling, whereby we model based on assumed system behaviour, being called into question. This epistemic uncertainty is related to the issue of Knightian uncertainty: the inherent indeterminacy of the system ("that which cannot be known" – Lane, 2014). In cases of epistemic and Knightian uncertainty, the use of adaptive management techniques (Garmestani, 2013) is an effective way of acting in a practical sense, but does not necessarily provide a solution to unknown unknowns. Modelling is a key part of the reduction of epistemic uncertainty: Di Baldassarre et al. (2015a) call for the iterative process of "new observations, empirical studies and conceptual modelling" to increase knowledge regarding human–water systems, in order to reduce these uncertainties.

4   How?

The final component of this paper covers the "how" of socio-hydrological modelling. Sivapalan and Blöschl (2015) give an excellent overview of how the overall modelling process should be carried out in socio-hydrology, which the reader is highly encouraged to read. This paper focuses on the different specific techniques available to modellers, the background to these techniques, how they would be developed, applied and used in socio-hydrology, as well as the difficulties that might be faced. The above "what?" and "why?" sections will be utilised to aid in these discussions. Table 1 shows some examples of modelling studies which involve some element of human–water interaction, including details of the technique that is used, the case studied and the reason for modelling. While some of the studies included would be deemed socio-hydrological in nature, many of them would not, but are present as the inclusion of some aspect of human–water interaction that they exhibit may be useful to future socio-hydrological modellers.

Table 1. Examples of studies that include some aspect of modelling human–water interaction.

Reference	Approach	Case studied	Reason for modelling
Barreteau et al. (2004)	ABM	Irrigation system, Senegal River Valley	Determining suitability of modelling approach to application
Becu et al. (2003)	ABM	Water management, northern Thailand	Analysis of policy approaches
Medellín-Azuara et al. (2012)	ABM	Prediction of farmer responses to policy options	Understanding behavioural processes
Schlüter and Pahl-Wostl (2007)	ABM	Amu Darya River basin, Central Asia	Determining origins of system resilience
Fabre et al. (2015)	CCM	Herault (France) and Ebro (Spain) catchments	Understanding supply–demand dynamics
Fraser et al. (2013)	CCM	Worldwide, areas of cereal production	Predicting areas of future vulnerability
Dougill et al. (2010)	SD	Pastoral drylands, Kalahari, Botswana	Predicting areas of future vulnerability
Elshafei et al. (2014)	SD	Murrumbidgee Catchment, Australia	System understanding
van Emmerik et al. (2014)	SD	Murrumbidgee Catchment, Australia	System understanding
Liu et al. (2015b)	SD	Water quality of Dianchi Lake, Yunnan Province, China	Decision support
Liu et al. (2015a)	SD	Tarim River basin, Western China	System understanding
Fernald et al. (2012)	SD	Acequia irrigation systems, New Mexico, USA	System understanding; stakeholder participation; prediction of future scenarios
Di Baldassarre et al. (2013b)	SD	Human–flood interactions, fictional catchment	System understanding
Viglione et al. (2014)	SD	Human–flood interactions, fictional catchment	System understanding
Garcia et al. (2015)	SD	Reservoir operation policies	System understanding
Madani and Hooshyar (2014)	GT	Multi-operator reservoir systems (no specific case)	Policy
van Dam et al. (2013)	BN	Nyando Papyrus Wetlands, Kenya	System understanding; evaluation of policy options
Srinivasan (2015)	Other	Water supply and demand, Chennai, India	System understanding; analysis of possible alternative historical trajectories
Srinivasan et al. (2015)	Other	Decreasing flows in the Arkavathy River, South India	Policy; focusing future research efforts
Odongo et al. (2014)	Other	Social, ecological and hydrological dynamics of the Lake Naivasha basin, Kenya	System understanding

ABM: agent-based modelling; CCM: coupled component modelling; SD: system dynamics; GT: game theory; BN: Bayesian network; POM: pattern-oriented modelling.

Liebman (1976) said that "modelling is thinking made public", and so models may be used to demonstrate the knowledge currently held in a community. Troy et al. (2015a) even state that socio-hydrological models at present may be thought of as hypotheses (rather than predictive tools), and so reinforce this view. With the current feeling in socio-hydrological circles being that the integration of the social and economic interactions with water is a vital component of study, this integration should be seen and should be included centrally in models in such a way that demonstrates the importance of these interactions to modellers (Lane, 2014). This should mean integration of the two disciplines in a holistic sense, including integrating the issues faced across hydrological, social and economic spheres, the integration of different processes from the different areas of study, integration of different levels of scale (hydrological processes will operate on a different scale to social and economic processes), as well as the integration of different stakeholders across the different disciplines (Kelly et al., 2013).

There are numerous ways to classify models, and so before each individual modelling technique is detailed, the more general classifications will be detailed.

4.1 Model classifications

4.1.1 Data-based vs. physics-based vs. conceptual

The distinction between these different types of model is fairly clear: physics-based models use mathematical rep-

resentations of physical processes to determine system response, data-based models seek to reproduce system behaviour utilising available data (Pechlivanidis and Jackson, 2011) (there also exist hybrid models using a combination of these two approaches), and conceptual models are based on a modeller's conceptual view of a system. The common criticisms of the two approaches are that physics-based model results are not always supported by the available data (Wheater, 2002) and are limited due to the homogenous nature of equations in a heterogeneous world (Beven, 1989), while metric models can represent processes that have no physical relevance (Malanson, 1999).

4.1.2 Bottom-up vs. top-down

There is a similar distinction between bottom-up and top-down models as between metric and physically based. Bottom-up modelling techniques involve the representation of processes (not necessarily physical) to develop system behaviour, whereas top-down approaches look at system outcomes and try to look for correlations to determine system behaviours. Top-down approaches have been criticised for their inability to determine base-level processes within a system, and so their inability to model the impact of implementing policies and technologies (Srinivasan et al., 2012). Bottom-up methods, while the message they present does not need to be "disentangled" (Lorenzoni et al., 2000), require a great deal of knowledge regarding specific processes and sites, which in social circumstances in particular can be very challenging (Sivapalan, 2015) and specific in both a spatial and temporal sense. More detail on bottom-up and top-down modelling approaches will be given in the sections on agent-based modelling and system dynamics modelling, since these are the archetypal bottom-up and top-down approaches respectively.

4.1.3 Distributed vs. lumped

The final distinction that is drawn here is that of distributed and lumped models. Distributed models include provisions for spatial, as well as temporal, heterogeneity, while lumped models concentrate study at discrete spatial points, where dynamics vary only in time. The advantages of distributed models are clear, particularly in a hydrological context where spatial heterogeneity is of such importance; however, the drawbacks of high-resolution data requirements, with high potential for uncertainty, and larger computational requirements (Sivapalan et al., 2003) mean that lumped models can be an attractive choice.

4.2 Approaches

Kelly et al. (2013) gives an excellent, critical overview of which modelling approaches may be used in modelling socio-ecological systems. As socio-hydrology is closely linked to socio-ecology, these modelling approaches are largely the same. The modelling techniques that will be discussed here are

- agent-based modelling (ABM),

- system dynamics (SD),

- pattern-oriented modelling (POM),

- Bayesian networks (BN),

- coupled-component modelling (CCM),

- scenario-based modelling, and

- heuristic/knowledge-based modelling.

While it is acknowledged that the modelling techniques detailed in this review are established, traditional techniques, this should certainly not be taken as implying that modellers in socio-hydrology should only use traditional techniques. As has been said, this review is not intended to be a review of socio-hydrological modelling thus far, but rather a review of current knowledge designed to guide future socio-hydrological modelling efforts. New or hybrid modelling techniques are likely to emerge to tackle the specific problems that socio-hydrology poses, but any new techniques are very likely to be based around existing methods. As such, these modelling processes for these approaches are detailed, with a critical view on their application in socio-hydrology taken.

In the discussions that follow, the factors that would affect the choice of modelling approach will also be used. These are

- model purpose

- data availability (quantity, quality and whether it is quantitative or qualitative),

- treatment of space,

- treatment of time,

- treatment of system entities,

- uncertainty, and

- model resolution.

Now that these pre-discussions have been included, a section on the importance of model conceptualisation is included, before each modelling approach is focused on.

4.3 The importance of model conceptualisation

The previously mentioned statement of modelling being "thinking made public" (Liebman, 1976) highlights the significance of the process behind model development for the distribution of knowledge. The conceptual basis on which a model is built defines the vision that a developer has of a system ("framing the problem" – Srinivasan, 2015), and is therefore both a vital step in model development and a way that understanding can be shared. Conceptualisations often involve "pictures", whether these be mental or physical pictures, and these pictures can be an excellent point of access for those who wish to understand a system, but who do not wish to delve into the potentially more quantitative or involved aspects. In some cases, a conceptual modelling study can also be an important first step towards the creation of a later quantified model (e.g. Liu et al., 2014, 2015a).

There are certain facets of socio-hydrology that should be captured in all SHS models, and so frameworks for socio-hydrological models should underlie conceptualisations. Two frameworks for socio-hydrological models that have been developed thus far are those of Carey et al. (2014) and Elshafei et al. (2014). The framework of Carey et al. (2014) highlights some key facets of the human side of the system that are important to capture:

- "Political agenda and economic development

- Governance: laws and institutions

- Technology and engineering

- Land and resource use

- Societal response".

Elshafei et al. (2014) present a framework for the whole system, which is composed of

- catchment hydrology,

- population dynamics,

- economics,

- ecosystem services,

- societal sensitivity, and

- behavioural response.

Both of these frameworks give a view of the key parts of socio-hydrological systems: the second gives a good base for modelling the entirety of the system, and has a very abstracted point of view of the societal dynamics, whereas the former takes a more detailed look at the societal constructs that lead to a particular response. Depending on the level of detail that is sought, either or both of these frameworks could be used as a basis for a socio-hydrological conceptualisation.

4.4 Agent-based modelling (ABM)

Having its origins in object-oriented programming, game theory and cognitive psychology (An, 2012), ABM is a bottom-up approach to the modelling of a system, in which the focus is on the behaviour and decision-making of individual "agents" within a system (Bousquet and Le Page, 2004). These agents may be individuals, groups of individuals, or institutions, but are defined by the attributes of being autonomous and self-contained, the presence of a state and the existence of interactions with other agents and/or the environment in which an agent exists (Macal and North, 2010). Decision rules are determined for agents (these may be homogeneous or heterogeneous), which determine the interactions and feedbacks that occur between agents (often agents on different organisational levels Valbuena et al., 2009), as well as between agents and the environment. ABMs are almost necessarily coupled in a socio-ecological sense (though they are often not necessarily termed as such), given that they use the decision-making processes of those within a society to determine the actions that they will take, and as such their impacts upon the environment and associated feedbacks, though they might not fully look at impacts that society has upon the environment, and rather look at human reactions to environmental changes.

Agent-based models themselves come in many forms, for example:

- Microeconomic: agent rules are prescribed to optimise a given variable, for instance profit, and make rational (or bounded rational) choices with regards to this (e.g. Becu et al., 2003; Filatova et al., 2009; Nautiyal and Kaechele, 2009).

- Evolutionary: agent decision-making processes change over time as agents "learn" (e.g.. Manson and Evans, 2007) and test strategies (e.g. Evans et al., 2006).

- Heuristic/experience-based: agents' rules are determined either through via either experience, or the examination of data (e.g. Deadman et al., 2004; An et al., 2005; Matthews, 2006; Gibon et al., 2010; Valbuena et al., 2010, 2009).

- Scenario-based: various environmental scenarios are investigated to see the impact upon behaviours, or different scenarios of societal behaviours are investigated to see impacts upon the environment (e.g. Murray-Rust et al., 2013).

The development of an ABM involves a fairly set method, the general steps of which are the following.

1. Problem definition

2. Determination of relevant system agents

3. Description of the environment in which agents exist

4. Elicitation of agent decision-making process and behaviours (Elsawah et al., 2015)

5. Determination of the interactions between agents

6. Determination of the interactions between agents and the environment

7. Development of computational algorithms to represent agents, environment, decision-making processes, behaviours and interactions

8. Model validation and calibration.

The results from ABMs will generally be spatially explicit representations of system evolution over time, and so lend themselves well to integration with GIS software (Parker et al., 2005).

ABMs may be used in socio-hydrological modelling in two contexts: firstly, the discovery of emergent behaviour (Kelly et al., 2013) in a system, and secondly determining the macro-scale consequences that arise from interactions between many individual heterogeneous agents and the environment. ABM may be used for a number of different reasons: in the context of system understanding, the elicitation of emergent behaviours and outcomes leads to an understanding of the system, and in particular decision-making mechanisms where they can represent important phenomena that may be difficult to represent mathematically (Lempert, 2002). ABMs are also very applicable in the area of policy-making, as the outcomes of different policy options may be compared when the impact of agent behaviours are accounted for; for instance, O'Connell and O'Donnell (2014) suggest that ABMs may be more useful in determining appropriate flood investments than current cost-benefit analysis (CBA) methods. In the area of resilience, the importance of human behaviours in creating adaptive capacity of socio-ecological systems (Elsawah et al., 2015) has meant that

ABMs have been used to look at the differing levels of resilience in different governance regimes (Schlüter and Pahl-
Wostl, 2007). The usage of ABM can be particularly strong
in participatory modelling (Purnomo et al., 2005), where
agents may be interviewed to determine their strategies, and
then included in subsequent modelling stages. While ABM
is seen by many as a technique with a wide range of uses,
others are less sure of its powers (Couclelis, 2001), particularly in predictive power at small scales (An, 2012), along
with the difficulties that can be present in validation and verification of decision-making mechanisms (An, 2012). One
study that has been carried out in the specific area of socio-
hydrology which incorporates agent-based aspects is that of
Srinivasan (2013). In this historical study, social and hydrological change in Chennai, India (Srinivasan, 2013) was investigated to determine the vulnerability of those within the
city to water supply issues. The model was successfully able
to incorporate different temporal scales, and was able to identify the possibility for vulnerability of water supplies on both
a macro- and micro-scale level; the adaptive decisions of
agents that the model was able to account for played a big
part in this success. This work has been carried on via another study (Srinivasan, 2015) in which alternative trajectories are investigated to examine how the system might now
be different had different decisions been made in the past.

Agent-based modelling may be particularly well placed to
investigate the role of changing norms and values in socio-
hydrology; by considering the decision-making processes of
individual agents, there is an ability to determine the implications of slow changes in these decision-making processes.
This does not, however, diminish the difficulty involved in
determining how to represent these changing norms.

4.4.1 Game theory

"Game theory asks what moves or choices or allocations are
consistent with (are optimal given) other agents' moves or
choices or allocations in a strategic situation." (Arthur, 1999),
and so is potentially very applicable to agent-based modelling in determining the decisions that agents make (Bousquet and Le Page, 2004). For a great deal of time, game theory has been used to determine outcomes in socio-ecological
systems (for example the tragedy of the commons – Hardin,
1968), and game theory has been used extensively in water resource management problems (Madani and Hooshyar,
2014), so there is the potential that game theory could be extended to problems in a socio-hydrological setting. However,
the uncertainties that will be dealt with in socio-hydrology
(which have been discussed earlier) would be beyond those
that are currently considered in game theory, and so special
attention would need to be paid to this area were game theory
to be applied.

4.5 System dynamics (SD)

System dynamics (and the linked technique of system analysis Dooge, 1973) takes a very much top-down view of a
system; rather than focusing on the individual processes that
lead to overall system behaviours, system dynamics looks at
the way a system converts inputs to outputs and uses this
as a way to determine overall system behaviour. In system
dynamics, describing the way a system "works" is the goal
rather than determining the "nature of the system" (Dooge,
1973) by examining the system components and the physical laws that connect them. System dynamics can, therefore, avoid the potentially misleading analysis of the interactions and scaling up of small-scale processes (potentially
misleading due to the complexity present in small-scale interactions not scaling up) (Sivapalan et al., 2003). Macro-scale
outcomes such as non-linearities, emergence, cross-scale interactions and surprise can all be investigated well using system dynamics (Liao, 2013), and its high-level system outlook
allows for holism in system comprehension (Mirchi et al.,
2012).

An important facet of the system dynamics approach is
the development procedure: a clear and helpful framework
that is integral in the development of a successful model,
and also provides an important part of the learning experience. As with other modelling techniques, this begins with
a system conceptualisation, which, in this case, involves the
development of a causal loop diagram (CLD). A CLD (see
examples in Figs. 4 and 5) is a qualitative, pictorial view of
the components of a system and the linkages between them.
This allows for a model developer to visualise the potential
feedbacks and interconnections that may lead to system-level
behaviours (Mirchi et al., 2012) from a qualitative perspective, without needing to delve into the quantitative identification of the significance of the different interconnections.
Depending on how a modeller wishes to represent a system,
different levels of complexity may be included in a CLD (this
complexity may then later be revisited during the more quantitative model development phases), and CLDs (and indeed
SD models) of different complexity may be useful in different circumstances. The differences in complexity between
Figs. 4 and 5 show very different levels of complexity that
modellers may choose to use (particularly since Fig. 4 is only
a CLD for one of four linked subsystems). Once a CLD has
been devised, the next stage in model development is to turn
the CLD into a stocks and flows diagram (SFD). This process is detailed in Table 2, and essentially involves a qualitative process of determining the accumulation and transfer of
"stocks" (the variables, or proxy variables used to measure
the various resources and drivers) in and around a system.
Figure 6 shows the SFD developed from a CLD. SFD formulation lends itself better to subsequent development into
a full quantitative model, though is still qualitative in nature
and fairly simple to develop, requiring little or no computer
simulation (a good thing, as Mirchi et al. (2012) says, "ex-

Table 2. Procedure for building SFD using CLD (from Mirchi et al., 2012).

Step	Purpose
Key variable recognition	Identify main drivers
Stock identification	Identify system resources (stocks) associated with the main drivers
Flow module development	Provide rates of change and represent processes governing each stock
Qualitative analysis	Identify (i) additional main drivers that may have been overlooked;
	(ii) causal relationships that require further analysing by specific methods;
	(iii) controllable variables and their controllers;
	(iv) systemic impact of changes to controllable variables;
	(v) system's vulnerability to changes in uncontrollable variables.

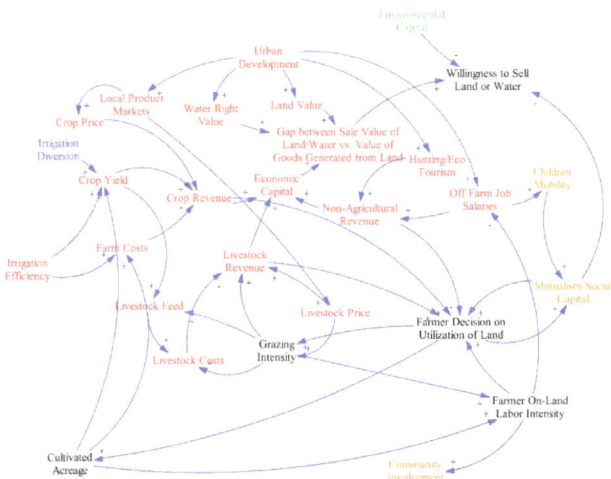

Figure 4. Fernald et al. (2012), reproduced under the CC Attribution License 3.0. An example of a complex CLD (this is approximately one quarter of the complete diagram).

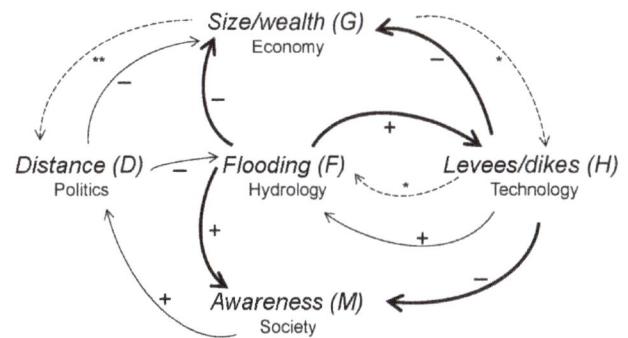

Figure 5. Di Baldassarre et al. (2013b), reproduced with permission under the CC Attribution License 3.0. An example of a simple CLD from Di Baldassarre et al. (2013b).

tensive computer simulations should be performed only after a clear picture...has been established"). Once a SFD has been developed, this then leads into the development of a full quantitative model, which will help "better understand the magnitude and directionality of the different variables within each subsystem (Fernald et al., 2012) and the overall impacts that the interactions between variables have. Turning the SFD into a quantitative model essentially involves the application of mathematical computations in the form of differential/difference equations to each of the interactions highlighted in the SFD. As with other modelling techniques, this quantitative model should go through full validation and calibration steps before it is used.

The application of a top-down modelling strategy, such as system dynamics, carries with it certain advantages. The impact that individual system processes and interactions thereof may be identified, as the root causes of feedbacks, time-lags and other non-linear effects can be traced. This trait makes system dynamics modelling particularly good in system understanding applications. The usefulness of SD in learning circumstances is increased by the different levels on which

system understanding can be generated: the different stages of model development, varying from entirely qualitative and visual to entirely quantitative, allow for those with different levels of understanding and inclination to garner insight at their own level, and during different stages of model development. As such, system dynamics is an excellent tool for use in participatory modelling circumstances. SD techniques also give a fairly good level of control over model complexity to the developer, since the level at which subsystems and interactions are defined by the model developer. There are clear outcomes that emerge in many socio-ecological and socio-hydrological systems, but the inherent complexity and levels of interaction of small-scale processes "prohibits accurate mechanistic modelling" (Scheffer et al., 2012), and so viewing (and modelling) the system from a level at which complexity is appreciated but not overwhelming allows for modelling and analyses. Another advantage that follows from this point is that system dynamics may be used in situations where the physical basis for a relationship is either unknown or difficult to represent, since correlative relationships may be used as a basis for modelling (Öztürk et al., 2013). The nature of SD models also makes it easy to integrate the important (Gordon et al., 2008) aspect of spatio-temporal scale integration, and the data-based typology of system dynamics means that the "opportunity" (Rosenberg and Madani,

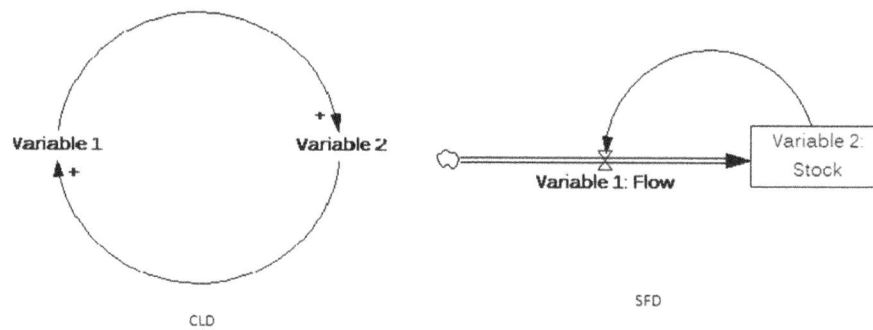

Figure 6. An example of a stocks and flows diagram (SFD) developed from a causal loop diagram (CLD).

2014) presented by big data can be harnessed in water resource management.

There are, of course, reasons why system dynamics would not be chosen as a modelling technique. The first of these is the fundamental issue that all models that view systems from a top-down perspective, inferring system characteristics from behaviours, can only produce deterministic results (Liu et al., 2006). Great care must also be taken with the level of complexity included in a system dynamics model, since very simplistic relationships between variables will fail to capture the complexity that is present (Kandasamy et al., 2014), while the inclusion of too much complexity is easy, and can result in relationships that do not occur in the real world (Kelly et al., 2013). In systems of evolution and co-evolution, using SD techniques may also be difficult, as the "very nature of systems may change over time" (Folke et al., 2010), and so time invariant equations may not properly model long-term dynamics. This is of particular importance in socio-hydrology, where changing (and so time invariant) social norms and values play a particularly important role. As such, for application in socio-hydrology, the use of time-variant equations in SD models may be useful.

Of all of the modelling techniques detailed in this review, system dynamics has perhaps seen the most explicit usage in socio-hydrology thus far. This is perhaps due to the usefulness of SD in developing system understanding (the stage that socio-hydrology would currently be characterised as being at), and the ease with which disciplines may be integrated. Models thus far have generally been fairly simple, involving five or so system components, using proxy measures for high-level system "parameters". Examples include the work of Di Baldassarre et al. (2013b) in which there are five system parameters with a total of seven difference equations governing the behaviour of a fictional system investigating the coupled dynamics of flood control infrastructure, development and population in a flood-prone area. The parameters used are proxies for the subsystems of the economy, politics, hydrology, technology and societal sensitivity. The usage of a fairly simple model has allowed for further work using this model, in which the impact of changing parameters

which represent the risk-taking attitude of a society, its collective memory and trust in risk-reduction strategies are investigated, alongside developments in which a stochastic hydrological input was used (Viglione et al., 2014), and a study in which control theory was used to investigate optimality in this context, and in which the stochastic elements of the model were replaced with periodic deterministic functions (Grames et al., 2015). The model was further developed, this time simplified in structure, by Di Baldassarre et al. (2015b); here, the core dynamics were focused on, and the number of parameters and variables reduced. This step of simplification is surely good in system dynamics models, isolating the core features and relationships which produce system-level outcomes, while reducing the risks of overparameterisation and excessive model complexity. The structure of the modelling framework allowed for the development of a fairly simple model that could show complex interactions between society and hydrology, producing emergent outcomes, and leading to development in thought around the subject. Another example of a system dynamics approach being taken in socio-hydrological study is the work of Kandasamy et al. (2014), where the co-evolution of human and water systems in the Murrumbidgee Basin (part of the Murray–Darling Basin) was investigated in a qualitative sense to form a system conceptualisation; this was then followed by work by van Emmerik et al. (2014) in which this conceptualised system view was turned into a quantitative model, formed from coupled differential equations capable of modelling past system behaviour. In this case, a slightly different set of variables are investigated (reservoir storage, irrigated area, human population, ecosystem health and environmental awareness), which provide indicators of the economic and political systems in a more indirect (e.g. the irrigated area giving an idea of economic agricultural production) but directly measurable way. Again, this fairly simple mathematical model was able to replicate the complex, emergent behaviours seen in the system, particularly the "pendulum swing" between behaviours of environmental exploitation and restoration. Studies investigating the Tarim Basin, Western China, have followed a similar development process, with a conceptual model de-

Table 3. Key advantages and disadvantages of top-down and bottom-up modelling techniques.

	Advantages	Disadvantages
Top-down	– Incomplete knowledge of system and/or processes acceptable – Complexity determined more by modeller	– Difficult to determine underlying processes – Correlations in data may be coincidental, rather than due to underlying processes
Bottom-up	– Processes properly represented (where they are understood) – Causal link between process and outcome discernable	– Large amount of system knowledge required – Model complexity determined in part by process complexities

veloped (Liu et al., 2014) first to examine the system from a qualitative, historical perspective, before a quantitative approach (Liu et al., 2015a), including proxy variables for hydrological, ecological, economic and social sub-systems, is taken to develop further understanding of how and why specific co-evolutionary dynamics have occurred; the focus in this study was on system learning, and so a simple model was developed to facilitate easy understanding. The final socio-hydrological study that explicitly takes a system dynamics approach looks at the dynamics of lake systems (Liu et al., 2015b); this study involves a slightly more complex SD model, but is an excellent example of the development path through conceptualisation, CLD formation, conversion to an SFD and subsequent quantitative analysis. The five feedback loops that exist within the model, and their significance in terms of system behaviour, are well explained. Again, similar (though a slightly higher number of) variables are used in the model, including population, economics, water demand, discharge, pollutant load and water quality. As is clear from the choice of variables, the hydrological system is viewed in more detail in this study, and the aspects of community sensitivity and behavioural responses are not included explicitly.

As is clear from the studies highlighted, system dynamics has been well applied to socio-hydrological studies. The ease with which SD facilitates system learning, the ability for relatively simple models to (re)produce emergent phenomena seen in socio-hydrological systems, and the clear model development process have led to this being a common choice of modelling framework in early socio-hydrological system study. The highlighted studies make clear the aspects of integrated socio-hydrological systems that should be included in all such studies (i.e. some inclusion of hydrological systems, impacts on livelihoods and societal responses), but also the importance of tailoring models to show in more detail those aspects that are pertinent to a particular case study.

4.6 Pattern-oriented modelling (POM)

The previously described techniques of agent-based modelling and system dynamics are archetypal examples of bottom-up and top-down modelling frameworks respectively. The advantages and disadvantages of these approaches have been detailed earlier, but are summed up in Table 3. Over-

coming these deficiencies is key in furthering the pursuit of accurate, useful modelling. One way of attempting to overcome the difficulties posed by top-down and bottom-up strategies is to attempt to "meet in the middle" (something that has been called for a long while; Veldkamp and Verburg, 2004), and this is where POM sits. Pattern-oriented models are essentially process-based (and so bottom-up) models where system results are matched to observed patterns of behaviour in the model calibration/validation stage (Grimm et al., 1996). The use of patterns in calibration, as opposed to exact magnitudes of output parameters, makes validation simpler (Railsback, 2001), since maximum use may be found for data that are available, and the often impracticable collection of data regarding all output parameters becomes less necessary. Also, imperfect knowledge of base-level processes may be overcome through emergent pattern identification (Magliocca and Ellis, 2013). The use of POM would allow for a simpler process-based model, with few parameters, overcoming the problems associated with the complexity in bottom-up models, whereby overparameterisation may lead to the tendency for models to be able to fit data despite potentially incorrect processes and structure, as well as reducing model uncertainty, while also being defined by processes, rather than data, and so overcoming the criticisms commonly levelled at top-down approaches. There are, of course, drawbacks to the use of POM: a model being able to fit patterns does not necessarily mean that the mechanisms included in the model are correct, and the data required for model validation may be quite different to those which are commonly required at present, and so using POM may require a different approach to data collection (Wiegand et al., 2003). Also, pattern-oriented models may still be significantly more complex than system dynamics models, due to the modelling of base-level processes. The very fact that they are pattern-oriented also leaves difficulties in dealing with surprise, a very important aspect of socio-hydrology.

The model development process in POM is the following (Wiegand et al., 2003).

1. Identification of processes and development of a process-based model

2. Model parameterisation

3. Aggregation of relevant data and identification of patterns

4. Comparison of observed patterns and those predicted by the model

5. Comparison of model results with other predictions (key model outputs may need to be validated against as well as patterns)

6. Necessary cyclical repetition of previous steps

Pattern-oriented models would be well applied in socio-hydrological situations. The various emergent characteristics and patterns that are created in coupled socio-ecological and socio-hydrological systems lend themselves perfectly to the integrated use of processes and patterns, particularly since there are sub-systems and processes which are well understood and the dynamics of which can be well modelled, but also those system components which are less well understood. In less well understood system sections, underlying processes may be uncovered by using the patterns which define the system (Grimm et al., 2005). POM has already found applications in socio-ecological investigations into land-use change (Evans and Kelley, 2008; Iwamura et al., 2014), though it has potential uses in many other areas.

4.7 Bayesian networks (BN)

Often, relationships between variables are stochastic, rather than deterministic, i.e. a given input does not always give the same output and instead there is a distribution of possible outputs. In such situations, Bayesian networks are well applied. The advantages of using Bayesian networks come directly from the modelling approach: uncertainties are directly and explicitly accounted for since all inputs and outputs are stochastic (Kelly et al., 2013), and the use of Bayes' theorem means that probability distributions of output variables may be "updated" as new knowledge and data become available (Barton et al., 2012). Using Bayes' theorem also allows the use of prior knowledge, since distributions of output parameters are required to be specified prior to model start-up (to then be changed and updated), and these prior distributions may be informed by the literature (Barton et al., 2012). The fact that there are relationships (albeit stochastic rather than deterministic) between variables also means that direct causal links between variables may be established (Jellinek et al., 2014). The drawbacks in using BNs are the difficulties present in modelling dynamic systems, since BNs tend to be set up as "acyclic" (Barton et al., 2012) (though object-oriented (Barton et al., 2012) and dynamic Bayesian networks (Nicholson and Flores, 2011), which can model dynamic feedbacks, are being developed and becoming more prevalent), and in the potential statistical complexities present. A Bayesian network may be seen as a stochastic version of a system dynamics model, and so many of the criticisms of SD models may also be applicable to BNs; in

particular, the fact that BNs are largely based around data-defined relationships (as opposed to physically determined or process-based relationships) between variables means that BNs can only yield deterministic (albeit stochastically deterministic) results that arise from data.

The model development process for a Bayesian network follows the following basic outline.

1. The model is conceptualised, with variables represented as "nodes" in the network and causal linkages between variables determined

2. "Parent" and "child" nodes are related with a conditional probability distribution determining how a "child" node changes in relation to parent nodes (Jellinek et al., 2014)

3. Data are collected and fed into the model.

4. These new data cause output probability distributions to be updated.

5. As new data and knowledge are accumulated, the network can be continually updated, and so the previous two points may be carried out cyclically.

Many uncertain relationships exist within hydrology and sociology, and indeed in the linkages between the two, so perhaps the use of stochastic relationships and the BN framework would be an appropriate technique in socio-hydrological studies. However adept BNs are at dealing with aleatory uncertainties, they still cannot include information about what we do not know we do not know, and so the issues of dealing with epistemic uncertainty and surprise are still prevalent. van Dam et al. (2013) has applied an acyclic BN to a wetlands scenario to determine how wetlands may be impacted by both natural and anthropogenic factors in an ecosystem functionality sense and how change in wetlands ecosystems may impact upon livelihoods; however, this model could not account for potentially significant dynamic feedbacks. The development of dynamic Bayesian networks in a socio-hydrological context should be a research priority in this area; the development of such models would be of value in contexts of system understanding, policy development and forecasting, due to the vital role that uncertainties play in all of these areas.

4.8 Coupled component modelling (CCM)

Coupled component models take specialised, disciplinary models for each part of a system and integrate them to form a model for the whole system. Kelly et al. (2013) describe how this may be "loose", involving the external coupling of models, or much more "tight", involving the integrated use of inputs and outputs. CCM therefore offers a flexibility of levels of integration (this is of course dependent on the degree to which models are compatible), and can be a very efficient method of model development, since it takes knowledge from models that already exist, and will already have some degree of validity in the system that they are modelling. The flexibility also extends into the fact that different modelling techniques may be integrated, and so those techniques that suit specific disciplines may be utilised. CCM can also be an excellent catalyst for interdisciplinary communication; models that experts from different disciplines have developed may be integrated, necessitating communication between modellers and leading to development in understanding of modelling in different disciplines.

However, there are of course drawbacks to using CCM; the models used may not be built for integration (Kelly et al., 2013), which may lead to difficulties and necessitate significant recoding. There may also be aspects of models that cannot be fully integrated, which could potentially lead to feedbacks being lost. Different treatments of space and time could potentially create difficulties in integration (though this could also be a positive, since aspects that do not require computationally intensive models may be coupled with those that do and result in savings). Uncertainties could also be an issue when coupling models directly: models will have been developed such that the outputs they generate have acceptable levels of uncertainty, though when integrated these uncertainties may snowball. When considering applications in socio-hydrology, the use of CCM raises other points. Using previously developed models means coupling together previously developed knowledge, which does have the capacity to generate new insights into coupled systems, but does not perhaps give the view of a totally integrated system. Some of the most important things in socio-hydrology occur at the interface between society and water, and so using models developed to explore each of these aspects separately may limit the capacity to learn about strictly socio-hydrological processes. New and unconventional data types, which will be important in socio-hydrology, will also struggle to be incorporated using coupled disciplinary models. The use of CCM could, however, be a good way to foster interdisciplinary communication between those in hydrology and those in the social sciences, and may be a way to improve transdisciplinary learning (a very important part of socio-hydrology).

Models have certainly been coupled between hydrology and other disciplines (for example economics e.g. Akter et al., 2014), and indeed different aspects of hydrology have been integrated using CCM (Falter et al., 2015). In socio-hydrology specifically, Hu et al. (2015) incorporates a multi-agent simulation model with a physical groundwater model to try to understand declining water table levels.

4.9 Scenario-based modelling

While perhaps not a "modelling technique" per se, and rather a method of resolution that can be applied, the usage of scenarios in analysis has important implications for modelling that warrant discussion. Scenario-based approaches fall into two main categories, those which investigate different policy implementation scenarios, and those which use scenarios of different initial conditions (within this, initial conditions could be for instance different socio-economic behavioural patterns, or future system states). This means that the impact that policies may have can be analysed from two angles; that of assuming knowledge of system behaviour and comparing decisions that may be made, as well as admitting lack of system knowledge and analysing how different system behaviour may impact the results that decisions have (indeed these may also be mixed). There are several issues that socio-hydrological modelling studies may encounter that will lead to scenario-based techniques being applicable. Firstly, long-term modelling of systems that will involve a large amount of uncertainty, particularly in terms of socio-economic development, is difficult due to the snowballing of uncertainties; as such, using likely scenarios of future development may be a more prudent starting point for modelling studies that go a long way into the future. In a similar way, scenarios that look at the occurrence of different surprising events would be useful in socio-hydrology. Even if uncertainties are deemed acceptable, the computational effort required to conduct integrated modelling studies far into the future may make such studies infeasible, and so the use of scenarios as future initial conditions may be necessary. Particularly in a policy context, policies are generally discrete options, and so the first use of scenario-based approaches mentioned (comparing options) certainly makes sense. Studies conducted on the subject of climate change tend to use a scenario-based approach for socio-economic development, and CHANS studies also sometimes use scenario-based approaches (e.g. Monticino et al., 2007). The usage of scenarios has been said to have improved recently (Haasnoot and Middelkoop, 2012), with more scenarios generally being used, and appropriate interpretation of the relative probabilities of different scenarios occurring being investigated. While the use of a scenario-based approach for analysing policy alternatives involves very few compromises, the use of scenarios as initial conditions for modelling future system states can involve compromise in that the "dynamic interactions" between social and hydrological systems will be lost (Carey et al., 2014) in the intervening period between model development and the time at which the model is analysing.

4.10 Heuristic/knowledge-based modelling

Heuristic modelling involves collecting knowledge of a system and using logic or rules to infer outcomes (Kelly et al., 2013). The process of model development here is quite clear, with an establishment of the system boundaries and processes, and simply gathering knowledge of system behaviour to determine outcomes. As with scenario-based modelling and coupled component modelling, the use of heurism in models allows the use of different modelling techniques within the tag of "heurism", for example Acevedo et al.

(2008); Huigen (2006) have used ABMs encoded with a great deal of heuristic knowledge. The advantage of heuristic modelling is in the heurism: experience and knowledge of systems is a valuable source of information, and if system processes are understood well enough that logic may be used to determine outcomes, then this is an excellent method. However, where system knowledge is incomplete, or imperfect in any way (as in socio-hydrology at present), then the usefulness of experience-based techniques falls down. Heuristic modelling is also not generally all that useful in system learning applications, though in cases where disciplinary models are integrated, new heurism may be generated in the interplay between subjects.

Gober and Wheater (2015) have identified that some current socio-hydrological models (that of Di Baldassarre et al., 2015b) may have "heuristic value" (Gober and Wheater, 2015), as opposed to practical, applicable value, in that some conceptualised models of socio-hydrological systems tend to assume relationships between variables, rather than define them via data. This gives a different value to the term heuristic, and implies the development of models of different structures via heuristic means. The challenge in taking this approach "is to avoid biasing the model to predict the social behaviour that we think should happen" (Loucks, 2015).

5 Conclusions

This paper has reviewed the literature surrounding the modelling of socio-hydrological systems, including concepts that underpin all such models (for example conceptualisation, data and complexity) and modelling techniques that have and/or could been applied in socio-hydrological study. It shows that there is a breadth of issues to consider when undertaking model-based study in socio-hydrology, and also a wide range of techniques and approaches that may be used. Essentially, however, in socio-hydrological modelling, there is a decision to be made between top-down and bottom-up modelling, which represents a choice between representing individual system processes (including the behaviours and decisions of people in this case) and viewing the system as a whole; both of these approaches have advantages and disadvantages, and the task of the modeller is to maximise the advantages and minimise the disadvantages. There are significant challenges in representing, modelling and analysing coupled human–water systems, though the importance of the interactions that now occur between humans and water means that these challenges should be the focus of significant research efforts. With regards to future research that could be conducted following the work that has been reviewed here, without resorting to the platitudes of improving predictions, reducing and managing uncertainties, increasing interdisciplinary integration and improving data, there are several examples of areas in which research would be of benefit. Some of these topics are common to other subjects; however, there

are specific aspects that are of particular importance in socio-hydrology.

- Conceptual models of stylised socio-hydrological systems, for example systems of inter-basin water transfer, drought or agricultural water use: the strength that socio-hydrology should bring is a greater understanding of how human–water interaction affects overall system behaviour. A great deal of understanding can be generated through conceptual studies of generalised systems, and so modelling of archetypal systems would be of benefit. The challenge here is to move beyond models developed to mimic behaviour that we expect, towards those capable of giving insight.

- Determining the appropriate complexity for models of highly interconnected socio-hydrological systems: the broadening of system boundaries brings issues regarding model complexity and trade-offs between deterministic uncertainty and uncertainty propagation. Quantifying these trade-offs in socio-hydrological circumstances, and so determining the appropriate level of abstraction for modelling would allow for more effective modelling efforts.

- Gathering data in socio-hydrological studies: as an interdisciplinary subject, data in socio-hydrological study will come from a variety of sources. While methods for collection of hydrological data are well established, the social data that will be required, and indeed the new, unconventional data that may be required to describe socio-hydrological processes, may pose issues in availability and collection. The challenge here is to maximise the utility of what is available and to develop models in an iterative fashion, allowing early stage, conceptual models to guide data collection, and adapting models to suit what data are available.

- Determining methods for calibration and validation in socio-hydrology: calibration and validation are issues in almost all modelling areas. However, as a new subject, there is no calibration/validation protocol for socio-hydrological modelling, and with the aforementioned issues with social science data, conducting formal calibration and validation may be difficult. As such, the development of guidelines regarding what constitutes "validation" in socio-hydrology would be worthy of investigation.

- Discussion of emergence in socio-hydrological systems, particularly emergence of more abstract properties, such as risk, vulnerability and resilience: the stochastic nature of hydrological drivers and the unpredictability of human responses renders any definite statement regarding system behaviour largely anecdotal (though often anecdotes of merit), and so acknowledging this stochasticity in analysis and discussion, using properties of more abstract meaning to describe the system may be useful in socio-hydrology.

– More in-depth socio-hydrological modelling studies across social, economic and hydrological gradients: while conceptual modelling can build understanding to a point, case-based models can often give a greater insight into specific system behaviours. Applying socio-hydrological models to a range of cases will help build understanding in this way, particularly if these cases are similar, but differentiated in some way (e.g. responses to drought across a range of levels of economic development). The challenge (and opportunity) that this presents is understanding the dynamics which are general across cases, those which vary across gradients and those which are place-specific.

– Determining how best to present and use findings from socio-hydrological studies in policy applications: the way that socio-hydrological understanding will likely be applied in the real world is via policy decisions. As such, understanding the best way to communicate findings in socio-hydrology is vital. The challenge here is to communicate the differences between the outcomes predicted by traditional analyses and socio-hydrological studies regarding the way that policy decisions may impact the system in the long term, while acknowledging the limitations in both approaches.

The unifying feature of these future research topics is the development of understanding regarding socio-hydrological systems. The most important way in which socio-hydrology differs from other water management subjects is in understanding the system as a whole, as opposed to focusing on problem solving. As such, the research priorities at this stage are focused on different ways of improving and communicating understanding.

Acknowledgements. This work was supported by the Natural Environment Research Council as part of the Science and Solutions for a Changing Planet Doctoral Training Programme NE/L002515/1, and UK Natural Environment Research Council projects NE-K010239-1 (Mountain-EVO) and NE/I022558/1 (Hydroflux India). The authors would also like to extend great thanks to Giuliano Di Baldassarre and two other anonymous referees for reviewing this paper and making valuable suggestions on how to improve it.

References

Acevedo, M., Baird Callicott, J., Monticino, M., Lyons, D., Palomino, J., Rosales, J., Delgado, L., Ablan, M., Davila, J., Tonella, G., Ramírez, H., and Vilanova, E.: Models of natural and human dynamics in forest landscapes: Cross-site and cross-cultural synthesis, Geoforum, 39, 846–866, doi:10.1016/j.geoforum.2006.10.008, 2008.

Adger, W.: Evolution of economy and environment: an application to land use in lowland Vietnam, Ecol. Econ., 31, 365–379, doi:10.1016/S0921-8009(99)00056-7, 1999.

AghaKouchak, A., Feldman, D., Hoerling, M., Huxman, T., and Lund, J.: Water and Climate: Recognize anthropogenic drought, Nature, 524, 409–411, 2015.

Akter, S., Quentin Grafton, R., and Merritt, W. S.: Integrated hydro-ecological and economic modeling of environmental flows: Macquarie Marshes, Australia, Agricult. Water Manage., 145, 98–109, doi:10.1016/j.agwat.2013.12.005, 2014.

An, L.: Modeling human decisions in coupled human and natural systems: Review of agent-based models, Ecol. Model., 229, 25–36, doi:10.1016/j.ecolmodel.2011.07.010, 2012.

An, L., Linderman, M., and Qi, J.: Exploring complexity in a human-environment system: an agent-based spatial model for multidisciplinary and multiscale integration, Ann. Assoc. Am Geograph., 95, 54–79, doi:10.1111/j.1467-8306.2005.00450.x, 2005.

Anderies, J. M., Janssen, M. A., and Ostrom, E.: A Framework to Analyze the Robustness of Social-Ecological Systems from an Institutional Perspective, Ecol. Soc., 9, 1–18, 2004.

Archer, M. S.: Realist Social Theory: The Morphogenetic Approach, Cambridge University Press, Cambridge, p. 184, 1995.

Arkesteijn, L. and Pande, S.: On hydrological model complexity, its geometrical interpretations and prediction uncertainty, Water Resour. Res., 49, 7048–7063, doi:10.1002/wrcr.20529, 2013.

Arthur, W. B.: Complexity and the Economy, Science, 284, 107–109, doi:10.1126/science.284.5411.107, 1999.

Barreteau, O., Bousquet, F., Millier, C., and Weber, J.: Suitability of Multi-Agent Simulations to study irrigated system viability: Application to case studies in the Senegal River Valley, Agricult. Syst., 80, 255–275, doi:10.1016/j.agsy.2003.07.005, 2004.

Barton, D. N., Kuikka, S., Varis, O., Uusitalo, L., Henriksen, H. J., Borsuk, M., de la Hera, A., Farmani, R., Johnson, S., and Linnell, J. D. C.: Bayesian networks in environmental and resource management, Int. Environ. Assess. Manage., 8, 418–429, doi:10.1002/ieam.1327, 2012.

Becu, N., Perez, P., Walker, A., Barreteau, O., and Le Page, C.: Agent based simulation of a small catchment water management in northern Thailand, Ecol. Model., 170, 319–331, doi:10.1016/S0304-3800(03)00236-9, 2003.

Berkes, F.: Understanding uncertainty and reducing vulnerability: Lessons from resilience thinking, Nat. Hazards, 41, 283–295, doi:10.1007/s11069-006-9036-7, 2007.

Beven, K.: Changing Ideas in Hydrology – the Case of Physically-Based Models, J. Hydrol., 105, 157–172, 1989.

Biggs, R., Carpenter, S. R., and Brock, W. A.: Turning back from the brink: detecting an impending regime shift in time to avert it, P. Natl. Acad. Sci. USA, 106, 826–831, doi:10.1073/pnas.0811729106, 2009.

Blomquist, W. and Schlager, E.: Political Pitfalls of Integrated Watershed Management, Soc. Nat. Resour., 18, 37–41, doi:10.1080/08941920590894435, 2005.

Blöschl, G. and Sivapalan, M.: Scale Issues in Hydrological Modelling: a Review, Hydrol. Process., 9, 251–290, 1995.

Boelens, R.: Cultural politics and the hydrosocial cycle: Water, power and identity in the Andean highlands, Geoforum, 57, 234–247, doi:10.1016/j.geoforum.2013.02.008, 2013.

Bohensky, E.: Learning dilemmas in a social-ecological system: An agent-based modeling exploration, JASSS, 17, doi:10.18564/jasss.2448, 2014.

Bourblanc, M. and Blanchon, D.: The challenges of rescaling South African water resources management: Catchment Management Agencies and interbasin transfers, J. Hydrol., 519, 2381–2391, doi:10.1016/j.jhydrol.2013.08.001, 2013.

Bousquet, F. and Le Page, C.: Multi-agent simulations and ecosystem management: a review, Ecol. Model., 176, 313–332, doi:10.1016/j.ecolmodel.2004.01.011, 2004.

Carey, M., Baraer, M., Mark, B. G., French, A., Bury, J., Young, K. R., and McKenzie, J. M.: Toward hydro-social modeling: Merging human variables and the social sciences with climate-glacier runoff models (Santa River, Peru), J. Hydrol., 518, 60–70, doi:10.1016/j.jhydrol.2013.11.006, 2014.

Chu, D., Strand, R., and Fjelland, R.: Theories of Complexity: Common Denominators of Complex Systems, Complexity, 8, 19–30, doi:10.1002/cplx.10059, 2003.

Cotter, M., Berkhoff, K., Gibreel, T., Ghorbani, A., Golbon, R., Nuppenau, E.-A., and Sauerborn, J.: Designing a sustainable land use scenario based on a combination of ecological assessments and economic optimization, Ecol. Ind., 36, 779–787, doi:10.1016/j.ecolind.2013.01.017, 2014.

Couclelis, H.: Why I no longer work with Agents, Tech. rep., Centre for Spatially Integrated Social Science, University of California, Santa Barbara, http://www.csiss.org/events/other/agent-based/papers/couclelis.pdf (last access: 13 October 2014), 2001.

Crépin, A.-S.: Using fast and slow processes to manage resources with thresholds, Environ. Resour. Econ., 36, 191–213, doi:10.1007/s10640-006-9029-8, 2007.

Crook, J. H.: Social organisation and the environment: Aspects of contemporary social ethology, Animal Behav., 18, 197–209, 1970.

Crutzen, P. J.: Geology of mankind, Nature, 415, p. 23, doi:10.1038/415023a, 2002.

Crutzen, P. J. and Stoermer, E. F.: The 'Anthropocene', IGBP Global Change Newsletter, 17–18, http://www.igbp.net/publications/globalchangemagazine/globalchangemagazine/globalchangenewslettersno4159.5.5831d9ad13275d51c098000309.html (last access: 14 March 2015), 2000.

Dakos, V., Carpenter, S. R., Nes, E. H. V., and Scheffer, M.: Resilience indicators: prospects and limitations for early warnings of regime shifts, Phil. Trans. Roy. Soc. B, 370, 20130263, doi:10.1098/rstb.2013.0263, 2015.

Deadman, P., Robinson, D., Moran, E., and Brondizio, E.: Colonist household decisionmaking and land-use change in the Amazon Rainforest: an agent-based simulation, EnviroN. Plan., 31, 693–709, doi:10.1068/b3098, 2004.

Destouni, G., Jaramillo, F., and Prieto, C.: Hydroclimatic shifts driven by human water use for food and energy production, Nature Clim. Change, 3, 213–217, doi:10.1038/nclimate1719, 2012.

Di Baldassarre, G., Kooy, M., Kemerink, J. S., and Brandimarte, L.: Towards understanding the dynamic behaviour of floodplains as human-water systems, Hydrol. Earth Syst. Sci., 17, 3235–3244, doi:10.5194/hess-17-3235-2013, 2013a.

Di Baldassarre, G., Viglione, A., Carr, G., Kuil, L., Salinas, J. L., and Blöschl, G.: Socio-hydrology: conceptualising human-flood interactions, Hydrol. Earth Syst. Sci., 17, 3295–3303, doi:10.5194/hess-17-3295-2013, 2013b.

Di Baldassarre, G., Brandimarte, L., and Beven, K.: The seventh facet of uncertainty: wrong assumptions, unknowns and surprises in the dynamics of human-water systems, Hydrol. Sci. J., doi:10.1080/02626667.2015.1091460, 2015a.

Di Baldassarre, G., Viglione, A., Carr, G., Kuil, L., Yan, K., Brandimarte, L., and Blöschl, G.: Debates-Perspectives on socio-hydrology: Capturing feedbacks between physical and social processes, Water Resour. Res., 51, 4770–4781, doi:10.1002/2014WR016416, 2015b.

Dinar, S.: Physical and political impacts: Complex river boundaries at risk, Nature Clim. Change, 4, 955–956, doi:10.1038/nclimate2421, 2014.

Dooge, J.: Linear theory of hydrologic systems: Technical Bulletin No. 1468, Tech. rep., Agricultural Research Service – United States Department of Agriculture, Washington, http://books.google.com/books?hl=en&lr=&id=iVgTfUhBi2gC&oi=fnd&pg=PA1&dq=Linear+Theory+of+Hydrologic+Systems&ots=dvGbEATLVP&sig=A5G0et_9hcEK7L08Z3nJT3CemrA (last access: 17 December 2014) 1973.

Dougill, A. J., Fraser, E. D. G., and Reed, M. S.: Anticipating vulnerability to climate change in dryland pastoral systems: Using dynamic systems models for the Kalahari, Ecol. Soc., 15, http://www.ecologyandsociety.org/vol15/iss2/art17/ (last access: 24 March 2015) 2010.

D'Odorico, P., Bhattachan, A., Davis, K. F., Ravi, S., and Runyan, C. W.: Global desertification: Drivers and feedbacks, Adv. Water Resour., 51, 326–344, doi:10.1016/j.advwatres.2012.01.013, 2013.

Ehret, U., Gupta, H. V., Sivapalan, M., Weijs, S. V., Schymanski, S. J., Blöschl, G., Gelfan, A. N., Harman, C., Kleidon, A., Bogaard, T. A., Wang, D., Wagener, T., Scherer, U., Zehe, E., Bierkens, M. F. P., Di Baldassarre, G., Parajka, J., van Beek, L. P. H., van Griensven, A., Westhoff, M. C., and Winsemius, H. C.: Advancing catchment hydrology to deal with predictions under change, Hydrol. Earth Syst. Sci., 18, 649–671, doi:10.5194/hess-18-649-2014, 2014.

Elsawah, S., Guillaume, J. H. A., Filatova, T., Rook, J., and Jakeman, A. J.: A methodology for eliciting, representing, and analysing stakeholder knowledge for decision making on complex socio-ecological systems: From cognitive maps to agent-based models, J. Environ. Manage., 151, 500–516, doi:10.1016/j.jenvman.2014.11.028, 2015.

Elshafei, Y., Sivapalan, M., Tonts, M., and Hipsey, M. R.: A prototype framework for models of socio-hydrology: identification of key feedback loops and parameterisation approach, Hydrol. Earth Syst. Sci., 18, 2141–2166, doi:10.5194/hess-18-2141-2014, 2014.

Epstein, G. and Vogt, J. M.: Missing ecology: integrating ecological perspectives with the social-ecological system framework, International J. Commons, 7, 432–453, 2013.

Ertsen, M. W., Murphy, J. T., Purdue, L. E., and Zhu, T.: A journey of a thousand miles begins with one small step - human agency, hydrological processes and time in socio-hydrology, Hydrol. Earth Syst. Sci., 18, 1369–1382, doi:10.5194/hess-18-1369-2014, 2014.

Evans, T. P. and Kelley, H.: Multi-scale analysis of a household level agent-based model of landcover change., Journal of environmental management, 72, 57–72, doi:10.1016/j.jenvman.2004.02.008, 2004.

Evans, T. P. and Kelley, H.: Assessing the transition from deforestation to forest regrowth with an agent-based model of land cover change for south-central Indiana (USA), Geoforum, 39, 819–832, doi:10.1016/j.geoforum.2007.03.010, 2008.

Evans, T. P., Sun, W., and Kelley, H.: Spatially explicit experiments for the exploration of land-use decision-making dynamics, Int. J. Geogr. Inf. Sci., 20, 1013–1037, doi:10.1080/13658810600830764, 2006.

Fabre, J., Ruelland, D., Dezetter, A., and Grouillet, B.: Simulating past changes in the balance between water demand and availability and assessing their main drivers at the river basin scale, Hydrol. Earth Syst. Sci., 19, 1263–1285, doi:10.5194/hess-19-1263-2015, 2015.

Falkenmark, M.: Water and Mankind: A Complex System of Mutual Interaction, Ambio, 6, 3–9, 1977.

Falkenmark, M.: Main Problems of Water Use and Transfer of Technology, GeoJournal, 3, 435–443, 1979.

Falkenmark, M.: Freshwater as shared between society and ecosystems: from divided approaches to integrated challenges., Phil. Trans. Roy. Soc. Lnd., 358, 2037–49, doi:10.1098/rstb.2003.1386, 2003.

Falkenmark, M.: Ecohydrosolidarity-towards better balancing of humans and nature, Waterfront, 4–5, 2009.

Falkenmark, M.: What's new in water, what's not, and what to do now, Rev. Environ. Sci. Bio/Technol., 10, 107–109, doi:10.1007/s11157-011-9238-7, 2011.

Falkenmark, M. and Folke, C.: The ethics of socio-ecohydrological catchment management: towards hydrosolidarity, Hydrol. Earth Syst. Sci., 6, 1–10, doi:10.5194/hess-6-1-2002, 2002.

Falter, D., Schröter, K., Dung, N. V., Vorogushyn, S., Kreibich, H., Hundecha, Y., Apel, H., and Merz, B.: Spatially coherent flood risk assessment based on long-term continuous simulation with a coupled model chain, J. Hydrol., 524, 182–193, doi:10.1016/j.jhydrol.2015.02.021, 2015.

Falvo, D. J.: On modeling Balinese water temple networks as complex adaptive systems, Human Ecol., 28, 641–649, 2000.

Fernald, A., Tidwell, V., Rivera, J., Rodríguez, S., Guldan, S., Steele, C., Ochoa, C., Hurd, B., Ortiz, M., Boykin, K., and Cibils, A.: Modeling sustainability of water, environment, livelihood, and culture in traditional irrigation communities and their linked watersheds, Sustainability, 4, 2998–3022, doi:10.3390/su4112998, 2012.

Fernald, A., Guldan, S., Boykin, K., Cibils, A., Gonzales, M., Hurd, B., Lopez, S., Ochoa, C., Ortiz, M., Rivera, J., Rodriguez, S., and Steele, C.: Linked hydrologic and social systems that support resilience of traditional irrigation communities, Hydrol. Earth Syst. Sci., 19, 293–307, doi:10.5194/hess-19-293-2015, 2015.

Filatova, T., van der Veen, A., and Parker, D. C.: Land Market Interactions between Heterogeneous Agents in a Heterogeneous Landscape-Tracing the Macro-Scale Effects of Individual Trade-Offs between Environmental Amenities and Disameni-

ties, Canad. J. Agr. Econ., 57, 431–457, doi:10.1111/j.1744-7976.2009.01164.x, 2009.

Fish, R. D., Ioris, A. A. R., and Watson, N. M.: Integrating water and agricultural management: collaborative governance for a complex policy problem., Sci. Total Eenviron., 408, 5623–5630, doi:10.1016/j.scitotenv.2009.10.010, 2010.

Folke, C.: Resilience: The emergence of a perspective for social-ecological systems analyses, Global Environ. Change, 16, 253–267, doi:10.1016/j.gloenvcha.2006.04.002, 2006.

Folke, C., Carpenter, S. R., and Walker, B.: Resilience thinking: integrating resilience, adaptability and transformability, Ecol. Soc., 15, 2010.

Foster, J.: From simplistic to complex systems in economics, Cambridge Journal of Economics, 29, 873–892, doi:10.1093/cje/bei083, 2005.

Fraser, E. D., Simelton, E., Termansen, M., Gosling, S. N., and South, A.: "Vulnerability hotspots: Integrating socio-economic and hydrological models to identify where cereal production may decline in the future due to climate change induced drought, Agr. Forest Meteorol., 170, 195–205, doi:10.1016/j.agrformet.2012.04.008, 2013.

Garcia, M., Portney, K., and Islam, S.: A question driven socio-hydrological modeling process, Hydrol. Earth Syst. Sci. Discuss., 12, 8289–8335, doi:10.5194/hessd-12-8289-2015, 2015.

Garmestani, A. S.: Sustainability science: accounting for nonlinear dynamics in policy and social-ecological systems, Clean Technol. Environ. Pol., 16, 731–738, doi:10.1007/s10098-013-0682-7, 2013.

Gibon, A., Sheeren, D., Monteil, C., Ladet, S., and Balent, G.: Modelling and simulating change in reforesting mountain landscapes using a social-ecological framework, Landsc. Ecol., 25, 267–285, doi:10.1007/s10980-009-9438-5, 2010.

Gober, P. and Wheater, H. S.: Socio-hydrology and the science-policy interface: a case study of the Saskatchewan River basin, Hydrol. Earth Syst. Sci., 18, 1413–1422, doi:10.5194/hess-18-1413-2014, 2014.

Gober, P. and Wheater, H. S.: Debates-Perspectives on socio-hydrology: Modeling flood risk as a public policy problem, Water Resour. Res., 51, 4782–4788, doi:10.1002/2015WR016945, 2015.

Gordon, L. J., Peterson, G. D., and Bennett, E. M.: Agricultural modifications of hydrological flows create ecological surprises., Trends Ecol. Evol., 23, 211–219, doi:10.1016/j.tree.2007.11.011, 2008.

Grames, J., Prskawetz, A., Grass, D., and Blöschl, G.: Modelling the interaction between flooding events and economic growth, Proc. Int. Assoc. Hydrol. Sci., 369, 3–6, doi:10.5194/piahs-369-3-2015, 2015.

Grimm, V., Frank, K., Jeltsch, F., Brandl, R., Uchmaski, J., and Wissel, C.: Pattern-oriented modelling in population ecology, Sci. Total Environ., 183, 151–166, doi:10.1016/0048-9697(95)04966-5, 1996.

Grimm, V., Revilla, E., Berger, U., Jeltsch, F., Mooij, W. M., Railsback, S. F., Thulke, H.-H., Weiner, J., Wiegand, T., and DeAngelis, D. L.: Pattern-oriented modeling of agent-based complex systems: lessons from ecology, Science, 310, 987–91, doi:10.1126/science.1116681, 005.

Haasnoot, M. and Middelkoop, H.: A history of futures: A review of scenario use in water policy studies in the Netherlands, Environ.

Sci. Pol., 19–20, 108–120, doi:10.1016/j.envsci.2012.03.002, 2012.

Haddeland, I., Heinke, J., Biemans, H., Eisner, S., Flörke, M., Hanasaki, N., Konzmann, M., Ludwig, F., Masaki, Y., Schewe, J., Stacke, T., Tessler, Z. D., Wada, Y., and Wisser, D.: Global water resources affected by human interventions and climate change, P. Natl. Acad. Sci., 111, 3251–3256, doi:10.1073/pnas.1222475110, 2014.

Hadfield, L. and Seaton, R.: A co-evolutionary model of change in environmental management, Futures, 31, 577–592, doi:10.1016/S0016-3287(99)00015-4, 1999.

Hardin, G.: The Tragedy of the Commons, Science, 162, 1243–1248, doi:10.1126/science.162.3859.1243, 1968.

Harou, J. J., Pulido-Velazquez, M., Rosenberg, D. E., Medellín-Azuara, J., Lund, J. R., and Howitt, R. E.: Hydro-economic models: Concepts, design, applications, and future prospects, J. Hydrol., 375, 627–643, doi:10.1016/j.jhydrol.2009.06.037, 2009.

Harte, J.: Toward a Synthesis of the Newtonian and Darwinian Worldviews, Phys. Today, 55, 29–34, doi:10.1063/1.1522164, 2002.

Hodgson, G. M.: Darwinism and institutional economics, J. Econ. Iss., 37, 85–97, 2003.

Hoekstra, A. and Hung, P.: Virtual Water Trade: A Quantification of Virtual Water Flows Between Nations in Relation to International Crop Trade, Tech. Rep. 11, UNESCO, IHE Delft, Delft, http://www.waterfootprint.org/Reports/Report12.pdf (last access: 27 March 2015), 2002.

Holling, C.: Resilience and stability of ecological systems, Annu. Rev. Ecol. Syst., 4, 1–23, 1973.

Hrachowitz, M., Savenije, H., Blöschl, G., McDonnell, J., Sivapalan, M., Pomeroy, J., Arheimer, B., Blume, T., Clark, M. P., Ehret, U., Fenicia, F., Freer, J. E., Gelfan, A., Gupta, H., Hughes, D., Hut, R., Montanari, A., Pande, S., Tetzlaff, D., Troch, P. A., Uhlenbrook, S., Wagener, T., Winsemius, H., Woods, R., Zehe, E., and Cudennec, C.: A decade of Predictions in Ungauged Basins (PUB)–a review, Hydrol. Sci. J., 58, 1198–1255, doi:10.1080/02626667.2013.803183, 2013.

Hu, Y., Garcia-Cabrejo, O., Cai, X., Valocchi, A. J., and DuPont, B.: Global sensitivity analysis for large-scale socio-hydrological models using Hadoop, Environ. Model. Softw., 73, 231–243, doi:10.1016/j.envsoft.2015.08.015, 2015.

Hughes, T. P., Linares, C., Dakos, V., van de Leemput, I. A., and van Nes, E. H.: Living dangerously on borrowed time during slow, unrecognized regime shifts, Trends Ecol. Evol., 28, 149–155, doi:10.1016/j.tree.2012.08.022, 2013.

Huigen, M. G. A.: Multiactor modeling of settling decisions and behavior in the San Mariano watershed, the Philippines: a first application with the MameLuke framework, Ecol. Soc., 11, 2006.

Hurford, A. P., Huskova, I., and Harou, J. J.: Using many-objective trade-off analysis to help dams promote economic development, protect the poor and enhance ecological health, Environ. Sci. Pol., 38, 72–86, doi:10.1016/j.envsci.2013.10.003, 2014.

Iwamura, T., Lambin, E. F., Silvius, K. M., Luzar, J. B., and Fragoso, J. M.: Agent-based modeling of hunting and subsistence agriculture on indigenous lands: Understanding interactions between social and ecological systems, Environ. Model. Softw., 58, 109–127, doi:10.1016/j.envsoft.2014.03.008, 2014.

Janssen, M. A. and Ostrom, E.: Governing social-ecological systems, in: Handbook of computational economics, edited by: Tes-

fatsion, L. and Judd, K., chap. 30, 1466–1502, Elsevier B.V., Amsterdam, Netherlands, doi:10.1016/S1574-0021(05)02030-7, 2006.

Jeffrey, P. and McIntosh, B. S.: Description, diagnosis, prescription: a critique of the application of co-evolutionary models to natural resource management, Environ. Conserv., 33, 281–293, doi:10.1017/S0376892906003444, 2006.

Jellinek, S., Rumpff, L., Driscoll, D. A., Parris, K. M., and Wintle, B. A.: Modelling the benefits of habitat restoration in socio-ecological systems, Biol. Conserv., 169, 60–67, doi:10.1016/j.biocon.2013.10.023, 2014.

Kain, J.-H., Kärrman, E., and Söderberg, H.: Multi-criteria decision aids for sustainable water management, Proc. ICE-Eng. Sustain., 160, 87–93, doi:10.1680/ensu.2007.160.2.87, 2007.

Kallis, G.: When is it coevolution?, Ecol. Econ., 62, 1–6, doi:10.1016/j.ecolecon.2006.12.016, 2007.

Kallis, G.: Coevolution in water resource development, Ecol. Econ., 69, 796–809, doi:10.1016/j.ecolecon.2008.07.025, 2010.

Kandasamy, J., Sountharajah, D., Sivabalan, P., Chanan, A., Vigneswaran, S., and Sivapalan, M.: Socio-hydrologic drivers of the pendulum swing between agricultural development and environmental health: a case study from Murrumbidgee River basin, Australia, Hydrol. Earth Syst. Sci., 18, 1027–1041, doi:10.5194/hess-18-1027-2014, 2014.

Karoly, D. J.: Climate change: Human-induced rainfall changes, Nature Geosci., 7, 551–552, doi:10.1038/ngeo2207, 2014.

Kelly (Letcher), R. A., Jakeman, A. J., Barreteau, O., Borsuk, M. E., ElSawah, S., Hamilton, S. H., Henriksen, H. J., Kuikka, S., Maier, H. R., Rizzoli, A. E., van Delden, H., and Voinov, A. A.: Selecting among five common modelling approaches for integrated environmental assessment and management, Environ. Model. Softw., 47, 159–181, doi:10.1016/j.envsoft.2013.05.005, 2013.

Kumar, P.: Typology of hydrologic predictability, Water Resour. Res., 47, W00H05, doi:10.1029/2010WR009769, 2011.

Ladyman, J., Lambert, J., and Wiesner, K.: What is a complex system?, Eur. J. Phil. Sci., 3, 33–67, doi:10.1007/s13194-012-0056-8, 2013.

Lane, S. N.: Acting, predicting and intervening in a socio-hydrological world, Hydrol. Earth Syst. Sci., 18, 927–952, doi:10.5194/hess-18-927-2014, 2014.

Lansing, J. S.: Complex Adaptive Systems, Annual Review of Anthropology, 32, 183–204, doi:10.1146/annurev.anthro.32.061002.093440, 2003.

Lansing, J. S. and Kremer, J. N.: Emergent Properties of Balinese Water Temple Networks: Coadaptation on a Rugged Fitness Landscape, Am. Anthropol., 95, 97–114, doi:10.1525/aa.1993.95.1.02a00050, 1993.

Lansing, J. S., Cox, M. P., Downey, S. S., Janssen, M. A., and Schoenfelder, J. W.: A robust budding model of Balinese water temple networks, World Archaeol., 41, 112–133, doi:10.1080/00438240802668198, 2009.

Lempert, R.: Agent-based modeling as organizational and public policy simulators., P. Natl. Acad. Sci. USA, 99, Suppl 3, 7195–7196, doi:10.1073/pnas.072079399, 2002.

Letcher, R. A., Croke, B. F. W., and Jakeman, A. J.: Integrated assessment modelling for water resource allocation and management: A generalised conceptual framework, Environ. Model. Softw., 22, 733–742, doi:10.1016/j.envsoft.2005.12.014, 2007.

Levin, S., Xepapadeas, T., Crépin, A.-S., Norberg, J., de Zeeuw, A., Folke, C., Hughes, T., Arrow, K., Barrett, S., Daily, G., Ehrlich, P., Kautsky, N., Mäler, K.-G., Polasky, S., Troell, M., Vincent, J. R., and Walker, B.: Social-ecological systems as complex adaptive systems: modeling and policy implications, Environ. Develop. Econ., 18, 111–132, doi:10.1017/S1355770X12000460, 2012.

Liao, K.-H.: From flood control to flood adaptation: a case study on the Lower Green River Valley and the City of Kent in King County, Washington, Nat. Hazards, 71, 723–750, doi:10.1007/s11069-013-0923-4, 2013.

Liebman, J. C.: Some Simple-Minded Observations on the Role of Optimization in Public Systems Decision-Making, Interfaces, 6, 102–108, 1976.

Linton, J. and Budds, J.: The hydrosocial cycle: Defining and mobilizing a relational-dialectical approach to water, Geoforum, 57, 170–180, doi:10.1016/j.geoforum.2013.10.008, 2013.

Liu, D., Tian, F., Lin, M., and Sivapalan, M.: A conceptual socio-hydrological model of the co-evolution of humans and water: case study of the Tarim River basin, western China, Hydrol. Earth Syst. Sci., 19, 1035–1054, doi:10.5194/hess-19-1035-2015, 2015a.

Liu, H., Benoit, G., Liu, T., Liu, Y., and Guo, H.: An integrated system dynamics model developed for managing lake water quality at the watershed scale, J. Environ. Manage., 155, 11–23, doi:10.1016/j.jenvman.2015.02.046, 2015b.

Liu, J., Dietz, T., Carpenter, S. R., Alberti, M., Folke, C., Moran, E., Pell, A. N., Deadman, P., Kratz, T., Lubchenco, J., Ostrom, E., Ouyang, Z., Provencher, W., Redman, C. L., Schneider, S. H., and Taylor, W. W.: Complexity of coupled human and natural systems, Science, 317, 1513–1516, doi:10.1126/science.1144004, 2007a.

Liu, J., Dietz, T., Carpenter, S. R., Folke, C., Alberti, M., Redman, C. L., Schneider, S. H., Ostrom, E., Pell, A. N., Lubchenco, J., Taylor, W. W., Ouyang, Z., Deadman, P., Kratz, T., and Provencher, W.: Coupled Human and Natural Systems, AMBIO, 36, 639–649, doi:10.1579/0044-7447(2007)36[639:CHANS]2.0.CO;2, 2007b.

Liu, X., Li, X., and Anthony, G.-O. Y.: Multi-agent systems for simulating spatial decision behaviors and land-use dynamics, Sci. China Ser. D-Earth Sci., 49, 1184–1194, doi:10.1007/s11430-006-1184-9, 2006.

Liu, Y., Tian, F., Hu, H., and Sivapalan, M.: Socio-hydrologic perspectives of the co-evolution of humans and water in the Tarim River basin, Western China: the Taiji-Tire model, Hydrol. Earth Syst. Sci., 18, 1289–1303, doi:10.5194/hess-18-1289-2014, 2014.

Lorenzoni, I.: A co-evolutionary approach to climate change impact assessment – Part II: A scenario-based case study in East Anglia (UK), Global Environ. Change, 10, 145–155, doi:10.1016/S0959-3780(00)00016-9, 2000.

Lorenzoni, I., Jordan, A., Hulme, M., Kerry Turner, R., and O'Riordan, T.: A co-evolutionary approach to climate change impact assessment: Part I. Integrating socio-economic and climate change scenarios, Global Environ. Change, 10, 57–68, doi:10.1016/S0959-3780(00)00012-1, 2000.

Loucks, D. P.: Debates-Perspectives on socio-hydrology: Simulating hydrologic-human interactions, Water Resour. Res., 51, 4789–4794, doi:10.1002/2015WR017002, 2015.

Ludy, J. and Kondolf, G. M.: Flood risk perception in lands protected by 100-year levees, Nat. Hazards, 61, 829–842, doi:10.1007/s11069-011-0072-6, 2012.

Lumbroso, D. M. and Vinet, F.: A comparison of the causes, effects and aftermaths of the coastal flooding of England in 1953 and France in 2010, Nat. Hazards Earth Syst. Sci., 11, 2321–2333, doi:10.5194/nhess-11-2321-2011, 2011.

Macal, C. M. and North, M. J.: Tutorial on agent-based modelling and simulation, J. Simul., 4, 151–162, doi:10.1057/jos.2010.3, 2010.

Madani, K. and Hooshyar, M.: A game theory-reinforcement learning (GT-RL) method to develop optimal operation policies for multi-operator reservoir systems, J. Hydrol., 519, 732–742, doi:10.1016/j.jhydrol.2014.07.061, 2014.

Magliocca, N. R.: Induced coupling: an approach to modeling and managing complex human-landscape interactions, Syst. Res. Behav. Sci., 25, 655–661, doi:10.1002/sres.938, 2009.

Magliocca, N. R. and Ellis, E. C.: Using Pattern-oriented Modeling (POM) to Cope with Uncertainty in Multi-scale Agent-based Models of Land Change, Trans. GIS, 17, 883–900, doi:10.1111/tgis.12012, 2013.

Malanson, G.: Considering complexity, Ann. Assoc. Am Geograph., 89, 746–753, 1999.

Manson, S. M.: Simplifying complexity: a review of complexity theory, Geoforum, 32, 405–414, doi:10.1016/S0016-7185(00)00035-X, 2001.

Manson, S. M.: Does scale exist? An epistemological scale continuum for complex human-environment systems, Geoforum, 39, 776–788, doi:10.1016/j.geoforum.2006.09.010, 2008.

Manson, S. M. and Evans, T.: Agent-based modeling of deforestation in southern Yucatan, Mexico, and reforestation in the Midwest United States, P. Natl. Acad. Sci., 104, 20678–20683, doi:10.1073/pnas.0705802104, 2007.

Marshall, G. R. and Stafford Smith, D. M.: Natural resources governance for the drylands of the Murray?Darling Basin, Rangeland J., 32, 267, doi:10.1071/RJ10020, 2013.

Matthews, R.: The People and Landscape Model (PALM): Towards full integration of human decision-making and biophysical simulation models, Ecol. Model., 194, 329–343, doi:10.1016/j.ecolmodel.2005.10.032, 2006.

McClain, M. E., Chícharo, L., Fohrer, N., Gaviño Novillo, M., Windhorst, W., and Zalewski, M.: Training hydrologists to be ecohydrologists and play a leading role in environmental problem solving, Hydrol. Earth Syst. Sci., 16, 1685–1696, doi:10.5194/hess-16-1685-2012, 2012.

McDonnell, R. A.: Circulations and transformations of energy and water in Abu Dhabi's hydrosocial cycle, Geoforum, 57, 225–233, doi:10.1016/j.geoforum.2013.11.009, 2013.

Medellín-Azuara, J., Howitt, R. E., and Harou, J. J.: Predicting farmer responses to water pricing, rationing and subsidies assuming profit maximizing investment in irrigation technology, Agricult. Water Manage., 108, 73–82, doi:10.1016/j.agwat.2011.12.017, 2012.

Merz, B., Vorogushyn, S., Lall, U., Viglione, A., and Blöschl, G.: Charting unknown waters – On the role of surprise in flood risk assessment and management, Water Resour. Res., 51, 6399–6416, doi:10.1002/2015WR017464, 2015.

Milly, P. C. D., Betancourt, J., Falkenmark, M., Hirsch, R. M., Zbigniew, W., Lettenmaier, D. P., and Stouffer, R. J.: Stationarity Is

Dead: Whither Water Management ?, Science, 319, 573–574, 2008.

Mirchi, A., Madani, K., Watkins, D., and Ahmad, S.: Synthesis of System Dynamics Tools for Holistic Conceptualization of Water Resources Problems, Water Resour. Manage., 26, 2421–2442, doi:10.1007/s11269-012-0024-2, 2012.

Mirchi, A., Watkins, D. J., Huckins, C., Madani, K., and Hjorth, P.: Water resources management in a homgenizing world: averting the growth and underinvestment trajectory, Water Resour. Res., 50, 7515–7526, doi:10.1002/2013WR015128.Received, 2014.

Molle, F.: Scales and power in river basin management: The Chao Phraya River in Thailand, Geograph. J., 173, 358–373, doi:10.1111/j.1475-4959.2007.00255.x, 2007.

Mollinga, P. P.: Canal irrigation and the hydrosocial cycle, Geoforum, 57, 192–204, doi:10.1016/j.geoforum.2013.05.011, 2014.

Montanari, A., Young, G., Savenije, H. H. G., Hughes, D., Wagener, T., Ren, L. L., Koutsoyiannis, D., Cudennec, C., Toth, E., Grimaldi, S., Blöschl, G., Sivapalan, M., Beven, K., Gupta, H., Hipsey, M., Schaefli, B., Arheimer, B., Boegh, E., Schymanski, S. J., Di Baldassarre, G., Yu, B., Hubert, P., Huang, Y., Schumann, A., Post, D. A., Srinivasan, V., Harman, C., Thompson, S., Rogger, M., Viglione, A., McMillan, H., Characklis, G., Pang, Z., and Belyaev, V.: "Panta Rhei–Everything Flows: Change in hydrology and society–The IAHS Scientific Decade 2013–2022, Hydrol. Sci. J., 58, 1256–1275, doi:10.1080/02626667.2013.809088, 2013.

Monticino, M., Acevedo, M., Callicott, B., Cogdill, T., and Lindquist, C.: Coupled human and natural systems: A multi-agent-based approach, Environ. Model. Softw., 22, 656–663, doi:10.1016/j.envsoft.2005.12.017, 2007.

Murray-Rust, D., Rieser, V., Robinson, D. T., Miličič, V., and Rounsevell, M.: Agent-based modelling of land use dynamics and residential quality of life for future scenarios, Environ. Model. Softw., 46, 75–89, doi:10.1016/j.envsoft.2013.02.011, 2013.

Nautiyal, S. and Kaechele, H.: Natural resource management in a protected area of the Indian Himalayas: a modeling approach for anthropogenic interactions on ecosystem., Environ. Monitor. Assess., 153, 253–71, doi:10.1007/s10661-008-0353-z, 2009.

Nicholson, A. E. and Flores, M. J.: Combining state and transition models with dynamic Bayesian networks, Ecol. Model., 222, 555–566, doi:10.1016/j.ecolmodel.2010.10.010, 2011.

Norgaard, R. B.: Sociosystem and ecosystem coevolution in the Amazon, J. Environ. Econ. Manage., 254, 238–254, 1981.

Norgaard, R. B.: Coevolutionary development potential, Land Econ., 60, 160–173, 1984.

Norgaard, R. B.: Beyond Materialism: A Coevolutionary Reinterpretation of the Environmental Crisis, Rev. Social Econ., 53, 475–492, doi:10.1080/00346769500000014, 1995.

O'Connell, P. E. and O'Donnell, G.: Towards modelling flood protection investment as a coupled human and natural system, Hydrol. Earth Syst. Sci., 18, 155–171, doi:10.5194/hess-18-155-2014, 2014.

Odongo, V. O., Mulatu, D. W., Muthoni, F. K., van Oel, P. R., Meins, F. M., van der Tol, C., Skidmore, A. K., Groen, T. A., Becht, R., Onyando, J. O., and van der Veen, A.: Coupling socio-economic factors and eco-hydrological processes using a cascade-modeling approach, J. Hydrol., 518, 49–59, doi:10.1016/j.jhydrol.2014.01.012, 2014.

Orth, R., Staudinger, M., Seneviratne, S. I., Seibert, J., and Zappa, M.: Does model performance improve with complexity? A case study with three hydrological models, J. Hydrol., 523, 147–159, doi:10.1016/j.jhydrol.2015.01.044, 2015.

Ostrom, E.: A diagnostic approach for going beyond panaceas., P. Natl. Acad. Sci. USA, 104, 15 181–7, doi:10.1073/pnas.0702288104, 2007.

Ostrom, E.: A general framework for analyzing sustainability of social-ecological systems., Science, 325, 419–422, doi:10.1126/science.1172133, 2009.

Öztürk, M., Copty, N. K., and Saysel, A. K.: Modeling the impact of land use change on the hydrology of a rural watershed, J. Hydrol., 497, 97–109, doi:10.1016/j.jhydrol.2013.05.022, 2013.

Paalvast, P. and van der Velde, G.: Long term anthropogenic changes and ecosystem service consequences in the northern part of the complex Rhine-Meuse estuarine system, Ocean Coast. Manage., 92, 50–64, doi:10.1016/j.ocecoaman.2014.02.005, 2014.

Pandey, V. P., Babel, M. S., Shrestha, S., and Kazama, F.: A framework to assess adaptive capacity of the water resources system in Nepalese river basins, Ecol. Ind., 11, 480–488, doi:10.1016/j.ecolind.2010.07.003, 2011.

Parker, D. C., Maguire, D., Goodchild, M., and Batty, M.: Integrating of Geographic Information Systems and Use: Prospects and Challenges, in: GIS, Spatial Analysis and Modeling, chap. 19, 403–422, ESRI Press, Redlands, CA, 2005.

Parveen, S., Winiger, M., Schmidt, S., and Nüsser, M.: Irrigation in Upper Hunza: evolution of socio-hydrological interactions in the Karakoram, northern Pakistan, Erdkunde, 69, 69–85, doi:10.3112/erdkunde.2015.01.05, 2015.

Pataki, D. E., Boone, C. G., Hogue, T. S., Jenerette, G. D., McFadden, J. P., and Pincetl, S.: Socio-ecohydrology and the urban water challenge, Ecohydrology, 4, 341–347, doi:10.1002/eco.209, 2011.

Pechlivanidis, I. G. and Jackson, B. M.: Catchment Scale hydrological modelling: a review of model types, calibration approaches and uncertainty analysis methods in the context of recent developments in technology, Global NEST J., 13, 193–214, 2011.

Peel, M. C. and Blöschl, G.: Hydrological modelling in a changing world, Prog. Phys. Geogr., 35, 249–261, doi:10.1177/0309133311402550, 2011.

Perdigão, R. A. P. and Blöschl, G.: Spatiotemporal flood sensitivity to annual precipitation: Evidence for landscape-climate coevolution, Water Resour. Res., 50, 5492–5509, doi:10.1002/2014WR015365.Received, 2014.

Postel, S. L.: Foreword–Sharing the benefits of water, Hydrol. Sci. J., 56, 529–530, doi:10.1080/02626667.2011.578380, 2011.

Purnomo, H., Mendoza, G. A., Prabhu, R., and Yasmi, Y.: Developing multi-stakeholder forest management scenarios: a multi-agent system simulation approach applied in Indonesia, Forest Pol. Econ., 7, 475–491, doi:10.1016/j.forpol.2003.08.004, 2005.

Railsback, S.: Getting Results: The Pattern-oriented Approach to Analyzing Natural Systems With Individual-Based Models, Nat. Resour. Model., 14, 465–475, 2001.

Rammel, C. and van den Bergh, J. C.: Evolutionary policies for sustainable development: adaptive flexibility and risk minimising, Ecol. Econ., 47, 121–133, doi:10.1016/S0921-8009(03)00193-9, 2003.

Rammel, C., Stagl, S., and Wilfing, H.: Managing complex adaptive systems – A co-evolutionary perspective on natural resource management, Ecol. Econ., 63, 9–21, doi:10.1016/j.ecolecon.2006.12.014, 2007.

Ratna Reddy, V. and Syme, G. J.: Social sciences and hydrology: An introduction, J. Hydrol., 518, 1–4, doi:10.1016/j.jhydrol.2014.06.022, 2014.

Reed, P. and Kasprzyk, J.: Water Resources Management: The Myth, the Wicked, and the Future, J. Water Resour. Plan. Manage., 135, 411–413, 2009.

Ren, L., Wang, M., Li, C., and Zhang, W.: Impacts of human activity on river runoff in the northern area of China, J. Hydrol., 261, 204–217, doi:10.1016/S0022-1694(02)00008-2, 2002.

Reyer, C. P. O., Brouwers, N., Rammig, A., Brook, B. W., Epila, J., Grant, R. F., Holmgren, M., Langerwisch, F., Leuzinger, S., Medlyn, B., Pfeifer, M., Verbeeck, H., and Villela, D. M.: Forest Resilience and tipping points at different spatio-temporal scales: approaches and challenges, J. Ecol., 103, 5–15, doi:10.1111/1365-2745.12337, 2015.

Rittel, H. and Webber, M.: Dilemmas in a general theory of planning, Policy Sci., 4, 155–169, 1973.

Roberts, C., Stallman, D., and Bieri, J.: Modeling complex human-environment interactions: the Grand Canyon river trip simulator, Ecol. Model., 153, 181–196, doi:10.1016/S0304-3800(01)00509-9, 2002.

Rodriguez-Iturbe, I.: Ecohydrology : A hydrologic perspective of climate-soil-vegetation dynamics, Water Resour. Res., 36, 3–9, 2000.

Rosenberg, D. E. and Madani, K.: Water Resources Systems Analysis: A Bright Past and a Challenging but Promising Future, J. Water Resour. Plan. Manage., 140, 407–409, doi:10.1061/(ASCE)WR.1943-5452.0000414, 2014.

Runyan, C. W., D'Odorico, P., and Lawrence, D.: Physical and biological feedbacks of deforestation, Rev. Geophys., 50, 1–32, doi:10.1029/2012RG000394.1.INTRODUCTION, 2012.

Sandker, M., Campbell, B. M., Ruiz-Pérez, M., Sayer, J. A., Cowling, R., Kassa, H., and Knight, A. T.: The role of participatory modeling in landscape approaches to reconcile conservation and development, Ecol. Soc., 15, 2010.

Savenije, H. H. G., Hoekstra, A. Y., and van der Zaag, P.: Evolving water science in the Anthropocene, Hydrol. Earth Syst. Sci., 18, 319–332, doi:10.5194/hess-18-319-2014, 2014.

Scheffer, M., Carpenter, S. R., Lenton, T. M., Bascompte, J., Brock, W., Dakos, V., van de Koppel, J., van de Leemput, I. A., Levin, S. A., van Nes, E. H., Pascual, M., and Vandermeer, J.: Anticipating critical transitions, Science, 338, 344–348, doi:10.1126/science.1225244, 2012.

Schlüter, M.: New Horizons for Managing the Environment: A Review of Coupled Social-Ecological Systems Modeling, Nat. Resour. Model., 25, 219–272, 2012.

Schlüter, M. and Pahl-Wostl, C.: Mechanisms of resilience in common-pool resource management systems: an agent-based model of water use in a river basin, Ecol. Soc., 12, 2007.

Showqi, I., Rashid, I., and Romshoo, S. A.: Land use land cover dynamics as a function of changing demography and hydrology, GeoJournal, 79, 297–307, doi:10.1007/s10708-013-9494-x, 2013.

Simelton, E., Fraser, E. D., Termansen, M., Forster, P. M., and Dougill, A. J.: Typologies of crop-drought vulnerability: an empirical analysis of the socio-economic factors that influence the sensitivity and resilience to drought of three major food crops in China (1961-2001), Environ. Sci. Pol., 12, 438–452, doi:10.1016/j.envsci.2008.11.005, 2009.

Sivakumar, B.: Socio-hydrology: not a new science, but a recycled and re-worded hydrosociology, Hydrol. Process., 26, 3788–3790, doi:10.1002/hyp.9511, 2012.

Sivapalan, M.: Debates-Perspectives on socio-hydrology: Changing water systems and the tyranny of small problems – Socio-hydrology, Water Resour. Res., 51, 4795–4805, doi:10.1002/2015WR017080, 2015.

Sivapalan, M. and Blöschl, G.: Time scale interactions and the co-evolution of humans and water, Water Resour. Res., 51, 6988–7022, doi:10.1002/2015WR017896, 2015.

Sivapalan, M., Blöschl, G., Zhang, L., and Vertessy, R.: Downward approach to hydrological prediction, Hydrol. Process., 17, 2101–2111, doi:10.1002/hyp.1425, 2003.

Sivapalan, M., Savenije, H. H. G., and Blöschl, G.: Socio-hydrology: A new science of people and water, Hydrol. Process., 26, 1270–1276, doi:10.1002/hyp.8426, 2012.

Sivapalan, M., Konar, M., and Srinivasan, V.: Socio-hydrology: Use-inspired water sustainability science for the Anthropocene Earth's Future, Earth's Future, 2, 225–230, doi:10.1002/2013EF000164.Received, 2014.

Srinivasan, V.: Reimagining the past – use of counterfactual trajectories in socio-hydrological modelling: the case of Chennai, India, Hydrol. Earth Syst. Sci., 19, 785–801, doi:10.5194/hess-19-785-2015, 2015.

Srinivasan, V.: Reimagining the past - use of counterfactual trajectories in socio-hydrological modelling: the case of Chennai, India, Hydrol. Earth Syst. Sci., 19, 785–801, doi:10.5194/hess-19-785-2015, 2015.

Srinivasan, V., Lambin, E. F., Gorelick, S. M., Thompson, B. H., and Rozelle, S.: The nature and causes of the global water crisis: Syndromes from a meta-analysis of coupled human-water studies, Water Resour. Res., 48, W10516, doi:10.1029/2011WR011087, 2012.

Srinivasan, V., Seto, K. C., Emerson, R., and Gorelick, S. M.: The impact of urbanization on water vulnerability: A coupled human-environment system approach for Chennai, India, Global Environ. Change, 23, 229–239, doi:10.1016/j.gloenvcha.2012.10.002, 2013.

Srinivasan, V., Thompson, S., Madhyastha, K., Penny, G., Jeremiah, K., and Lele, S.: Why is the Arkavathy River drying? A multiple-hypothesis approach in a data-scarce region, Hydrol. Earth Syst. Sci., 19, 1905–1917, doi:10.5194/hess-19-1905-2015, 2015.

Steffen, W., Crutzen, P. J., and McNeill, J. R.: The Anthropocene: Are Humans Now Overwhelming the Great Forces of Nature, AMBIO, 36, 614–621, doi:10.1579/0044-7447(2007)36[614:TAAHNO]2.0.CO;2, 2007.

Steffen, W., Grinevald, J., Crutzen, P., and McNeill, J.: The Anthropocene: conceptual and historical perspectives, Phil. Trans A, 369, 842–867, doi:10.1098/rsta.2010.0327, 2011.

Swyngedouw, E.: The Political Economy and Political Ecology of the Hydro-Social Cycle, J. Contemp. Water Res. Edu., 142, 56–60, doi:10.1111/j.1936-704X.2009.00054.x, 2009.

Thompson, S. E., Sivapalan, M., Harman, C. J., Srinivasan, V., Hipsey, M. R., Reed, P., Montanari, A., and Blöschl, G.: Developing predictive insight into changing water systems: use-inspired hydrologic science for the Anthropocene, Hydrol. Earth Syst. Sci., 17, 5013–5039, doi:10.5194/hess-17-5013-2013, 2013.

Troy, T. J., Konar, M., Srinivasan, V., and Thompson, S.: Moving sociohydrology forward: a synthesis across studies, Hydrol. Earth Syst. Sci., 19, 3667–3679, doi:10.5194/hess-19-3667-2015, 2015a.

Troy, T. J., Pavao-Zuckerman, M., and Evans, T. P.: Debates-Perspectives on socio-hydrology: Socio-hydrologic modeling: Tradeoffs, hypothesis testing, and validation, Water Resour. Res., 51, 4806–4814, doi:10.1002/2015WR017046, 2015b.

Underdal, A.: Complexity and challenges of long-term environmental governance, Global Environ. Change, 20, 386–393, doi:10.1016/j.gloenvcha.2010.02.005, 2010.

Valbuena, D., Verburg, P. H., Bregt, A. K., and Ligtenberg, A.: An agent-based approach to model land-use change at a regional scale, Landsc. Ecol., 25, 185–199, doi:10.1007/s10980-009-9380-6, 2009.

Valbuena, D., Bregt, A. K., McAlpine, C., Verburg, P. H., and Seabrook, L.: An agent-based approach to explore the effect of voluntary mechanisms on land use change: a case in rural Queensland, Australia, J. Environ. Manage., 91, 2615–2625, doi:10.1016/j.jenvman.2010.07.041, 2010.

van Dam, A. A., Kipkemboi, J., Rahman, M. M., and Gettel, G. M.: Linking Hydrology, Ecosystem Function, and Livelihood Outcomes in African Papyrus Wetlands Using a Bayesian Network Model, Wetlands, 33, 381–397, doi:10.1007/s13157-013-0395-z, 2013.

Van den Bergh, J. C. J. M. and Gowdy, J. M.: Evolutionary theories in environmental and resource economics: approaches and applications, Environ. Resour., 17, 37–57, 2000.

van Emmerik, T. H. M., Li, Z., Sivapalan, M., Pande, S., Kandasamy, J., Savenije, H. H. G., Chanan, A., and Vigneswaran, S.: Socio-hydrologic modeling to understand and mediate the competition for water between agriculture development and environmental health: Murrumbidgee River Basin, Australia, Hydrol. Earth Syst. Sci., 18, 4239–4259, doi:10.5194/hess-18-4239-2014, 2014.

Veldkamp, A. and Verburg, P. H.: Modelling land use change and environmental impact, J. Environ. Manage., 72, 1–3, doi:10.1016/j.jenvman.2004.04.004, 2004.

Viglione, A., Di Baldassarre, G., Brandimarte, L., Kuil, L., Carr, G., Salinas, J. L., Scolobig, A., and Blöschl, G.: Insights from socio-hydrology modelling on dealing with flood risk – Roles of collective memory, risk-taking attitude and trust, J. Hydrol., 518, 71–82, doi:10.1016/j.jhydrol.2014.01.018, 2014.

Wada, Y., van Beek, L. P. H., Wanders, N., and Bierkens, M. F. P.: Human water consumption intensifies hydrological drought worldwide, Environ. Res. Lett., 8, 034036, doi:10.1088/1748-9326/8/3/034036, 2013.

Wagener, T. and Montanari, A.: Convergence of approaches toward reducing uncertainty in predictions in ungauged basins, Water Resour. Res., 47, 1–8, doi:10.1029/2010WR009469, 2011.

Wagener, T., Sivapalan, M., Troch, P. A., McGlynn, B. L., Harman, C. J., Gupta, H. V., Kumar, P., Rao, P. S. C., Basu, N. B., and Wilson, J. S.: The future of hydrology: An evolving science for a changing world, Water Resour. Res., 46, W05 301, doi:10.1029/2009WR008906, 2010.

Wanders, N. and Wada, Y.: Human and climate impacts on the 21st century hydrological drought, J. Hydrol., 526, 208–220, doi:10.1016/j.jhydrol.2014.10.047, 2015.

Wang, S. and Huang, G.: An integrated approach for water resources decision making under interactive and compound uncertainties, Omega, 44, 32–40, doi:10.1016/j.omega.2013.10.003, 2014.

Welsh, W. D., Vaze, J., Dutta, D., Rassam, D., Rahman, J. M., Jolly, I. D., Wallbrink, P., Podger, G. M., Bethune, M., Hardy, M. J., Teng, J., and Lerat, J.: An integrated modelling framework for regulated river systems, Environ. Model. Softw., 39, 81–102, doi:10.1016/j.envsoft.2012.02.022, 2013.

Wheater, H. S.: Progress in and prospects for fluvial flood modelling., Philos. Transactions A, 360, 1409–1431, doi:10.1098/rsta.2002.1007, 2002.

White, G. F.: Human adjustment to floods, Doctoral thesis, The University of Chicago, http://agris.fao.org/agris-search/search.do?recordID=US201300257437 (last access: 14 October 2014), 1945.

Wiegand, T., Jeltsch, F., Hanski, I., and Grimm, V.: Using pattern-oriented modeling for revealing hidden information: a key for reconciling ecological theory and application, Oikos, 65, 209–222, 2003.

Wilson, N. J.: Indigenous water governance: Insights from the hydrosocial relations of the Koyukon Athabascan village of Ruby, Alaska, Geoforum, 57, 1–11, doi:10.1016/j.geoforum.2014.08.005, 2014.

Winder, N., McIntosh, B. S., and Jeffrey, P.: The origin, diagnostic attributes and practical application of co-evolutionary theory, Ecol. Econ., 54, 347–361, doi:10.1016/j.ecolecon.2005.03.017, 2005.

Zeitoun, M.: Global environmental justice and international transboundary waters: An initial exploration, Geograph. J., 179, 141–149, doi:10.1111/j.1475-4959.2012.00487.x, 2013.

Zeitoun, M. and Allan, J. A.: Applying hegemony and power theory to transboundary water analysis, Water Pol., 10, 3–12, doi:10.2166/wp.2008.203, 2008.

Zeitoun, M. and Mirumachi, N.: Transboundary water interaction I: Reconsidering conflict and cooperation, Int. Environ. Agreem.-P., 8, 297–316, doi:10.1007/s10784-008-9083-5, 2008.

Zeitoun, M. and Warner, J.: Hydro-hegemony – A framework for analysis of trans-boundary water conflicts, Water Pol., 8, 435–460, doi:10.2166/wp.2006.054, 2006.

Zlinszky, A. and Timár, G.: Historic maps as a data source for socio-hydrology: A case study of the Lake Balaton wetland system, Hungary, Hydrol. Earth Syst. Sci., 17, 4589–4606, doi:10.5194/hess-17-4589-2013, 2013.

9

Spatio-temporal variability of snow water equivalent in the extra-tropical Andes Cordillera from distributed energy balance modeling and remotely sensed snow cover

E. Cornwell[1], N. P. Molotch[2,3], and J. McPhee[1,4]

[1]Advanced Mining Technology Center, Facultad de Ciencias Físicas y Matemáticas, Universidad de Chile, Santiago, Chile
[2]Department of Geography and Institute of Arctic and Alpine Research, University of Colorado, Boulder, USA
[3]Jet Propulsion Laboratory, California Institute of Technology, Pasadena, California, USA
[4]Departamento de Ingeniería Civil, Facultad de Ciencias Físicas y Matemáticas, Universidad de Chile, Santiago, Chile

Correspondence to: J. McPhee (jmcphee@u.uchile.cl)

Abstract. Seasonal snow cover is the primary water source for human use and ecosystems along the extratropical Andes Cordillera. Despite its importance, relatively little research has been devoted to understanding the properties, distribution and variability of this natural resource. This research provides high-resolution (500 m), daily distributed estimates of end-of-winter and spring snow water equivalent over a 152 000 km^2 domain that includes the mountainous reaches of central Chile and Argentina. Remotely sensed fractional snow-covered area and other relevant forcings are combined with extrapolated data from meteorological stations and a simplified physically based energy balance model in order to obtain melt-season melt fluxes that are then aggregated to estimate the end-of-winter (or peak) snow water equivalent (SWE). Peak SWE estimates show an overall coefficient of determination R^2 of 0.68 and RMSE of 274 mm compared to observations at 12 automatic snow water equivalent sensors distributed across the model domain, with R^2 values between 0.32 and 0.88. Regional estimates of peak SWE accumulation show differential patterns strongly modulated by elevation, latitude and position relative to the continental divide. The spatial distribution of peak SWE shows that the 4000–5000 m a.s.l. elevation band is significant for snow accumulation, despite having a smaller surface area than the 3000–4000 m a.s.l. band. On average, maximum snow accumulation is observed in early September in the western Andes, and in early October on the eastern side of the continental divide. The results presented here have the potential of inform-
ing applications such as seasonal forecast model assessment and improvement, regional climate model validation, as well as evaluation of observational networks and water resource infrastructure development.

1 Introduction

Accurately predicting the spatial and temporal distribution of snow water equivalent (SWE) in mountain environments remains a significant challenge for the scientific community and water resource practitioners around the world. The Andes Cordillera, a formidable mountain range that constitutes the backbone of the South American continent, remains one of the relatively least studied mountain environments due to its generally low accessibility and complex topography. The extratropical stretch of the Andes, extending south from approximately latitude 27° S, is a snow-dominated hydrological environment that provides key water resources for a majority of the population in Chile and Argentina. Until now, a very sparse network of snow courses and automated snow measuring stations (snow pillows) has been the only source of information about this key resource. In a context of sustained climate change characterized by warming trends and likely future precipitation reductions (Vera et al., 2006; Vicuña et al., 2011), it becomes ever more relevant to understand the past dynamics of the seasonal snowpack in order to validate predictive models of future snow water resources. This research presents the first spatially and temporally explicit high-resolution SWE reconstruction over the snow-dominated extratropical Andes of central Chile and Argentina based on a physical representation of the snowpack energy balance (Kustas et al., 1994) and remotely sensed snow extent (Dietz et al., 2012) between years 2001 and 2014. A key advantage of the presented product is its independence from notoriously scarce and unreliable precipitation measurements at high elevations. Estimates of maximum SWE accumulation and depletion curves are obtained at 500 m resolution, coincident with the MODIS MOD10A1 Fractional Snow Cover product (Hall et al., 2002).

Patterns of hydroclimatic spatio-temporal variability in the extratropical Andes have been studied with increased intensity over the last couple of decades, as pressure for water resources has mounted while at the same time rapid changes in land use and climate have highlighted the societal need for increased understanding of water resource variability and trends under present and future climates. The vast majority of studies have relied on statistical analyses of instrumental records and regional climate models to present synoptic-scale summaries of precipitation (e.g., Aravena and Luckman, 2009; Falvey and Garreaud, 2007; Garreaud, 2009), temperature (Falvey and Garreaud, 2009), snow accumulation (Masiokas et al., 2006) and streamflow variability (Cortés et al., 2011; Núñez et al., 2013). Currently, no high-resolution, large-scale distributed assessments of snow water equivalent are available for the Andes region.

The SWE reconstruction method seeks to estimate end-of-winter accumulation by back accumulating melt energy fluxes during the depletion season. The methods and assumptions required for SWE reconstruction have been tested and refined since initial development (Cline et al., 1998). Applications across a variety of scales have been presented in recent years. In the Sierra Nevada, Jepsen et al. (2012) compared SWE reconstructions to distributed snow surveys in a 19.1 km^2 basin ($R^2 = 0.79$), while Guan et al. (2013) obtained good correlation with SWE observations from an operational snow sensor network across the entire Sierra Nevada ($R^2 = 0.74$). In the Rocky Mountains, Jepsen et al. (2012) obtained an R^2 value of 0.61 when comparing reconstructed SWE to spatial regression from snow surveys, and Molotch (2009) estimated SWE with a mean absolute error (MAE) of 23 % compared to intensive study areas. A useful discussion on the uncertainties of the SWE reconstruction method – albeit one based on temperature-index melt equations – was presented by Slater et al. (2013), who demonstrated that errors in forcing data are at least, if not more, important than snow-covered area data availability. The vast majority, if not all, of SWE reconstruction exercises have been developed in the northern hemisphere, under environmental conditions quite different from those predominant in the extratropical Andes Cordillera. Here, snow distribution and properties have been analyzed in a few local studies (e.g., Ayala et al., 2014; Cortés et al., 2014b; Gas-

coin et al., 2013), but no large-scale estimations at a relevant temporal and spatial resolution for hydrologic applications have been presented. In fact, the Andes of Chile and Argentina display near-ideal conditions for the SWE reconstruction approach due to (1) the near absence of forest cover over a large fraction of the domain where snow accumulation is hydrologically significant; (2) the sharp climatological distinction between wet (winter: June through August) and dry (spring/summer: September through March) seasons, with most of annual precipitation falling during the former; and (3) the low prevalence of cloudy conditions during the spring and summer months over the mountains, which afford a high availability of remotely sensed snow cover information. Conversely, the SWE reconstruction presented here is certainly subject to a series of uncertainty sources, such as the sparseness of the hydrometeorological observational network, which limits both the availability of forcing and validation data.

However, this is the first estimation of peak SWE and snow depletion distribution at this scale and spatial resolution for the extratropical Andes, and the information shown here can be useful for several applications such as understanding year-to-year differential accumulation patterns that may impact the performance of seasonal streamflow forecast models that rely on point-scale data only. Also, the SWE reconstruction can be used to validate output from global or regional climate models and reanalysis, which are being increasingly employed to estimate hydrological states and fluxes in ungauged regions. By analyzing the spatial correlation of snow accumulation and hydrometeorological variables, distributed SWE estimates can inform the design of improved climate observation networks. Likewise, from analyzing the obtained SWE estimates in light of the necessary modeling assumptions and data availability we are able to highlight future research directions aimed at quantifying and reducing these uncertainties.

The objectives of this research include the following: (1) to assess the dominant patterns of spatio-temporal variability in snow water equivalent of the snow-dominated extratropical Andes Cordillera; and (2) to explicitly evaluate the strengths and weaknesses of the SWE reconstruction approach in different sub-regions of the extratropical Andes using snow sensors and distributed snow surveys.

2 Study area

Figure 1 shows the study area, which includes headwater basins in the Andes Mountains of central Chile and Argentina, between 27 and 38° S. These basins supply freshwater to low valleys located on both sides of the cordillera, a topographic barrier more than 5 km high that strongly controls the spatial variability in atmospheric processes (Garreaud, 2009; Montgomery et al., 2001). In Chile, runoff from the Andes Mountains benefits 75 % of the popula-

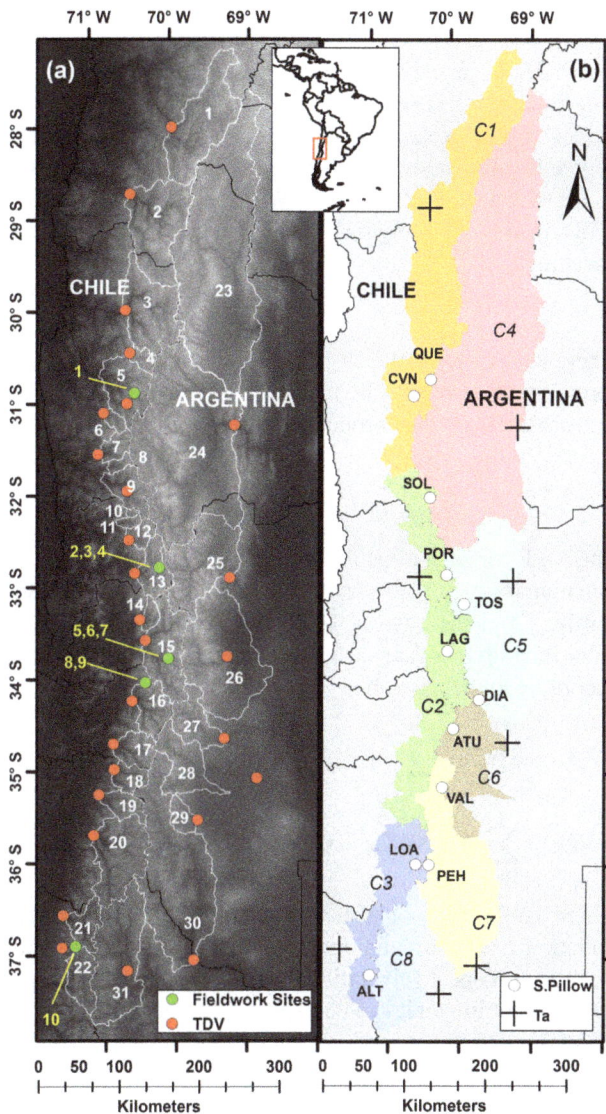

Figure 1. Study area and model domain: **(a)** river basins, stream gages (red circles) and sites where snow survey data are available (green circles); **(b)** hydrologic units (C1 to C8) and snow-pillow stations (white circles).

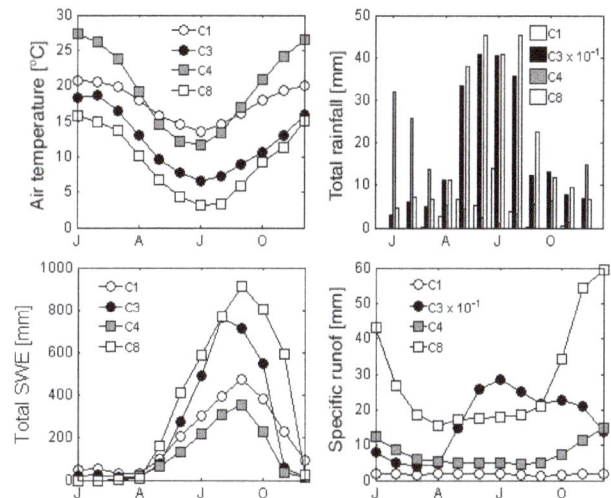

Figure 2. Summarized hydro-climatology of the model domain. Data from meteorological stations located within zones C1, C4, C3 and C8 summarized the hydro-climatological regime of the northwestern, northeastern, southwestern and southeastern zones, respectively. Total SWE is SWE measured at selected snow-pillow stations.

tion (http://www.ine.cl) as well as most of the country's agricultural output, hydropower and industrial activities. In the case of Argentina, 7 % of the population is located in the provinces of La Rioja, San Juan, Mendoza and northern Neuquén (http://www.indec.gov.ar/), with primary water uses in agriculture and hydropower. The selected watersheds have unimpeded streamflow observations and a snow-dominated hydrologic regime (Fig. 2). River basins included in this study have been grouped in eight clusters, or hydrologic response units, based on the seasonality of river flow, numbered C1 to C8 in Fig. 1b. Due to differences in topography and locations of stream gages, the number of headwater basins contained within clusters differs markedly on both

sides of the cordillera, with larger watersheds on the Argentinean side.

The hydro-climate is mostly controlled by orographic effects on precipitation (Falvey and Garreaud, 2007) and interannual variability associated with the Pacific Ocean through the El Niño–Southern Oscillation and Pacific Decadal Oscillation (Masiokas et al., 2006; Newman et al., 2003; Rubio-Álvarez and McPhee, 2010). Precipitation is concentrated in winter months on the western slope (Aceituno, 1988) and sporadic spring and summer storms occur on the mountain front plains of the eastern slope. The vegetation cover presents a steppe-type condition on the western slope up to 33° S, transitioning to the south into tall bushes and sparse mountain forest. On the eastern slope the steppe vegetation prevails until 37° S, with an intermittent presence of mountain forests in the Patagonian plains (Eva et al., 2004).

Figure 2 summarizes the dominant climatology and associated hydrological regime of rivers in the study region. The temperature seasonality (upper left panel) is typical of a temperate, Mediterranean climate, and precipitation is strongly concentrated in the fall–winter months of May through August (upper right panel). The hydrological regime is markedly snow-dominated in the northern part of the domain, which can be seen from the sharp increase in river flow from October and into the summer months of Dec, Jan and Feb (lower right panel) that follows the seasonal melt of snow (lower left panel). Only rivers in the southern subregion display a significant rainfall-dominated seasonal hydrograph. The importance of SWE for the region is demonstrated by the fact that for the studied basins, ablation-season (September–March) river flow accounts for two-thirds of average annual streamflow. Maximum SWE accumulation is reached between the months of August and September on the western side and between late September and early October on the eastern side (Fig. S4 in the Supplement). Scattered snow showers in mid-spring (September through November) affect the study area, but they do not affect significantly the decreasing trend of snow-covered area during the melt season (see timing of peak SWE and fractional snow-covered area (fSCA) analysis in online supplementary material). This feature is essential for choosing the SWE reconstruction methodology used in this work, which is most applicable to snow regimes with distinct snow accumulation and snow ablation seasons.

By and large, the existing network of high-elevation meteorological stations does not include appropriately shielded solid precipitation sensors. Some climate reanalysis products exist, but their representation of Andean topography is crude, and their spatial resolution is not readily amenable to hydrological applications without significant bias correction (Krogh et al., 2015; Scheel et al., 2011). Previous attempts at estimating precipitation amounts at high-elevation reaches in the Andes suggest uncertainties on the order of 50 % (Castro et al., 2014; Falvey and Garreaud, 2007; Favier et al., 2009). In some basins, runoff is partially dictated by glacier contributions, which occur in summer. According to the Randolph Glacier Inventory (http://www.glims.org/RGI/), the central Andes Cordillera has a glacier area of 2245 km^2 between 27 and 38° S, which is equivalent to 1.5 % of the modeling domain surface area ($\sim 152\,000$ km^2).

3 Methods

3.1 SWE reconstruction model

A retrospective SWE reconstruction model based on the convolution of the fSCA depletion curve and time-variant energy inputs for each domain pixel is implemented. For each year, the model is run at a daily time step between 15 August (end of winter) and 15 January (mid-summer). This time window ensures that the most likely time at which peak SWE occurs is captured – which itself is variable from year to year – and the almost complete depletion of the seasonal snowpack. Isolated pixels with non-negative fSCA values may remain after 15 January at glacier and perennial snowpack sites. However, the relative area that these pixels represent with respect to the entire model domain is very low (< 1.5 %), and can be neglected in the context of this work.

The energy balance model adopted here derives from the formulation proposed by Brubaker et al. (1996), which considers explicit net shortwave and longwave radiation terms and a conceptual, pseudo-physically based formulation for turbulent fluxes that depends only on the degree-day air temperature:

$$M_{\mathrm{p}} = \max\left\{\left(Q_{\mathrm{nsw}} + Q_{\mathrm{nlw}}\right) f_B + T_{\mathrm{d}} a_{\mathrm{r}},\, 0\right\}, \tag{1}$$

where M_{p} is potential melt; Q_{nsw} is the net shortwave energy flux; Q_{nlw} is the net longwave energy flux; T_{d} is the degree-day temperature, a_{r} (mm °C^{-1} day^{-1}) is the restricted degree-day factor, and f_B is the energy-to-mass conversion factor with a value of 0.26 (mm W^{-1} m^2 day^{-1}). Actual melt is obtained by multiplying potential melt by fractional snow cover area:

$$M = M_{\mathrm{p}} \mathrm{fSCA}^{\mathrm{fc}}, \tag{2}$$

where fSCA^{fc} is the fSCA MOD10A1 estimate adjusted to forest cover correction by a vegetation fractional f_{veg} (0 to 1) from the MOD44B product (Hansen et al., 2003):

$$\mathrm{fSCA}^{\mathrm{fc}} = \frac{\mathrm{fSCA}^{\mathrm{obs}}}{\left(1 - f_{\mathrm{veg}}\right)}. \tag{3}$$

The SWE for each pixel is computed for each year by accumulating the melt fluxes back in time during the melt season, starting from the day on which fSCA reaches a minimum value, and up to a date such that winter fSCA has plateaued, according to the relations

$$\mathrm{SWE}_t = \mathrm{SWE}_0 - \sum_1^t M = M_{t+1} + \mathrm{SWE}_{t+1}, \tag{4}$$

$$\mathrm{SWE}_0 = \sum_{t=1}^{n} M_t;\ \mathrm{SWE}_n = 0, \tag{5}$$

where SWE_0 is end-of-winter or initial maximum SWE accumulation, SWE_n is a minimum or threshold value. The model was run retrospectively until 15 August, an adequate date before which little melt can be expected for most of the winter seasons within the modeling period in this region (please see Fig. S5).

3.2 Fractional snow-covered area and land use data

Spatio-temporal evolution of snow-covered area was estimated using the fSCA product from the Moderate Resolution Imaging Spectroradiometer (MODIS) on-board the Terra satellite (MOD10A1 C5 Level 3). The MOD10A1 product provides daily fSCA estimates at 500 m resolution. Percentages of snow extent (i.e., 0 to 100 %) are derived from an empirical linearization of the Normalized Difference Snow Index (NDSI), considering the total MODIS reflectance in the visible range (0.545–0.565 μm; band 4) and shortwave infrared (1.628–1.652 μm; band 6) (Hall et al., 2002; Hall and Riggs, 2007).

Binary and fractional MODIS fSCA estimates are limited by the use of an empirical NDSI-based method. These errors are notoriously sensitive to surface features such as fractional vegetation and surface temperature (Rittger et al., 2013). Arsenault et al. (2014) reviewed MODIS fSCA accuracy estimates from several studies under different climatic conditions, and report a range between 1.5 and 33 % in terms of absolute error with respect to ground observations and operational snow cover data sets. Errors stem mainly from cloud masking and detection of very thin snow (<10 mm depth), forest cover and terrain complexity. In general, commission and omission errors are greatest in the early and late portions of the snow cover season (Hall and Riggs, 2007) and decrease with increasing elevation (Arsenault et al., 2014). Molotch and Margulis (2008) compared MODIS and Landsat Enhanced Thematic Mapper performance in the context of SWE reconstruction, showing that significant differences in SWE estimates were a result of SCA estimation accuracy and less so of model spatial resolution. The latter conclusion supports the feasibility of using the snow-covered area products at a 500 m spatial resolution for regional-scale studies. In order to minimize the effect of cloud cover on the temporal continuity and extent of the fSCA estimates, the MOD10A1 fSCA product was post-processed by a modified algorithm for non-binary products, based on the algorithm proposed by Gafurov and Bárdossy (2009). Their method is adapted here to the fractional snow cover product, applying a three-step correction consisting of: (1) a pixel-specific linear temporal interpolation over 1, 2 or 3 days prior and posterior to a cloudy pixel; (2) a spatial interpolation over the eight-pixel kernel surrounding the cloudy pixel, retaining information from lower-elevation pixels only; and (3) assigning the 2001–2014 fSCA pixel specific average when steps (1) and (2) where not feasible. This step minimized the effect of cloud cover on data availability over the spatial domain, yielding cloud cover percentages ranging from 21 % in September to 8 % in December.

The Normalized Difference Vegetation Index (NDVI) (Huete et al., 2002) derived from the MOD13Q1 v5 MODIS Level 3 product (16 days – 250 m) is used to classify forest presence for each model pixel. For pixels classified as forested, both fSCA and energy fluxes where corrected: fractional SCA was modified on the basis of percentage forest cover (Molotch, 2009; Rittger et al., 2013), using the average of the forest percentage product from MOD44B V51. Forest attenuation (below canopy) of energy fluxes at the snow surface was estimated from forest cover following the method from Ahl et al. (2006) assuming invariant NDVI over each melt season. The selected NDVI pattern is obtained by averaging the four NDVI scenes available in the December–January time window through 14 study years. This time window displays the average state of evergreen forest with the maximum amount of data.

3.3 Model forcings

Spatially distributed forcings are required at each grid element in order to run the SWE reconstruction model. In order to ensure the tractability of the extrapolation process, we divided the model domain into sub-regions or clusters, composed of one or more river basins. The river basins were grouped using a clusterization algorithm (please see Sect. S2

in the Supplement) based on melt-season river flow volume as described in Rubio-Álvarez and McPhee (2010). Then, spatially distributed variables (surface temperature, fSCA, global irradiance) are combined with homogeneous variables for each cluster (e.g., cloud cover index) and point data from meteorological stations in order to obtain a distributed product as described below. A further benefit of the clustering process is that it allows us to analyze distinct regional features of the SWE reconstruction parameters, input variables and output estimates.

Net shortwave radiation, Q_{nsw} is estimated as a function of incoming solar radiation based on the equation

$$Q_{nsw} = (1 - \alpha_s)(G_\downarrow)\tau_a, \tag{6}$$

where α_s is snow surface broadband albedo; $G_\downarrow$ is incoming solar radiation (global irradiance); and τ_a is the shortwave transmissivity as a function of LAI for mixed forest cover (Pontailler et al., 2003; Sicart et al., 2004), which in turn is estimated as

$$\tau_a = e^{(-\kappa \mathrm{LAI})}; \quad \mathrm{LAI} = -1.323 \ln\left(\frac{0.88 - \mathrm{NDVI}}{0.72}\right), \tag{7}$$

with $\kappa = 0.52$ for mixed forest species (DeWalle and Rango, 2008). Equation (7) is valid for NDVI values between 0.16 and 0.87. Global irradiance under cloudy sky conditions is estimated considering a daily distributed spatial pattern of clear sky irradiance $G_{c\downarrow}$ derived by the r.sun GRASS GIS module (Hofierka and Suri, 2002; Neteler et al., 2012) and the clear sky index K_c derived from the insolation incident on a horizontal surface from the "Climatology Resource for Agroclimatology" project in the NASA Prediction Worldwide Energy Resource "POWER" (http://power.larc.nasa.gov/) $1° \times 1°$ gridded product.

$$G_\downarrow = K_c G_{c\downarrow}; \quad K_c = \overline{(G_{r\downarrow}/G_{c\downarrow})} \tag{8}$$

In Eq. (8), $\overline{G_{r\downarrow}}$ and $\overline{G_{c\downarrow}}$ are spatial averages over each hydrologic response unit (cluster) of the POWER and r.sun-derived products, respectively.

A snow-age decay function based on snowfall detection is implemented to estimate daily snow surface albedo (Molotch and Bales, 2006) constrained between values of 0.85 and 0.40 (Army Corps of Engineers, 1960). Snowfall events were diagnosed using a unique minimum threshold for fSCA increments of 2.5 % for each hydrologic unit area.

Net longwave radiation estimates are derived using

$$Q_{nlw} = L_\downarrow f_{sv}\varepsilon_s + \sigma T_a^4(1 - f_{sv})\varepsilon_{sf} - \sigma T_s^4 \varepsilon_s, \tag{9}$$

$$L_\downarrow = 0.575 e_a^{1/7} \sigma T_a^4 \left(1 + a_c C^2\right), \tag{10}$$

where T_a is air temperature, T_s is the snow surface temperature, ε_s is the snow emissivity (i.e., 0.97), ε_{sf} is the canopy emissivity (i.e., 0.97), f_{sv} is the sky-view factor (i.e., assumed equal to shortwave transmissivity; Pomeroy et al.,

2009; Sicart et al., 2004), σ is the Stefan–Boltzmann constant, and $L_\downarrow$ is the incoming longwave radiation. Air vapor pressure (e_a) required for longwave radiation estimates was derived from air temperature and relative humidity, which in turn was assumed constant throughout the melt period and equal to 40 % based on observations at selected high-elevation meteorological stations. The multiplying factor $(1 + a_c C^2)$ represents an increase in energy input relative to clear sky conditions due to cloud cover, where a_c equals 0.17 and $C = 1 - K_c$ is an estimate of the cloud cover fraction (DeWalle and Rango, 2008).

Spatially distributed air temperature is generated by combining daily air temperature recorded at index meteorological stations and a weekly spatial pattern of skin temperature derived from the MODIS Land Surface Temperature product (MOD11A1.V5) (Wan et al., 2002, 2004). The product MOD11A1 V5 Level 3 estimates surface temperature from thermal infrared brightness temperatures under clear sky conditions using daytime and nighttime scenes and has been shown to adequately represent measurements at meteorological stations ($R^2 \geq 0.7$), displaying moderate overestimation in spring and underestimation in fall (Neteler, 2010). Other studies have reported similar accuracies, with RSME values around 4.5 °K in cold mountain environments (Williamson et al., 2014). Taking into account the high correlation between air temperature and LST (Benali et al., 2012; Colombi et al., 2007; Williamson et al., 2014), we define

$$T_a = T_{a\,base} + \Delta T_a = T_{a\,base} + \mu\,(LST - LST_{base}) + \nu, \quad (11)$$

where $T_{a\,base}$ is daily air temperature at an index station for each cluster and ΔT_a is the difference in air temperature between any pixel and the pixel where the index station is located. To determine ΔT_a we use a linear regression between MODIS LST data and ΔT_a considering pairs of stations located at high-altitude and valley (base) sites, taking into account the melt season average values over the 2001–2014 period. In Eq. (11), $LST - LST_{base}$ denotes the difference between skin temperatures from any pixel and the index station pixel. The linear regression between skin temperature and air temperature differences has a slope μ of 0.65, an intersect ν of -0.5 and R^2 of 0.93 (Fig. S3). Estimation of LST during cloudy conditions is done as follows: (1) a pixel-specific linear temporal interpolation is performed over 1 and 2 days prior and posterior to the cloudy pixel; and (2) estimation of remaining null values by an LST-elevation linear regression (Rhee and Im, 2014).

This spatial extrapolation method was preferred over more traditional methods – for example, based on vertical lapse rates (Minder et al., 2010; Molotch and Margulis, 2008) – after initial tests showed that the combined effect of the relatively low elevation of index stations and the large vertical range of the study domain resulted in unreasonably low air temperatures at pixels with the highest elevations. Likewise, the scarcity of high-elevation meteorological stations and the large spatial extent of the model domain precluded us from adopting more sophisticated temperature estimation methods (e.g., Ragettli et al., 2014).

Snow surface temperature and degree-day temperature are estimated (Brubaker et al., 1996) as

$$T_d = \max\,(T_a, 0)\,;\;\; T_s = \min\,(T_a - \Delta_T, 0), \quad (12)$$

where Δ_T is the difference between air and snow surface temperature. To the best of our knowledge, no direct, systematic values of snow surface temperature exist in this region, so for the purposes of this paper we adopt an average value $\Delta_T = 2.5$ [°C], following the suggestion in Brubaker et al. (1996). Slightly higher values ranging from 3 to 6 °C are shown for continental and alpine snow types (Raleigh et al., 2013), indicating an additional source of uncertainty over net longwave radiation computations. More sophisticated parametrizations for T_s, for example based on heat flow through the snowpack, have been proposed (e.g., Rankinen et al., 2004; Tarboton and Luce, 1996) but those require explicit knowledge about the snowpack temperature profile and/or more complex model formulations to estimate the internal snowpack heat and mass budgets simultaneously.

The a_r coefficient in the restricted degree-day energy balance equation was computed using a combination of station and reanalysis data, and assumed spatially homogeneous within each of the clusters that subdivide the model domain. Brubaker et al. (1996) propose a scheme in which this parameter can be explicitly computed from air and snow surface temperature, air relative humidity, and atmospheric pressure and wind speed. Wind speed was obtained from the NASA POWER reanalysis described previously. A correction for atmospheric stability is applied on the bulk transfer coefficient C_h according to the formulation presented by Kustas et al. (1994), assuming a surface roughness of 0.0005 m:

$$C_h = \left\{ \begin{array}{ll} (1 - 58Ri)^{0.25} & \text{for } Ri < 0 \\ (1 + 7Ri)^{-0.1} & \text{for } Ri > 0 \end{array} \right\};$$
$$Ri = \frac{gz\,(T_a - T_s)}{u^2 T_a} \quad (13)$$

where Ri is the Richardson number, g is the gravity acceleration (9.8 [m s^{-2}]), z is the standard air temperature measurement height (2 m) and u is wind speed. The calculation of Ri and a_r is based on the standard assumptions of T_s at the freezing point and a water vapor saturated snow surface over all high-elevation meteorological stations with available air temperature and relative humidity records (Molotch and Margulis, 2008). Further in the text, we discuss some implications of these assumptions and of the input data used on the ability of the model of simulating relevant components of the snowpack energy exchange.

Table 1 shows the main cluster characteristics and regionalized model parameters. It can be seen that for those clusters located in the southern and middle reaches of the model domain, the a_r parameter values range from 0.10 to 0.23

Table 1. Study area subdivision, relevant characteristics and model parameters.

Cluster	Area $\times 10^3$ (km^2)	Average elevation (m a.s.l.)	Average cluster latitude (°)	Clear sky index (K_c)	Avg. a_r (cm °C^{-1} day^{-1})	T_a (°C)	Forest cover (%)
C1	26.5	3300	−29.4	0.78	0.02	18.3	2.0
C2	17.9	2760	−33.7	0.89	0.11	16.1	5.5
C3	9.20	1890	−36.4	0.83	0.18	12.2	13.8
C4	49.3	3520	−30.1	0.8	0.04	20.4	1.4
C5	18.5	2855	−33.4	0.83	0.15	15.6	3.0
C6	7.60	2807	−34.8	0.83	0.21	13.9	2.3
C7	14.8	2167	−36.1	0.85	0.20	16.7	2.5
C8	8.30	1840	−37.0	0.82	0.23	15.7	4.9
Total/ average	152.1	2320	−	0.83	0.14	−	3.3

Table 2. Snow-pillow measurements available within the study domain.

ID	SWE data	Symbol	Lat. (S)	Long. (W)	Elevation (m a.s.l.)	Reference cluster
			Chile			
1	Quebrada Larga	QUE	30°43′	70°16′	3500	C1
2	Cerro Vega Negra	CVN	30°54′	70°30′	3600	C1
3	El Soldado	SOL	32°00′	70°19′	3290	C2
4	Portillo	POR	32°50′	70°06′	3000	C2
5	Laguna Negra	LAG	33°39′	70°06′	2780	C2
6	Lo Aguirre	LOA	35°58′	70°34′	2000	C3
7	Alto Mallines	ALT	37°09′	70°14′	1770	C3
			Argentina			
8	Toscas	TOS	33°09′	69°53′	3000	C5
9	Laguna Diamante	DIA	34°11′	69°41′	3300	C6
10	Laguna Atuel	ATU	34°30′	70°02′	3420	C6
11	Valle Hermoso	VAL	35°08′	70°12′	2250	C7
12	Paso Pehuenches	PEH	35°08′	70°23′	2545	C7

(cm °C^{-1} day^{-1}), which is similar to values reported in previous studies performed in other mountain ranges in the Northern Hemisphere (0.20–0.25 in Martinec, 1989; 0.17 in Kustas et al., 1994; 0.20 in Brubaker at al., 1996; 0.15 in Molotch and Margulis, 2008). However, values associated with the northernmost clusters of our study area are quite low, reaching under 0.02 for the C1 cluster in northern Chile.

Clear sky index (K_c) values range between 0.78 and 0.89, which is similar to values reported by Salazar and Raichijk (2014), who estimate K_c values on the order of 0.90 for a single location at 1200 m a.s.l. in northern Argentina. A 5 to 6 °C difference can be observed in mean air temperature at index stations between the northern and southern edge of the domain. Temperatures for the C4 cluster are subject to greater uncertainty, because no high-elevation climate station data were available for this study (Fig. S4). Forest cover values are lower than 6 % throughout the model domain, with

the exception of cluster C3, with a value of 13.8 %. The difference in forest cover between clusters C3 and C8 can be attributed to the precipitation shadow effect induced by the Andes ridge. Forest corrections applied to MODIS fSCA resulted in a 17 % increase with respect to the original values over the southern sub-domain (C3).

3.4 Evaluation data: SWE, snow depth and river flow observations

Operational daily snow-pillow data from stations maintained by government agencies in Chile and Argentina were available for this study (Table 2). Only stations with 10 or more years of record were included, and manual snow course data were neglected because of their discontinuous nature. Approximately 10 % of observed maximum SWE accumulation values were discarded due to obvious measurement er-

Table 3. Summary of snow depth and density intensive study campaigns.

Year	ID (Fig. 1)	Symbol	Field site	Date	Snow-pit density (kg m^{-3})	SWE average (mm)	SWE SD (mm)	SWE range (mm)	Sample size
2010	2	ODA	Ojos de Agua	25 Sep	352	450	163	848–0	134
2011	2	ODA	Ojos de Agua	30 Aug	341	705	199	1194–136	374
	5	MOR	Morales	1 Sep	367	642	282	1101–0	171
	8	OBL	Olla Blanca DET	31 Aug	333	539	217	1032–79	289
2012	1	CVN	Cerro Vega Negra	28 Aug	308	296	115	700–40	166
	3	MAR	Juncal–Mardones	30 Aug	373	530	230	1120–40	163
	5	MOR	Morales	12 Sep	412	590	360	1240–150	152
	8	OBL	Olla Blanca DET	3 Sep	411	590	260	1230–0	309
	4	POR	Portillo	15 Sep	410	170	180	1230–0	181
2013	1	CVN	Cerro Vega Negra	21 Aug	356	405	165	1040–10	282
	2	ODA	Ojos de Agua	23 Aug	355	540	220	1310–100	300
	10	CHI	Nevados Chillán[a]	27 Aug	416	980	240	1270–30	104
	10	CHI	Nevados Chillán[b]	27 Aug	416	600	240	1230–70	216
	4	POR	Portillo	23 Aug	392	340	210	1120–0	91
	6	LAG	Laguna Negra	30 Aug	455	480	250	1770–0	32
2014	1	CVN	Cerro Vega Negra	5 Aug	321	163	85	620–0	326
	5	MOR	Morales	12 Aug	401	510	250	1190–0	329
	7	LVD	Lo Valdez	13 Aug	365	710	290	1260–0	186
	8	OBL	Olla Blanca DET	12 Sep	363	420	240	1210–0	334
	9	RBL	Río Blanco DET	6 Sep	354	620	290	1210–0	99
	10	CHI	Nevados Chillán[a]	26 Sep	504	830	400	380–1510	18
	10	CHI	Nevados Chillán[b]	26 Sep	504	980	250	530–1500	87
	4	POR	Portillo	19 Aug	436	170	140	850–0	73
	6	LAG	Laguna Negra	30 Aug	365	300	110	540–0	117

[a] Without forest cover (upper part of basin). [b] With forest cover (lower part of basin).

rors and data gaps. An analysis of the seasonal variability of snow-pillow records on the western and eastern slopes of the Andes suggests that the peak-SWE date is somewhat delayed on the latter, by approximately 1 month. Therefore, peak-SWE estimates for Chilean and Argentinean stations are evaluated on 1 September and 1 October, respectively, although in the results section we show values for 15 September in order to use a unique date for the entire domain. Manual snow depth observations were taken in the vicinity of selected snow-pillow locations in order to evaluate the representativeness of these measurements at the MODIS grid scale during the peak-SWE time window. These depth observations were obtained in regular grid patterns within an area the approximate size of a MODIS pixel (500 m), centered about the snow-pillow location. On average, 120 depth observations spaced at approximately 50 m increments were obtained at each snow-pillow site. Snow density was estimated by a depth-weighted average of snow densities measured in snow pits with a 1000-cc snow cutter. Samples where obtained either at regular 10 cm depth intervals along the snow pit face, or at the approximate mid depth of identifiable snow strata for very shallow snow pack conditions. Weights were computed as the fraction of total depth represented by each snow sample.

Distributed snow depth observations were available from snow surveys carried out during late winter between 2010 and 2014 at seven study catchments on the western side of the Andes, between latitudes 30 and 37° S (Fig. 1, Table 3). Snow depths were recorded with 3 m graduated avalanche probes inserted vertically into the snow pack. Depending on the terrain conditions, between three and five individual point snow depth measurements were obtained at each location, from which a mean snow depth and standard error are calculated; i.e., three-point observations are made forming a line with a spacing of 1 m and five-point observations are made forming a cross with an angle of 90° and a spacing of 1 m. Pixel-scale SWE estimates are obtained by averaging all depth observations within the limits of MODIS pixels and multiplying them by density observations from snow pits excavated at the time of each snow survey, i.e., two or three snow pits per field campaign. After this, individual depth observations are converted into SWE for model validation. Modeled SWE values are averaged at all MODIS pixels where manual depth observations are available, and their summary statistics are compared to those of SWE estimated from manual depth observations at the same pixels, multiplied by average density from snow pits.

Spring and summer season (September to March) total river flow volume (SSRV) for the 2001–2014 period is obtained from unimpaired (no human extractions) streamflow records at river gauges located in the mountain front along the model domain. Data were pre-selected leaving out series that showed too many missing values, and verified through the double mass curve method (Searcy and Hardison, 1960) in order to discard anomalous values and to ensure homogeneity throughout the period of study. Regional consistency was verified through regression analysis, only including streamflow records with R^2 values greater than 0.5 among neighboring catchments. Missing values constituted about 3.7 % of the entire period and were filled through linear regression.

4 Results

4.1 Model validation

Figure 3 compares reconstructed peak SWE (gray circles) to observed values at three snow-pillow locations (black diamonds) where additional validation sampling at the MODIS pixel scale was conducted (box plots). At the Cerro Vega Negra site (CVN), located in cluster C1, the model overestimates peak SWE (1 September) with respect to the snow-pillow value by 97 % in 2013 and by 198 % in 2014. At the Portillo site (POR, cluster C2), reconstructed SWE underestimates recorded values by 51 % in 2013 and 72 % in 2014. At the Laguna Negra site (LAG, also C2), reconstructed peak SWE slightly overestimates recorded values (8 %) (Table 4). However, reconstructed SWE compares favorably to distributed manual SWE observations obtained in the vicinity of the snow pillows at the POR and LAG sites. At POR, model estimates approach upper (2012) and lower (2013 and 2104) quartiles, while at LAG the model estimates are closer to the minimum value observed in 2013 and very similar to the observed mean in 2014.

Figure 4 depicts the comparison between reconstructed SWE and snow surveys carried out at pilot basins throughout the model domain. From left to right, it can be seen that the model slightly overestimates SWE with respect to observations at CVN (i.e., 18 % overestimation). Further south, there is a very good agreement at ODA-MAR (i.e., 4 % underestimation), with less favorable results at MOR-LVD (i.e., 39 % underestimation) and OB-RBL (i.e., 36 % underestimation). At CHI the model significantly underestimates SWE (i.e., by 67 %); note that this site is heavily forested. For the 2013a and 2014a boxes (Fig. 4) – which correspond to clearing sites – there is still underestimation, but of lesser magnitude (20 %). Summarizing, we detect model overestimation with respect to snow survey medians in four cases and underestimation in fifteen cases. In 11 out of 19 cases, reconstructed SWE lies within the snow survey data uncertainty bounds (standard deviation).

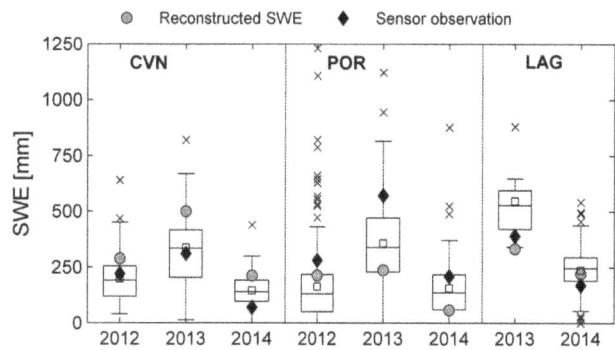

Figure 3. Reconstructed SWE validation at selected snow-pillow sites. Black diamonds are instrumental records, gray circles are model estimates, and box plots summarize the manual verification data set around the pillow site. Upper and lower box limits are the 75 and 25 % quartiles, the horizontal line is the median, the white box is the mean, upper and lower dashes represent plus and minus 2.5 SD from the mean, and crosses are outlying values.

Figure 5 shows a comparison between model estimates of peak (15 September) SWE and corresponding observations at snow-pillow sites. In general, directly contrasting pixel-based estimates with sensor observations should be attempted with caution. In areas with complex topography, slight variations in the position of the sensor with respect to the model grid, combined with high spatial variability in snow accumulation could lead to large differences between model estimates and observations. Also, small-scale variations in snow accumulation near the sensor, for example induced by protective fences, could introduce bias to the results (e.g., Meromy et al., 2013; Molotch and Bales, 2006; Rice and Bales, 2010). Taking the above into consideration, Fig. 5 suggests that the model tends to overestimate observed peak SWE at the two northernmost sites on the Chilean side (QUE and CVN); the equivalent cluster on the Argentinean side (C4) lacks SWE observations. The R^2 values indicated below refer to the best linear fit; regression line slope and intercept coefficients are provided in Table 4. Overall, we find a better agreement at the eastern slope sites (i.e., $R^2 = 0.74$) than at their western counterparts (i.e., $R^2 = 0.43$), with a combined R^2 value of 0.61. Individually, the worst and best linear agreements are obtained at POR ($R^2 = 0.32$) and LOA ($R^2 = 0.88$), respectively. Time series of observed SWE and model estimates for these two extreme cases are shown in the supplementary online material, and indicate a significant degree of inter-annual variability in model discrepancies in terms of peak SWE, but less in terms of, for instance, snow cover duration. Average standard error, $SE_{\bar{x}}$ is 284 mm ($SE_{\bar{x}} = 242$ mm at the western slope; $SE_{\bar{x}} = 302$ mm at the eastern slope), with a range between 72 mm (TOS) and 378 mm (ATU) (Table 4). Relative errors display some variability, with overestimation higher than 30 % at the two northernmost (QUE and CVN) and at the southernmost (PEH) snow pillows. For all other snow

Table 4. Model validation statistics against intensive study area observations around snow pillows and at catchment scale.

	Reconstructed SWE vs. MODIS pixel (grid) sampling (selected snow pillows)						
	Avg. sampling (mm) (1)	SD sampling (mm) (2)	Avg. model (mm) (3)	SP (sensor) (mm) (4)	RE% (avg.) (1) vs. (3)	RE% (avg.) (1) vs. (4)	RE% (avg.) (3) vs. (4)
CVN	223	110	334	200	49 %	−10 %	98 %
POR	227	177	170	353	−25 %	35 %	−51 %
LAG	395	180	283	280	−28 %	−30 %	8 %

	Reconstructed SWE vs. snow surveys (pilot basins)					
	Avg. sampling (mm) (1)	AD sampling (mm) (2)	Avg. model (mm) (3)	SD model (mm) (4)	RE% (avg.) (1) vs. (3)	RE% (SD) (2) vs. (4)
CVN	253	133	298	63	18 %	−53 %
ODA-MAR	556	203	535	128	−4 %	−37 %
MOR-LVD	613	295	375	115	−39 %	−61 %
OBL-RBL	497	252	317	89	−36 %	−65 %
CHI (forest)	790	245	257	46	−67 %	−81 %
CHI (clear)	905	320	724	170	−20 %	−47 %

	Reconstructed SWE vs. snow pillows (1 Sep – Chile and 1 Oct – Argentina)							
	R^2	Slope	Intercept (mm)	$SE_{\bar{x}}$ (mm)	RE%	RMSE (mm)	Mod. SWE average (mm)	Mod. SWE SD (mm)
QUE	0.71	1.39	131	208	79	335	529	350
CVN	0.78	0.92	247	140	56	251	609	281
SOL	0.68	0.85	−16	112	−19	127	401	241
POR	0.32	0.52	87	277	−36	398	437	324
LAG	0.42	0.76	16	217	−21	230	424	263
LOA	0.88	0.79	101	123	−5	171	734	316
ALT	0.83	0.56	5	89	−41	332	489	296
TOS	0.78	0.41	26	72	−52	251	120	141
DIA	0.76	0.85	38	141	−4	137	455	291
ATU	0.56	1.04	77	378	9	349	1263	496
VAL	0.72	0.74	11	211	−24	273	457	371
PEH	0.74	1.01	303	334	32	436	1302	580
Average	0.68	–	–	192	−2	274	602	330

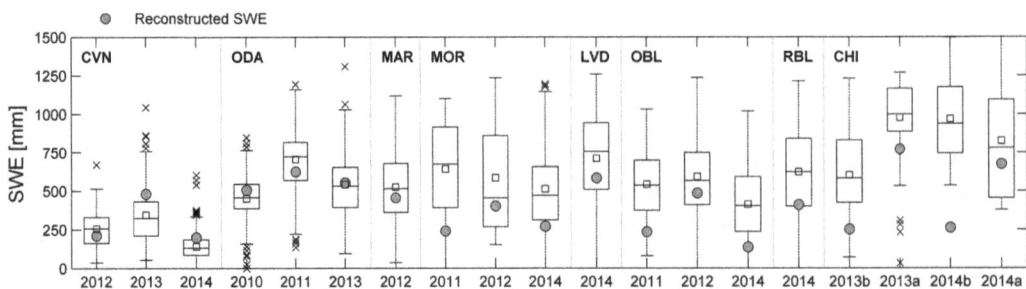

Figure 4. Reconstructed SWE validation at pixels with snow survey data. Box plots summarize all individual measurements at pixels co-located with SWE reconstruction. Symbology analogous to Fig. 3.

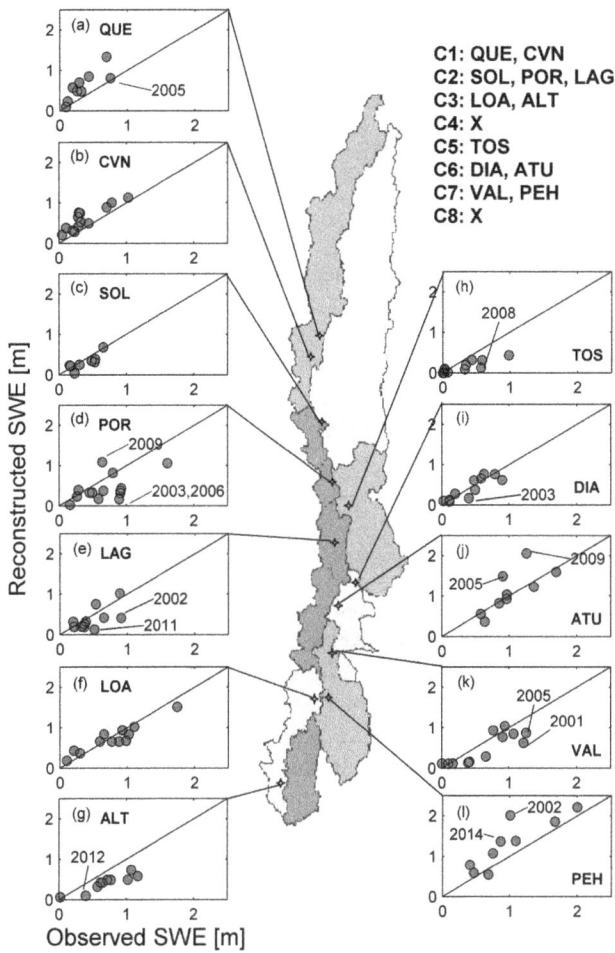

Figure 5. Comparison between peak reconstructed and observed SWE at snow-pillow sites. Solid line represents the 1 : 1 line.

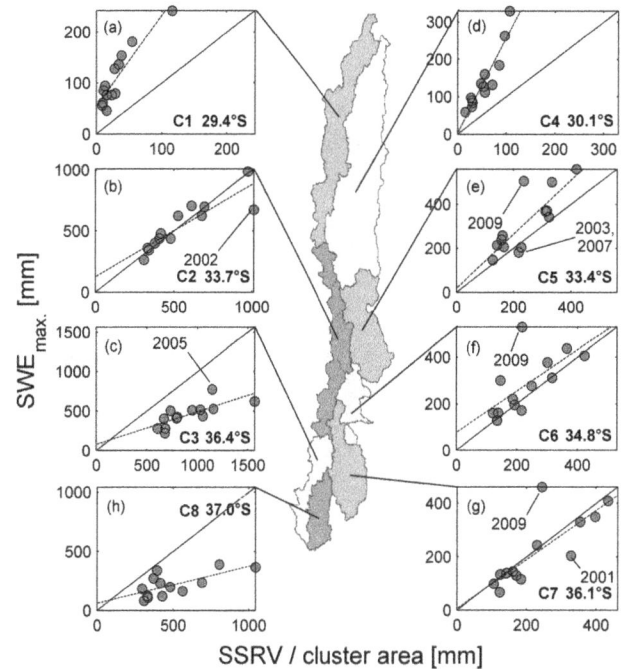

Figure 6. Area-specific spring–summer runoff volume (SSRV) versus peak SWE. Clusters 1 through 3 include rivers on the Chilean (western) slope of the Andes range; clusters 4 through 8 correspond to Argentinean (eastern) rivers. Solid line represents the 1 : 1 line. C4 and C8 SSRV were estimated by the area-transpose method.

pillows, the model estimates are lower than the sensor observation; the range of relative errors for those sites with underestimation goes from -52 to $-5\,\%$.

4.2 Correlation with melt-season river flows

Under the assumption of unimpaired flows, peak SWE and seasonal flow volume should show some degree of correlation, even though no assumptions can be made here about other relevant hydrologic processes, such as flow contributions from glaciated areas, subsurface storage carryover at the basin scale and influence of spring and summer precipitation. Differences can be expected due to losses to evapotranspiration and sublimation affecting the snowpack and soil water throughout the melt season. Hence, basin-averaged peak SWE should always be higher than melt season river volume. A clear regional pattern emerges when inspecting the results of this comparison in Fig. 6. Correlation between peak SWE and melt season river flow is higher in clusters C1 and C4 with R^2 values of 0.84 and 0.86, respectively. The re-

sult for Cluster C4 indicates that liquid precipitation during the melt season (Fig. 2) does not result in decreased correlation between peak SWE and river flow. Clusters C2, C5, C6 and C7 display a somewhat lower correlation, with some individual years departing more significantly from the overall linear trend. R^2 values range between 0.46 and 0.78 in these cases. Finally, not only are correlation coefficients lower for the southern clusters C3 ($R^2 = 0.56$) and C8 ($R^2 = 0.48$), but also estimated peak SWE is always lower than river flow, which indicates the importance of spring and summer precipitation in determining streamflow variability. In fact, Castro et al. (2014) analyze patterns of daily precipitation in this area and document average spring and summer rainfall amounts of approximately 520 mm in C3 and 85 mm in C8. A promising avenue for further research in this region emerges when comparing the correlation between melt-season river flow and the spatially distributed reconstructed product versus that of river flows and snow-pillow data. Table 5 shows values of R^2 for the linear regression between these variables. It can be seen that for two of the three clusters on the western side of the continental divide, the end-of-winter distributed reconstruction has more predictive power than observed SWE. Only for central Chile the *Laguna Negra* (LAG, with a value of 0.82) site has a better correlation with river flows, but the reconstructed product has a value of 0.78, which lies in between those found for LAG and for *Portillo*

Table 5. Coefficient of determination R^2 between river melt season flows (SSRV) and estimated and observed SWE (end-of-winter).

| | R^2 value-specific SSRV vs. estimated SWE per cluster | | R^2 value-specific SSRV vs. SWE at snow pillows (2001–2013)* | |
	2001–2014	Neglecting 2009 at Argentinean clusters**	Best	Second best
C1	0.84	–	0.74 (CVN)	0.69 (QUE)
C2	0.78	–	0.82 (LAG)	0.68 (POR)
C3	0.57	–	0.17 (LOA)	0.16 (ALT)
C4	0.87	–	–	–
C5	0.66	0.82	0.81 (TOS)	–
C6	0.45	0.76	0.87 (ATU)	0.77 (DIA)
C7	0.64	0.89	0.77 (VAL)	0.41 (PEH)
C8	0.48	0.64	–	–

* 2014 flows in Argentina unavailable to us at the time of writing. ** 2009 is considered an outlier year for the reconstruction at Argentinean sites.

(POR, with a value of 0.68). For the eastern side of the continental divide, the distributed product shows similar skill than that of snow pillows except for Atuel, which has a very high correlation (R^2 of 0.87) with cluster C6 river flows, and for cluster C7, in which the reconstruction shows higher predictive power (R^2 of 0.89) than the available SWE observations (VAL and PEH).

4.3 Regional SWE estimates

Figure 7 shows the 15 September SWE average over the 2001–2014 period obtained from the reconstruction model, and the percent annual deviations (anomalies) from that average. Steep elevation gradients can be inferred from the climatology, as well as the latitudinal variation expected from precipitation spatial patterns. For the northern clusters (C1 and C4), the peak SWE averaged over snow-covered areas is on the order of 300 mm, while in the middle of the domain (C2, C5, C6), it averages approximately 750 mm. The southern clusters (C3, C7, C8) do show high accumulation averages (≈ 650 mm), despite the sharp decrease in the Andes elevation south of latitude 34° S. The anomaly maps convey the important degree of inter-annual variability, as well as distinct spatial patterns associated with it. Between 2001 and 2014, years 2002 and 2005 stand out for displaying large positive anomalies throughout the entire mountainous region of the model domain, with values 2000 mm and more above the simulation period average. Other years prior to 2010 show differential accumulation patterns, where either the northern or southern parts of the domain are more strongly affected by positive or negative anomalies. Overall, the northern clusters (C1 and C4) show above-average accumulation in only 3 (2002, 2005 and 2007) of the 14 simulated years, whereas the other clusters show above-average accumulation for 6 years (2001, 2002, 2005, 2006, 2008 and 2009). In particular, years 2007 and 2009 show a bimodal spatial structure, with excess

accumulation (deficit) in the northern (southern) clusters during the former, and the inverse pattern in the latter year.

A longitudinal pattern in the distribution of negative anomalies can be discerned from Fig. 7, whereby drought conditions tend to be more acute on one side of the divide versus the other. Conversely, during positive anomaly years, both sides of the Andes seem to show similar behavior. Further research on the mechanisms of moisture transport during below-average precipitation years may shed light on this result.

Figure 8 provides a different perspective on the region's peak SWE climatology by presenting our results aggregated into elevation bands for each hydrologic unit. Elevation bands are defined at 1000 m increments starting from 1000 m a.s.l. Crosses indicate average peak SWE for each band (mm), and circle areas are proportional to the surface area covered by each elevation band. From north to south, hydrologic unit C4 shows slightly higher SWE than C1 between 3000 and 5000 m a.s.l., but much larger surface areas ($\sim 32\,000$ vs. $\sim 17\,000$ km^2), indicating a larger water resource potential. C2 stands out as having the greatest area-weighted cluster SWE and the greatest SWE for each elevation band. Compared to its counterpart on the eastern side of the Andes range (C5), C2 shows higher accumulations (up to ~ 1800 mm) at all elevations. The area included between 2000 and 4000 m a.s.l. ($\sim 13\,000$ km^2), which shows an estimated peak SWE accumulation on the order of 600 mm, represents the most predominant snow volume accumulation zone. Although the 4000–5000 m a.s.l. elevation band contributes approximately half the 2000–4000 band surface area in C2, its average peak SWE is roughly twice that of the 3000–4000 band (~ 6000 km^2). This makes this subregion interesting for future research, because most snow observations in the area are obtained below 4000 m a.s.l.; the same is true for unit C5. Further to the south, the barrier effect of the Andes is also suggested by the displacement of the SWE-

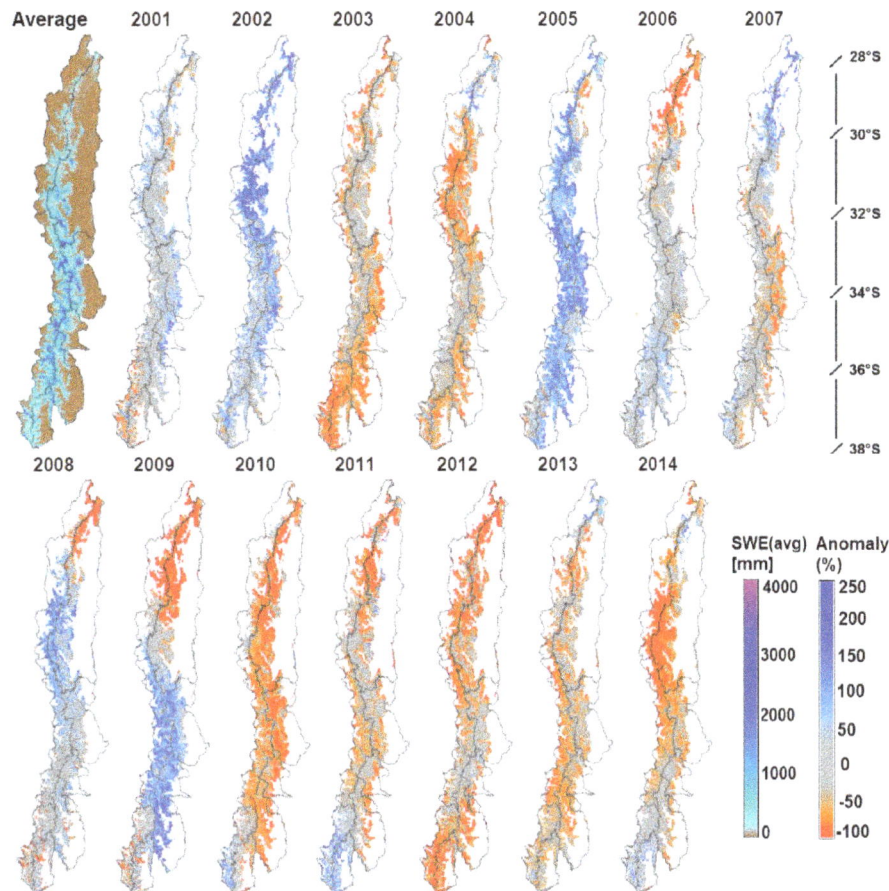

Figure 7. Regional peak (15 September) SWE climatology for the 2001–2014 period (upper left panel), and annual peak SWE anomalies.

elevation distribution in C6 and C7 when compared to C3. On the eastern side of the model domain, it is interesting to see a steepening of the average peak SWE elevation profile between C6 and C8, suggesting that C8 is less affected by Andes blockage than its northern counterparts.

Estimated net energy inputs (Fig. 9) shows a decrease from the northern (C1 and C4) into the mid-range clusters (C2, C5 and C6), with increases again in the southern reaches of the domain (C3, C7 and C8). This is a result of a combination of an increasing trend in net shortwave radiation in the south–north direction and a reverse spatial trend in net longwave radiation exchange, which increases (approaches less negative values) in the north–south direction. Modeled turbulent energy fluxes (Eq. 1) are negligible in the northern clusters, but their contribution to the net energy exchange increases with latitude as a result of the spatial variation in the a_r parameter.

Figure 10 shows the temporal (seasonal) variation in average fSCA and SWE for each cluster, and Table 6 shows peak SWE at the watershed scale, averaged both over the entire basin and over the snow-covered area. Maximum fSCA increases in the north–south direction, consistent with the climatological increase in winter precipitation and decrease in

temperature. A dramatic increase in snow coverage is observed between the northern (i.e., C1 and C4) and adjacent southern clusters (i.e., C2 and C5), with average peak fSCA increasing from 20 to 50 %. The highest average snow coverage is observed for cluster C8, with more than 60 %. Snow water equivalent displays a similar regional variability with lower seasonal variability than snow cover for all clusters except for C2, where fSCA and SWE variability throughout the melt season are identical. Mean peak SWE in northern Chile is the lowest among the eight clusters, with approximately 100 mm SWE over the 2001–2014 period. The largest estimate is for cluster C2, central Chile, where mean peak SWE exceeds 500 mm. The rain shadow effect of the Andes range is apparent in the comparison of SWE and fSCA in C2 and C5–C6–C7. Fractional snow-covered area is lower on the eastern side because of the larger basin sizes, which increases the proportional area of lower elevation terrain. In addition, peak SWE is approximately 25 % lower on the eastern side, with less than 400 mm SWE for the eastern clusters. Cluster C4 is not affected by this phenomenon, showing higher snow coverage and water equivalent accumulation than its counterpart, C1. Cluster C8 represents an interesting exception in that its average fSCA is the largest within the model

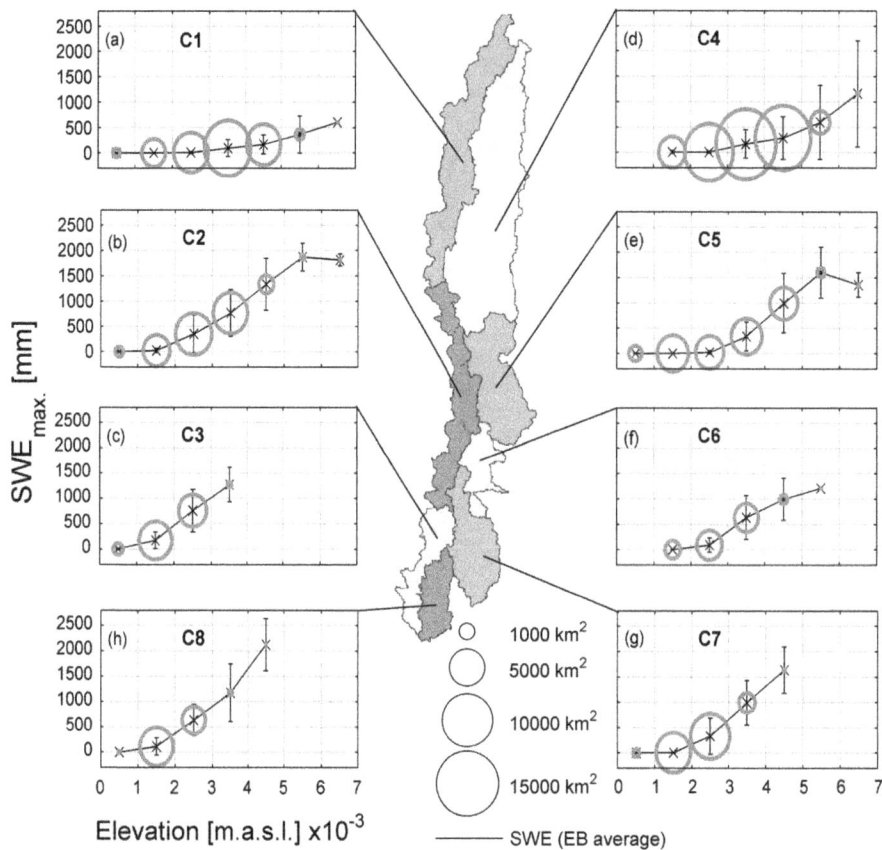

Figure 8. Maximum SWE through 1000 m elevation bands (EB). Crosses are mean values within EB, lines are the estimated SWE-elevation profile. Circle radius indicates EB area (km^2) scaled by 0.05 and takes values from the SWE axis.

domain, but peak SWE is not significantly higher than the estimates in the other clusters on the Argentinean side of the Andes.

5 Discussion

5.1 Sensitivity analysis

The Andes Cordillera, on the one hand, displays ideal conditions for SWE reconstruction, including low cloud cover, infrequent snowfall during spring and summer, and very low forest cover. On the other hand, the scarcity of basic climate data poses challenges that would affect any modeling exercise. A local sensitivity analysis is implemented in order to gain insights regarding the influence of some of the assumptions required for SWE modeling (Fig. 11). The influence of the clear sky factor (K_c), snow surface albedo (α_s), the slope of the Δ_{LST} vs. Δ_{T_a} relationship (μ), the a_r parameter, and the difference between air and snow surface temperature are explored. Results are shown for the model pixels corresponding to two of the snow-pillow sites, each located at the northern and southern sub-regions of the model domain respectively. The clear sky factor, snow albedo and Δ_{LST} vs. Δ_{T_a}

slope are the most sensitive parameters at the northern (CVN) site. Increasing the slope in the Δ_{LST} vs. Δ_{T_a} relationship results in decreasing temperature at pixels with higher elevations than the index station, thus lowering longwave cooling and resulting in higher SWE estimates. The impact of increasing slope values decreases progressively, because an increasing slope results in increased pixel air temperature, but snow surface temperature cannot exceed 0 °C. The influence of snow albedo is analyzed by perturbing the entire albedo time series for each season from the values predicated by the USACE model. Increasing albedo values restricts the energy available for melt therefore decreasing peak SWE estimates. Again, a nonlinear effect is observed, constrained by a minimum albedo value of 0.4. The sensitivity of the clear sky factor, on the other hand, is monotonic, with increasing values generating more available solar energy, resulting in higher SWE estimates. At the southern site (ALT), the shape of the sensitivity functions is the same as at CVN, but the magnitude of SWE variations as a function of parameter perturbations is smaller. This is likely related to the fact that turbulent fluxes constitute a larger fraction of the simulated overall energy balance at the southern sites; a_r parameter values are greater in the southern portions of the domain. There-

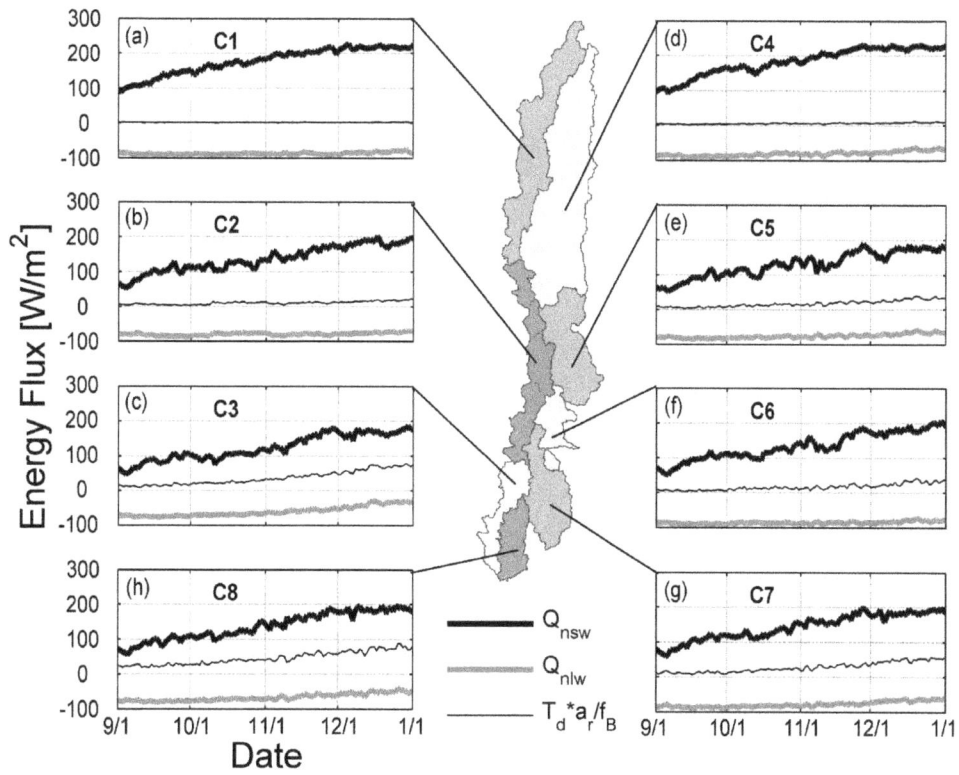

Figure 9. Time series of energy fluxes over snow surface (average over 14 years) and global average per cluster. Unique axes scale for all plots.

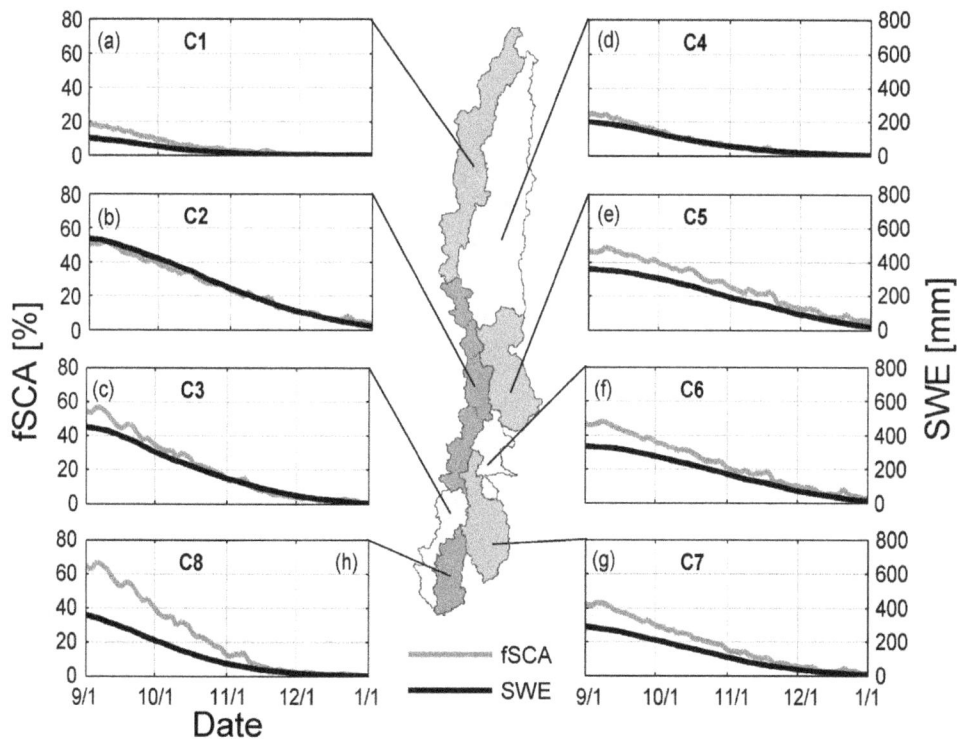

Figure 10. Average seasonal evolution of fSCA and SWE in the study region. Lower right panel shows the spatial correlation between time-averaged fSCA, SWE and specific melt-season river discharge.

Table 6. Peak SWE 2001–2014 climatology for river basins within the study region. Basin-wide averages, SCA-wide averages and basin-wide water volumes shown.

ID	Basin – gauge station	Lat. S	Long. W	Outlet elev. (m a.s.l.)	Area (km^2)	SWE Basin-wide (mm)	SWE Over SCA (mm)	SWE Basin-wide (m$^3 \times 10^{-6}$)
					Chile			
1	Copiapó en Pastillo	27°59′	69°58′	1300	7470	45	120	336
2	Huasco en Algodones	28°43′	70°30′	750	7180	68	161	488
3	Elqui en Algarrobal	29°59′	70°35′	760	5710	151	269	862
4	Hurtado en San Agustín	30°27′	70°32′	2050	676	302	325	204
5	Grande en Puntilla San Juan	30°41′	70°55′	2140	3545	137	306	486
6	Cogotí en La Fraguita	31°06′	70°53′	1021	491	182	335	89
7	Illapel en Huintil	31°33′	70°57′	650	1046	180	305	188
8	Chalinga en San Agustín	31°41′	70°43′	920	437	142	332	62
9	Choapa en Salamanca	31°48′	70°55′	560	2212	214	356	473
10	Sobrante en Piñadero	32°12′	70°42′	2057	126	172	198	22
11	Alicahue en Colliguay	32°18′	70°44′	852	344	92	184	32
12	Putaendo en Resg. Los Patos	32°30′	70°34′	1218	890	273	346	243
13	Aconcagua en Chacabuquito	32°51′	70°30′	950	2110	609	692	1285
14	Mapocho en Los Almendros	33°22′	70°27′	970	640	269	342	172
15	Maipo en El Manzano	33°35′	70°22′	850	4840	692	760	3349
16	Cachapoal en Puente Termas	34°15′	70°34′	700	2455	700	814	1719
17	Tinguiririca en Los Briones	34°43′	70°49′	560	1785	532	677	950
18	Teno en Claro	34°59′	70°49′	650	1210	438	524	530
19	Lontué en Colorado–Palos	35°15′	71°02′	600	1330	656	759	872
20	Maule en Armerillo	35°42′	70°10′	470	5465	525	554	2869
21	Ñuble en San Fabián	36°34′	71°33′	410	1660	376	430	624
22	Polcura en Laja	37°19′	77°32′	675	2088	358	378	748
					Argentina			
23	Jachal en Pachimoco	30°12′	68°49′	1563	24 266	79	175	1917
24	San Juan en km 101	31°15′	69°10′	1129	23 860	308	569	7349
25	Mendoza en Guido	32°54′	69°14′	1479	7304	460	672	3360
26	Tunuyán en Zapata	33°46′	69°16′	852	11 230	289	592	3245
27	Diamante en La Jaula	34°40′	69°18′	1451	2832	395	489	1118
28	Atuel en Loma Negra	35°15′	69°14′	1353	3696	338	525	1249
39	Malargue en La Barda	35°33′	69°40′	1568	1055	171	284	180
30	Colorado en Buta Ranquil	37°04′	69°44′	817	14 896	288	495	4290
31	Neuquén en Rahueco	37°21′	70°27′	870	8266	356	446	2943

fore, perturbations of the other terms account for a smaller fraction of the energy exchange at the southern sites.

5.2 Model performance and conceptual energy balance representation

Among the many factors that influence model performance, the sub-region delineation involves the selection of index meteorological stations for extrapolating input data at the domain level. Thus, for example, two adjacent pixels that are part of different sub-regions may be assigned input data derived from two different meteorological stations that are many kilometers apart. It would be preferable to use distributed inputs only, but these were not available for this do-

main. Future research is needed to explore alternative strategies for domain clustering.

Overall, the model performance, evaluated against SWE observations, is comparable to that achieved in other mountain regions of the world. Our average coefficient of determination R^2 of 0.68 is lower than that obtained by Guan et al. (2013) in the Sierra Nevada (0.74) when comparing operational snow-pillow observations, although this value is affected by three stations with much lower agreement (POR, LAG, ATU); the median R^2 in our study, on the other hand, is 0.73, which we consider satisfactory in light of the scarcity of forcing data and direct snow properties observations available in this region. The overall relative error is -2% for observations from snow pillows within our study region, but

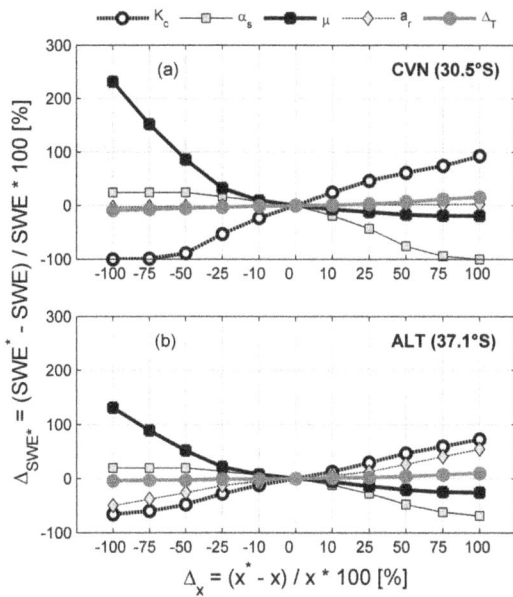

Figure 11. Sensitivity of peak SWE estimates to model forcings and parameters. Average over the 2001–2014 period at selected snow-pillow sites. Δ_x represents the percentage change over each parameter studied with respect to the base case.

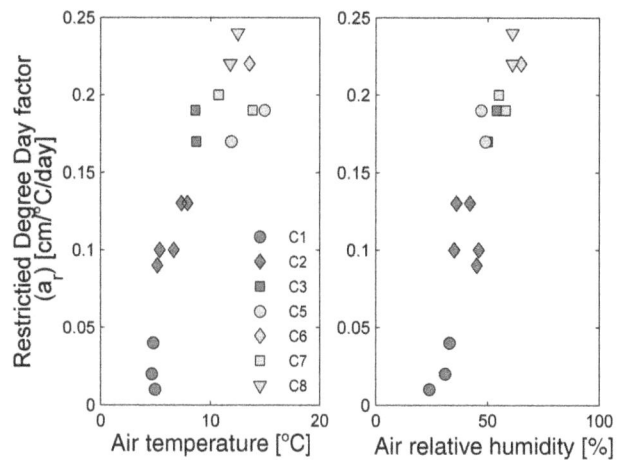

Figure 12. Restricted degree-day factor as a function of space (basin cluster) and climatological properties. Bowen (β) coefficient shown between parentheses in the legend.

this value is strongly affected by two stations where we observed significant overestimation (QUE and CVN). When including the remaining ten snow pillows only, relative error increases to -16%. Given that forest cover is minimal in our modeling domain, we can attribute this bias to either weaknesses in the simplified energy balance model formulation or to errors in the MOD10A1 fSCA product. Previous work in the northern hemisphere (Rittger et al., 2013) has shown that MODIS can underestimate fractional snow cover during the snowmelt season. On the one hand, land cover heterogeneity at spatial resolutions lower than the MODIS scale (i.e., 500 m) results in mixed-pixel detection problems. On the other hand, spectral unmixing based on the NDSI approach tends to underestimate fSCA under patchy snow distributions. In addition, surface temperatures greater than $10\,^\circ\text{C}$ – more likely to exist during late spring – induce MODIS fSCA underestimation. Molotch and Margulis (2008) tested the SWE reconstruction model using Landsat ETM and MOD10A1 and found that maximum basin-wide mean SWE estimates were significantly lower when using MOD10A1. More recently, Cortés et al. (2014a) showed that a similar pattern can be seen for the extratropical Andes, whereby MODIS fSCA consistently underestimated LANDSAT TM fSCA retrievals. MODIS fSCA underestimation during spring combined with increased net energy fluxes over the snowpack can result in a marked underestimation ($\sim 20\%$) for available energy flux for snowpack melting and consequently ($\sim 45\%$) for maximum SWE (Molotch and Margulis, 2008).

Comparisons against spatial interpolations from intensive-study areas in the Sierra Nevada or Rocky Mountains (e.g., Erxleben et al., 2002; Jepsen et al., 2012) are not directly applicable, because in this study we do not employ interpolation methods to derive our manual snow survey SWE estimates. However, the average overestimation found with respect to snow survey data could be explained by the fact that manual surveys are limited by site accessibility and sampling procedures. For example, snow probes utilized are only 3.0 m long, which precludes observation of deeper snowpack; likewise, deep snow is expected in sites exposed to avalanching, which were generally avoided in snow survey design due to safety considerations. On the other hand, manual snow surveys do not visit steep snow-free areas where snow depth is expected to be lower than the 500 m pixel reconstruction. The combined effect of these two contrasting effects is the subject of further research in this region.

Another possible explanation for model errors is the simplified formulation of the energy balance equation, which may be problematic when applied over a large, climatically variable model domain. To explore the implications of the simplified energy balance with respect to model errors, we focus on the representation of turbulent energy fluxes, represented here through a linear temperature-dependent term. Figure 12 describes the spatial distribution of the a_r parameter, and its dependence on air temperature and relative humidity observed at index meteorological stations. The implication for energy balance modeling is that turbulent fluxes would account for a very small portion of the snowpack energy and mass balance in the northern area (C1 and C4), which is characterized by low air temperatures and relative humidity, which yield very low a_r values. The reader must recall that a_r values were computed based on index station data and assumed spatially homogeneous over each cluster. The simplified model formulation used in this research, however, although pseudo-physically based – compared to degree-day or fully calibrated models – allows only for positive net turbulent fluxes, because both the a_r and the degree-day temperature index are positive values. However, previous studies in this region (Corripio and Purves, 2005; Favier et al., 2009) have suggested that latent heat fluxes have a relevant role because of high sublimation rates favored by high winds and low relative humidity conditions predominant in the area.

In order to diagnose differential performance of the model across the hydrologic units defined in this study, we compute the Bowen ratio (β) at the point scale from data available only at the few high-elevation meteorological stations in the region with recorded relative humidity. The calculations show that at stations located within cluster C1, latent heat fluxes are opposite in sign and larger in magnitude than sensible heat fluxes (Fig. S6). While this results in net turbulent cooling of the snowpack, this energy loss is not considered in our simplified energy balance approach. Note that for the clusters C5, C6, C7 and C8, all located on the eastern (Argentinean) slope of the Andes, sensible and latent heat fluxes are positive, compared to negative latent heat fluxes for all the index stations within clusters C2 and C3 on the Chilean side. This result is consistent with Insel et al. (2010), who applied a regional circulation model (RegCM3) in the area and showed a significant difference in relative humidity ($\sim 70\,\%$ on the eastern side vs. $\sim 40\,\%$ on the western side). The fact that we extrapolate the a_r parameter value based on relatively low elevation meteorological observations throughout the southern Argentinean hydrologic units may result in a yet not quantified overestimation of seasonal energy inputs and peak SWE for those clusters.

6 Conclusions

Snow water equivalent is the foremost water source for the extratropical Andes region in South America. This paper presents the first high-resolution distributed assessment of this critical resource, combining instrumental records with remotely sensed snow-covered area and a physically based snow energy balance model. Overall errors in estimated peak SWE, when compared with operational station data, amount to $-2.2\,\%$, and correlation with observed melt-season river flows is high, with an R^2 value of 0.80. MODIS fractional SCA data proved adequate for the goals of this study, affording high temporal resolution observations and an appropriate spatial resolution given the extent of the study region. These results have implications for evaluating seasonal water supply forecasts, analyzing synoptic-scale drivers of snow accumulation, and validating precipitation estimates from regional climate models. In addition, the strong correlation between peak SWE and seasonal river flow indicates that our results could be useful for the evaluation of alternative water resource projects as part of development and climate change adaptation initiatives. Finally, the regional SWE and

anomaly estimates illustrate the dramatic spatial and temporal variability of water resources in the extratropical Andes, and provide a striking visual assessment of the progression of the drought that has affected the region since 2009. These results should motivate further research looking into the climatic drivers of this spatially distributed phenomenon.

Acknowledgements. This research was conducted with support from CONICYT, under grants FONDECYT 1121184, SER-03, FONDEF CA13I10277 and CHILE-USA2013. The authors wish to thank everybody involved in field data collection, including brothers Santiago and Gonzalo Montserrat, Mauricio Cartes, Alvaro Ayala, and many others. Gonzalo Cortés provided insightful comments to working drafts of this paper.

References

Aceituno, P.: On the functioning of the Southern Oscillation in the South American sector. Part I: Surface climate, Mon. Weather Rev., 116, 505–524, 1988.

Ahl, D. E., Gower, S. T., Burrows, S. N., Shabanov, N. V., Myneni, R. B., and Knyazikhin, Y.: Monitoring spring canopy phenology of a deciduous broadleaf forest using MODIS, Remote Sens. Environ., 104, 88–95, 2006.

Aravena, J.-C. and Luckman, B. H.: Spatio-temporal rainfall patterns in southern South America, Int. J. Climatol., 29, 2106–2120, 2009.

Army Corps of Engineers: Engineering and design: runoff from snowmelt, Washington, 1960.

Arsenault, K. R., Houser, P. R., and De Lannoy, G. J. M.: Evaluation of the MODIS snow cover fraction product, Hydrol. Process., 28, 980–998, doi:10.1002/hyp.9636, 2014.

Ayala, A., McPhee, J., and Vargas, X.: Altitudinal gradients, midwinter melt, and wind effects on snow accumulation in semiarid midlatitude Andes under La Niña conditions, Water Resour. Res., 50, 3589–3594, doi:10.1002/2013WR014960, 2014.

Benali, A., Carvalho, A. C., Nunes, J. P., Carvalhais, N., and Santos, A.: Estimating air surface temperature in Portugal using MODIS LST data, Remote Sens. Environ., 124, 108–121, 2012.

Brubaker, K., Rango, A., and Kustas, W.: Incorporating Radiation Inputs into the Snowmelt Runoff Model, Hydrol. Process., 10, 1329–1343, doi:10.1002/(SICI)1099-1085(199610)10:10<1329::AID-HYP464>3.0.CO;2-W, 1996.

Castro, L. M., Gironás, J., and Fernández, B.: Spatial estimation of daily precipitation in regions with complex relief and scarce data using terrain orientation, J. Hydrol., 517, 481–492, 2014.

Cline, D. W., Bales, R. C., and Dozier, J.: Estimating the spatial distribution of snow in mountain basins using remote sensing and energy balance modeling, Water Resour. Res., 34, 1275–1285, 1998.

Colombi, A., De Michele, C., Pepe, M., and Rampini, A.: Estimation of daily mean air temperature from MODIS LST in Alpine areas, EARSeL EProceedings 6, 38–46, 2007.

Corripio, J. G. and Purves, R. S.: Surface energy balance of high altitude glaciers in the central Andes: The effect of snow penitentes, in: Clim. Hydrol. Mt. Areas, edited by: Collins, D., de Jong, C., and Ranzi, R., Wiley, London, 15–27, 2005.

Cortés, G., Vargas, X., and McPhee, J.: Climatic sensitivity of streamflow timing in the extratropical western Andes Cordillera, J. Hydrol., 405, 93–109, doi:10.1016/j.jhydrol.2011.05.013, 2011.

Cortés, G., Cornwell, E., McPhee, J. P., and Margulis, S. A.: Snow Cover Quantification in the Central Andes Derived from Multi-Sensor Data, in: AGU Fall Meeting Abstracts, San Francisco, p. 0410, 2014a.

Cortés, G., Girotto, M., and Margulis, S. A.: Analysis of sub-pixel snow and ice extent over the extratropical Andes using spectral unmixing of historical Landsat imagery, Remote Sens. Environ., 141, 64–78, doi:10.1016/j.rse.2013.10.023, 2014b.

DeWalle, D. and Rango, A.: Principles of snow hydrology, Cambridge University Press, New York, 2008.

Dietz, A. J., Kuenzer, C., Gessner, U., and Dech, S.: Remote sensing of snow – a review of available methods, Int. J. Remote Sens., 33, 4094–4134, 2012.

Erxleben, J., Elder, K., and Davis, R.: Comparison of spatial interpolation methods for estimating snow distribution in the Colorado Rocky Mountains, Hydrol. Process., 16, 3627–3649, 2002.

Eva, H. D., Belward, A. S., De Miranda, E. E., Di Bella, C. M., Gond, V., Huber, O., Jones, S., Sgrenzaroli, M., and Fritz, S.: A land cover map of South America, Global Change Biol., 10, 731–744, 2004.

Falvey, M. and Garreaud, R.: Wintertime precipitation episodes in central Chile: Associated meteorological conditions and orographic influences, J. Hydrometeorol., 8, 171–193, 2007.

Falvey, M. and Garreaud, R. D.: Regional cooling in a warming world: Recent temperature trends in the southeast Pacific and along the west coast of subtropical South America (1979–2006), J. Geophys. Res.-Atmos., 114, D04102, doi:10.1029/2008JD010519, 2009.

Favier, V., Falvey, M., Rabatel, A., Praderio, E., and López, D.: Interpreting discrepancies between discharge and precipitation in high-altitude area of Chile's Norte Chico region (26–32° S). Water Resour. Res., 45, W02424, doi:10.1029/2008WR006802, 2009.

Gafurov, A. and Bárdossy, A.: Cloud removal methodology from MODIS snow cover product, Hydrol. Earth Syst. Sci., 13, 1361–1373, doi:10.5194/hess-13-1361-2009, 2009.

Garreaud, R. D.: The Andes climate and weather, Adv. Geosci., 22, 3–11, doi:10.5194/adgeo-22-3-2009, 2009.

Gascoin, S., Lhermitte, S., Kinnard, C., Bortels, K., and Liston, G. E.: Wind effects on snow cover in Pascua-Lama, Dry Andes of Chile, Adv. Water Resour., 55, 25–39, doi:10.1016/j.advwatres.2012.11.013, 2013.

Guan, B., Molotch, N. P., Waliser, D. E., Jepsen, S. M., Painter, T. H., and Dozier, J.: Snow water equivalent in the Sierra Nevada: Blending snow sensor observations with snowmelt model simulations, Water Resour. Res., 49, 5029–5046, doi:10.1002/wrcr.20387, 2013.

Hall, D. K. and Riggs, G. A.: Accuracy assessment of the MODIS snow products, Hydrol. Process., 21, 1534–1547, 2007.

Hall, D. K., Riggs, G. A., Salomonson, V. V., DiGirolamo, N. E., and Bayr, K. J.: MODIS snow-cover products, Remote Sens. Environ., 83, 181–194, 2002.

Hansen, M. C., DeFries, R. S., Townshend, J. R. G., Carroll, M., Dimiceli, C., and Sohlberg, R. A.: Global percent tree cover at a spatial resolution of 500 meters: First results of the MODIS vegetation continuous fields algorithm, Earth Interact., 7, 1–15, 2003.

Hofierka, J. and Suri, M.: The solar radiation model for Open source GIS: implementation and applications, in: Proceedings of the Open Source GIS-GRASS Users Conference, Trento, Italy, 1–19, 2002.

Huete, A., Didan, K., Miura, T., Rodriguez, E. P., Gao, X., and Ferreira, L. G.: Overview of the radiometric and biophysical performance of the MODIS vegetation indices, Remote Sens. Environ., 83, 195–213, 2002.

Insel, N., Poulsen, C. J., and Ehlers, T. A.: Influence of the Andes Mountains on South American moisture transport, convection, and precipitation, Clim. Dynam., 35, 1477–1492, 2010.

Jepsen, S. M., Molotch, N. P., Williams, M. W., Rittger, K. E., and Sickman, J. O.: Interannual variability of snowmelt in the Sierra Nevada and Rocky Mountains, United States: Examples from two alpine watersheds, Water Resour. Res., 48, W02529, doi:10.1029/2011WR011006, 2012.

Krogh, S. A., Pomeroy, J. W., and McPhee, J.: Physically Based Mountain Hydrological Modeling Using Reanalysis Data in Patagonia, J. Hydrometeorol., 16, 172–193, doi:10.1175/JHM-D-13-0178.1, 2015.

Kustas, W. P., Rango, A., and Uijlenhoet, R.: A simple energy budget algorithm for the snowmelt runoff model., Water Resour. Res., 30, 1515–1527, 1994.

Martinec, J.: Hour-to-hour snowmelt rates and lysimeter outflow during an entire ablation period, Snow Cover Glacier Var., in: Glacier and Snow Cover Variations, IAHS Publ. no. 183, edited by: Colbeck, S. C., Proceedings of the Baltimore Symposium, Maryland, 19–28, 1989.

Masiokas, M. H., Villalba, R., Luckman, B. H., Le Quesne, C., and Aravena, J. C.: Snowpack variations in the central Andes of Argentina and Chile, 1951–2005: Large-scale atmospheric influences and implications for water resources in the region, J. Climate, 19, 6334–6352, 2006.

Meromy, L., Molotch, N. P., Link, T. E., Fassnacht, S. R., and Rice, R.: Subgrid variability of snow water equivalent at operational snow stations in the western USA, Hydrol. Process., 27, 2383–2400, 2013.

Minder, J. R., Mote, P. W., and Lundquist, J. D.: Surface temperature lapse rates over complex terrain: Lessons from the Cascade Mountains, J. Geophys. Res.-Atmos., 115, 1984–2012, 2010.

Molotch, N. P.: Reconstructing snow water equivalent in the Rio Grande headwaters using remotely sensed snow cover data and a spatially distributed snowmelt model, Hydrol. Process., 23, 1076–1089, doi:10.1002/hyp.7206, 2009.

Molotch, N. P. and Bales, R. C.: Comparison of ground-based and airborne snow surface albedo parameterizations in an alpine watershed: Impact on snowpack mass balance, Water Resour. Res., 42, W05410, doi:10.1029/2005WR004522, 2006.

Molotch, N. P. and Margulis, S. A.: Estimating the distribution of snow water equivalent using remotely sensed snow cover data and a spatially distributed snowmelt model: A multi-resolution, multi-sensor comparison, Adv. Water Resour., 31, 1503–1514, doi:10.1016/j.advwatres.2008.07.017, 2008.

Montgomery, D. R., Balco, G., and Willett, S. D.: Climate, tectonics, and the morphology of the Andes, Geology, 29, 579–582, 2001.

Neteler, M.: Estimating Daily Land Surface Temperatures in Mountainous Environments by Reconstructed MODIS LST Data, Remote Sens., 2, 333–351, doi:10.3390/rs1020333, 2010.

Neteler, M., Bowman, M. H., Landa, M., and Metz, M.: GRASS GIS: A multi-purpose open source GIS, Environ. Model. Softw., 31, 124–130, 2012.

Newman, M., Compo, G. P., and Alexander, M. A.: ENSO-forced variability of the Pacific decadal oscillation, J. Climate, 16, 3853–3857, 2003.

Núñez, J., Rivera, D., Oyarzún, R., and Arumí, J. L.: Influence of Pacific Ocean multidecadal variability on the distributional properties of hydrological variables in north-central Chile, J. Hydrol., 501, 227–240, 2013.

Pomeroy, J. W., Marks, D., Link, T., Ellis, C., Hardy, J., Rowlands, A., and Granger, R.: The impact of coniferous forest temperature on incoming longwave radiation to melting snow, Hydrol. Process., 23, 2513–2525, 2009.

Pontailler, J.-Y., Hymus, G. J., and Drake, B. G.: Estimation of leaf area index using ground-based remote sensed NDVI measurements: validation and comparison with two indirect techniques, Can. J. Remote Sens., 29, 381–387, 2003.

Ragettli, S., Cortés, G., McPhee, J., and Pellicciotti, F.: An evaluation of approaches for modelling hydrological processes in high-elevation, glacierized Andean watersheds, Hydrol. Process., 28, 5674–5695, doi:10.1002/hyp.10055, 2014.

Raleigh, M. S., Landry, C. C., Hayashi, M., Quinton, W. L., and Lundquist, J. D.: Approximating snow surface temperature from standard temperature and humidity data: New possibilities for snow model and remote sensing evaluation, Water Resour. Res., 49, 8053–8069, 2013.

Rankinen, K., Karvonen, T., and Butterfield, D.: A simple model for predicting soil temperature in snow-covered and seasonally frozen soil: model description and testing, Hydrol. Earth Syst. Sci., 8, 706–716, doi:10.5194/hess-8-706-2004, 2004.

Rhee, J. and Im, J.: Estimating high spatial resolution air temperature for regions with limited in situ data using MODIS products, Remote Sens., 6, 7360–7378, 2014.

Rice, R. and Bales, R. C.: Embedded-sensor network design for snow cover measurements around snow pillow and snow course sites in the Sierra Nevada of California, Water Resour. Res., 46, W03537, doi:10.1029/2008WR007318, 2010.

Rittger, K., Painter, T. H., and Dozier, J.: Assessment of methods for mapping snow cover from MODIS, Adv. Water Resour., 51, 367–380, 2013.

Rubio-Álvarez, E. and McPhee, J.: Patterns of spatial and temporal variability in streamflow records in south central Chile in the period 1952–2003, Water Resour. Res., 46, W05514, doi:10.1029/2009WR007982, 2010.

Salazar, G. and Raichijk, C.: Evaluation of clear-sky conditions in high altitude sites, Renew. Energy, 64, 197–202, 2014.

Scheel, M. L. M., Rohrer, M., Huggel, C., Santos Villar, D., Silvestre, E., and Huffman, G. J.:. Evaluation of TRMM Multi-satellite Precipitation Analysis (TMPA) performance in the Central Andes region and its dependency on spatial and temporal resolution, Hydrol. Earth Syst. Sci., 15, 2649–2663, doi:10.5194/hess-15-2649-2011, 2011.

Searcy, J. K. and Hardison, C. H.: Double-mass curves, in: Manual of Hydrology: Part 1, General Surface Water Techniques, US Geol. Surv. Water-Supply Pap. 1541-B, US Geological Survey, Washington, D.C., 31–59, 1960.

Sicart, J. E., Essery, R. L., Pomeroy, J. W., Hardy, J., Link, T., and Marks, D.: A sensitivity study of daytime net radiation during snowmelt to forest canopy and atmospheric conditions, J. Hydrometeorol., 5, 774–784, 2004.

Slater, A. G., Barrett, A. P., Clark, M. P., Lundquist, J. D., and Raleigh, M. S.: Uncertainty in seasonal snow reconstruction: Relative impacts of model forcing and image availability, Adv. Water Resour., 55, 165–177, doi:10.1016/j.advwatres.2012.07.006, 2013.

Tarboton, D. G. and Luce, C. H.: Utah energy balance snow accumulation and melt model (UEB), Citeseer, Computer model technicaldescription and users guide, Utah Water Research Laboratory and USDA Forest Service Intermountain Research Station, 1996.

Vera, C., Silvestri, G., Liebmann, B., and González, P.: Climate change scenarios for seasonal precipitation in South America from IPCC-AR4 models, Geophys. Res. Lett., 33, L13707, doi:10.1029/2006GL025759, 2006.

Vicuña, S., Garreaud, R. D., and McPhee, J.: Climate change impacts on the hydrology of a snowmelt driven basin in semiarid Chile, Climatic Change, 105, 469–488, doi:10.1007/s10584-010-9888-4, 2011.

Wan, Z., Zhang, Y., Zhang, Q., and Li, Z.: Validation of the land-surface temperature products retrieved from Terra Moderate Resolution Imaging Spectroradiometer data, Remote Sens. Environ., 83, 163–180, 2002.

Wan, Z., Zhang, Y., Zhang, Q., and Li, Z.-L.: Quality assessment and validation of the MODIS global land surface temperature, Int. J. Remote Sens., 25, 261–274, 2004.

Williamson, S. N., Hik, D. S., Gamon, J. A., Kavanaugh, J. L., and Flowers, G. E.: Estimating temperature fields from MODIS land surface temperature and air temperature observations in a sub-Arctic Alpine environment, Remote Sens., 6, 946–963, 2014.

10

The Hydrological Open Air Laboratory (HOAL) in Petzenkirchen: a hypothesis-driven observatory

G. Blöschl[1,2], A. P. Blaschke[1,2,8], M. Broer[1], C. Bucher[1,3], G. Carr[1], X. Chen[1], A. Eder[1,4], M. Exner-Kittridge[1,5], A. Farnleitner[1,6,8], A. Flores-Orozco[7], P. Haas[1,2], P. Hogan[1], A. Kazemi Amiri[1], M. Oismüller[1], J. Parajka[1,2], R. Silasari[1], P. Stadler[1,5], P. Strauss[4], M. Vreugdenhil[1], W. Wagner[1,7], and M. Zessner[1,5]

[1]Centre for Water Resource Systems, TU Wien, Karlsplatz 13, 1040 Vienna, Austria

[2]Institute of Hydraulic Engineering and Water Resources Management, TU Wien, Karlsplatz 13/222, 1040 Vienna, Austria

[3]Institute of Building Construction and Technology, TU Wien, Karlsplatz 13/206, 1040 Vienna, Austria

[4]Institute for Land and Water Management Research, Federal Agency for Water Management, Pollnbergstraße 1, 3252 Petzenkirchen, Austria

[5]Institute for Water Quality, Resource and Waste Management, TU Wien, Karlsplatz 13/226, 1040 Vienna, Austria

[6]Institute of Chemical Engineering, TU Wien, Gumpendorfer Straße 1a, 1060 Vienna, Austria

[7]Department for Geodesy and Geoinformation, TU Wien, Gußhausstraße 25-29/120, 1040 Vienna, Austria

[8]Interuniversity Cooperation Centre for Water & Health, TU Wien, Vienna, Austria

Correspondence to: G. Blöschl (bloeschl@hydro.tuwien.ac.a)

Abstract. Hydrological observatories bear a lot of resemblance to the more traditional research catchment concept, but tend to differ in providing more long-term facilities that transcend the lifetime of individual projects, are more strongly geared towards performing interdisciplinary research, and are often designed as networks to assist in performing collaborative science. This paper illustrates how the experimental and monitoring set-up of an observatory, the 66 ha Hydrological Open Air Laboratory (HOAL) in Petzenkirchen, Lower Austria, has been established in a way that allows meaningful hypothesis testing. The overarching science questions guided site selection, identification of dissertation topics and the base monitoring. The specific hypotheses guided the dedicated monitoring and sampling, individual experiments, and repeated experiments with controlled boundary conditions. The purpose of the HOAL is to advance the understanding of water-related flow and transport processes involving sediments, nutrients and microbes in small catchments. The HOAL catchment is ideally suited for this purpose, because it features a range of different runoff generation processes (surface runoff, springs, tile drains, wetlands), the nutrient inputs are known, and it is convenient from a logistic point of view as all instruments can be connected to the power grid and a high-speed glassfibre local area network (LAN). The multitude of runoff generation mechanisms in the catchment provides a genuine laboratory where hypotheses of flow and transport can be tested, either by controlled experiments or by contrasting sub-regions of different characteristics. This diversity also ensures that the HOAL is representative of a range of catchments around the world, and the specific process findings from the HOAL are applicable to a variety of agricultural catchment settings. The HOAL is operated jointly by the Vienna University of Technology and the Federal Agency for Water Management and takes advantage of the Vienna Doctoral Programme on Water Resource Systems funded by the Austrian Science Funds. The paper presents the science strategy of the set-up of the observatory, discusses the implementation of the HOAL, gives examples of the hypothesis testing and summarises the lessons learned. The paper concludes with an outlook on future developments.

1 Introduction

Understanding water-related flow and transport processes in catchments and their interactions with other environmental processes across space scales and timescales forms essential research issues in the context of environmental technology, planning and management. From a water quantity perspective, understanding runoff generation mechanisms is very important for better estimating floods that may occur in small catchments, in particular if one is interested in extrapolating from small to large floods (Merz and Blöschl, 2008). Water yield under different management options as well as land–atmosphere feedbacks are particularly relevant when addressing issues related to climate change. From chemical and sediment perspectives, understanding the relevant mechanisms is important in the context of land management practices that aim at reducing sediment production (e.g. Yeshaneh et al., 2015), and for water resource management where the interest resides in understanding the fate of nutrients and designing relevant management practices (Schilling et al., 2005; Zessner et al., 2005; Strauss and Klaghofer, 2006; Kovacs et al., 2012). From a human-health-related perspective, characterising microbial faecal hazards in water and identifying contamination sources contribute to more reliable hazard characterisation and risk estimation in the context of water safety management, for example by allowing target-oriented protection measures in the catchment and delineating effective and site-specific protection zones (Reischer et al., 2011; Farnleitner et al., 2011). While these research issues are relevant individually, they are also closely connected to each other through process interactions. Integrated research into these processes is therefore needed to shed light on the interactions and fully explore the causal relationships of the catchment system.

Experimental research addressing these issues differs from experiments in many other fields of science in at least two ways. First, the processes related to water flow in the landscape are strongly controlled by the forcing of the weather. It is therefore difficult, if not impossible, to conduct controlled experiments where one varies the boundary conditions in a prescribed way. As a consequence, the processes associated with water flow are intrinsically non-repeatable and require particular care when hypothesis testing (Blöschl et al., 2014). Second, the processes occur at the catchment scale (where much of the interesting process interactions occur) and may not be present at the small laboratory scale. As a result, the experimental set-up must be designed at the catchment scale which, again, involves a number of scientific and logistic challenges.

Experimental catchments have a long tradition in hydrology. Some corner stones include the Coweeta hydrologic laboratory (Southern Appalachians) in the early 1930s where the focus was on forest management practices (Swank and Crossley, 1988; Elliott and Vose, 2011), the Plynlimon catchment (Wales) in the late 1960s where pollution was the main interest (Kirby et al., 1991; Robinson et al., 2013), the Weiherbach (Germany) and Löhnersbach (Austria) catchments in the 1990s where a broader, interdisciplinary approach was taken (Plate and Zehe, 2008; Zehe et al., 2001; Kirnbauer et al., 2005); and the Tarrawarra catchment (Australia) in the 1990s where the focus was specifically on spatial process patterns (Western et al., 1998, 1999, 2001). An overview of some of the European experimental catchments is given in Schumann et al. (2010) and Holko et al. (2015).

More recently, the concept of *environmental observatories* has been developed and implemented. Examples are the Critical Zone Observatories (CZO) in the US where the starting point was geochemical processes (e.g. Anderson et al., 2008; Lin and Hopmans, 2011), and the Terrestrial Environmental Observatories (TERENO) in Germany where the starting point was processes at the hydrological–ecological interface (Zacharias et al., 2011). While these observatories bear a lot of resemblance to the more traditional research catchments, they differ in three important ways. (a) Similar to astronomical and meteorological observatories, their objective is to provide long-term facilities that transcend the lifetime of individual projects. (b) Even more so than their more traditional counterparts, they are geared towards performing interdisciplinary research. (c) Often they are designed as networks to assist in performing collaborative science within the research community. Indeed, long-term interdisciplinary research in networks may be the hallmark of catchment-scale experimental research in an era where "Humans may no longer be treated as boundary conditions but should be seen as an integral part of the coupled human-nature system... [and] the coupling between the geoscience disciplines ... gets more important." (Blöschl et al., 2015, p. 17).

Establishment of research catchments or hydrological observatories may be either driven by management questions as was the case with much of the early experimental work, or by fundamental research questions, and the two aims may feed into each other. In both instances, the experimental or monitoring set-up must be designed in a way that enables the critical research questions to be tested. The classical example are paired catchment studies (e.g. Brown et al., 2005) where the effects of forest management on the hydrological cycle are studied with a similar, untreated catchment used as a control. Differences in the observations between these two catchments are then used to test hypotheses on, e.g. the effects of forest on water yield. Again, a classical hypothesis to be tested by this set-up is that forest cutting will increase water yield from the catchment. In the Coweeta, for example, "the largest water yield increases occurred the first year after cutting when evapotranspiration (Et) was most reduced due to minimal leaf area index (LAI). As vegetation regrew, LAI and Et increased and streamflow declined logarithmically, until it returned to the pre-treatment level by five to six years." (Elliott and Vose, 2011; p. 906). For more complex hypotheses, the experimental or monitoring set-up must be more elaborate in order to allow the hypothesis testing in a meaningful way.

The purpose of this paper is to illustrate how the experimental or monitoring set-up of an observatory can be established in a way that allows meaningful hypothesis testing, and to communicate the lessons learned from the experiences with the Hydrology Open Air Laboratory (HOAL) in Petzenkirchen, Austria. We will first present the science strategy of the set-up of such an observatory, discuss the implementation of the HOAL, give examples of the hypothesis testing and summarise the lessons learned.

2 Science strategy of the HOAL

The success of a research programme hinges on whether new, cutting-edge scientific findings are achieved. The HOAL observatory is designed to facilitate cutting-edge research by providing long-term experimental infrastructure, fostering interdisciplinary collaboration and encouraging networking within the science community. All three aspects are considered through the prism of the hypotheses to be addressed.

2.1 Long-term experimental infrastructure

Some of the most interesting science questions require long-term observation. These include questions related to hydrological change where one aims at detecting differences of hydrological fluxes and/or processes between decades. Another such question relates to hydrological extremes, since the likelihood of observing extreme events increases with the observation period. At the same time, long-term infrastructure can most efficiently be used if a range of complementary research questions is addressed that all build on that infrastructure, i.e. where the synergies of different questions are exploited. To cater for a range of questions, a nested approach was therefore adopted for the HOAL related to overarching science questions and specific hypotheses (Fig. 1).

2.1.1 Overarching science questions

First, overarching science questions were identified that were relevant for advancing the fundamental understanding of water-related flow and transport processes at the catchment scale. These were defined in a broad way and included the following.

- What are the space–time patterns of flow paths and evaporation in a small agricultural catchment?

- What are the space–time patterns of erosion and sediment transport processes in the catchment and what are their driving forces?

- What are the processes controlling nutrient and faecal pollution dynamics in the catchment?

These questions are aligned with the interests of the individuals and institutions involved in the context of prior experience, societal relevance and funding opportunities. The site location, the research student dissertation topics, and the base monitoring were selected based on the overarching questions.

Site location: selection of the site was guided by the ability to address the overarching science questions. Importantly, much of the research is related to runoff generation. It was therefore deemed important to select an area with many different runoff generation mechanisms in the same catchment to make the scientific findings applicable to a wide spectrum of catchments around the world. Questions such as erosion and nutrient dynamics are usually associated with agricultural practices, which was another criterion for selecting the site.

Dissertation topics: the topics of the dissertations were chosen in a way that a number of generations of research students can build on each other. The topics of the first generation students (2009–2013) were geared towards the more fundamental processes of water and matter flow in the catchment as well as soil moisture. The second generation (2012–2016) had more elaborate topics such as microbial processes, land–atmosphere interactions and linkages to the deep subsurface. The third generation of students will, again, build on these findings and address upscaling and hydrological change more explicitly. The fourth generation of students will be concerned with how all of these findings can be generalised to other climatic and management conditions around the world.

Base monitoring: all overarching research questions require an understanding of the hydrological fluxes with high spatial and temporal detail. Consequently, a substantial number of high-resolution raingauges and stream gauges were chosen as the base monitoring set-up. Locations of runoff-related measurements were carefully considered to sample different runoff mechanisms. At the catchment outlet, basic chemical and physical parameters were monitored by online sensors and regular grab sampling. To complement these, a weather station was set up to monitor the energy fluxes at the land–atmosphere interface. Spatial sampling to characterise the catchment included Lidar for high-definition topography, soil mapping and sampling.

2.1.2 Specific hypotheses

Nested into the overarching science questions, specific hypotheses were defined, dedicated monitoring and sampling was performed, and individual experiments were conducted, some of which were repeated with controlled boundary conditions.

Dedicated monitoring and sampling: a soil moisture network within the catchment was set up to understand the spatial soil moisture distribution and link it to remotely sensed soil moisture. Three eddy-correlation stations were set up to

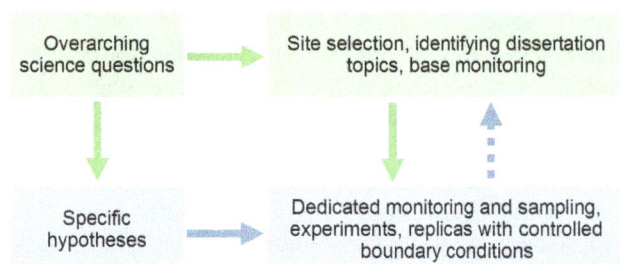

Figure 1. Interplay of hypotheses and experimental planning in the HOAL.

understand the spatial distribution of land–atmosphere interactions. Faecal indicators were monitored to test alternative measurement methods and understand the dynamics of faecal contamination, and water quality characteristics were monitored at a number of locations to understand nutrient fluxes (Exner-Kittridge et al., 2013).

Individual experiments: field campaigns were conducted over limited periods of time to obtain more in-depth understanding of the processes at the field scale. Examples include tracer tests in the stream to elucidate stream aquifer interactions and a field campaign dedicated to measuring transpiration and bare soil evaporation separately in a field of maize.

Repeated experiments with controlled boundary conditions: a small number of experiments were conducted with controlled boundary conditions. Examples include resuspension experiments were sediment-free water was pumped into the stream to understand the sources of suspended sediments at the beginning of events (Eder et al., 2014) and an experiment where soil plots were prepared to a prescribed roughness and moisture, which were then measured by Lidar to understand the controls on Lidar response.

New instruments and new data transmission technologies are of particular interest in the HOAL, as detailed in Sect. 3.2.2 of this paper. More detailed examples of how instrumentation and experimental set-up were selected on the basis of the specific hypotheses are given in Sect. 4.

2.2 Interdisciplinary collaboration

One of the hallmarks of an observatory is its ability for fostering cooperation across the disciplinary boundaries. In the case of the HOAL much of the research is conducted within the frame of the Vienna Doctoral Programme on Water Resource Systems (Blöschl, et al., 2012). The programme is funded by the Austrian Science Funds and aims at producing top graduates capable of conducting advanced, independent research of the highest international standards which cuts across multiple disciplines. The HOAL is therefore a natural platform for the Programme and benefits from its integration strategy. The Programme enables integration between disciplines that ensures that students can address more complex science questions than is possible through individ-

ual dissertations. The main strategy for achieving this consists of organising the research through joint groups, joint research questions, and joint study sites. One of the joint study sites is the HOAL.

As an example, the concept of integration between the research of the nine doctoral students currently working in the HOAL is illustrated for one of the overarching science questions, i.e. "Space time patterns of flow paths and evaporation". Atmospheric scientist Patrick Hogan is investigating the soil moisture and land use controls on spatial evaporation patterns within the catchment. One specific hypothesis Patrick Hogan is testing is that the relative importance of soil moisture controls exceeds that of topographic controls at all times of the year. As evaporation is an important flux in the HOAL it will directly affect soil moisture (of interest to remote sensing specialist Mariette Vreugdenhil) and indirectly affect the flow paths (of interest to hydrogeologist Michael Exner-Kittridge who deals with nutrient fluxes). Structural engineer Abbas Kazemi Amiri is taking advantage of the eddy-correlation systems and conducts measurements of the dynamic wind loading of the mast structure to understand the interactions of water resource structures with wind, and specifically the role of fatigue. Conversely, Patrick Hogan can make use of the expertise and research progress of other students by testing the spatial distribution of evaporation obtained by his eddy-correlation instrumentation against observed runoff volumes in different parts of the catchment. Hydrologist Rasmiaditya Silasari's thesis quantifies the spatial organization of the flow patterns. One specific hypothesis she is testing is that spatial connectivity is a major determinant of the flow rates and flow dynamics. The numerical hydrological simulations she conducts for testing her hypotheses are directly relevant to Mariette Vreugdenhil for interpreting spatial soil moisture.

2.3 Networking within the science community and beyond

Another key characteristic of observatories is that they are embedded into a network of scientists to maximise the opportunities of producing novel and societally relevant research. Networking of the HOAL has therefore been designed at a number of levels.

The TU Wien – IKT collaboration: at the centre of the HOAL stands the collaboration between a number of institutes and centres of the Vienna University of Technology (TU Wien) and the Institute for Land and Water Management Research (IKT) of the Federal Agency for Water Management. The expertise of a number of TU Wien institutes is brought together through their affiliation with the Centre for Water Resource Systems at TU Wien, involving professors from structural mechanics, remote sensing, hydrology, hydrogeology and water quality. Each institute operates their own in-house laboratories in their area of specialisation. In addition, the IKT has a long standing expertise in measuring and modelling soil water, sediments and nutrients with a focus on field work. They have operated experimental sites for decades and also operate a physical and chemical soil laboratory and workshop.

Collaborations with instrument companies: a second level of networking and collaboration takes place with some of the providers of the instrumentation. Although most of the instrumentation has simply been purchased from the vendors, for a number of providers a joint venture has been embarked upon to test new instrumentation and methods. One such collaboration is with the Microtronics company regarding telemetering data from the catchment to the central server and data management. Another is with the VWM (Vienna Water Monitoring) company regarding testing novel devices for automated measurements of a proxy parameter of microbial faecal pollution in the stream of the HOAL under field conditions (Farnleitner et al., 2002; Ryzinska-Paier et al., 2014).

Collaborations with other research institutions: a range of collaborations with both national and international research institutes and agencies are under way, most of which focus on testing a particular hypothesis. A collaboration with the Austrian Institute of Technology (AIT) focuses on stable isotope analyses to understand water age, a collaboration with the International Atomic Energy Agency (IAEA) is geared to testing a cosmic ray soil moisture sensor against the soil moisture network, and a collaboration with the Helmholtz Centre for Environmental Research (UFZ) deals with understanding water isotopic signatures in a regional context. HOAL is one of the ground truthing sites of the NASA's SMAP (soil moisture active passive satellite) mission. Collaborations with additional institutes are being planned. The doctoral students working in the HOAL are entitled to spend a semester abroad with a research institution of their choice. This provides further opportunity to knit a strong network of collaborations with leading groups around the world in their field of expertise.

Communication and outreach: visibility of the research output hinges on suitable dissemination of the research results at a range of scales. Dissemination has therefore been designed as a multi-scale process involving the university (e.g. workshops and seminars within the university, email and website communication), the national and international scientific communities (through journal papers, conference presentations, and a guest scientist programme) and the general public through a range of outreach activities (e.g. newspaper, television and radio interviews with scientists working in the HOAL, as well as regular meetings with the local community).

3 Implementation

3.1 Site selection and hydrological characteristics

3.1.1 Site selection

Since many of the questions are related to runoff generation it was considered important to select an area with many different runoff generation mechanisms in the same catchment. Also, as the interest was on experimental hydrology, a catchment scale of a square kilometre or less was envisaged. A small catchment near Petzenkirchen, Lower Austria, was found to be ideally suited. In this catchment a wide range of runoff generation mechanisms occurs, including infiltration excess overland flow, re-infiltration of overland flow, saturation excess runoff from wetlands, tile drainage flow, shallow aquifer seepage flow and groundwater discharge from springs. The multitude of runoff generation mechanisms in the catchment provides a genuine laboratory where hypotheses of flow and transport can be tested, either by controlled experiments or by contrasting sub-regions of contrasting characteristics. This diversity also ensures that the HOAL is representative of a range of catchments around the world and the specific process findings from the HOAL are applicable to a variety of agricultural catchment settings.

As many of the overarching science questions are related to erosion and nutrients, it was considered an advantage that most of the catchment is used for agricultural purposes where sediment and nutrient fluxes tend to be bigger than for forested or urban settings. The crops include winter wheat and maize, which allows examination of the effect of different crops on the hydrological processes. Manure and fertiliser application are accurately known from farmers' bookkeeping, which is useful for estimating nutrient and faecal pollution inputs. Part of the catchment is pasture and part of it is forested, which opens up more comparative research opportunities.

The catchment selected also had other, more practical, advantages over other catchments. Importantly, it is very convenient from a logistic point of view. It is located within walking distance of the premises of the Institute for Land and Water Management Research, which greatly facilitates the day-to-day maintenance of the instruments and experimental set-ups. Because of the proximity to the institute, the instruments can be connected to the power grid which, again, has major advantages as it avoids battery failures – a frequent cause of data loss. Finally, the instruments can be connected to a high-speed glassfibre local area network which is very useful for data management and remote monitoring of the functioning of the instruments and the short-term planning of experiments. Alternative potential site locations such as the Löhnersbach, a previous research catchment of the TU Wien (Kirnbauer et al., 2005), while interesting hydrologically, did not meet the criteria of logistic convenience.

An additional bonus for the selection of the site is that runoff measurements at the catchment outlet started in 1945 (Blümel und Klaghofer, 1977; Turpin et al., 2006; Strauss et al., 2007), which helps put the recent observations into a longer-term context.

3.1.2   Catchment description

The Petzenkirchen HOAL (Hydrology Open Air Laboratory) catchment is situated in the western part of Lower Austria (48°9′ N, 15°9′ E) (Fig. 2). The catchment area at the outlet (termed MW) is 66 ha. The elevation of the catchment ranges from 268 to 323 m a.s.l. with a mean slope of 8 %. At present, 87 % of the catchment area is arable land, 5 % is used as pasture, 6 % is forested and 2 % is paved. The crops are mainly winter wheat and maize.

The climate can be characterised as humid with a mean annual temperature of 9.5 °C and a mean annual precipitation of 823 mm yr^{-1} from 1990 to 2014. Precipitation tends to be higher in summer than in winter (Fig. 3, Appendix A). Crop evapotranspiration (ET$_c$) estimated by the FAO (1998) method using local climate data and crop growth information for this period was 471 mm yr^{-1}. Annual evapotranspiration estimated by the water balance ranged from 435 to 841 with a mean of 628 mm yr^{-1} (1990–2014) (assuming deep percolation is negligible). The natural surface water outlet of the catchment is known as the Seitengraben stream. Mean annual flow from the catchment in this stream is 4.1 L s^{-1} (or 195 mm yr^{-1}) (1990–2014). Mean flows tend to peak in the spring (Fig. 3). The largest flood events on record occurred in 1949 and 2002 with estimated peak discharges of 2.8 and 2.0 m^3 s^{-1}, respectively. The highest discharge in recent times occurred in summer 2013 with 0.66 m^3 s^{-1}. The subsurface consists of Tertiary sediments of the Molasse zone and fractured siltstone. The dominant soil types are Cambisols and Planosols with medium to poor infiltration capacities. Gleysols occur close to the stream (Fig. 4).

The HOAL is special in that many runoff generation mechanisms can be observed simultaneously in different parts of the catchment (Fig. 5). Due to shallow, low permeable soils and the use of the catchment area as agricultural land, the concave part of the catchment was tile drained in the 1940s in an effort to reduce water logging. The estimated drainage area from the tile drains is about 15 % of the total catchment and can be divided into two bigger systems in the south-western part of the catchment and four smaller drainage systems in the north-eastern part. The pipes drain into the main stream at four locations. Two tile drain systems (Sys1, Sys2) do not dry out during the year, while two are ephemeral (Frau1, Frau2) (see Fig. 7). The uppermost 25 % of the stream was piped in the 1940s to enlarge the agricultural production area. The pipe enters the main stream at inlet Sys4. Its flow dynamics and chemistry are similar to those of the permanent tile drains as it drains the surrounding soil.

There are two clearly visible springs that directly discharge into the stream. These are Q1 and K1. The water from Q1 originates from a fractured siltstone aquifer with distinct hydrologic and chemical characteristic from those of other point sources along the stream. The hydrologic dynamics and chemical characteristics of K1 are more similar to the perennial tile drainages. Q1 is perennial, while K1 is not.

In the south-eastern part of the catchment is a small wetland close to the stream which permanently seeps into the stream via two rivulets (A1, A2). The wetland is fed by springs at the upper part of the wetland and usually responds very quickly to all types of rainfall due to its high saturation state.

During low-intensity events in summer, the flow in the main stream responds to rainfall with substantial delay as the soil usually offers a lot of storage capacity, depending on soil moisture. A mixture of tile drainage water, diffusive inflow from the shallow aquifer, spring water, and surface water from the wetland tends to feed the stream. During major storms, saturation overland flow occurs across the fields (mainly in the depression areas along the talweg and close to the stream) which enters the stream at two (E1, E2) or three locations, depending on the magnitude of the event. The overland flow causes gully erosion.

During high-intensity thunderstorms in summer and spring, infiltration excess overland flow tends to occur with a very substantial, fast contribution from the tile drainage system. During infiltration excess overland flow events, all forms of erosion from interrill to gully erosion may occur on the fields that are poorly covered by crops (such as bare soil after soil management), but sedimentation immediately occurs when the sediment laden water enters a field with better cover (such as wheat). During very dry periods in summer, the high clay contents will cause shrinking cracks which act as macropores for re-infiltration during subsequent events.

In winter, rain-on-snow runoff may occur as saturation overland flow during large events leading to gully erosion. In fact, this is when most of the overland flow occurs during the year. However, the main runoff generation mechanism in winter is through lateral subsurface pathways (shallow subsurface preferential flow paths, drainage pipes). Even minor events (of, say, 5 mm) will lead to a significant increase in streamflow due to high soil moisture during the winter. After freezing periods, when the soil is still frozen, infiltration excess overland flow may occur.

3.2   Setting up the HOAL and instrumentation

Setting up the base monitoring and the dedicated monitoring and sampling was guided by the overarching science questions and the specific hypotheses.

Figure 2. View of the Petzenkirchen HOAL catchment looking south (trees in the centre of the photo constitute the riparian zone of the Seitengraben stream).

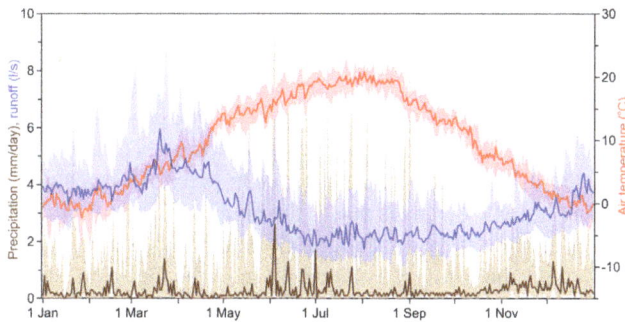

Figure 3. Precipitation and air temperature at the weather station, and runoff at the catchment outlet (MW) of the HOAL. Lines show medians of the period 1990–2014, shaded areas the 25 and 75 % percentiles based on the data aggregated to daily values.

3.2.1 Basic infrastructure and monitoring

Planning of the HOAL started in 2008. In September 2009 the Vienna Doctoral Programme on Water Resource Systems started and the financial resources for the base instrumentation were made available through the TU Wien. In line with the overarching science questions, the instrumentation was designed for a high spatial and temporal resolution which involves substantial power consumption. Consequently, a mains cable was run from the nearest connector a few hundred meters outside the catchment along the stream to the weather station to enable 380 V electric power supply to the instruments. To facilitate maintenance of the instruments, data storage and the short-term planning of experiments, a high-speed glassfibre cable was run from the premises of the Institute for Land and Water Management Research into the HOAL to provide a local area network (LAN) for data transmission. The glassfibre network allows fast streaming of the data and is less susceptible to damage due to lightening than electrical transmission lines. Subsequently, a range of instruments was installed as the basic monitoring set-up to measure dynamic data. All are operated at a temporal resolution of 1 min with the exception of the

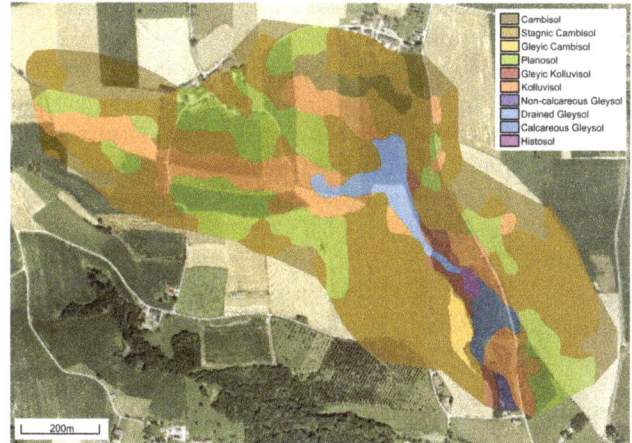

Figure 4. Soil types in the HOAL.

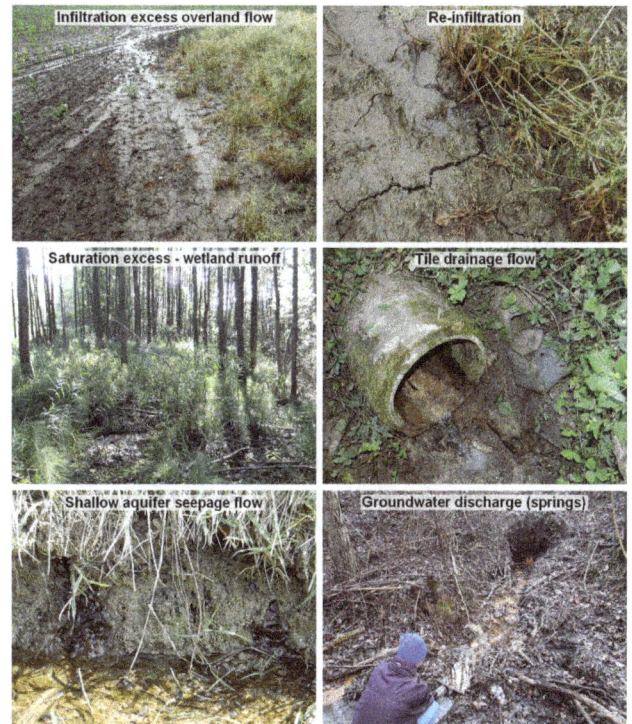

Figure 5. Runoff generation mechanisms in the HOAL.

piezometers, where groundwater levels are recorded at temporal resolutions of 5 to 30 min.

Atmospheric processes: four raingauges were installed to monitor spatial rainfall patterns which were strategically placed to cover spatial rainfall patterns well.

Atmospheric and soil processes: monitoring at the weather station located approximately in the centre of gravity of the catchment includes air temperature, air humidity, wind speed and direction (all at three heights), incoming and outgoing solar and long-wave radiation, wind load on the construction, raindrop size distribution, snow depths, soil heat flux and soil temperatures at different depths.

Surface water: a total of 12 flumes were installed within the catchment to monitor discharge at 1 min resolution from the inlet piped stream, tile drains, erosion gullies, springs and tributaries from wetlands. These flumes are the backbone of the HOAL. All flumes were calibrated in the Hydraulic Laboratory of the TU Wien to obtain a reliable stage–discharge relationship.

Surface water: at the catchment outlet, the existing H-flume (dating from 1945 with a number of changes since) was upgraded in 2009. The maximum discharge capacity was increased and a number of additional sensors were installed including water temperature, electrical conductivity, turbidity (two probes from different makes), chloride, pH, and nitrate. Grab samples are taken weekly for a range of chemical analyses including suspended solids and various compounds of nutrients. Additionally, autosamplers take water samples during events. A video camera was installed to monitor the water level in the flume and the functioning of the instruments.

Groundwater: 23 piezometers were installed within the catchment where groundwater level and water temperature are monitored. Most of the piezometers are located along transects perpendicular to the stream to help understand stream–aquifer interactions. Two additional air pressure sensors were needed to correct the readings of the pressure transducers for the air pressure fluctuations.

Table 1 and Appendix B give more details of the instruments. All of these instruments are connected to data loggers (some of them through interfaces) where the data are stored temporarily. Most of the data loggers are then directly connected to a computer at IKT through the glassfibre LAN. These include the loggers of the discharge pressure transducers, the turbidity measurements, the water chemical parameters and the instruments at the weather station. The raingauges are connected through a GSM (mobile phone) module. The data of the piezometers and the movable eddy-correlation stations are stored locally and read out manually at regular intervals.

To complement the base monitoring of the dynamics of the hydrological flow and transport processes at specific locations, a number of spatial surveys were conducted after setting up the HOAL, which included a Lidar survey, aerial photographs, soil mapping and sampling, and collection of agricultural data (Table 2). Further details are given in Appendix B, Figs. 6 and 7.

3.2.2 Dedicated monitoring and experiments

Dedicated monitoring and experiments were more specifically geared towards the testing of individual hypotheses (see Sect. 4 for examples) and involve new instruments and new data transmission technologies in addition to proven technology. Three eddy-correlation stations were set up in 2012 and 2013 to understand the spatial distribution of land–atmosphere interactions. As evaporation is an important flux in the HOAL, it will directly affect soil moisture and flow paths of interest to other HOAL research questions. One set of instruments has been set up at the weather station location, using a closed path device. Additionally, two mobile stations are deployed (using open path devices) based on a site rotation plan to optimise the locations for each sensor relating to the factors of interest: topography, soil type and moisture and vegetation. The data are processed offline using dedicated software (Mauder and Foken, 2011) to provide 30 min values for the sensible, latent heat and CO_2 fluxes. Soil heat flux and net radiation sensors are also installed to complete the energy balance. Scintillometer measurements of aggregated fluxes over a line of about 150 m are made for comparison to obtain momentum flux, sensible heat flux and information on the turbulent parameters of the air. Acceleration sensors (accelerometers) are installed on the guyed mast of the weather station to evaluate the fatigue of water related structures caused by the fluctuating components of wind. For the elements of steel structures (such as poles of water supply towers) fatigue damages due to the high cyclic wind-induced vibration are a relevant failure mechanism. Another step is to identify the wind loads inversely from the measured structural response and correlate them to the wind statistics from eddy-correlation measurements. The wind load identification follows the general lines of the experiments already accomplished at the TU Wien laboratory of structural model dynamics (Kazemi and Bucher, 2015).

A soil moisture monitoring network was set up to understand the effect of small-scale variability of landscape characteristics on the microwave response of satellite sensors (see Sect. 4.3). Since soil moisture is such a key parameter, a better understanding of its space–time patterns will also be useful to other processes in the HOAL including evapotranspiration and runoff generation. The network uses wireless transmission technology based on the ZigBee protocol (Bogena et al., 2010). Time domain transmission sensors were installed at four depths below ground surface. One of the difficulties with measuring soil moisture in agricultural catchments is that it is not possible to install the sensors permanently in the field. As a consequence, 20 stations were installed permanently on pasture, while 11 stations are temporary, which are removed and replaced once or twice a year in accordance with the agricultural manipulations on the fields. Monitoring of saturation patterns within the catchment is conducted using a video camera to understand the space–time patterns and connectivity of surface flow following the pattern comparison paradigm of Blöschl et al. (1991) and Parajka et al. (2012). This is complementary to the soil moisture network by providing visual observation with better spatial resolution. Pictures of overland flow generation provide a visual assessment of saturation patterns and are useful for validating distributed models of surface runoff (Grayson et al., 2002; Horvath et al., 2015).

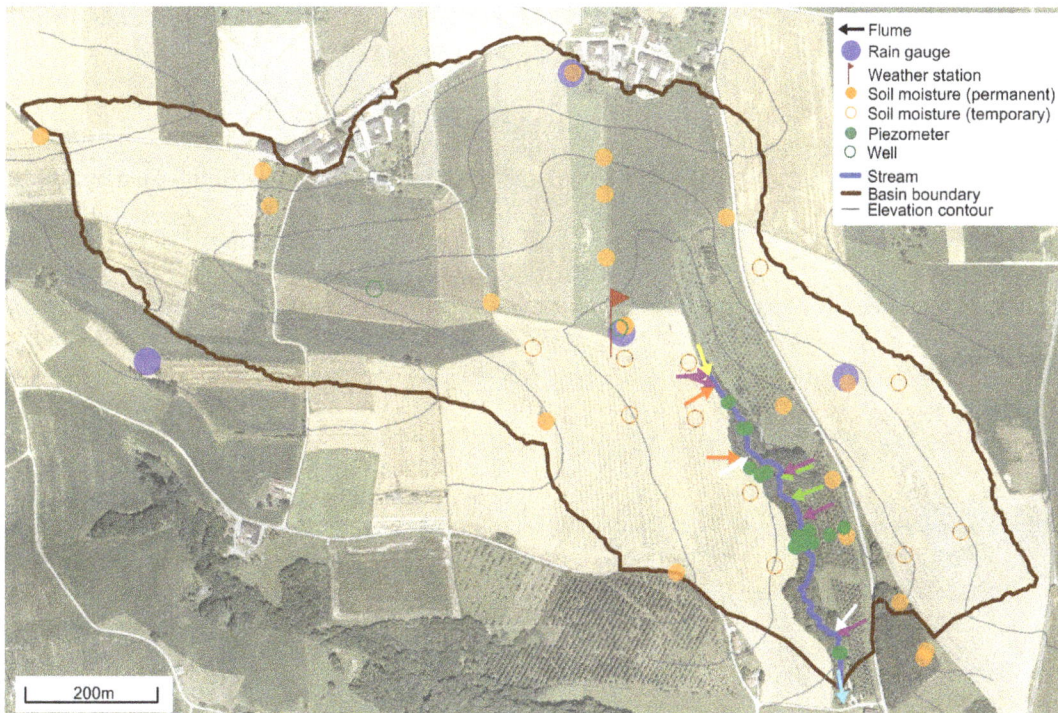

Figure 6. Instrumentation in the HOAL catchment (see Tables 1 and B1).

To understand the dynamics of nutrients such as phosphorus and nitrogen and their compounds, a water quality monitoring network was installed at the tributaries within the HOAL to complement the base monitoring at the catchment outlet (see Sect. 4.2). Flumes for overland flow from the fields (erosion gullies), the wetland, tile drains and the springs were equipped with in situ sensors for electrical conductivity, temperature, turbidity, pH, nitrate and chloride sampling at 1 min intervals. While the high sediment concentrations in the HOAL facilitated the sediment process analyses, they turned out to be a challenge for monitoring the water quality parameters, as the stilling wells in which sensors are usually placed tended to silt up quickly. A new device was developed, termed the Water Monitoring Enclosure (WME), which allows in situ monitoring of water quality parameters for highly dynamic, sediment-laden streams (Exner-Kittridge et al., 2013). The WME ensures a minimum internal water level which keeps the monitoring equipment submerged even when there is no flow into the enclosure. Four WME and six autosamplers were installed throughout the catchment for event sampling. Grab sampling is performed monthly at the tributaries, in addition to the weekly sampling at the catchment outlet, and analysed for a range of parameters including stable isotopes. Four enzymatic analysers were set up at the catchment outlet to understand the dynamics and pathways of faecal pollution and to test the instruments for real-time surface water monitoring (see Sect. 4.4). The devices sample stream water at intervals of 1 or 3 h. The results from the devices are compared for different set-ups with laboratory analyses of water samples to understand the strengths and limitations of the instruments in an on-line mode, and interpreted in the context of a range of physical and chemical parameters for events with contrasting characteristics (e.g. fast and short response times, dry and wet antecedent soil moisture) to shed light on the processes of microbial pollution.

Sediment monitoring and experimentation were conducted to understand the sources and pathways of sediments within the catchment (see Sect. 4.1). Turbidity is monitored at both erosion gullies along with autosamplers to be able to calibrate the sediment–turbidity relationships for each event separately (Eder et al., 2010). Further autosamplers are located at the inlet of the piped stream and some of the tile drains on the right bank to investigate subsurface sediment transport. Aerial photographs are taken to identify erosion patterns and calculate eroded soil volumes after erosive rainfall events. To understand sediment deposition and resuspension in the stream, flushing experiments were conducted, where sediment-free water was pumped into the source of the stream and flow rates, and sediment and solute concentrations as well as grain size distributions were measured (Eder et al., 2014).

To explore the stream–aquifer interactions, several stream tracer tests were performed in the main stream. One set of tracer experiments was performed during winter baseflow conditions (where evaporation can be assumed to be negligible). Bromide was injected as a tracer and bromide con-

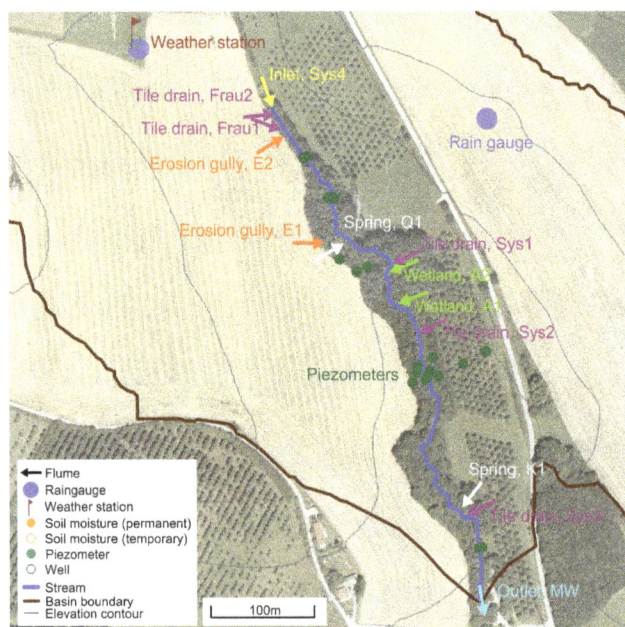

Figure 7. Detail of instrumentation in the HOAL catchment (see Tables 1 and B1).

centrations and flow were measured for five locations along the stream. This allowed the estimation of stream bank fluxes (Exner-Kittridge et al., 2014). An infrared camera was used to identify hotspots of groundwater recharge into the stream. Mass balances over sections of the stream were used to determine the role of near-stream riparian trees in the daily fluctuations of the stream flow during low-flow conditions. A number of geophysical surveys were conducted to improve the delineation of hydrogeological heterogeneities and processes in the subsurface. Initially, a series of measurements using ground-penetrating radar were performed in 2010 for the characterisation of drainage pipes. Imaging surveys are being started with the induced polarization method for delineating the aquifer geometry and hydrogeological structures such as preferential flow paths. Application of the spectral-induced polarization method at different frequencies is planned to gain information on hydraulic conductivity and changes in the subsurface associated with microbial activity (e.g. Flores Orozco et al., 2011, 2013, 2015). Low-induction number electromagnetic induction methods will permit the collection of data at extensive areas with reduced acquisition times. Tables B1 and B2 give details on the instrumentation and the associated laboratory analyses.

3.3 Managing the HOAL

Meaningful hypothesis testing in an observatory not only requires careful planning of installation of instruments and conducting the monitoring and the experimentation, but also coordination of the research between the groups involved,

maintenance of the instruments, dealing with landowners, and data management.

3.3.1 Coordination of research

One of the main strengths of this kind of observatory comes from the synergies between a critical group of people conducting related research. In the HOAL, currently more than 20 researchers are involved plus support staff. Nine dissertation projects focussed on the HOAL are being conducted. While observatories sometimes adopt a top-down approach where the individual research activities are subsidiary to the main goal, a slightly different approach has been adopted in the HOAL. A general master plan for the research to be conducted was defined as the overarching sciences questions. These were specified in the research proposals of the Doctoral Programme on Water Resource Systems that were submitted to the Austrian Science Funds (Blöschl et al., 2012). The research proposals also included more specific hypotheses. When actually implementing the research, the individual doctoral students were given considerable freedom in specifying their own hypotheses and their experimental/monitoring set-ups. This then led to an iterative network structure of the interactions between the research of the students. Figure 8 illustrates the general concept of implementation. For each hypothesis, the individual steps of implementation consisted of (i) planning of the dedicated monitoring and experiments, (ii) conducting monitoring and experiments, (iii) data analysis and hypothesis testing, and (iv) research write up. Depending on the outcomes of the experiments, these steps would be repeated in an iterative way. At the same time other hypotheses are tested in the HOAL (by the same or other students). These interact, as indicated by the double arrows in Fig. 8. The interactions occur at all four steps of the research, from the planning to the write up. The main advantages of this iterative, network-based process of conducting hypothesis testing are its flexibility and the encouragement of creative thinking by the students.

The Doctoral Programme on Water Resource Systems is an ideal setting for this exchange as it is specifically geared towards fostering collaboration between students, including from different disciplines. As part of the doctoral programme, each student is encouraged to develop collaborations through joint supervision (each student has two supervisors), regular research cluster meetings focusing on research themes, and annual and 6-monthly symposia that bring all research students and supervisors together for 1 or 2 days for research presentations, posters and discussion sessions.

3.3.2 Maintenance of instruments

The overall responsibility of coordinating the maintenance of the instruments lies with the HOAL manager who draws the maintenance plans and coordinates or supports any repairs and replacements. The manager also coordinates the instal-

Table 1. Instrumentation in the HOAL (most of which has a 1 min time resolution). Most data are transmitted to the server at the institute by glassfibre cable. For details, see Appendix B; for locations, see Figs. 6 and 7.

Compartment	Variables	Locations	Number of stations	Basic/ dedicated
Atmosphere	Precipitation intensity	Within (or close to) catchment	4	B
Atmosphere	Air temperature, humidity, wind speed and direction (three heights); atmospheric pressure, incoming and outgoing short-wave and long-wave radiation, raindrop size distribution	Weather station	1	B
Atmosphere	Carbon dioxide flux, latent heat flux, sensible heat flux, momentum flux (eddy correlation)	Weather station and other locations	3	D
Atmosphere	Sensible heat flux (scintillometer)	Within catchment	1	D
Atmosphere	Wind load	Weather station	1	D
Ground surface	Snow depth	Weather station	1	B
Ground surface	Saturation patterns (photos, video)	Within catchment	1	B
Surface water	Discharge, electric conductivity, temperature, pH, chloride, nitrate	Inlet: piped stream (Sys4)	1	B/D
Surface water	Discharge; partly electric conductivity, temperature, pH, chloride, nitrate	Tile drains (Frau1, Frau2, Sys1, Sys2, Sys3)	5	B/D
Surface water	Discharge, turbidity	Erosion gullies (E1, E2)	2	B
Surface water	Discharge	Springs (Q1, K1)	2	B
Surface water	Discharge; partly electric conductivity, temperature, pH, chloride, nitrate	Wetland runoff (A1, A2)	2	B/D
Surface water	Discharge, electric conductivity, temperature, turbidity, pH, chloride, nitrate, enzymatic activity, UV-Vis fingerprint, video images	Catchment outlet (MW)	1	B/D
Surface water	Runoff water samples (automatic samplers, 24 bottles each, event triggered)	Inlet (Sys4), tile drain (Frau2), erosion gullies (E1, E2), catchment outlet (MW)	6	B/D
Soil	Soil heat flux, soil temperatures	Weather station	1, 2	B
Soil	Soil moisture, soil temperature (in four depths, partly five depths)	Within catchment	31	D
Soil	Soil moisture (cosmic ray)	Weather station	1	D
Groundwater	Groundwater level, temperature, partly air pressure	Within catchment	24	B

lation of new instrumentation and the set-up of experiments. An important part of the maintenance work relates to the base monitoring, and in particular the cleaning of the H-flumes at the stream tributaries. Some of the water quality sensors need regular cleaning to avoid biofilm formation and calcification. The sensors on the weather station are checked regularly for

level position and cleanliness. The soil moisture sensors and the other sensors that are not connected to the power grid are checked regularly for power supply (change of batteries, cutting out grass to prevent solar panels from being overgrown). A regular schedule of checking the instrumentation is operated. In these tasks, the HOAL manager is assisted by a

Table 2. Spatial surveys of catchment characteristics. For details see Appendix B.

Survey	Variables	Spatial resolution	Date of survey	Basic/ dedicated
Lidar	Digital elevation model	0.5 m	March 2010	B
Soil mapping through auger holes	Soil type	50 m grid	Spring 2010	B
Soil sampling by profiles	Soil horizons, photos, colour, texture, organic carbon, anorganic carbon, plant available phosphorus and potassium and pH of each soil horizon	50 m grid	Summer 2010	B
Soil sampling by profiles	Saturated and unsaturated hydraulic conductivity, pF and bulk density of each soil horizon	50 m grid	Summer 2012– ongoing	B
Geophysics	Georadar profiles	Four profiles	August 2010	D
Geophysics	Seismic profiles	Seven profiles	March 2011	D
Soil moisture survey	Soil moisture	100 pts	Spring 2014	D
Aerial photographs from powered paraglider	Digital surface model, surface roughness, soil loss volumes, erosion patterns	Depending on flight height	bimonthly	D
Agricultural data interviews with farmers	Crops, cultivation period, seeding, fertilization, plant protection, harvest times, harvested biomass, fertiliser and manure application	By field	Annually	B
Water withdrawal interviews with farmers	Water withdrawal from wells	Two wells in catchment	Annually	B

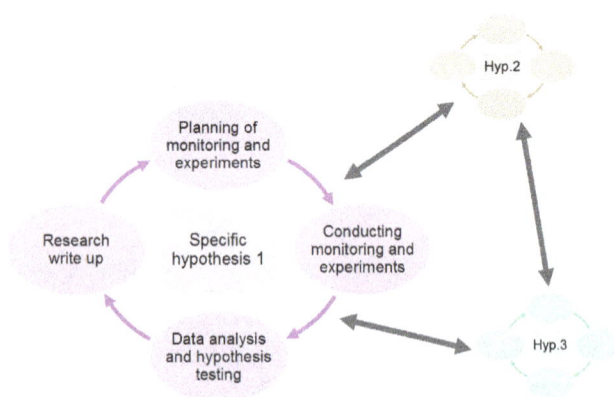

Figure 8. Network-based coordination of hypothesis-guided research.

number of local technicians with diverse expertise, including electronics.

Generally, each student is responsible for the proper set-up and operation of any dedicated monitoring and experimentation for their PhD research. There is, again, a set maintenance schedule. Maintenance and regular checking of the stations are coordinated with the HOAL manager and carried out by the students and the local technicians.

One of the main advantages of the HOAL is its location within walking distance of the premises of the Institute for Land and Water Management Research, which vastly facilitates the day-to-day maintenance of the instruments and experimental set-ups. Both the HOAL manager and the local technicians are based at the institute. Heavy rainfall events can be observed live and reference measurements can be taken during events. The operation of the auto samplers can be checked during events, to maximise the number of water samples from an event. After events associated with lightening the entire system is checked for operation (e.g. power outages).

To facilitate the exchange of information between the team members, a web-logbook has been specifically created for the HOAL. All activities within the HOAL are entered into the logbook including installation and maintenance of instrumentation, all sampling and surveying activities, and any other activities that are relevant to the operation of the HOAL. The web-logbook is a web application that allows access anywhere anytime by simply using a web browser. The main advantage of the logbook is that it sets a minimum standard protocol for all the information relevant to operating the HOAL and its easy, instantaneous accessibility to all team members. The logbook is often accessed in the field during manual measurements. The software also features user management, search and import/export facilities.

3.3.3 Landowners

Observatories in most other geoscience disciplines, such as astronomy and meteorology, require relatively modest space on the land. Typically, the land is purchased by the operators of the observatory. In contrast, hydrology is about water and matter fluxes at the landscape scale, so the requirements regarding space are invariably more extensive, and purchasing the entire catchment of interest is rarely an option. Arrangements have therefore to be made between the operators of the observatories and the landowners. The arrangements in the HOAL involve

- permissions to use the land, and

- information on agricultural management practices.

Permissions to use land are needed for the permanent instrumentation (such as the weather station) as well as to access the fields for sampling and for the temporary sensors of the soil moisture network. Information on agricultural management practices is particularly important for estimating nutrient input and it is also very relevant for estimating other fluxes such as transpiration. In a number of instances specific tillage practices are part of hillslope experiments.

Agreements have been drawn up between the HOAL management and the landowners to make arrangements for both aspects. About half the land is privately owned by a total of nine farmers. The remaining land is state owned and managed by the Austrian agricultural research agency, which facilitates the collaboration with the HOAL team due to similar objectives. A small fee is paid as part of the agreement but, more importantly, a good working relationship is always sought. Any maintenance or experimentation activities in the field are planned in agreement with the landowners, in order to avoid obstructions of the daily agricultural routines. The HOAL manager makes an effort to introduce the doctoral students and their research to the farmers on site, e.g. when they meet by chance during sensor installations or field work. The farmers are given access to the weather data, which is generally appreciated. They also get Christmas presents and there is an annual open day where the students explain their recent research to the local community. The main source of income of the farms in the catchment is crop production for pig fattening. Fertiliser costs and fertiliser leaching as well as the problem of soil loss by erosion are important tasks farmers have to deal with. This makes them additionally interested in the research and the cooperation.

3.3.4 Data management

As indicated in Appendix B, most sensors are connected to a computer (IKT server, HOAL PC, Soil Net PC) at the IKT via a fast glassfibre LAN. A database, known as Mydatanet, is run on the IKT server and hosts most of the data. Mydatanet imports the data at 1 min intervals from the data loggers along the stream (discharge and water chemistry parameters) and the raingauges. Mydatanet features online access and a web-based graphical interface (Fig. 9) to the database which allows a regular check of data and fast identification of specific hydrologic situations and instrumentation failures. Mydatanet also provides for easy importing and downloading, user management, device administration and reporting.

Some sensors are connected through fast glassfibre LAN to dedicated computers. For example, the sensors of the weather station are connected to the HOAL PC, the sensors of the soil moisture network to the Soil Net PC where they are stored as files. Some data (such as the eddy-correlation data) are read out manually from the data loggers and uploaded on the data bases on the dedicated computers.

All measured data are stored as two separate layers. The first contains raw data as directly obtained by the instruments. These data are regularly screened for errors and inconsistencies. They are corrected or labelled as missing data according to a set protocol. The corrected data are stored in the second layer with data flagging and a processing report. Data quality check is an important step in data management not only for scientific usage of the data, but also for providing a direct feedback to maintaining and updating instrumentation configurations. All raw and processed data are exported from the various databases and uploaded in consistent CSV file format to an ftp server at TU Wien at daily and bimonthly (raw data) and 6-monthly (processed data) intervals. A backup of all data is performed on a daily basis by the grandfather–father–son method. Monthly backups of all databases are kept for 1 year.

The HOAL manager is responsible for the overall data management process. Two IT professionals (one at IKT, one at TU Wien) are responsible for the back up of the data and hardware maintenance. The quality check and the correction of the data are carried out by the research students as part of their PhD work. The data correction protocols are stored on the ftp server in simple readme text format.

4   Examples of specific hypotheses

Currently, nine research students are conducting their PhD in the HOAL. Based on the literature and previous work in the HOAL the students identify specific hypotheses within their research programmes. Typically, one hypothesis conforms to one research paper they are planning to prepare, but sometimes the hypotheses are more specific. The following steps

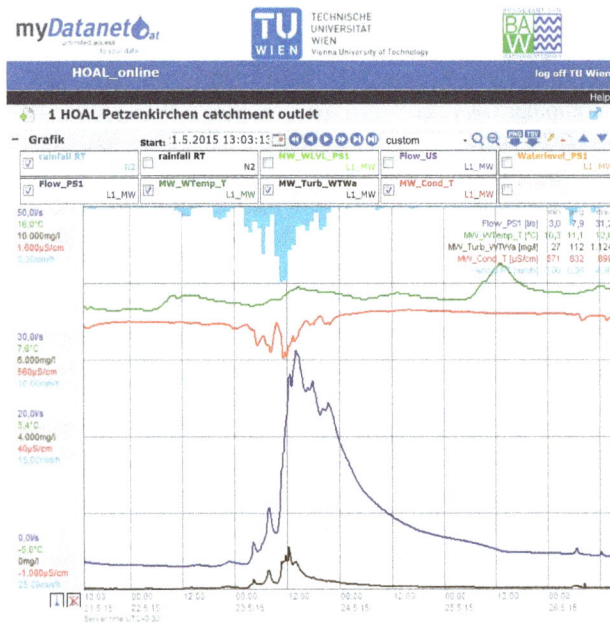

Figure 9. Screen shot of web-based real-time monitoring of the data collected in the HOAL.

were adopted in inferring the instrumentation or experimental set-up from the hypothesis to be tested.

– Background: importance of the research issue, prior knowledge of the issue and specific research question. In many instances the specific hypotheses are formulated and tested as a collaboration among students (joint science questions) building on previous work.

– Hypothesis: stating the hypothesis from knowledge of the processes in the literature and prior analyses in the catchment.

– Test: anticipating alternative test results and their implications for rejecting (or not rejecting) the hypothesis. If possible, more than one test is performed to test the same hypothesis, preferably based on different data and/or different rationales.

– Experiment: performing the experiment or the monitoring with required sensitivity.

– Outcomes: testing the hypothesis against the results of the experiment or the monitoring in the context of the assumptions involved and implications for the overarching science questions.

Below a number of examples of hypothesis testing are presented to illustrate the approach adopted in the HOAL. They relate to repeatable experiments (Example 1), temporal monitoring (Example 2), spatial monitoring (Example 3) and testing of instruments (Example 4). All of them use both basic (Sect. 3.2.1) and dedicated (Sect. 3.2.2) infrastructure.

4.1 Example 1: what is the source of early stream sediment concentrations?

– Background: understanding the sources of sediments is very relevant for managing contaminants such as phosphorus and for controlling soil loss from agricultural landscapes. During rainfall events, an early peak in the suspended sediment concentration is often observed (Eder et al., 2010). The sediments may either stem from erosion from hillslopes close to the stream or from reactivation of sediments on the stream bed that have been deposited during previous events. Observations of sediment concentrations during natural events are inconclusive, as sediment inputs may occur in a diffuse way along the stream which are difficult to measure. Alternative experiments are needed to test the origin of early suspended sediments in the stream.

– Hypothesis: early suspended sediment concentration peaks in the stream are a result of resuspension of sediments in the stream bed deposited during previous events, rather than a result of erosion from the catchment.

– Test 1: does sediment-free water pumped into the stream produce suspended sediment concentrations similar to those observed for natural events? Yes: cannot reject hypothesis. No: reject hypothesis.

– Test 2: do suspended sediment loads decrease for repeated experiments? Yes: cannot reject hypothesis. No: reject hypothesis.

– Experiment: two flushing experiments were conducted by pumping sediment-free water into the stream and measuring flow and sediment concentrations at three sites with high temporal resolution. The discharges were similar to those of early stages of natural events with comparable bed shear stresses.

– Outcomes: at the most upstream section (site 360) of the stream, significant sediment was resuspended from the stream bed with concentrations similar to those of natural events, so the first hypothesis was not rejected. Sediment concentrations and loads decreased along the stream as the flow velocities decreased as a result of the dispersion of the hydrograph (Fig. 10). During the second experiment the sediment load was much smaller than during the first experiment, so hypothesis 2 was not rejected either. This finding was interpreted as the result of the depletion of stream bed sediments during the first experiment. Comparison with natural events supported stream bed resuspension as the source of early sediment peaks.

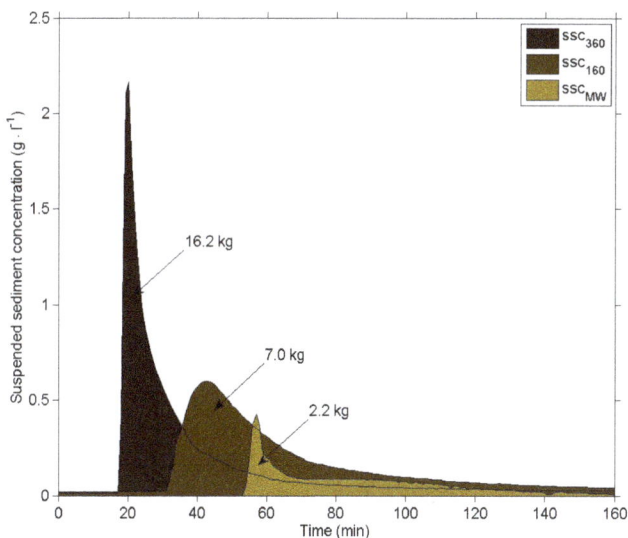

Figure 10. Hypothesis testing example 1: sediment concentrations for a flushing experiment in August 2011 at the three monitoring locations. 360 is the most upstream location at 360 m from the catchment outlet, MW. From Eder et al. (2014).

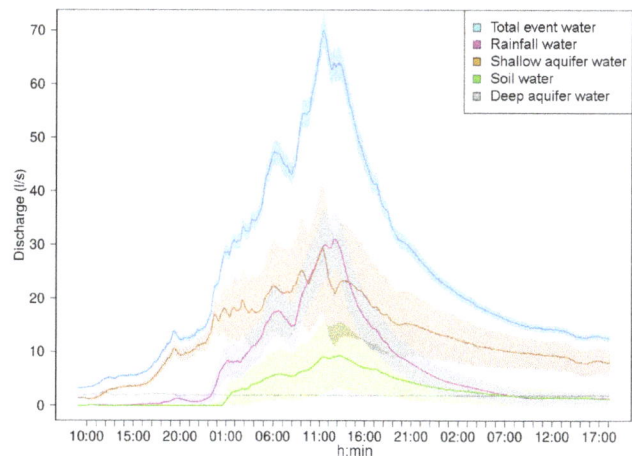

Figure 11. Hypothesis testing example 2: flow contributions for the event on 20 January 2012 at MW including uncertainties. End-member mixing analysis (EMMA) was applied in a Monte Carlo mode with given error distributions on the concentrations and discharges which translate into the uncertainty distributions shown as shaded areas in the graph. See Exner-Kittridge et al. (2016) for details.

4.2 Example 2: what are the sources and flow paths of event runoff?

– Background: agricultural runoff into surface waters during rainfall events can originate from many different sources (e.g. multiple aquifers, unsaturated zone, event rainfall) and can take multiple interconnected flow paths (e.g. overland flow, macropore flow, matrix flow, tile drainage systems, etc.). Cost-effective mitigation measures of excess nutrients are harmful to the aquatic environment should be targeted on the sources and flow paths that conduct the bulk of the nutrient load rather than all sources and flow paths. Additionally, specific sources and flow paths may dominate during different periods within a runoff event throughout the entire length of the stream. Methods are needed to identify both sources and flow paths.

– Hypothesis: the shallow aquifer contributes the majority of the total discharge at MW during rainfall events.

– Test: does the shallow aquifer contribute less than 50 % to the total event discharge volume as compared to the event rain water and the unsaturated soil water? Yes: reject hypothesis. No: cannot reject hypothesis.

– Experiment: monitor discharge, chloride (Cl) and nitrogen (N) at MW over several years. Perform end-member mixing analysis (EMMA) based on the chemical signatures of the end-member reservoirs (i.e. event rainfall: low Cl, low N; soil water: medium Cl, high N; shallow aquifer: medium Cl, medium N) and assess the uncertainties.

– Outcomes: EMMA suggests that, over the period 2011-2012, the shallow aquifer contributes between 10 % and 70 % of the event discharge volume with an average of 45 %, depending on the event magnitude. During small to average events in summer, the shallow aquifer water dominates the total volume of the hydrograph, while the unsaturated soil water tends to contribute very little. Both preferential flow and pressure displacement appear to be the dominant pathways during these periods. During the winter months and events with high rainfall volumes, the contribution of unsaturated soil water and rain water can increase substantially (Fig. 11). This is attributed to high soil saturation conditions during these periods.

4.3 Example 3: how do spatial soil moisture patterns change during rainfall events?

– Background: understanding the controls of spatial soil moisture patterns in small catchments is essential for upscaling soil moisture from point to catchment scales and linking ground data to satellite data. The relative importance of the factors driving the spatial distribution of soil moisture was found to change during the season, e.g. topography may control the soil moisture distribution during wet periods, and vegetation and soil properties may be more dominant during dry conditions (Grayson et al., 1997). The changes in the patterns during rainfall events are less well documented and it has been hypothesised that the relative patterns remain

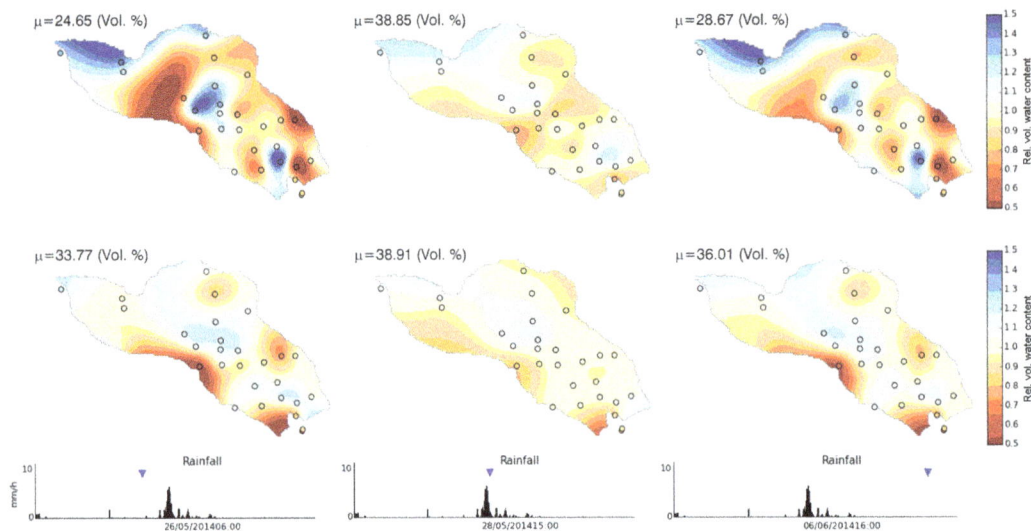

Figure 12. Hypothesis testing example 3: soil moisture patterns scaled by the mean catchment soil moisture μ before, during and after an event in May 2014. Top and bottom panels show 5 and 20 cm soil moisture, respectively. Circles show measurement locations, patterns are interpolations. Time series at the very bottom show rainfall with the time of the soil moisture patterns indicated by blue triangles.

stable (Grayson and Western, 1998). Observations are needed to test whether this is actually the case.

– Hypothesis: during spring rainfall events, the relative spatial soil moisture pattern remains stable throughout the events.

– Test: is a clear change in relative soil moisture patterns observable over the catchment? Yes: reject hypothesis. No: cannot reject hypothesis.

– Experiment: soil moisture was monitored at many locations within the catchment before, during and after a large rain event.

– Outcomes: the spatial patterns do change during the rainfall event examined in this particular catchment, both at 5 and 20 cm depth (Figure 12), so the hypothesis is rejected. Relative soil moisture is more evenly distributed during the event than before, although the centre and the north-eastern part of the catchment are consistently wetter. After the event the soil dries out and the patterns return to a similar state as before the event. The main difference in the patterns is their variance, so a different scaling (rather than by the spatial mean) may produce greater similarity. On the other hand, one would expect bigger changes than those in Fig. 12 for drier antecedent soil moisture as is typical of summer events.

4.4 Example 4: can faecal indicators be consistently monitored on an on-line basis?

– Background: on-line detection of enzymatic beta-D-Glucurondiase (GLUC) activity has been suggested as

a potential surrogate for microbiological faecal pollution monitoring with a capacity for near-real-time applications in the context of water safety management (Farnleitner et al., 2001, 2002). Such measurements will also allow shedding of light on microbial transport processes at the catchment scale. While automated measuring devices have already been tested for groundwater (Ryzinska-Paier et al., 2014), so far no evaluation exists for surface water. Surface water may involve additional challenges due to higher sediment concentrations and bacterial contamination levels which may contaminate or block inlet pipes and other system components. The HOAL is an ideal test bed for the method due to its highly dynamic runoff, sediment concentrations and bacterial contamination. Devices with two different designs (BACTcontrol and ColiMinder) are available in the HOAL (Stadler et al., 2016).

– Hypothesis: GLUC activity in surface water can be consistently measured by devices differing in construction (consistent meaning $R^2 > 0.9$ and p value < 0.001).

– Test 1: are measurements of devices with identical constructions consistent? Yes: cannot reject hypothesis. No: reject hypothesis.

– Test 2: are measurements of devices with different constructions consistent? Yes: cannot reject hypothesis. No: reject hypothesis.

– Experiment: four devices for automated GLUC measurements were installed at the catchment outlet and operated in parallel for a period of 12 months (two sets of two identical devices, BACTcontrol and ColiMinder).

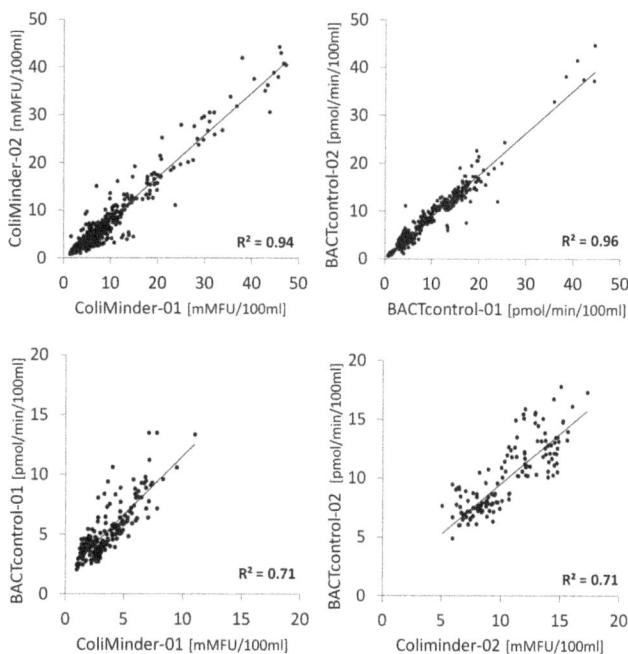

Figure 13. Hypothesis testing example 4: consistency of online monitoring of microbial faecal pollution by GLUC activity as a surrogate. Top panels (Test 1) compare devices with the same design and assay (BACTcontrol vs. BACTcontrol, ColiMinder vs. ColiMinder). Bottom panels (Test 2) compare devices with different designs and assays (BACTcontrol vs. ColiMinder). The smaller range of the bottom panels is related to differences in the measurement period. All correlations are significant at p values < 0.001. See Stadler et al. (2016) for details.

– Outcomes: results from Test 1 (Fig. 13, top) show that devices with identical constructions are indeed extremely consistent ($R^2 > 0.9$). Test 2 (Fig. 13, bottom), however, shows that different designs lead to less consistent results ($R^2 = 0.71$), so hypothesis 2 was rejected. The lower correlations in the latter case are mainly due to the different designs and partly related to slightly different intake locations (about 2 m separation) and measurement times (up to 60 min time offset). Overall, the experiments suggest that the instruments are indeed useful for near-real-time monitoring of GLUC activity.

5 Lessons learned and outlook

5.1 Lessons regarding science strategy of the HOAL

5.1.1 Long-term experimental infrastructure

The research since the inception of the HOAL has demonstrated that the strategy of base monitoring related to overarching science questions and dedicated monitoring related to specific hypotheses indeed works well. Substantial synergies

were realised between the dissertation studies that shared the base monitoring. For example, most students used the runoff measurements at high temporal and spatial resolution in the context of their own specific science questions such as runoff generation, flow paths, nutrient budgeting, sediment transport and evaporation estimation. On the other hand, the dedicated monitoring allowed collection of exactly the information needed to test specific hypotheses and thus maximise the efficiency of the HOAL. Two generations of research students have so far worked in the HOAL. The overall, structured set-up geared towards long-term research assisted students in building on the findings of the previous generation. When students left, there was sufficient expertise among the team members for a smooth transition to new students. Practical aspects such as the HOAL manager and the web logbook turned out to be valuable in this transition.

5.1.2 Interdisciplinary collaboration

Interdisciplinarity is both a consequence of the type of societally relevant research questions being addressed in the HOAL, and it also provides an opportunity to address more complex research questions than would be possible by researchers from only one discipline. Students have clearly recognised that through collaboration with others they are able to gain knowledge and understanding that enables them to delve deeper into their own research topic (see Carr et al., 2015). Additionally, they often also see immediate benefits to their collaboration in the form of a data set, which provides further motivations for continuing to work collaboratively. However, they also recognised that collaboration across the disciplines can bring additional challenges as time and effort is needed to understand and incorporate knowledge from other research fields. Study sites, such as the HOAL, provide a focal point where researchers from different disciplines can interact, develop joint hypotheses together and work collaboratively on data collection or experimentation. As such, this can be seen to raise the efficiency of interdisciplinary collaboration because research students have greater clarity on who and why they need to collaborate with to overcome specific research challenges in answering their joint research questions.

5.1.3 Networking within the science community and beyond

The collaboration between TU Wien and IKT fully realised the potential of the complementary expertise. Similarly, collaboration with some of the providers of the instrumentation turned out to be very useful and allowed science questions to be addressed (e.g. comparative testing of monitoring microbial pollution proxies) that would be difficult to address otherwise. Collaborations with other research institutions sometimes posed an issue regarding the time axes. Joint projects usually turned out to take longer (and consume more resources) than anticipated. The joint projects were not always the top priority of the project partners which added to delays. A more rigorous planning of joint projects in the future, including set deadlines, deliverables and clear budgets, may help increase the efficiency of such activities. Communication and outreach activities were received well, although there is probably potential for additional activities from local to global scales.

5.2 Lessons regarding implementation

5.2.1 Site selection and hydrological characteristics

The HOAL site turned out to be an excellent choice for the same reasons it was selected in the first place. The different runoff generation mechanisms indeed allowed some very interesting and unique hypothesis testing associated with flow paths and water sources. The proximity of the HOAL to the Institute for Land and Water Management Research was probably one of the most fortunate choices of the entire project. The logistic benefits for maintenance and connection to the power grid and high-speed glassfibre LAN turned out to be immense. This is certainly an important lesson learned and we can warmly recommend a similar set-up for other hydrological laboratories.

5.2.2 Setting up the HOAL and instrumentation

While the overall science strategy and site selection clearly worked well, the implementation of the instrumentation was not always easy. All instrumentation was finally installed and functional in a similar way as planned but, on the way, there were considerable challenges, even though there was substantial expertise within the team members with field experimentation. With the benefit of hindsight the HOAL team would probably approach some of the installations differently.

Challenges with the flumes: H-flumes were planned for runoff measurements for some of the tributaries within the HOAL. The main motivation for choosing H-flumes over V-notch weirs was the hope that they will be less prone to siltation although, ultimately, siltation was not completely avoided. The choice of H-flumes came at a cost of lower measurement accuracy at low flows. Initially the main scientific interest was on large floods, but soon it became clear that the entire runoff spectrum is of interest. Additionally, the H-flumes were overdesigned. This was partly due to the early focus on floods and partly due to internal communication issues where each of the team members added a "safety margin" to the maximum design flow to ensure that it is never exceeded. Finally, for simplicity only three size classes were constructed and in this step most flumes were additional increased to fit a class. After a year, when the problems became evident, the cross sections of the flumes were narrowed down to improve their accuracy and tipping buckets were added,

but the lesson learned is that some extra time and coordination in the initial planning of the flumes would probably have paid off.

Another problem with the flumes was freezing in winter. In the first winter, the team lost a number of pressure transducers (although the same make had worked fine during winter in a different catchment). Later, a heating system was installed and the flumes were insulated but freezing remains a problem in some situations. It was not always easy to seal the flumes to the ground because of erosion processes and leakage occurred repeatedly. One of the practical fixes were lateral metal sheets attached to the flumes and dug into the ground. Regular maintenance is needed to ensure that no leakage occurs. Finally, the positions of the pressure transducer changed due to the technical interventions. Regular checking of the positions and manual discharge measurements on a weekly basis for quality assurance purposes were found to be very important.

Similarly, operation of the soil moisture sensor network turned out to be more time consuming than anticipated. Part of the problem is the agricultural use of the HOAL requires temporary sensors to be removed and replaced once or twice a year in accordance with the agricultural manipulations on the fields. When replacing the sensors that soil profile had obviously been disturbed, so some of the continuity of the measurements was lost. Forested catchments (Bogena et al., 2010) or pastures (Western et al., 1999) would allow more straightforward instrumentation and more consistent soil moisture data. There were also leakage issues with the housing of the equipment, and the sensors were not always straightforward to calibrate. Clearly, soil moisture sensing needs utmost care in order to obtain accurate results.

Overall, there were a range of setbacks which is not surprising given the enormous number of sensors installed in the HOAL, a total of more than 300, in addition to sampling, experiments and surveys. Much of the research focused on high spatial detail which, in many instances, was realised by a large number of sensors at different locations, but any sensor failures during limited periods of time translated to missing data and reduced spatial detail. Other issues included faulty parts of the instruments, faulty software updates and damage to the glass fibre cable during construction of a nearby building. Additionally, compatibility between different sensors–data-logger systems was not always straightforward, for example due to different preset measurement intervals. The lesson learned is that such issues are difficult to avoid and it is important to be flexible and vigilant. A dedicated person (the HOAL manager) responsible for the overall management certainly turned out to be very valuable in minimising the loss of data and maximising their accuracy.

On the other hand, there were a number of things that worked better than anticipated. One of them is the web access to the data in real time, which allows an online-check of the measurements. There were numerous instances where measurement or connectivity problems were quickly identified and repaired. The web access also allows changing the sampling discs of the autosampler when needed and to control the sampling intervals remotely according to the current weather and streamflow conditions, which turned out to be useful. Permanent electrical supply throughout the entire length of the stream was a great benefit for the easy installation, testing, and long-term monitoring of equipment. Without a permanent power supply, certain types of equipment would have not been possible to have been installed, while others would have been very difficult to maintain.

5.2.3 Managing the HOAL

The HOAL manager position was filled in early 2013 and the benefits of a manager quickly became apparent. Previously, communication with the landowners was complicated as there was no single contact point of the HOAL team. Yet, a good system of co-operation is necessary for the installation and operation of instrumentation such as tripod-mounted eddy-covariance devices and the soil moisture sensors. The addition of the HOAL manager position to the project had a very positive effect on this process. With the manager recruited from the locality, communication with the landowners was now immediate, as was seen for example from the planning of an evaporation field campaign, as the manager works directly with the research students on a daily basis and hence has detailed knowledge of their research, much more effective and efficient than previously. A good working relationship with the landowners was facilitated by sharing some of the findings of the HOAL, e.g. on erosion sources and potential protection measures, groundwater protection, and fertiliser management, as well as on the weather data.

Similarly, the manager was invaluable for coordinating the maintenance. The implementation of a weekly maintenance plan by the manager ensured that no important work was overlooked. The plan also facilitated the communication between the group members, e.g. when research students were on temporary leave during their research semester abroad. Similarly the web-logbook was instrumental in maximising data quality and ensuring a realistic interpretation of the data. On the other hand, the maintenance turned out to be quite time consuming. For example, the cleaning of the H-flumes at the stream tributaries consumes considerable time resources. Additional maintenance is needed to clean out dead leaves in autumn. Maintenance works also involves mowing the grass around the instrumentation to avoid shading of solar panels. The soil moisture network required substantial maintenance, in particular the end devices buried a few centimetres below ground, which tended to get wet and had to be cleaned.

The main advantages of this iterative, network-based process of conducting hypothesis testing are its flexibility and the encouragement of creative thinking by the students. There were a number of instances where this flexibility allowed exploiting collaborative opportunities. Examples include a field campaign on identifying the relative contributions of transpiration and soil evaporation together with the IAEA and the validation of NASA's SMAP satellite based on soil moisture data in the HOAL.

At the beginning of the project, the data management was not an easy process. The main challenge were in organising and checking data from a large number of different sensors, communication and a consistent protocol between all the people involved. The raw data correction process consumed more resources than anticipated. Eventually, the overall data management approach did run smoothly. Regular backups and simple and robust data formats (such as the csv file format) for flexible data exchange proved to be useful.

5.3 Lessons regarding hypothesis testing

The general philosophy of a hypothesis-based observatory was considered by the HOAL team members to work well. The hypotheses provided guidance for the dedicated monitoring and experimentation and they facilitated the transformation of the research findings into publications. In particular, thinking in terms of hypotheses was found to be useful, as it is directly linked to the research questions addressed in individual papers.

However, hypothesis testing was not always as clear-cut as one would hope (Chamberlin, 1965; Srinivasan et al., 2015). There were two issues which were related to (a) setting up the hypothesis and (b) the outcomes of the hypothesis testing.

(a) Setting up of the hypothesis: setting up of the hypothesis was constrained by the available resources. Once equipment had been purchased, other hypotheses were also constrained by the available infrastructure. One of the issues is the sensitivity of the measurements with respect to the hypothesis. For example, soil moisture sensor pairs installed in the field at the same location gave very consistent results but comparisons with the gravimetric method (oven-drying of samples) did not. The main difficulty was the small sampling volume of the sensors and the immense spatial soil moisture variability, particularly near the surface due to burrows, roots, cracks and soil characteristics. Additional site-specific calibration of any soil moisture sensor prior to installation turned out to be essential. It is hence not clear how sensitive the soil moisture network is to catchment-scale hypotheses. In the near future, validation will be based on a portable TDR sensor with a sampling volume similar to that of the sensors. In a similar vein, the observations of the saturation and/or overland flow patterns by video monitoring were focussed on a particular 2 ha area that is usually cultivated with maize or winter wheat. With young crops the patterns can be clearly observed but later in the season this is no longer possible. However, since the field patches are cultivated at different times, it is possible to switch the observation area to an alternative bare patch to maximise the period of pattern observations within the HOAL. To decide about the best patches, prior planning is needed.

Clearly, the more complex the processes are, the more difficult it is to set up clear-cut hypotheses (see e.g. Reischer et al., 2011, for a complex case). As Knorr-Cetina (2013, 4–5) noted, "the products of science are contextually specific constructions which bear the mark of the situational contingency and interest structure of the process by which they are generated,", and "If there is a principle which seems to govern laboratory action, it is the scientists' concern with making things "work", which points to a principle of success rather than one of truth. [...] Thus, it is success in making things work which is reinforced as a concrete and feasible goal of scientific action, and not the distant ideal of truth which is never quite attained." It is therefore clear that the outcomes of the hypothesis testing will not be independent of the preconceptions and interests of all those involved. Holländer et al. (2014) illustrates this notion for the more specific case of hydrological predictions.

Some of the examples in this paper do illustrate the difficulty with reducing complex science questions to hypotheses with binary outcomes. Hypothesis example 2 in this paper is quite specific, while a more relevant question would perhaps enquire about the causal mechanisms driving the sources and flow paths of event runoff. Similarly, one is usually more interested in the more complex questions of the mechanisms driving spatial soil moisture patterns rather than simply whether they change during events or not (example 3). The issue of complex process interactions that cannot be easily disentangled by individual hypotheses permeates much of hydrology, similar to the non-repeatability of experiments. More work on framing relevant and yet testable hypotheses in hydrology is needed (Sivapalan, 2009).

(b) Outcomes of the hypothesis testing: the second issue was that the outcomes of the hypothesis testing were not always fully conclusive. This is illustrated by the four examples, where rejection (or non-rejection) of the hypothesis was usually associated with some qualifications. Perhaps more importantly, the issue is that, with the exception of a few repeatable experiments performed in the HOAL (such as example 1), most testing was not repeatable due to the randomness in weather and other boundary and initial conditions which cannot be fully specified (Zehe and Blöschl, 2004; Zehe et al., 2007). The lack of repeatability of experiments and the associated limits to predictability (Blöschl and Zehe, 2005) are a real issue in hydrology. Ideally, one would like to have at least three replicas of the same experiments. There are a few examples where this is already put into practice, such as the Biosphere 2 hillslope experiments performed in a green house (Hopp et al., 2009) but, usually, even dedicated large-scale experiments such as the Chicken Creek artificial catchment (Holländer et al., 2009, 2014) do not involve multiple replicas.

A similar question that arises is how representative the HOAL of other (experimental or larger) catchments around the world is, i.e. the question of whether and how the findings of the HOAL can be generalised. The diversity of runoff generation processes encountered in the HOAL is considered instrumental in making the findings more generally applicable to a variety of agricultural catchment settings. The students are forced from the beginning of their PhD projects to carefully think about, formulate and interpret their experiments with respect to such broader settings. On the other hand, care needs to be taken in building models that are based on the specifics of the HOAL.

Notwithstanding these caveats related to hypothesis testing, experiences in the HOAL also showed that there were a number of unforeseen opportunities to test hypotheses and acquire knowledge that were not anticipated, i.e. positive surprises. Some of them occurred through collaborations with partners. The HOAL has been shown to numerous guest scientists, it has been used for field training during the Meeting of the European Geosciences Union and it is used as a site for TU Wien courses on field work. Unexpected opportunities that arose from these collaborations were a field campaign on separating transpiration and bare soil evaporation based on isotopic measurements, operated in collaboration with the International Atomic Energy Agency (IAEA) in June 2014. A Picarro isotope analyser system was installed in the field and both institutions benefited from the shared expertise during this field test.

Another example of an unforeseen opportunity was an unplanned, yet very interesting observation in February 2015 when the research students conducted field work to sample stream water quality to test hypotheses regarding diurnal fluctuations relative to summer conditions. A period without snow and rain was selected to ensure no surface water input into the stream. As it turned out, temperatures rose slight above zero which melted frozen soil water in the catchment and produced a quite significant discharge into the stream, more than doubling streamflow. The event was neither rain nor snowmelt driven, which is not commonly observed. The comprehensive instrumentation in the HOAL allowed a detailed analysis of the hydrological situation of this interesting event. It was speculated that such mechanisms may occur more often than usually assumed, yet they are rarely considered in rainfall runoff models.

5.4 Outlook

The ambition of the Hydrological Open Air Laboratory (HOAL) in Petzenkirchen is to advance the understanding of water-related flow and transport processes involving sediments, nutrients and microbes in small catchments. Overall, the concept of long-term facilities that transcend the lifetime of individual projects, a commitment to interdisciplinary research, and the involvement in networks to assist in performing collaborative science has worked well so far.

Two generations of research students have so far pursued research in the HOAL. The third generation of students will address issues of upscaling and hydrological change. The fourth generation of students will be concerned with how all of these findings can be generalised to other climatic and management conditions around the world. More extensive collaborations will allow addressing a wider range of science questions and realise the full potential of an important network within the scientific community. Future societal water challenges revolve around sustainable water management in a changing world (Leibundgut et al., 2014). Hydrological change, hydrological and microbial risks, nutrients and emerging substances are all issues that will likely increase in importance in the near future. Innovative technology and approaches for better understanding water-related processes in the environment are needed. There are numerous opportunities to further develop new technology (e.g. new tracers, sensors, protocols) in the framework of the HOAL to address some of the knowledge gaps and issues discussed above.

This paper has focused on the scientific and implementation aspects of the HOAL. There are also financial aspects that are fundamental to the safe operation of the laboratory. Currently, much of the funding comes from the Austrian Sciences Funds, the TU Wien and the Federal Agency for Water Management. As the activities branch out to a larger number of collaboration partners, care needs to be taken to ensure the long-term funding of the Hydrological Open Air Laboratory.

The HOAL is becoming a hub for hosting guest scientists, through a closely knit network with other academic institutions and observatories. The HOAL is special in that many runoff generation processes (surface runoff, spring runoff, tile drainage, runoff from wetlands) can be observed simultaneously and at the high spatial and temporal resolution with which the processes are monitored. This particular profile opens exciting opportunities for complementary, comparative research with different hydrological observatories and experimental catchments in different environments (e.g. Schumann et al., 2010) to foster progress in the interdisciplinary water sciences.

Appendix A: Catchment details

Table A1. Catchment details.

Location	Near Petzenkirchen, in the western part of Lower Austria Weather station: $48°09'17.7''$ N, $15°08'54.0''$ E Catchment outlet: $48°09'00.9''$ N, $15°09'10.9''$ E
Catchment size:	65.8 ha.
Climate and runoff:	Mean annual air temperature (1990–2014): $9.5\,°C$ Range (1990–2014): $7.9\,°C$ (1996) to $10.6\,°C$ (1994) Mean monthly air temperature (1990–2014): Jan $-0.4\,°C$, Feb $0.9\,°C$, Mar $5.1\,°C$, Apr $9.6\,°C$, May $14.3\,°C$, Jun $17.4\,°C$, Jul $19.2\,°C$, Aug $18.9\,°C$, Sep $14.3\,°C$, Oct $9.4\,°C$, Nov $4.4\,°C$, Dec $0.0\,°C$
	Mean annual precipitation (1990–2014): $823\,\mathrm{mm\,yr^{-1}}$ Range (1990–2014): $591\,\mathrm{mm\,yr^{-1}}$ (2003) to $1090\,\mathrm{mm\,yr^{-1}}$ (2002) Mean monthly precipitation (1990–2014): Jan 48.7 mm, Feb 46.7 mm, Mar 64.9 mm, Apr 50.0 mm, May 78.4 mm, Jun 99.4 mm, Jul 89.5 mm, Aug 95.5 mm, Sep 79.6 mm, Oct 54.8 mm, Nov 62.4 mm, Dec 53.6 mm
	Mean catchment evaporation (1990–2014) based on the water balance: $628\,\mathrm{mm\,yr^{-1}}$
	Mean annual runoff at catchment outlet (MW): mean (1990–2014): $4.07\,\mathrm{L\,s^{-1}}$ Range (1990–2014): $1.91\,\mathrm{L\,s^{-1}}$ (2004) to $6.99\,\mathrm{L\,s^{-1}}$ (2013) Mean monthly runoff (1990–2014): Jan $4.91\,\mathrm{L\,s^{-1}}$, Feb $5.72\,\mathrm{L\,s^{-1}}$, Mar $5.74\,\mathrm{L\,s^{-1}}$, Apr $5.04\,\mathrm{L\,s^{-1}}$, May $4.08\,\mathrm{L\,s^{-1}}$, Jun $3.92\,\mathrm{L\,s^{-1}}$, Jul $2.88\,\mathrm{L\,s^{-1}}$, Aug $3.12\,\mathrm{L\,s^{-1}}$, Sep $2.82\,\mathrm{L\,s^{-1}}$, Oct $3.07\,\mathrm{L\,s^{-1}}$, Nov $3.85\,\mathrm{L\,s^{-1}}$, Dec $4.12\,\mathrm{L\,s^{-1}}$ Maximum runoff (1990–2014): $2000\,\mathrm{L\,s^{-1}}$ (1 Sep 2002, estimate), $656\,\mathrm{L\,s^{-1}}$ (25 June 2013)
Soils:	The soil types are Cambisols (56 %), Planosols (21 %), Anthrosols (17 %), Gleysols (6 %) and Histosols (<1 %). Infiltration capacities tend to be medium to low, water storage capacities tend to be high, and shrinking cracks may occur in summer due to high clay contents.
Geology and aquifers:	The subsoil consists of Tertiary sediments of the Molasse zone and fractured siltstone. The shallow aquifer is associated with the water draining the shallow subsurface soil, while the deep aquifer is within the fractured siltstone unit.
Topography:	Elevation range: 268 to 323 m a.s.l. Mean slope: 8 %.
Vegetation/land use:	At present, 87 % of the catchment area is arable land, 5 % is used as pasture, 6 % is forested and 2 % is paved. The crops are mainly maize, winter wheat, rape and barley.
Fertiliser input:	Nitrogen fertiliser input (2010–2013, catchment average): Range $140\,\mathrm{kg\,N\,ha^{-1}}$ (2013) to $210\,\mathrm{kg\,N\,ha^{-1}}$ (2012) Phosphorus fertiliser input (2010–2013, catchment average): Range 12 kg P ha (2013) to 26 kg P ha (2012)
Seitengraben stream:	Length: 620 m, slightly meandering Continuous shaded by deciduous trees and bushes in riparian zone. Biologically active ecosystem with small water animals and plants. Discharges into Hauptgraben river which discharges into the Erlauf and finally the Danube).

Appendix B: Instrumentation

Table B1. Instrumentation in the HOAL. Compartments (Comp): GW = groundwater, SW = surface water, A = atmosphere. Location: MW = catchment outlet, WS = Weather station (also see Figs. 6 and 7). Data connectivity: LAN = glassfibre LAN, M = local storage and manual data transfer, P = GSM phone, S = satellite modem. Main data storage: Cosmos = Cosmos project server, HP = HOAL PC, IKT = IKT server (plain file system), MN = IKT server (MydataNet), Soil = Soil Net PC, TU = TU server, VWM = VWM server.

Variable	Units	Comp.	No. of stations × sensors	Location	Sensor	Sensor type	Temporal resolution (min)	Data connectivity	Main data storage	Data from (year)
Precipitation	mm min^{-1}	A	4	Catchment	Precipitation gauge	OTT Pluvio	1	P	MN	2010
Air temperature	°C	A	1 × 3	WS	Temperature sensor at 2, 5, 10 m	HMP 155	30	LAN	HP	2012
Air humidity	%	A	1 × 3	WS	Humidity sensor at 2, 5, 10 m	HMP 155	30	LAN	HP	2012
Wind speed and direction	m s^{-1}, °	A	1 × 3	WS	Wind sensor at 2, 5, 10 m	Gill Wind-Sonic	30	LAN	HP	2012
Atmospheric pressure	hPa	A	1	WS	Barometer	EC100	1	LAN	HP	2012
Radiation (incoming short-wave, incoming long-wave, outgoing short-wave, outgoing long-wave)	W m^{-2}	A	1 × 4	WS	Four-component net radiometer	Kipp & Zonen CNR 4	1	LAN	HP	2012
Raindrop distribution, air temperature, relative humidity	Number of drops, °C, %	A	1 × 3	WS	Present weather sensor at 1.7 m	Campbell PWS 100	1	LAN	HP	2012
Carbon dioxide flux, latent heat flux, sensible heat flux, momentum flux	mmol m^{-2} s^{-1}, W m^{-2}, W m^{-2}, kg m^{-1} s^{-2}	A	2	Catchment (movable device)	Open-path eddy covariance (3-D wind speed, water vapour, carbon dioxide density)	Campbell IR-GASON	10 Hz, 30 min aggregation	M	HP	2012/2013
Carbon dioxide flux, latent heat flux, sensible heat flux, momentum flux	mmol m^{-2} s^{-1}, W m^{-2}, W m^{-2}, kg m^{-1} s^{-2}	A	1	WS	Closed-path eddy covariance (3-D wind speed, water vapour, carbon dioxide mixing ratio)	Campbell EC155	10 Hz, 30 min aggregation	M	HP	2013
Momentum and sensible heat flux	W m^{-2}	A	1	Catchment (movable device)	Scintillometer	Scintec SLS-20	1	LAN	HP	2012
Wind load (acceleration)	m s^{-2}	A	1 × 7	WS	Triaxial DC Accelerometer	3713B1110G, MEMS Capacitive	100 Hz	M	HP	2015
Snow depth	m	Surface	1	WS	Snow depth US sensor	SR50AT	1	LAN	HP	2012
Saturation patterns on land surface	–	Surface	1	WS	Camera, timelapse pictures, recorded video (on detected motion)	Sanyo VCC-MCH5600P	1	LAN	HP	2013
Discharge	L s^{-1}	SW	1	Sys4 (inlet, piped stream)	H-flume, pressure transducer (water level)	Druck PTX1830	1	LAN	MN	2011
Electrical conductivity and water temperature	µS cm^{-1}, °C	SW	1 × 2	Sys4 (inlet, piped stream)	Electric conductivity probe	WTW TetraCon	1	LAN	MN	2011
pH, Cl, NO$_3$–N	–, mg L^{-1}, mg L^{-1}	SW	1 × 3	Sys4 (inlet, piped stream)	Multiparameter probe	Nadler pH electrode, ion selective electrodes	1	LAN	MN	2011
Discharge	L s^{-1}	SW	1	Frau1 (tile drain)	H-flume, pressure transducer (water level)	Druck PTX1830	1	LAN	MN	2011
Discharge (low flows)	L s^{-1}	SW	1	Frau1 (tile drain)	Tipping bucket (counts)	Reed sensor	1	LAN	MN	2011
Discharge	L s^{-1}	SW	1	Frau2 (tile drain)	H-flume, pressure transducer (water level)	OTT PS1	1	LAN	MN	2011
Electrical conductivity and water temperature	µS cm^{-1}, °C	SW	1 × 2	Frau2 (tile drain)	Electric conductivity probe	WTW TetraCon	1	LAN	MN	2012
pH, Cl, NO$_3$–N	–, mg L^{-1}, mg L^{-1}	SW	1 × 3	Frau2 (tile drain)	Multiparameter probe	Nadler pH electrode, ion selective electrodes	1	LAN	MN	2012
Discharge	L s^{-1}	SW	1	Sys1 (tile drain)	H-flume, pressure transducer (water level)	OTT PS1	1	LAN	MN	2011
Discharge (low flows)	L s^{-1}	SW	1	Sys1 (tile drain)	H-flume, Tipping bucket (counts)	Reed sensor	1	LAN	MN	2011
Discharge	L s^{-1}	SW	1	Sys2 (tile drain)	H-flume, pressure transducer (water level)	OTT PS1	1	LAN	MN	2011
Electrical conductivity and water temperature	µS cm^{-1}, °C	SW	1 × 2	Sys2 (tile drain)	Electric conductivity probe	WTW TetraCon	1	LAN	MN	2011
pH, Cl, NO$_3$–N	–, mg L^{-1}, mg L^{-1}	SW	1 × 3	Sys2 (tile drain)	Multiparameter probe	Nadler pH electrode, ion selective electrodes	1	LAN	MN	2011
Discharge	L s^{-1}	SW	1	Sys3 (tile drain)	H-flume, pressure transducer (water level)	Druck PTX1830	1	LAN	MN	2011
Discharge (low flows)	L s^{-1}	SW	1	Sys3 (tile drain)	H-flume, Tipping bucket (counts)	Reed sensor	1	LAN	MN	2011
Discharge	L s^{-1}	SW	1	E1 (erosion gully)	H-flume, pressure transducer (water level)	Druck PTX1830	1	LAN	MN	2011

Table B1. Continued.

Variable	Units	Comp.	No. of stations × sensors	Location	Sensor	Sensor type	Temporal resolution (mins)	Data connectivity	Main data storage	Data from (year)
Turbidity	$mg\,L^{-1}$	SW	1	E1 (erosion gully)	Turbidity probe	WTW ViSolid	1	LAN	MN	2011
Discharge	$L\,s^{-1}$	SW	1	E2 (erosion gully)	H-flume, pressure transducer (water level)	OTT PS1	1	LAN	MN	2011
Turbidity	$mg\,L^{-1}$	SW	1	E2 (erosion gully)	Turbidity probe	WTW ViSolid	1	LAN	MN	2011
Discharge	$L\,s^{-1}$	SW	1	Q1 (spring)	V-notch weir, pressure transducer (water level)	Druck PTX1830	1	LAN	MN	2011
Discharge	$L\,s^{-1}$	SW	1	K1 (spring)	V-notch weir, pressure transducer (water level)	Druck PTX1830	1	LAN	MN	2011
Discharge	$L\,s^{-1}$	SW	1	A1 (wetland runoff)	H-flume, pressure transducer (water level)	OTT PS1	1	LAN	MN	2011
Electrical conductivity and water temperature	$\mu S\,cm^{-1}$, °C	SW	1 × 2	A1 (wetland runoff)	Electric conductivity probe	WTW TetraCon	1	LAN	MN	2011
pH, Cl, NO_3–N	–, $mg\,L^{-1}$, $mg\,L^{-1}$	SW	1 × 3	A1 (wetland runoff)	Multiparameter probe	Nadler pH electrode, ion selective electrodes	1	LAN	MN	2011
Discharge	$L\,s^{-1}$	SW	1	A2 (wetland runoff)	H-flume, pressure transducer (water level)	Druck PTX1830	1	LAN	MN	2011
Discharge (low flows)	$L\,s^{-1}$	SW	1	A2 (wetland runoff)	H-flume, tipping bucket (counts)	Reed sensor	1	LAN	MN	2011
Discharge	$L\,s^{-1}$	SW	1	MW	Thomson weir, stage recorder (water level)	Ott stage recorder	1	M	IKT	1945–2002
Discharge and water temperature	$L\,s^{-1}$, °C	SW	1	MW	H-flume, pressure transducer (water level)	Ott PS1	1	M	MN	2002–2009
Discharge and water temperature	$L\,s^{-1}$, °C	SW	1	MW	H-flume, pressure transducer (water level)	Druck PTX1830	1	LAN	MN	2009
Discharge	$L\,s^{-1}$	SW	1	MW	H-flume, Ultrasonic probe (water level)	Endress/Hauser	1	LAN	MN	2010
Electrical conductivity and water temperature	$\mu S\,cm^{-1}$, °C	SW	1 × 2	MW	Electric conductivity probe	WTW TetraCon	1	LAN	MN	2009
Turbidity	$mg\,L^{-1}$	SW	1	MW	Turbidity probe	WTW ViSolid	1	LAN	MN	2009
Turbidity	$mg\,L^{-1}$	SW	1	MW	Turbidity probe	Hach Lange SOLITAX ts-line sc	1	LAN	MN	only 2010
pH, Cl, NO_3–N	–, $mg\,L^{-1}$, $mg\,L^{-1}$	SW	1 × 3	MW	Multiparameter probe	Nadler pH electrode, ion selective electrodes	1	LAN	MN	2011
beta-D-glucuronidase activity	pmol/min/100 ml	SW	1 × 2	MW	Fluorescence analyser	Photometric cuvette (Coliguard 0025, 0035)	180	P	TU	2011, 2012
beta-D-glucuronidase activity	mMFU/100 ml	SW	1 × 2	MW	Fluorescence analyser	Photometric cuvette (Coliminder A, B)	60	P	VWM	2014
TSS, NO_3–N, COD, BOD, TOC, DOC, turbidity, UV254	$mg\,L^{-1}$, FNU, $Abs\,m^{-1}$	SW	1 × 2	MW	Spectrolyser	s::can spectrolyser UV-Vis, 15/35 mm pathlength, 220–700 nm range	10	M	TU	2013
Visual images of flume	–	SW	1	MW	Camera	Axis P5512-E	1	LAN	HP	2014
Soil heat flux	$W\,m^{-2}$	Soil	1 × 2	WS	Soil heat flux −30 cm	Huskeflux HFP01SC	30; 1	LAN	HP	2012
Soil temperature	°C	Soil	2 × 5	WS	Soil temperature at −5 cm, −10 cm, −15 cm, −20 cm, −30 cm	PT107	30; 1	LAN	HP	2012
Soil moisture, soil temperature	% volumetric soil water content, °C	Soil	18 × 4, 2 × 5	Catchment	Permanent soil moisture sensors −5, −10, −20, −50, (2–100 cm)	Spade-TDT (Jülich)	30	LAN	Soil	2013
Soil moisture, soil temperature	% volumetric soil water content, °C	Soil	11 × 4	Catchment	Temporary soil moisture sensors −5, −10, −20, −50 cm	Spade-TDT (Jülich)	30	LAN	Soil	2013
Soil moisture	% volumetric soil water content	Soil	1	Catchment	Cosmic ray soil moisture neutron probe (680 m footprint, 12–76 cm depth)	CRS 1000/B	60	S	Cosmos	2013
Groundwater level and temperature	$cm\,H_2O$, °C	GW	23	Catchment near stream (BP01, BP02, …)	Groundwater data loggers	SWS Mini-Diver	5–30	M	IKT	2011
Atmospheric pressure and temperature	$cm\,H_2O$, °C	A	2	near stream (Baro1, Baro2)	Groundwater data loggers	SWS Baro-Diver	5–30	M	IKT	2011

Table B2. Laboratory analyses from samples taken in the HOAL. Main data storage: IKT = IKT server (plain file system), TU = TU server, UFZ = UFZ server. * Note: physical and chemical analysis (potentiometric, conductometric, filtering, ion chromatographic, photometric, spectrometric).

Variable	Units	Comp.	No. of stations × sensors	Location	Sampling	Analysis	Temporal resolution	Main data storage	Data from (year)
pH, EC, SSC, Cl, NO_3, NH_4, P	$-$, $\mu S\,cm^{-1}$, $mg\,L^{-1}$	SW	2	MW	Autosampler Isco 6712	Physical and chemical analysis* (IKT Lab)	Within event	IKT	2009
pH, EC, SSC, Cl, NO_3, NH_4, P	$-$, $\mu S\,cm^{-1}$, $mg\,L^{-1}$	SW	$-$	MW	Grab samples	Physical and chemical analysis* (IKT Lab)	Weekly	IKT	2010
TOC	$mg\,L^{-1}$	SW	1	MW	Grab samples	Thermal catalytic oxidation (IKT Lab)	Within event or weekly	IKT	2013
pH, EC, SSC, Cl, NO_3, NH_4, P	$-$, $\mu S\,cm^{-1}$, $mg\,L^{-1}$	SW	4×1	E1, E2 (erosion gullies), Frau2 (tile drain), Sys4 (inlet, piped stream)	Autosampler Isco 6712	Physical and chemical analysis* (IKT Lab)	Within event	IKT	2011 or 2013
DOC, TOC, PO4-P, NH_4-N, NO_3-N, TP, TN, SS, HCO_3, Cl, SO_4, pH, EC, Na, K, Ca, Mg	$-$, $\mu S\,cm^{-1}$, $mg\,L^{-1}$	SW	$-$	All tributaries	Grab samples	Physical and chemical analysis* (IKT Lab)	Monthly	IKT	2010
Precipitation ^{18}O, ^{2}H		A	1	Close to catchment (IKT)	Autosampler Manning S-4040, adapted	Laser spectroscopy (Picarro L1115-i, AIT Tulln	Event-based	IKT	2009
Discharge ^{18}O, ^{2}H		SW	$-$	MW, all tributaries	Grab samples	Laser spectroscopy (Picarro L1115-i, AIT Tulln	Within event or monthly	IKT	2009
Discharge ^{3}H		SW	$-$	Q1	Grab samples	Laser spectroscopy (Picarro L1115-i, AIT Tulln	Monthly	IKT	2013
Precipitation ^{18}O, ^{15}N		A	$-$	Close to catchment (IKT)	Autosampler Manning S-4040, adapted	Mass spectrometry (DELTA V Plus + GasBench II, Thermo Scientific; L1102-I, Picarro)	Event based	UFZ	2013
Discharge ^{18}O, ^{15}N		SW	$-$	All tributaries	Grab samples	Mass spectrometry (DELTA V Plus + GasBench II, Thermo Scientific; L1102-I, Picarro)	Monthly	UFZ	2013
E. coli, coliforms	MPN/100 ml	SW	$-$	MW, partly tributaries	Grab samples	Colilert-18 Quanti-Tray (IKT Lab)	Monthly	TU	2012
E. coli, aerobic spores, clostridium perfringens spores, total cell count	CFU/100 ml	SW	$-$	MW	Grab samples	TBX Agar ISO 16649-1 (Med Univ Vienna)	Monthly	TU	2012

Appendix C: Photos of stream gauges

Figure C1. Photos of the 13 stream gauges in the HOAL.

Acknowledgements. The authors would like to thank the Austrian Science Funds (Doctoral Programme on Water Resource Systems, W 1219-N22), the TU Wien (three innovative projects, Lion equipment pool, startup funds, ERC FloodChange commitment, staff base funding) as well as the Federal Agency for Water Management for financial support. We would also like to thank Vienna Water Monitoring (VWM) for the cooperation in the ColiMinder 2014–2015 project funded by the Austrian Research Promotion Agency. Financial support has also been provided by the private Land is Water association.

References

Anderson, S. P., Bales, R. C., and Duffy, C. J.: Critical zone observatories: Building a network to advance interdisciplinary study of Earth surface processes, Mineralogical Magazine, 72, 7–10, 2008.

Blöschl, G. and Zehe, E.: On hydrological predictability, Invited commentary, Hydrol. Process., 19, 3923–3929, 2005.

Blöschl, G., Gutknecht, D., and Kirnbauer, R.: Distributed snowmelt simulations in an Alpine catchment. 2. Parameter study and model predictions, Water Resour. Res., 27, 3181–3188, 1991.

Blöschl, G., Carr, G., Bucher, C., Farnleitner, A. H., Rechberger, H., Wagner, W., and Zessner, M.: Promoting interdisciplinary education – the Vienna Doctoral Programme on Water Resource Systems. Hydrology and Earth System Sciences, 16, 457–472, doi:10.5194/hess-16-457-2012, 2012.

Blöschl, G., Bárdossy, A., Koutsoyiannis, D., Kundzewicz, Z. W., Littlewood, I., Montanari, A., and Savenije, H.: Joint Editorial "On the future of journal publications in hydrology", Hydrol. Earth Syst. Sci., 18, 2433–2435, doi:10.5194/hess-18-2433-2014, 2014.

Blöschl, G., Thybo, H., Savenije, H., and Lovejoy, S.: A Voyage, in: A Voyage Through Scales – The Earth System in Space and Time, edited by: Blöschl, G., Thybo, H., and Savenije, H., Edition Lammerhuber, Vienna, 13–18, 2015.

Blümel, F. and Klaghofer, E.: Bericht über den Stand der Oberflächenabflußmessungen im Jahr 1976. Sonderdr. aus: Sitzungsberichte der Österr. Akademie der Wissenschaften, Math.-naturw. Kl., Abt. I, 186. Bd., 1. bis 3. Heft, Wien, 1977.

Bogena, H. R., Herbst, M., Huisman, J. A., Rosenbaum, U., Weuthen, A., and Vereecken, H.: Potential of wireless sensor networks for measuring soil water content variability, Vadose Z. J. 9, 1002–1013, 2010.

Brown, A. E., Zhang, L., McMahon, T. A., Western, A. W., and Vertessy, R. A.: A review of paired catchment studies for determining changes in water yield resulting from alterations in vegetation, J. Hydrol., 310, 28–61, 2005.

Carr, G., Loucks, D. P., Blanch, A. R., Blaschke, A. P., Brouwer, R., Bucher, C., Farnleitner, A. H., Fürnkranz-Prskawetz, A., Morgenroth, E., Parajka, J., Pfeifer, N., Rechberger, H., Wagner, W. W., Zessner, M., and Blöschl, G.: Evaluating multi-disciplinary collaboration in a doctoral programme on water resource systems, to be submitted, 2015.

Chamberlin, T. C.: The method of multiple working hypotheses, Science, 15, 754–759, doi:10.1126/science.148.3671.754, 1965.

Eder, A., Strauss, P., Krueger, T., and Quinton, J. N.: Comparative calculation of suspended sediment loads with respect to hysteresis effects (in the Petzenkirchen catchment, Austria), J. Hydrol. 389, 168–176, 2010.

Eder, A., Exner-Kittridge, M., Strauss, P., and Blöschl G.: Resuspension of bed sediment in a small stream – results from two flushing experiments, Hydrol. Earth Syst. Sci. 18, 1043–1052, doi:10.5194/hess-18-1043-2014, 2014.

Elliott, K. J. and Vose, J. M.: The contribution of the Coweeta Hydrologic Laboratory to developing an understanding of long-term (1934–2008) changes in managed and unmanaged forests, Forest Ecol. Manage., 261, 900–910, 2011.

Exner-Kittridge, M., Niederreiter, R., Eder, A., and Zessner, M.: A simple and flexible field tested device for housing water monitoring sensors at point discharges, Water Sci. Technol., 67, 1026–1033, 2013.

Exner-Kittridge, M., Salinas, J. L., and Zessner, M.: An evaluation of analytical stream to groundwater exchange models: a comparison of gross exchanges based on different spatial flow distribution assumptions, Hydrol. Earth Syst. Sci., 18, 2715–2734, doi:10.5194/hess-18-2715-2014, 2014.

Exner-Kittridge, M., Blöschl, G., Eder, A., Strauss, P., and Zessner, M.: Source contribution dynamics of runoff events in a headwater agricultural catchment, submitted to Water Resour. Res., 2016.

FAO (Food and Agriculture Organization of the United Nations): Crop evapotranspiration: guidelines for computing crop water requirements, no. 56 in FAO irrigation and drainage paper, Rome, 1998.

Farnleitner, A. H., Hocke, L., Beiwl, C., Kavka, G. G., Zechmeister, T., Kirschner, A. K. T., and Mach, L. R.: Rapid enzymatic detection of Escherichia coli contamination in polluted river water, Lett. Appl. Microbiol., 33, 246–250, 2001.

Farnleitner, A. H., Hocke, L., Beiwl, C., Kavka, G. G., and Mach, R. L.: Hydrolysis of 4-methylumbelliferyl-β-D-glucuronide in differing sample fractions of river waters and its implication for the detection of fecal pollution, Water Res., 36, 975–981, 2002.

Farnleitner, A. H., Reischer, G. H., Stadler, H., Kollanur, D., Sommer, R., Zerobin, W., Blöschl, G., Barrella, K. M., Truesdale, J. A., Casarez, E. A., and Di Giovanni, G. D.: Microbial Source Tracking: Methods, Applications and Case Studies. Chapter 18 – Agricultural and Rural Watersheds, edited by: Hagedorn, C., Haarwood, J., and Blanch A., Springer, New York, 399–432, 2011.

Flores Orozco, A., Williams, K. H., Long, P. E., Hubbard, S. S., and Kemna, A.: Using complex resistivity imaging to infer biogeochemical processes associated with bioremediation of a uranium-contaminated aquifer, J. Geophys. Res., 116, G03001, doi:10.1029/2010JG001591, 2011.

Flores Orozco, A., Williams, H. K., and Kemna, A.: Time-lapse spectral induced polarization imaging of stimulated uranium bioremediation, Near Surf. Geophys., 11, 531–544, doi:10.3997/1873-0604.2013020, 2013.

Flores Orozco, A., Velimirovic, M., Tosco, T., Kemna, A., Sapion, H., Klaas, N., Sethi, R., and Leen, B.: Monitoring the injection of microscale zero-valent iron particles for groundwater remediation by means of complex electrical conductivity imaging, Environ. Sci. Technol., 49, 5593–5600, doi:10.1021/acs.est.5b00208, 2015.

Grayson, R. B. and Western, A. W.: Towards areal estimation of soil water content from point measurements: time and space stability of mean response, J. Hydrol., 207, 68–82, 1998.

Grayson, R. B., Western, A. W., Chiew, F. H., and Blöschl, G.: Preferred states in spatial soil moisture patterns: Local and nonlocal controls, Water Resour. Res., 33, 2897–2908, 1997.

Grayson, R., Blöschl, G., Western, A., and McMahon, T.: Advances in the use of observed spatial patterns of catchment hydrological response, Adv. Water Resour., 25, 1313–1334, 2002.

Holko, L., Holzmann, L., de Lima, M. I. P., and de Lima, J. L. M. P.: Hydrological research in small catchments – an approach to improve knowledge on hydrological processes and global change impacts, J. Hydrol. Hydromech., 63, 181–182, 2015.

Holländer, H. M., Blume, T., Bormann, H., Buytaert, W., Chirico, G. B., Exbrayat, J.-F., Gustafsson, D., Hölzel, H., Kraft, P., Stamm, C., Stoll, S., Blöschl, G., and Flühler, H.: Comparative predictions of discharge from an artificial catchment (Chicken Creek) using sparse data, Hydrology and Earth System Sciences, 13, 2069–2094, 2009.

Holländer, H. M., Bormann, H., Blume, T., Buytaert, W., Chirico, G. B., Exbrayat, J.-F., Gustafsson, D., Hölzel, H., Krauße, T., Kraft, P., Stoll, S., Blöschl, G., and Flühler, H.: Impact of modellers' decisions on hydrological a priori predictions, Hydrol. Earth Syst. Sci., 18, 2065–2085, doi:10.5194/hess-18-2065-2014, 2014.

Hopp, L., Harman, C., Desilets, S. L. E., Graham, C. B., McDonnell, J. J., and Troch, P. A.: Hillslope hydrology under glass: confronting fundamental questions of soil-water-biota co-evolution at Biosphere 2, Hydrol. Earth Syst. Sci., 13, 2105–2118, doi:10.5194/hess-13-2105-2009, 2009.

Horvath, Z., Waser, J., Perdigao, R. A. P., Konev, A., and Blöschl, G.: A two-dimensional numerical scheme of dry/wet fronts for the Saint-Venant system of shallow water equations, Int. J. Num. Meth. Fluids, 77, 159–182, 2015.

Kazemi Amiri, A. and Bucher, C.: Derivation of a new parametric impulse response matrix utilized for nodal wind load identification by response measurement, J. Sound Vibr., 344, 101–113, 2015.

Kirby, C., Newson, M., and Gilman, K. (eds.): Plynlimon research: the first two decades – IH Report 109, www.ceh.ac.uk/products/publications/Plynlimonresearchthefirsttwodecades.html, (last access 11 July 2015) 1991.

Kirnbauer, R., Blöschl, G., Haas, P., Müller, G., and Merz, B.: Identifying space-time patterns of runoff generation – A case study from the Löhnersbach catchment, Austrian Alps, in: Global Change and Mountain Regions, edited by: Huber, U., Bugmann, H., and Reasoner, M., Springer, Series on Advances in Global Change Research, 23, 309–320, 2005.

Knorr-Cetina, K. D.: The manufacture of knowledge: An essay on the constructivist and contextual nature of science, Elsevier, 204 pp., 2013.

Kovacs, A., Honti, M., Zessner, M., Eder, A., Clement, A., and Blöschl, G.: Identification of phosphorus emission hotspots in agricultural catchments, Sci. Total Environ., 433, 74–88, 2012.

Leibundgut, Ch., Blöschl, G., Borchardt, D., Bundi, U., Hansjürgens, B., Merz, B., and Nobilis, F.: Nachhaltige Wassernutzung in der Schweiz. Gesamtsynthese: NFP61 weist Wege in die Zukunft (Sustainable water management in Switzerland: NRP 61 shows the way ahead), Aqua Gas, 11, 10–16, 2014.

Lin, H. and Hopmans, J. W.: Interdisciplinary sciences in a global network of critical zone observatories, Vadose Z. J., 10, 781–785, 2011.

Mauder, M. and Foken, T.: Documentation and instruction manual of the eddy covariance software package TK3. Universität Bayreuth, Abt. Mikrometeorologie, Arbeitsergebnisse, 60 pp., 2011.

Merz, R. and Blöschl, G.: Flood frequency hydrology: 1. Temporal, spatial, and causal expansion of information, Water Resour. Res., 44, W08432, doi:10.1029/2007WR006744, 2008.

Parajka, J., Haas, P., Kirnbauer, R., Jansa, J., and Blöschl, G.: Potential of time-lapse photography of snow for hydrological purposes at the small catchment scale, Hydrol. Process., 26, 3327–3337, 2012.

Plate, E. and Zehe, E.: Hydrologie und Stoffdynamik kleiner Einzugsgebiete- Prozesse und Modelle, Schweizerbart, Stuttgart 366 pp., 2008.

Reischer, G. H., Kollanur, D., Vierheilig, J., Wehrspaun, C., Mach, R., Stadler, H., Sommer, R., and Farnleitner, A. H.: A hypothesis-driven approach for the identification of fecal pollution sources in water resources, Environ. Sci. Technol., 45, 4038–4045, 2011.

Robinson, M., Rodda, J. C., and Sutcliffe, J. V.: Long-term environmental monitoring in the UK: origins and achievements of the Plynlimon catchment study, Transactions of the Institute of British Geographers, 38, 451–463, 2013.

Ryzinska-Paier, G., Lendenfeld, T., Correa, K., Stadler, P., Blaschke, A. P., Mach, R. L., Stadler, H., Kirschner, A. K. T., and Farnleitner, A. H.: A sensitive and robust method for automated on-line monitoring of enzymatic activities in water and water resources, Water Sci. Technol., 69, 1349–1358, 2014.

Schilling, C., Behrendt, H., Blaschke, A., Danielescu, S., Dimova, G., Gabriel, O., Heinecke, U., Kovacs, A., Lampert, C., Postolache, C., Schreiber, H., Strauss, P., and Zessner, M.: Lessons learned from investigations on case study level for modelling of nutrient emissions in the Danube Basin, Water Sci. Technol., 51, 183–191, 2005.

Schumann S., Schmalz, B., Meesenburg, H., and Schröder, U.: Status and Perspectives of Hydrology in Small Basins. IHP/HWRP-Berichte Heft 10, German Federal Institute of Hydrology, Koblenz, 69 pp., 2010.

Sivapalan, M.: The secret to "doing better hydrological science": change the question!, Hydrol. Process., 23, 1391–1396, 2009.

Srinivasan, V., Thompson, S., Madhyastha, K., Penny, G., Jeremiah, K., and Lele, S.: Why is the Arkavathy River drying? A multiple-hypothesis approach in a data-scarce region, Hydrol. Earth Syst. Sci., 19, 1905–1917, doi:10.5194/hess-19-1905-2015, 2015.

Stadler, P., Blöschl, G., Vogl, W., Koschelnik, J., Epp, M., Lackner, M., Oismüller, M., Kumpan, M., Nemeth, L., Strauss, P., Sommer, R., Ryzinska-Paier, G., Farnleitner, A. H., and Zessner, M.: Real-time monitoring of beta-D-glucuronidase activity in sediment laden streams: A comparison of instruments, submitted to Environ. Sci. Technol., 2016.

Strauss, P. and Klaghofer, E.: Soil Erosion in Europe: 1.17 Austria, edited by: Boardman J. and Poesen, J., Soil Erosion in Europe, John Wiley, 205–212, 2006.

Strauss, P., Leone, A., Ripa, M. N., Turpin, N., Lescot, J. M., and Laplana, R.: Using critical source areas for targeting cost-effective best management practices to mitigate phosphorus and sediment transfer at the watershed scale, Soil Use Manage., 23, 144–153, 2007.

Swank, W. T. and Crossley Jr., D. A.: Forest Hydrology and Ecology at Coweeta, Springer, 469 pp., 1988.

Turpin, N., Laplana, R., Strauss, P., Kaljonen, M., Zahm, F., and Bégué, V.: Assessing the cost, effectiveness and acceptability of best management farming practices: a pluri-disciplinary approach. International journal of agricultural resources, governance and ecology, 5, 272–288, 2006.

Western, A. W., Blöschl, G., and Grayson, R. B.: How well do indicator variograms capture the spatial connectivity of soil moisture?, Hydrol. Process., 12, 1851–1868, 1998.

Western, A. W., Grayson, R. B., Blöschl, G., Willgoose, G. R., and McMahon, T. A.: Observed spatial organisation of soil moisture and its relation to terrain indices, Water Resour. Res., 35, 797–810, 1999.

Western, A. W., Blöschl, G., and Grayson, R. B.: Towards capturing hydrologically significant connectivity in spatial patterns, Water Resour. Res., 37, 83–97, 2001.

Yeshaneh E., Eder, A., and Blöschl, G.: Effects of conservation measures on soil loss in the Koga catchment, Northwestern Ethiopia, Land Degr. Develop., accepted, 2015.

Zacharias, S., Bogena, H., Samaniego, L., Mauder, M., Fuß, R., Pütz, T., Frenzel, M., Schwank, M., Baessler, C., Butterbach-Bahl, K., Bens, O., Borg, E., Brauer, A., Dietrich, P., Hajnsek, I., Helle, G., Kiese, R., Kunstmann, H., Klotz, S., Munch, J. C., Papen, H., Priesack, E., Schmid, H. P., Steinbrecher, R., Rosenbaum, U., Teutsch, G., Vereecken, H.: A network of terrestrial environmental observatories in Germany, Vadose Z. J., 10, 955–973, 10.2136/vzj2010.0139, 2011.

Zehe, E. and Blöschl, G.: Predictability of hydrologic response at the plot and catchment scales: Role of initial conditions, Water Resour. Res., 40, W10202, doi:10.1029/2003WR002869, 2004.

Zehe, E., Elsenbeer, H., Lindenmaier, F., Schulz, K., and Blöschl, G.: Patterns of predictability in hydrological threshold systems, Water Resour. Res., 43, W07434, 10.1029/2006WR005589, 2007.

Zehe, E., Maurer, T., Ihringer, J., and Plate, E.: Modeling water flow and mass transport in a loess catchment, Phys. Chem. Earth, 6, 487–507, 2001.

Zessner, M., Postolache, C., Clement, A., Kovacs, A., and Strauss, P.: Considerations on the influence of extreme events on the phosphorus transport from river catchments to the sea, Water Sci. Technol., 51, 193–204, 2005.

PERMISSIONS

LIST OF CONTRIBUTORS

M. Garcia
Civil & Environmental Engineering Department, Tufts University, 200 College Avenue, Medford, MA 02155, USA

K. Portney
Bush School of Government & Public Service, Texas A&M University, 4220 TAMU, College Station, TX 77843, USA

S. Islam
Civil & Environmental Engineering Department, Tufts University, 200 College Avenue, Medford, MA 02155, USA
The Fletcher School of Law and Diplomacy, Tufts University, 160 Packard Avenue, Medford, MA 02155, USA

Y. Y. Zhang and J. Xia
Key Laboratory of Water Cycle and Related Land Surface Processes, Institute of Geographic Sciences and Natural Resources Research, Chinese Academy of Sciences, Beijing, 100101, China

Q. X. Shao
CSIRO Digital Productivity Flagship, Leeuwin Centre, 65 Brockway Road, Floreat Park, WA 6014, Australia

A. Z. Ye
College of Global Change and Earth System Science, Beijing Normal University, Beijing, 100875, China

H. T. Xing
CSIRO Agriculture Flagship, GPO BOX 1666, Canberra, ACT 2601, Australia

Y. Chen and J. Li
Department of Water Resources and Environment, Sun Yat-sen University, Room 108, Building 572, Guangzhou 510275, China

H. Xu
Bureau of Hydrology and Water Resources of Fujian Province. Fuzhou, Fujian, China

M. W. Scown and M. C. Thoms
Riverine Landscapes Research Laboratory, University of New England, Armidale, Australia

N. R. De Jager
Upper Midwest Environmental Sciences Center, United States Geological Survey, La Crosse, Wisconsin, USA

L. Foresti, M. Reyniers and L. Delobbe
Royal Meteorological Institute of Belgium, Brussels, Belgium

A. Seed
Bureau of Meteorology, Melbourne, Australia

C. Duvert
Queensland University of Technology, Brisbane, QLD 4001, Australia

M. K. Stewart
Aquifer Dynamics Ltd & GNS Science, P.O. Box 30368, Lower Hutt, 5040, New Zealand

D. I. Cendón
Australian Nuclear Science and Technology Organisation, Kirrawee DC, NSW 2232, Australia
School of Biological, Earth & Environmental Sciences, University of New South Wales, Sydney, NSW 2052, Australia

M. Raiber
5CSIRO Land & Water, Dutton Park, Brisbane, QLD 4102, Australia

B. C. Si
College of Hydraulic and Architectural Engineering, Northwest A&F University, Yangling 712100, China
University of Saskatchewan, Department of Soil Science, Saskatoon, SK S7N 5A8, Canada

W. Hu
University of Saskatchewan, Department of Soil Science, Saskatoon, SK S7N 5A8, Canada
New Zealand Institute for Plant&Food Research Limited, Private Bag 4704, 8140 Christchurch, New Zealand

P. Blair and W. Buytaert
Grantham Institute and Department of Civil and Environmental Engineering, Skempton Building, Imperial College London, SW7 2AZ, UK

E. Cornwell
Advanced Mining Technology Center, Facultad de Ciencias Físicas y Matemáticas, Universidad de Chile, Santiago, Chile

N. P. Molotch
Department of Geography and Institute of Arctic and Alpine Research, University of Colorado, Boulder, USA

Jet Propulsion Laboratory, California Institute of Technology, Pasadena, California, USA

J. McPhee
Advanced Mining Technology Center, Facultad de Ciencias Físicas y Matemáticas, Universidad de Chile, Santiago, Chile
Departamento de Ingeniería Civil, Facultad de Ciencias Físicas y Matemáticas, Universidad de Chile, Santiago, Chile

M. Broer, G. Carr, X. Chen, A. Kazemi Amiri, M. Oismüller, R. Silasari and M. Vreugdenhil Centre for Water Resource Systems, TU Wien, Karlsplatz 13, 1040 Vienna, Austria

G. Blöschl, P. Haas and J. Parajka
Centre for Water Resource Systems, TU Wien, Karlsplatz 13, 1040 Vienna, Austria
Institute of Hydraulic Engineering and Water Resources Management, TU Wien, Karlsplatz 13/222, 1040 Vienna, Austria

A. P. Blaschke
Centre for Water Resource Systems, TU Wien, Karlsplatz 13, 1040 Vienna, Austria
Institute of Hydraulic Engineering and Water Resources Management, TU Wien, Karlsplatz 13/222, 1040 Vienna, Austria
Interuniversity Cooperation Centre for Water & Health, TU Wien, Vienna, Austria

C. Bucher
Centre for Water Resource Systems, TU Wien, Karlsplatz 13, 1040 Vienna, Austria
Institute of Building Construction and Technology, TU Wien, Karlsplatz 13/206, 1040 Vienna, Austria

A. Eder
Centre for Water Resource Systems, TU Wien, Karlsplatz 13, 1040 Vienna, Austria
Institute for Land and Water Management Research, Federal Agency for Water Management, Pollnbergstraße 1, 3252 Petzenkirchen, Austria

P. Strauss
Institute for Land and Water Management Research, Federal Agency for Water Management, Pollnbergstraße 1, 3252 Petzenkirchen, Austria

M. Exner-Kittridge, M. Zessner and P. Stadler
Centre for Water Resource Systems, TU Wien, Karlsplatz 13, 1040 Vienna, Austria
Institute for Water Quality, Resource and Waste Management, TU Wien, Karlsplatz 13/226, 1040 Vienna, Austria

A. Farnleitner
Centre for Water Resource Systems, TU Wien, Karlsplatz 13, 1040 Vienna, Austria
Institute of Chemical Engineering, TU Wien, Gumpendorfer Straße 1a, 1060 Vienna, Austria
Interuniversity Cooperation Centre for Water & Health, TU Wien, Vienna, Austria

A. Flores-Orozco
Department for Geodesy and Geoinformation, TU Wien, Gußhausstraße 25-29/120, 1040 Vienna, Austria

W. Wagner
Centre for Water Resource Systems, TU Wien, Karlsplatz 13, 1040 Vienna, Austria
Department for Geodesy and Geoinformation, TU Wien, Gußhausstraße 25-29/120, 1040 Vienna, Austria

Index

www.ingramcontent.com/pod-product-compliance
Lightning Source LLC
Chambersburg PA
CBHW080622200326
41458CB00013B/4472